VERTEBRATES: Physiology

Readings from

SCIENTIFIC AMERICAN

VERTEBRATES: Physiology

With Introductions by

Norman K. Wessells

Stanford University

W. H. Freeman and Company
San Francisco

Most of the Scientific American articles in *Vertebrates: Physiology* are available as separate Offprints. For a complete list of articles now available as Offprints, write to W. H. Freeman and Company, 660 Market Street, San Francisco, California, 94104.

Library of Congress Cataloging in Publication Data

Main entry under title:

Vertebrates: physiology.

Bibliography: p.
Includes index.
1. Vertebrates—Physiology—Addresses, essays, lectures.
I. Wessells, Norman K. II. Scientific American.
QP71.V44 596′.01 79-28748
ISBN 0-7167-1198-2
IISBN 0-7167-1199-0 pbk.

Printed in the United States of America

9 8 7 6 5 4 3 2 1

PREFACE

The unprecedented triumphs of molecular biology since the 1950s tend to make us forget the broad and ancient base on which modern life sciences are built. As we succeed in constructing mechanistic interpretations of life processes, it is sobering to realize how perceptive and accurate were the observations of those phenomena made hundreds of years ago. Although the following quotation was written by Pliny the Elder 1900 years ago, nearly every topic in it is discussed in this book and is still under active investigation.

> The swiftest of all animals, not only those of the sea, is the dolphin: it is swifter than a bird and darts faster than a javelin, and were not its mouth much below its snout, almost in the middle of its belly, not a single fish would escape its speed. But nature's foresight contributes delay, because they cannot seize their prey except by turning over on their backs. . . . They have a habit of sallying out on to the land for an unascertained reason, and they do not die at once after touching earth—in fact they die more quickly if the gullet is closed up. . . . For a voice they have a moan like that of a human being. . . . The dolphin is an animal that is not only friendly to mankind but is also a lover of music, and it can be charmed by . . . the sound of the water-organ.
> (Pliny, *Natural History,* IX, vii, viii.)

Today we believe the dolphin's swiftness is due to laminar flow of water over its skin, so that eddies and turbulence do not generate drag. The sallies of whales or dolphins onto beaches may stem from a failure of the echo-locating navigational system to detect shallow, sloping sea bottoms, or may be due to ear parasites that interfere with the hearing of echoes. The dolphin's "voice" is of course an integral part of the echo-locating system and is also used for communication between these remarkably "brainy" cetaceans. And, what else but "nature's forethought" (natural selection in our terms) could explain so well the origin of these adaptations or the equally marvelous ones described throughout this collection of articles from *Scientific American*?

This book and its companion volume—*Vertebrates: Adaptation* (W. H. Freeman and Company, 1980)—are about the structures, functions, and adaptations of vertebrates. We shall study molecules, cells, organs, and organisms in these two books, and our purpose will be to see how the parts are orchestrated to yield the functioning individual vertebrate, adapted for its particular niche. Any overview of the kind endeavored in these books must be based on certain biases. The prejudice observed in this book is that it is just as sterile to study only bones and morphology as to concentrate only upon molecules or cells. A student can best achieve real understanding by sampling at all levels or organization and integrating the knowledge thus obtained to gain a fuller picture of a trout, a toad, or a titmouse.

This book concentrates upon physiological systems, their properties, and their controls. Several of the articles describe special uses of physiological systems so that the student may begin to gain a flavor of the remarkable versatility of the vertebrate organism in adaptating to its environment. The companion book, *Vertebrates: Adaptation*, concentrates upon the myriad ways that vertebrates are adapted to the physical and biological world in which they live.

These two collections of articles are designed to supplement courses in introductory biology, vertebrate biology, comparative anatomy, and physiology. As such, the books are meant to fill some of the gaps and answer some of the questions that might arise from a general survey of vertebrate biology. Orientation to the field and articles is provided by Introductions to the Sections of the book. Although *Scientific American* has not of course published articles on all the subjects that might interest a student, the scope is nevertheless quite broad; therefore references to Offprints on related subjects are included. In addition, a special list of recent references for each Section is found in the Bibliography at the end of each book.

I have assumed that teachers and students will be able to consult the *Annual Review of Physiology, Biological Reviews*, or *Physiological Reviews*, for many useful papers that are not cited specifically here. The important book by P. W. Hochachka and G. N. Somero, *Strategies of Biochemical Adaptation*, and *Animal Physiology* by R. Eckert and D. Randall, are to my mind, the most useful means of amplifying the many intriguing aspects of vertebrate life treated in these *Scientific American* articles. For those needing background in ancient vertebrates and paleontology, the book by B. J. Stahl, *Vertebrate History: Problems in Evolution*, is by far the best balanced evaluation and summary.

My thanks go to Charlea Massion, a friend who, while a student in my Vertebrate Biology course in 1972, wrote the poem that is placed at the close of "An Essay on Vertebrates." Thanks also go to Lois Wessells for aid and comfort in busy times.

November 1979 *Norman K. Wessells*

CONTENTS

V HORMONES AND INTERNAL REGULATION

Note on cross-references: References to articles included in this book are noted by the title of the article and the page on which it begins; references to articles that are available as Offprints, but are not included here, are noted by the article's title and Offprint number; references to articles published by SCIENTIFIC AMERICAN, but which are not available as Offprints, are noted by the title of the article and the month and year of its publication.

GENERAL INTRODUCTION: AN ESSAY ON VERTEBRATES

AN ESSAY ON VERTEBRATES

Adaptations are specializations in structure and function that permit animals to survive under a given set of environmental conditions. Whether they are physiological, biochemical, or structural, adaptations arc variations upon pre-existing themes: a change in a metabolic pathway occurs, causing a new form of nitrogen excretion; or the shape of a pelvic bone alters, and erect, bipedal locomotion becomes possible. Adaptations develop in the course of generations as responses of continually varying, sexually reproducing organisms to ever changing environments. Nowhere is this process better illustrated than in the remarkable evolution of vertebrates from filter feeder to abstract thinker.

Many adaptations seem so marvelously complex that to view one individually is, at first glance, to strain credulity. But as one studies the evolution of living forms, one acquires a perspective which makes it possible to see how the individual parts of an organism have altered to yield an animal specialized or "adapted" for some special circumstance. In this introductory essay we will attempt to gain this perspective from a brief, historical survey of the vertebrates. We begin with vertebrate origins.

One of the oldest characteristics unique to those organisms that gave rise to vertebrates was a filter-feeding apparatus with gill slits and cilia. Such a feeding apparatus is found in some organisms alive today—adult tunicates, amphioxi (see Figure 1), and even more primitive, worm-like creatures called Hemichordates—all of which are considered to be close relatives of the vertebrates. These marine organisms fed by ingesting water, detritus, and suspended food particles through the mouth. The food and detritus were then trapped in mucus, as the water left the body through gill slits located in each lateral wall of the anterior gut (the pharynx). Beating hairlike cilia propelled the stream of water.

Interestingly, the mucus used for trapping the food was secreted by cells that are thought to be a portion of the precursor of the vertebrate endocrine gland, the thyroid. Other cells of the same part of the pharynx apparently had the ability to bind iodine into organic molecules, one of which was the hormone thyroxine. This chemical has widespread effects upon both the developmental processes and the adult metabolic functions of today's fishes. Because

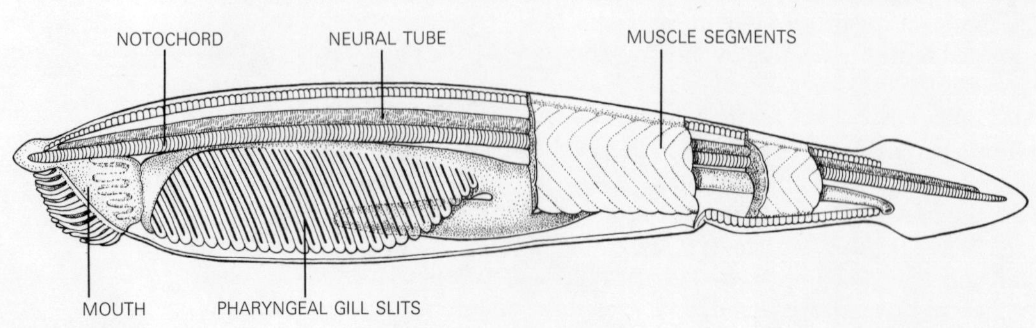

Figure 1. A simplified view of the organs in an amphioxus. In an adult animal a large number of gill slits may be found on each side of the body, although only a small number is shown here for clarity. The stiff notochord and segmental muscle masses make up the swimming apparatus, and the nerve cord lies directly above the notochord. [After S. Wischnitzer, _Atlas and Dissection Guide for Comparative Anatomy_, Second Edition, W. H. Freeman and Company. Copyright © 1972.]

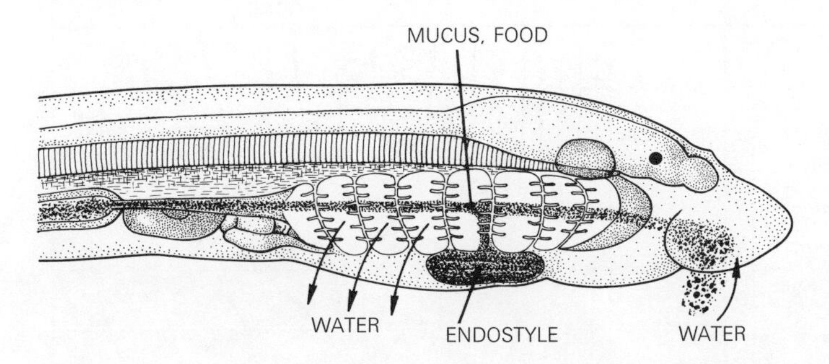

Figure 2. The anterior end of a lamprey ammocoete, showing the feeding apparatus. Water, food, and detritus enter the mouth and proceed posteriorly. The water leaves through the gill slits while the food is trapped in the mucus strand rising from the ventral endostyle.
[After J. Z. Young, *The Life of Vertebrates*, Oxford, 1962.]

animals as large and complex as the earliest vertebrates probably required a substantial set of control machinery, we should not be surprised that parts of the endocrine system evolved early in vertebrate history. In fact, the probable evolutionary precursor of the anterior pituitary gland (the so-called "master" endocrine gland of vertebrates) is seen in the form of special, glandlike structures located anterior to the gut in each of the vertebrate relatives mentioned above.

A second ancient characteristic of the ancestors of vertebrates was a locomotor complex that differed from that of any other organism. It consisted of segmental masses of muscle arranged down the length of the body, a stiff central rod (the notochord or, in later forms, the vertebral column) that prevented shortening of the body when muscle contraction occurred, and a nervous control system for coordination of the muscle masses. This combination of structures is seen today in larval tunicates, adult amphioxi, and all vertebrates, and is shown in Figure 1. Even in the most ancient forms there is a nerve tube, located above the notochord, that shows expansion at the anterior end into a brain. The brain had areas of specialization associated with the various senses (seeing, smelling, tasting); in addition, the brain served as a control center that could override the basic reflex activity of the spinal cord of the central nervous sytem which was responsible for the locomotory pattern itself.

The next major vertebrate adaptation is a muscular, rather than ciliary, filter-feeding apparatus. It is found in the earliest vertebrate fossils, as well as in some organisms alive today—the larvae of lampreys and hagfish (cyclostomes). Lamprey larvae (ammocoetes), for example, do not have cilia, but propel the feeding current by expanding and contracting the pharyngeal cavity (see Figure 2). The action thus produced is analogous to the regular movements of the gill cover (operculum) on the sides of common higher fish, such as trout or goldfish. The "suction pump" of ammocoetes is driven by muscles in the wall of the pharynx that pull upon the skeletal support rods between the gill slits. A substantial blood supply serves the gill muscles. One might guess that, in the ancient relatives of the cyclostomes, the presence of this blood near a point of rapid water flow over a body surface (the gill slits) could well provide conditions leading to the formation of the first vertebrate respiratory organ, the "gill."

Although the transition from ciliary to muscular filter feeding does not sound very astounding, its consequences were momentous. Feeding in the new way was apparently much more efficient: larger organisms begin to appear in the fossil record (see Figure 3), and a smaller percentage of the body is devoted to food gathering by filtering (for example, the pharynx-gill slit complex has ten or fewer pairs of gill slits, as opposed to fifty or more in amphioxi and other ciliary filter feeders). In the Ordovician period (see Figure 3), there were a variety of types of fishes and all of them probably fed by the muscular filter-feeding method or a variation of it. Unfortunately, we have no fossils of the intermediate forms between the ciliary filter feeders and these primitive agnathans (jawless vertebrates). In addition to the muscular filter-feeding apparatus, agnathans possessed many other "vertebrate" characteristics: bone, a skeletal material not found in invertebrates; an internal ear used in sensing

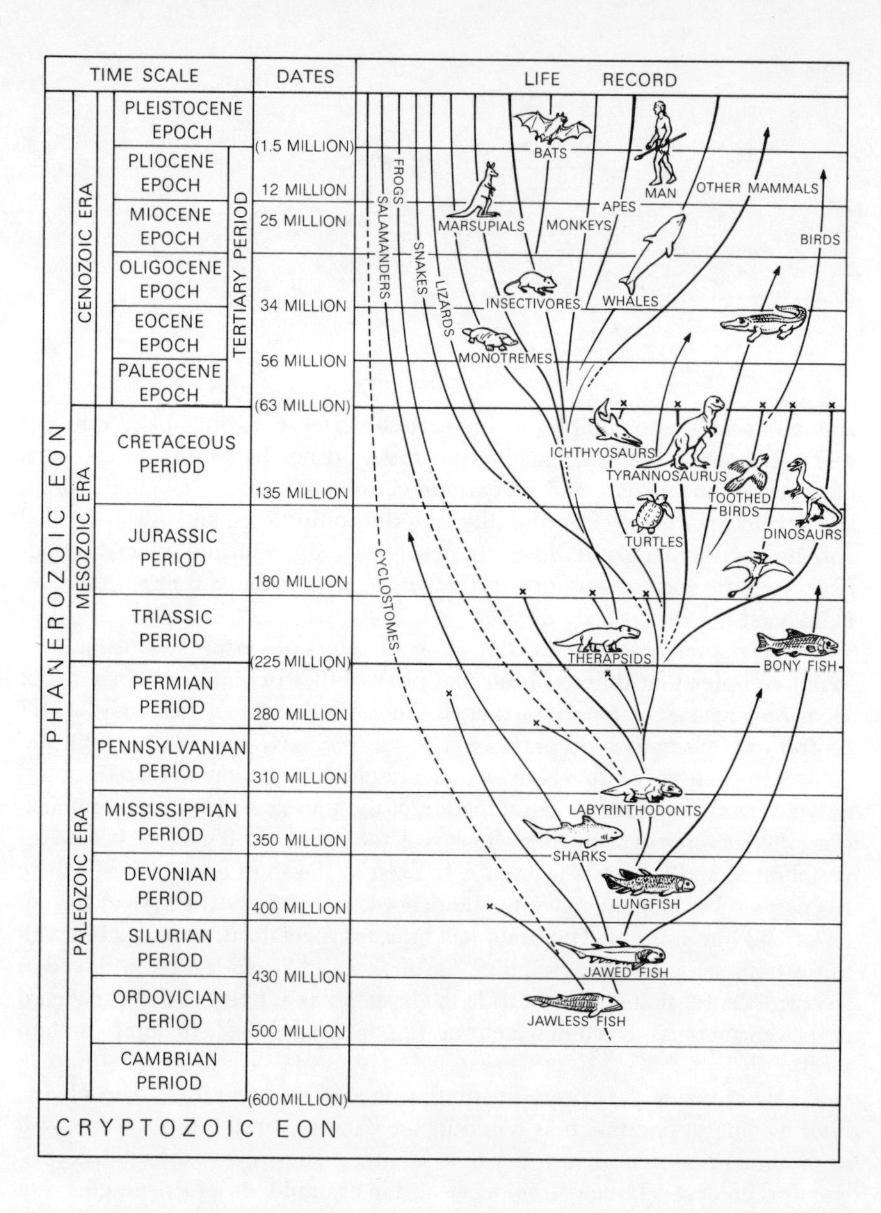

Figure 3. **The geological time scale and the appearance of various vertebrate types. The probable relationship between most of the organisms mentioned in the text is shown. Lines ending in a cross indicate extinction; those ending in an arrow indicate that descendants of that line are alive today. [After Carl O. Dunbar, *Historical Geology,* Second Edition, Wiley, 1960.]**

orientation in space or body movement; eyes very like those of modern fishes; nasal openings, indicating the organism's ability to smell; lateral-line canals like those of the higher fishes; a light-sensing pineal organ (see "The Pineal Gland" by Richard J. Wurtman and Julius Axelrod, in Section V); and cranial nerves, some of which led to the gill regions.

Interestingly, these ancient "vertebrates" did not possess a real vertebral column; instead, a notochord was present in adults and acted as the main longitudinal skeletal strut. The circulatory system of these fishes was probably "closed"; that is, it probably consisted of an interconnected set of tubular arteries, capillaries, veins, and heart. Because of the large body size, it seems likely that the vertebrate respiratory pigment, hemoglobin, had to be present. We can only guess about other systems composed of soft parts that were not preserved in the fossil record, but there is no reason not to suppose that endocrine and excretory systems, much like those of modern fish groups, were already present.

Bone deserves special comment as a vertebrate innovation. For many years, it has been assumed that bone arose as the result of selective pressures favoring development of hard protective armor or skeletal support materials. The British biologist L. B. Halstead has pointed out, however, that a more likely original use of bone was simply as a reservoir for phosphate, the crucial ion required for ATP production. Presence of such stored phosphate would have protected our ancient ancestors from variations in the phosphate cycle in the sea. Summer or autumn phytoplankton "blooms," for instance, can drastically reduce dissolved phosphate. It may have been fortunate happenstance that the storage form of phosphate, the apatites, was hard and so could be employed secondarily as armor or endoskeleton.

The next major advance for the vertebrate stock was also in feeding: it was the transition from filtering food to biting it. Moveable jaws probably arose as modifications of the anterior gill support bones and their musculature. The importance of this step can hardly be exaggerated—for the first time a vertebrate could eat a large invertebrate, or large plant material or another vertebrate! An added complexity of food chains resulted: big fish could eat little fish, or be eaten by bigger fish. No longer were most vertebrates bottom-dwellers, scooping up mud and food. Instead, a remarkable radiation into a variety of early fishes that had jaws took place (the acanthodians and placoderms). A reconstruction of an acanthodian jaw is shown in Figure 4. These fishes and their descendants came to dominate the oceans and the fresh waters of the earth. Their bodies were highly variable in structure, and many new vertebrate organs evolved. Since food was ingested in morsels, a storage and digestion reservoir, the stomach, evolved. The pursuit of prey and the escape from predators required mechanisms for controlling orientation in the three-dimensional world of water. Thus paired fin systems anterior and posterior to the center of gravity developed, as did changes in body and tail shape. Mechanisms for increasing buoyancy also appeared. As a result of such alterations, an extraordinary diversity of fishes lived in both the marine and fresh waters of the earth during the Devonian period (the "Ages of Fishes") which lasted from about 400 to 350 million years ago. Among them were the ancestors of today's sharks, the elasmobranchs, those of the higher bony fish, the teleosts, and those of the lines that gave rise to terrestrial vertebrates.

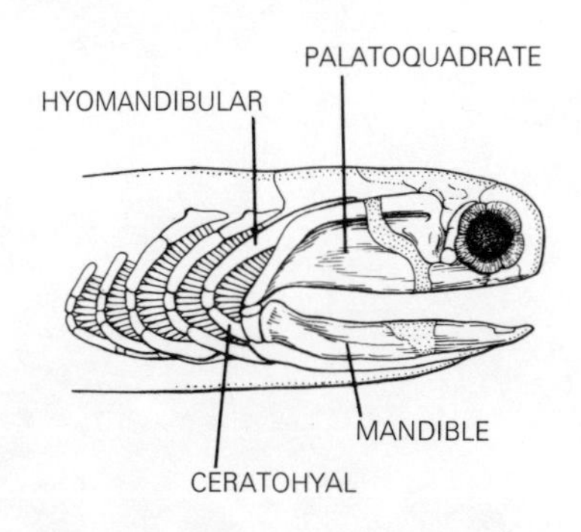

***Figure 4.* The jaws of an acanthodian. The upper and lower jaw bones are thought to be modified forms of the gill support bones that are shown posterior to the jaws.**
[After A. S. Romer, *The Vertebrate Body*, Third Edition, Saunders, 1962.]

One of the important features of the early history of bony fishes was the development of sacs extending from the anterior gut. Presumably these sacs were vascularized and were one of the respiratory organs of fresh-water fishes. In some salt-water fishes, a sac extending from the gut and similar to a lung in structure became modified into the swim bladder. This is a gas-filled organ that allows bony fishes to remain in neutral buoyancy so that they can hover in the water without expending muscular energy. Other types of fishes had primitive lungs, like those of the first amphibians who made the transition to life on land.

The next major advancements were several that occurred when vertebrates left the water and invaded the land. Adaptations to resist desiccation, to support the body, for locomotion on land, and to improve carbon dioxide elimination evolved. Probably, most attempts at meeting the challenge of life on land failed. Indeed, most of the surviving descendants of the earliest land dwellers—the frogs, toads, and salamanders—are even today only marginally emancipated from water.

Frogs and salamanders have moist skins and so are subject to evaporative water loss. Their kidneys are like those of fresh-water fish—they are ideally suited to eliminate large volumes of dilute urine but hardly adapted for water conservation. Moreover, because the eggs, sperm, and embryos have no protection from drying, at the time of reproduction most amphibians return to the water for breeding. But certain advances toward control of desiccation can be seen in these amphibians. A new type of endodermal bladder, a storage site for urine from which water can be resorbed, is present for the first time among vertebrates. Coincidentally, the structure and function of a portion of the pituitary gland changed, causing the release of a new hormone, vasopressin. This hormone affects the permeability of certain tissues, allowing water to enter the body spaces (through the skin, bladder wall, or kidney tubules). Ready availability of vasopressin is assured because of anatomical changes that have (1) affected the localization of the nerve endings so that release of the hormone occurs in a discrete region, the pars nervosa of the posterior pituitary gland; and (2) caused the appearance of a special "portal" blood supply that drains the portion of the pituitary that stores vasopressin. Finally, though most amphibians still reproduce in bodies of fresh water, behavioral adaptations permit a few to remain away from lakes and streams during breeding. The Javanese tree frog, for example, seals its eggs between two leaves; the egg jelly liquefies and the embryos and larvae live in the pool between the leaves until metamorphosis produces a young frog.

Transition to land posed great mechanical problems for the vertebrate body. No longer was the body supported and pressed on all sides by water: instead, it rested on the ground or was held up by paired derivatives of the pectoral and pelvic fins, the legs. The vertebral column, formerly a simple strut for preventing compression, now tended to move dorsally so that the bulk of the body mass hung below it. Finally, the points at which weight was transferred from the vertebral column to the legs—the pectoral and pelvic girdles—were highly modified and greatly strengthened. We see all these bones in a transitional condition in fossilized early amphibians or in today's salamanders.

The problem of gas exchange on land was especially serious. Primitive lungs simply don't do the job very well. In frogs and salamanders, the work of the simple saccular lungs is supplemented by respiratory exchange across the moist skin. The reason for this arrangement relates to the physical properties of oxygen and carbon dioxide in water and air. In water, oxygen tends to be limiting for a vertebrate, whereas the great solubility of carbon dioxide results in ready elimination of that waste product from the body. On land, the large proportion of oxygen in air means that amphibians need take fewer breaths to acquire sufficient oxygen. However, the relatively infrequent breaths generate a problem inasmuch as carbon dioxide accumulates in the blood and body fluids (with resultant changes in pH). Most amphibians eliminate much of the carbon dioxide across their moist skins, which in effect function as subsidiary respiratory organs. Incidentally, the necessity of having a layer of water on the skin (through which carbon dioxide can diffuse) is the reason why evaporative water loss occurs in most modern amphibians. In addition to these aspects of gas exchange, the Amphibia lack an efficient mechanism for filling and emptying the lungs. A frog swallows bubbles of air much as a lungfish does; unlike all the terrestrial vertebrates, it cannot expand its chest cavity to draw air into the lungs.

Thus we see that the first land vertebrates and most of their surviving amphibian relatives were only partly successful in adapting to the demands of the terrestrial habitat. It was the reptiles, descendants of the ancient amphibians, that succeeded where their predecessors had failed.

Reptiles have a dry and relatively impermeable skin. Their kidneys excrete a small volume of concentrated urine. Their nitrogen metalolism has been modified so that uric acid, rather than urea, is the main end product. Since a molecule of uric acid contains four nitrogen atoms, instead of the two present in urea, only half as many molecules (i.e., osmotically active particles) are produced for a given amount of protein catabolism, as is shown in Figure 5. Therefore, only half as much osmotic water need be lost from the body as wastes are carried away. In addition, uric acid precipitates out of solution as its concentration is raised in the special storage organs of the reptilian body, so that even more water is freed for resorption into the body.

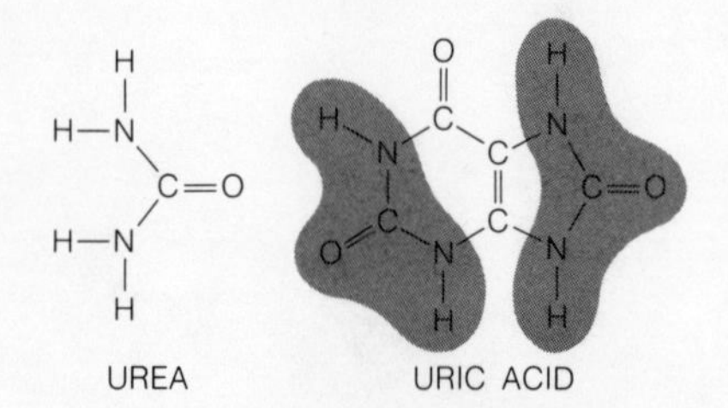

Figure 5. **Structure of urea and uric acid. Note that two molecules of urea occur in the uric acid. Despite its larger size, uric acid is equivalent to urea as an osmotic particle.**

In the Reptilia, an egg that was enclosed in a protective shell evolved. The female reproductive ducts covered the fertilized zygote with various protective layers that effectively sealed it from the atmosphere. Then, once the eggs had been laid in a terrestrial environment, they would develop if appropriately incubated so that evaporation was reduced. The male acquired a copulatory organ, which obviated the exposure of sperm and eggs to drying conditions, and insured fertilization of the egg far enough within the female reproductive tract so that there would be time for the zygote to be sealed within the protective coverings.

Other reptilian adaptations centered on the respiratory system. Because the surface of the skin, the epidermis, had become thick, dry, and dead, gases could no longer traverse it, and gas exchange necessarily became restricted to the lungs. This was made possible by the increased structural complexity and surface area of the lungs, and by the newly evolved, expandable rib cage, which filled and emptied the lungs by pumping large volumes of air to and from the respiratory surfaces.

These adaptations freed the reptile from dependence on water—except for the need to drink it—at every stage in its life cycle. Other organ systems also changed radically. The limbs and girdles were modified so that the legs were located more directly beneath the body, raising it higher off the ground. Bends and rotations of the bones at the elbow and knee allowed the limbs to move in an anterior-posterior direction, parallel with the long axis of the body, instead of in sweeping arcs to each side as did the primitive amphibian limbs that protruded outward much like fish fins. With the advent of the reptiles, there occurred for the first time great variation in the shapes and the uses of the legs as these organisms diversified and lived in new ways on the land.

Just as the fishes diverged into an extensive variety of forms after they acquired jaws, so did the reptiles as they overcame the major problems of terrestrial life. During much of the Mesozoic era (from about 225 to 63 million years ago) they were, in anthropomorphic terms, the "dominant" forms of animal life on land.

Nevertheless, reptiles had one significant limitation: their susceptibility to fluctuations in temperature. A terrestrial animal, because it is directly exposed to the atmosphere, is subject to much greater environmental variation than is a marine animal: temperature alters markedly from day to night; seasonal temperature variation in higher latitudes is dramatic; and daily changing weather conditions can produce rapid alterations in the environment. The aquatic vertebrates are not beset by such sudden variations because they are immersed in water, which warms and cools much more slowly than air. For these animals, the only rapid temperature changes would be those encountered in going from one depth of water to another. This basic difference between the two habitats directly affected the adaptations of aquatic and terrestrial vertebrates. A particular lizard so adapted that it could catch insects efficiently on warm summer afternoons might be unable to move rapidly, or function efficiently, in the cool evening or the cold winter. But fishes, never having been exposed to the high temperature of the atmosphere, are physiologically adapted to their relatively constant cold surroundings and can thus function year round.

The relative inability to tolerate environmental extremes, then, was one reason why the reptiles' utilization of the terrestrial habitat was limited. Add to it the fact that large regions of the earth were, because of the cold, largely beyond their reach, and one can imagine the selective pressure that must have operated to give rise to the two temperature-regulating groups alive today, the birds and the mammals (and perhaps to "warm-blooded" dinosaurs).

All biochemical reactions are somewhat inefficient in that they generate a small amount of heat. Therefore heat is continually being produced by normal animal metabolism. To conserve this heat, birds and mammals acquired feathers or fur to insulate their bodies. Both of these types of insulators consist of dead epidermal cells that require no blood supply. (The presence of superficial vascular networks in the skin is an important cause of heat loss in all vertebrates.) The feathers or hairs are arranged in overlapping layers that trap and immobilize a layer of insulating air next to the skin. To complement these means for conserving heat, three separate control systems evolved—one for heat production, another for heat dissipation, and another for heat retention. And, interestingly, even though birds and mammals arose at different times from different types of reptiles, there developed in all of them the same basic control features that permit maintenance of a constant high body temperature.

In both of these "warm-blooded" groups, the ability to regulate temperature was dependent upon key changes in the blood vascular system and its control machinery. A high-pressure system with rapid blood flow resulted from such changes as the complete separation of the pulmonary (lung) circulatory pathway from the systemic (body) pathway. Equally vital alterations occurred in the protein (globin) portion of hemoglobin, the respiratory pigment of the blood, so that avian and mammalian hemoglobin could bind and carry oxygen at high temperatures (37 degrees centigrade); the respiratory pigment of fishes or amphibians can carry oxygen only at lower temperatures. The most important alteration in avian and mammalian hemoglobin, however, was its new lower affinity for oxygen; this allows a higher oxygen content in the body tissues, which is essential for the high rates of metabolism required by life at 37 to 40 degrees centigrade.

As a result of the many adaptations that had made temperature control possible, a variety of birds and mammals appeared. Most could be active for longer periods of the day and, in the temperate regions, for greater parts of the year; some could even live in the polar regions of the earth.

Although feathers of primitive birds were important heat insulators, we don't know whether they originated for this purpose or for that of increasing wing surface area to permit short, gliding flights. Certainly, the feathers contributed to the great difference between avian and mammalian locomotion. They provide a light surface area that makes up much of the aerofoils of wings, the control surface of the tail, and the contouring that assures efficient airflow over the body during flight.

Most organ systems of the avian body have come to resemble those of mammals, but one, the reproductive system, is very different. A bird lays a shelled egg like that of the reptiles; this type of reproduction often necessitates certain procedures for nesting and for feeding of the young by both parents. In mammals, a new means of embryo incubation developed. The female reproductive tract and the hormonal system that controls its activity were altered so that the embryo would be retained and nourished within the body. Some advantages of this type of incubation are that the female is mobile during the gestation period, and that the mates do not have to remain together in order to take turns at incubating eggs. (A few species of birds are like mammals in that the female incubates the eggs and cares for the nestlings.) For the mammalian embryo, incubation within the mother's body provided a constant, high temperature. A much longer potential period of gestation became possible because of the availability of an essentially unlimited source of food—components of the mother's blood—for the embryo. In contrast, all the food for a bird's embryonic development must be put into the yolk before the egg shell is sealed off. An important adaptation associated with mammalian embryos maturing in a uterus (and in fact with avian ones in the cleidoic egg) is the evolution of various types of "fetal" hemoglobin. This embryonic respiratory pigment invariably has a greater affinity for oxygen than does adult hemoglobin of the same species; in this way, net transfer of oxygen from maternal to fetal blood is assured.

A crucial aspect of the evolution of mammalian reproduction was the development of a means of "insulating" the embryo and fetal tissues from the mother's immune system. Fetal tissue cells of the placenta are apparently covered or protected in some way so that they are not attacked and destroyed by maternal lymphocytes; thus, they are not recognized as "foreign" tissue (as would be a graft of skin or an implanted heart). In fact, breakdown of the immune protection of the young embryo may be a major cause of aborted or resorbed fetuses. (A. E. Beer and R. E. Billingham deal with this topic in their book *The Immunobiology of Mammalian Reproduction,* which is recommended for further reading.)

Mammals acquired still another convenience associated with reproduction—the mammary gland and its hormonal control network. This gland provided a ready source of a constant food type (milk, a species-specific balance of protein, fats, and carbohydrates) and eliminated the necessity of special food gathering to support early life of offspring. We think that mammary glands evolved from sweat glands, structures normally used as a means of lowering elevated body temperature. Thus an organ originally developed for one purpose, temperature regulation, was modified to serve a completely different function.

Other adaptations for life on land affected the ability of animals to gather information from the environment. Changes in the cornea and lens of the eye were particularly important because they enabled the animal living in air to focus light. In fishes the outermost structure of the eye, the cornea, has a refractive index close to that of water (the refractive index is the ratio of the velocity of light in a vacuum to the velocity of light in another medium, such as water or the cornea). Hence the cornea is of little use in bending and focusing light on the retina; consequently, it tends simply to be shaped like the side of the fish so that it offers the least resistance to movement through the water. Fishes do almost all focusing with the lens, by moving it out or in, or occasionally by changing its shape. But in terrestrial vertebrates, the cornea is the critical structure for focusing light because its refractive index (1.376) is so much greater than that of the air (1.00). As C. Ladd Prosser has pointed out, in air the cornea functions as the "coarse" adjustment on a microscope, and the lens as the "fine" adjustment.

Terrestrial adaptations of the ear affected primarily one of the chambers (the lagena) of the internal ear. The lagena became greatly elongated in most terrestrial vertebrates (in which it is termed the cochlea), and it contains large numbers of sensory hairs that respond to shearing movements of certain membranes. In addition, a "middle" ear developed to amplify and transmit sound from the external ear drum to the membrane system of the internal ear. In fact, these alterations in the ear are among the most complex morphological changes that took place in the transition from aquatic to terrestrial life. A comparison of the internal ears of the shark and the mammal is made in Figure 6.

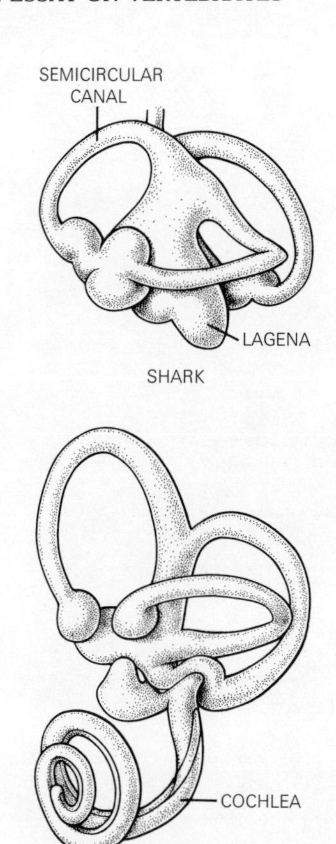

Figure 6. **The internal ears of a shark and of a mammal. Note how the lagena is expanded to form the long, coiled cochlea of the terrestrial vertebrate. The semicircular canals are similar in structure and in function; in these and other vertebrates the canals are part of the apparatus that senses acceleration, deceleration, and relative position of the body in the water.**

The most striking way in which mammals differ from birds is, of course, in the larger mammalian brain in which the most complex cognitive processes are carried out in the cerebral cortex. For unknown reasons, the most complex types of bird behavior are controlled by an entirely different region of the brain, the hyperstriatum (see Laurence J. Stettner and Kenneth A. Matyniak, "The Brain of Birds," Offprint 515). In seeking to understand why the mammalian brain, and particularly that of the primates, became so large, we must examine many aspects of mammalian biology.

One factor that may have contributed to the expansion of the primate brain is the unique alteration in the reproductive activity of these animals. Virtually all other vertebrates are seasonal breeders: all females mate only when they are in a condition of estrus, or heat, and a condition in which ovulation or release of fertilizable eggs is likely; females of a given population of a species come into estrus together, and mating occurs at such a time that the young will be born at a season propitious for their survival. In nonhuman primates (and in some other animals), we see an initial alteration in this pattern: females are asynchronous in estrus. They still mate only when ovulation is likely to occur, but as a result of the asynchrony, breeding within a population takes place throughout the year.

Only in humans do we see the additional step that dissociates mating from ovulation. The nervous and endocrine systems have been modified in such a way that the individual female is a potential breeder all year long. It has been argued, in fact, that this change is the most significant physiological difference between ourselves and the apes. The "competitive" societal organization of the apes could have resulted in large part from the sexual asynchrony of the females. Males repeatedly fight to establish their position in the dominance hierarchy month after month as new females enter estrus and become sexually receptive. Polygamy and harems are the rule in some such societies. Mature males tend not to take part in rearing the young, and the oldest, most experienced males are often driven from the group when they are no longer dominant. In humans, the dissociation of copulation from reproduction has eliminated at least one condition fostering competition between males—the constant competition for a new sex partner; stable relations between one male and one female can be established, since the female may be continuously active sexually. It has been proposed that this change established the conditions in which the "cooperative" society of humans could appear. Thus, one might argue that the unique feature of human physiology—the behavior that most distinguishes us from other animals—is sex for pleasure alone, rather than merely for reproduction.

Had such an alteration occurred in the structure of society of prehuman anthropoids, the resultant permanent relationships between males and females, and the cooperative living between old, reproductive, and young nonreproductive animals, could have greatly stimulated development of communication systems, transmittal of experience, and other factors conducive to expanded learning capacity.

Another possible cause contributing to the expansion of the brain might be the evolution of the hand. Indeed, the development of the hand is one of the most intriguing aspects of the recent evolutionary history of primates. Several reptiles and mammals (such as *Tyrannosaurus* and the kangaroo) abandoned the usual tetrapod stance for a bipedal one. The forelimbs of these animals became much smaller in a relative sense than typical forelimbs of tetrapods (see Figure 7). The only bipedal vertebrates in which this tendency did not occur were those in which the forelimbs had special functions; for instance, birds, flying dinosaurs (pterosaurs), and bats employed their enlarged forelimbs as wings. The other major exceptions to the tendency for small forelimbs among bipedal vertebrates are humans and our immediate ancestors. This we take to be a strong piece of indirect evidence that our primate ancestors lived in the trees. Monkeys and small primates have small, light bodies that they can balance easily above the limbs of trees. As a result, these creatures use four limbs to walk and run on *top* of the branches. It is probable that some of the ancestors of today's pongids developed a new type of locomotion—brachiation—in which they swung by their forelimbs *beneath* the branches. These creatures certainly gave rise to the gibbons of today. It is not clear whether the ancestors of other apes also passed through a brachiating condition or whether they became "knuckle walkers" like today's chimpanzees and gorillas.

Perhaps because of brachiation which put new stresses and strains on the forelimbs and pectoral girdle, and perhaps because of knuckle walking and

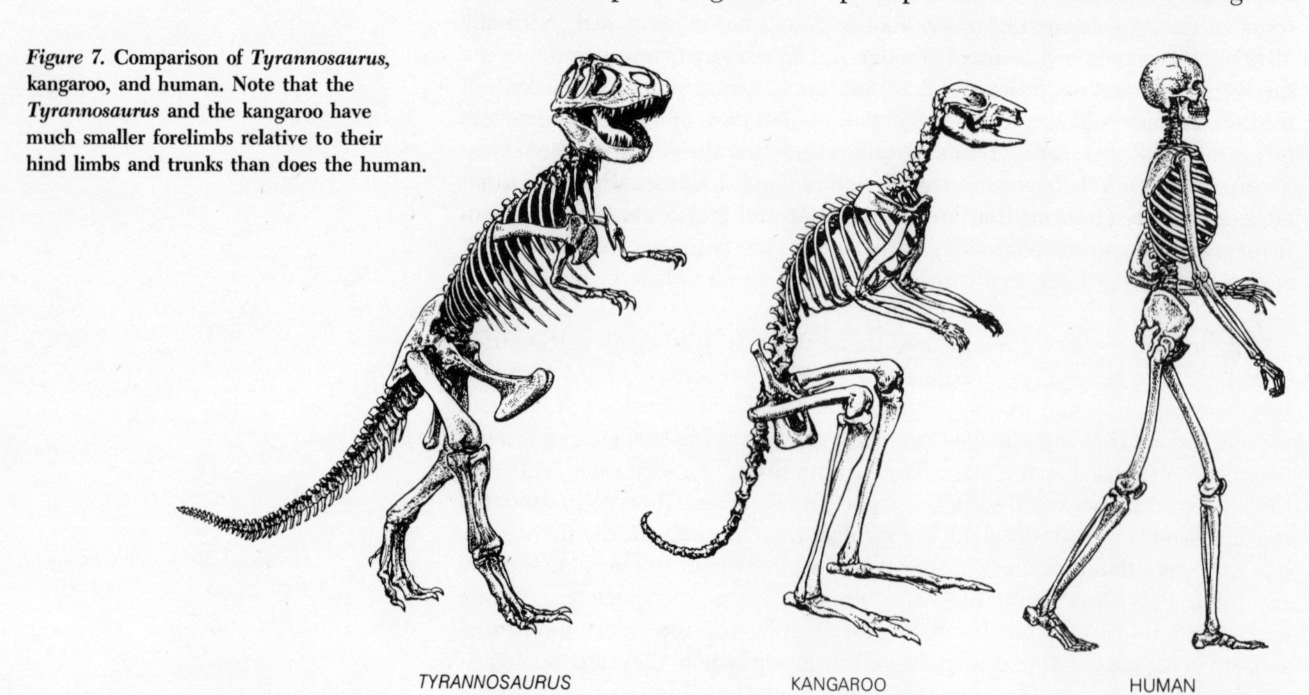

***Figure 7.* Comparison of *Tyrannosaurus*, kangaroo, and human. Note that the *Tyrannosaurus* and the kangaroo have much smaller forelimbs relative to their hind limbs and trunks than does the human.**

semi-erect locomotion on the ground, the forelimbs became accentuated in development. Witness the result in a chimp or a gibbon: the forelimbs are longer than the hindlimbs; the shoulder-girdle bones and muscles are expanded in size (perhaps to support total body weight hanging beneath the branches); and the typical mammalian "paw" has been modified for grasping branches. Several million years ago, when our ancestors gradually left the trees to become bipedal terrestrial dwellers, they probably already possessed large forelimbs and hands, which, because of the ways they were used, tended to be preserved as such, rather than shrinking in relative size as did those of *Tyrannosaurus*. With these forelimbs and hands the immediate ancestors of *Homo sapiens* acquired their skills in manipulating objects, and it is very possible that this increased dexterity contributed to the expansion of the brain.

So far we have argued that endocrinological and behavioral alterations in sexual activity made development of a cooperative society possible, and that the altered skeleton of hominids permitted bipedal locomotion and use of the hands in new ways. Ecological factors, too, were probably partly responsible for the changes in the societies of prehumans. It is thought that, when such creatures left the trees to live in the savannah forest-edge habitat, they ate both plant and animal food. Dentition, animal remains, and occasional tools or weapons that have been found support this idea. Some anthropologists have speculated that the practice of hunting large game was the key to cooperation between early hominids. Obvious advantages and consequences of hunting in groups can be imagined. Probably we shall never know whether this type of cooperative food gathering predated and precipitated the altered sex behavior and family structure.

Perhaps the main conclusion to be drawn from this discussion is that cultural and biological evolution of humans are intimately linked—feedback from each affects the other so that both change in time. The focus of change, of course, is the brain and its increase in size and complexity. In particular, the number of brain associational neurons (intermediate neurons between the sensory input cells and the motor output nerve cells) grew until ultimately a new property—selfawareness or abstract thought—appeared. We cannot say, as yet, at what point the number and complexity of interactions between neurons became great enough to generate this property, but there is no reason to invoke any nonbiological element to explain it.

Our survey, then, ends with humans. We see how an incredibly intricate series of changes, occurring in the course of millions of years, has led to the diversity of vertebrate life and to modern human beings. In ourselves we can see traces of many of those steps: filter feeding, swimming and its control, feeding mechanisms, the invasion of land, the control of high body temperature, and finally the cultural-biological interplay that led to thought; all have left their marks. Yet these are merely focal points for further study. We have looked at some questions that are still unanswered and at some speculations that are still controversial. As long as they remain so, our survey is not really at an end, but only just begun.

N.K.W.

SPECULATIONS ON A THEORY BY WESSELLS

"Thus, one might argue that the unique feature of human physiology—the behavior that most distinguishes us from other animals—is sex for pleasure alone, rather than merely for reproduction."

Although in cities we appear
to stand apart from eel and bear,
our heart is chambered like the hawk's;
our parts, connected limb to bone,
are analogs of antelopes';
and eyes and ears, your sweet tongue too,
supply the heads of skink and newt.
Placentas fed the shark and you.

Shiny fish can shift their prism;
the tree toad, switch from green to brown;
the bat and dolphin see by sound.
While geese, by light, detect direction,
by night the owl can hear to hunt.
Young chimps choose grass to capture ants.
And you, love, caught on the rim of sleep,
ask if humans are unique.

Objective, naked inspection shows,
oh not huge heads, odd thumbs or stance,
divide us from the beasts below—
instead: our vibrant body-dance.
You smile, sex for pleasure alone,
without intent, cycle or season;
the fire, caught to cross the pair,
becomes our reason dreaming here.

Charlea Massion

I

VASCULAR SYSTEM BIOLOGY

The arteries have no sensation, for they even are without blood, nor do they all contain the breath of life; and when they are cut only the part of the body concerned is paralysed . . . the veins spread underneath the whole skin, finally ending in very thin threads, and they narrow down into such an extremely minute size that the blood cannot pass through them nor can anything else but the moisture passing out from the blood in innumerable small drops which is called sweat.

Pliny
NATURAL HISTORY, XI, lxxxix

I VASCULAR SYSTEM BIOLOGY

INTRODUCTION

Each of the Sections in this book deals with one of the major physiological systems in the bodies of vertebrates. This first Section is concerned with the system that is central to the life process in large multicellular animals such as vertebrates: the circulatory system. The flowing blood is the source of oxygen and nutrients for tissue cells, the vehicle for removal and transport of cellular wastes, and the mode of transmittal of chemical regulators (hormones). It is involved in pH control, ionic balance, and the movement of heat. Consequently, the regulation of blood flow, overall and at specific sites in the body, is a crucial feature of the hierarchy of interconnected control circuits. Thus, the control of respiration, of body temperature, and of kidney function, among other systems, are dependent upon the vascular system and its proper regulation and function.

Since the time of classical studies by comparative anatomists, we have known much about evolutionary changes in the arterial arches, the veins, and the heart among various types of vertebrates. But these changes would be of little value if concomitant alterations had not occurred in the control machinery and in hemoglobin, the respiratory pigment. Let us consider, first, some properties of that pigment.

Each molecule of hemoglobin is a complex of the protein globin and four heme groups; each heme is a porphyrin molecule containing an iron atom. Oxygen binds reversibly to the iron in a manner that varies with the partial pressure of the gas in the neighborhood of the hemoglobin; thus, as partial pressure rises, binding rises, and vice versa. This relation is described by a sigmoid curve (an oxygen dissociation curve, such as those shown in Figure 1) that has a characteristic shape for each vertebrate species.

The position of the dissociation curve at various partial pressures of oxygen defines the hemoglobin's affinity for oxygen. Curves for most fish bloods show complete loading at relatively low partial pressures. Such hemoglobin is said to have a "high" affinity for oxygen because it binds the oxygen avidly at low partial pressure. Terrestrial vertebrates (and, in particular, birds and mammals) have curves far to the right of those of fish; their hemoglobin has a "low" affinity for oxygen and loads only at the high partial pressure of oxygen in normal air. The low affinity of avian and mammalian hemoglobin for oxygen is of vital importance for high rates of metabolism and elevated body temperature. Low affinity in effect means that the hemoglobin *unloads* its oxygen in the tissues at relatively high partial pressures (pressures at which fish hemoglobin would remain completely loaded). This allows more oxygen to be present around and in cells, a necessity for the high rates of metabolism required for body temperatures of 37 to 40 degrees centigrade.

Dissociation curves are affected by such things as carbon dioxide, pH, and temperature. An increase in carbon dioxide dissolved in blood lowers the pH (some other agents also produce hydrogen ions and lower the pH) and thus decreases hemoglobin's affinity for oxygen; consequently, dissociation occurs more readily in tissue spaces where carbon dioxide is produced by metabolic activity. But in lung alveoli, where carbon dioxide is lost to the air and pH rises, affinity is increased and oxygen is bound more readily to hemoglobin (the sensitivity of hemoglobin to carbon dioxide or pH is termed the Bohr effect).

Oxygen dissociation curves are also affected by the presence of an organic molecule produced by erythrocytes themselves. This 2,3-diphosphoglycerate decreases the affinity of hemoglobin for oxygen. As one might predict, larger quantities of the substance are present in red cells as they pass through body tissues; smaller quantities are present as blood passes through the lungs. Thus, this intracellular substance acts like hydrogen ions and produces a shift in the dissociation curve similar to that of the Bohr effect. It is interesting that, even in the first day or two of acclimatization to high altitude, the overall quantities of 2,3-diphosphoglycerate rise in human blood, thereby shifting the dissociation curve to the right and allowing oxygen to be surrendered more easily in the tissues.

Alterations in hemoglobins of different vertebrates provide several interesting lessons about the mechanisms of evolution. In terrestrial mammals the degree of the Bohr effect is inversely proportional to the body weight: mouse hemoglobin responds dramatically to increased carbon dioxide by giving up its oxygen very easily and, as a result, it is a more "efficient" respiratory pigment; the larger the mammal, the less the hemoglobin responds to increased partial pressures of carbon dioxide.

These differences are attributable to the problem of temperature control. The surface area of a mouse is large in relation to its mass; hence, the area from which heat can be lost is great in proportion to the quantity of tissue producing heat. Because the animal cannot remain active (and warm) if oxygen levels fall, the exaggerated Bohr effect works to insure that this emergency does not arise. The proportions are the opposite in an elephant, whose body mass is large in relation to surface area. As a result, there is less chance of precipitous heat loss in the event of subnormal oxygen supply to the tissues. Interestingly, a large aquatic mammal capable of prolonged diving has a much greater Bohr effect than would be predicted on the basis of mass. Humpback whale hemoglobin is the Bohr equivalent of a guinea pig's, for instance, and shows significant decrease in affinity for oxygen under conditions of increasing carbon dioxide. This is obviously a special adaptation that permits rapid oxygen loading while the animal is at the surface, complete unloading as carbon dioxide accumulates during long dives, and tolerance of hydrogen ion production during such dives.

Quite a different and unexpected alteration has taken place in hemoglobins of extremely active pelagic fishes. So much muscular activity goes on in these creatures that large quantities of lactic acid continually tend to lower blood pH to the point where normal fish hemoglobin with its Bohr effect would be unable to load oxygen in the gills. As emphasized by Peter Hochachka and George Somero in their important book *Strategies of Biochemical Adaptation*, the

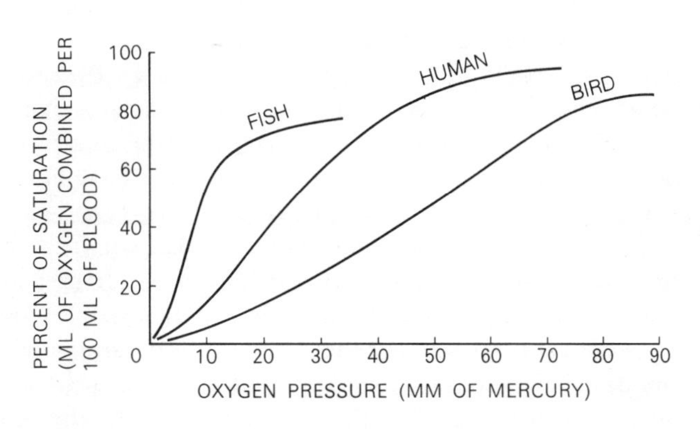

Figure 1. **Oxygen dissociation curves for various vertebrates. Fish hemoglobin loads at low partial pressures of oxygen, whereas the pigment of the homeotherms requires much more oxygen in the environment before high percentages of saturation are reached.**

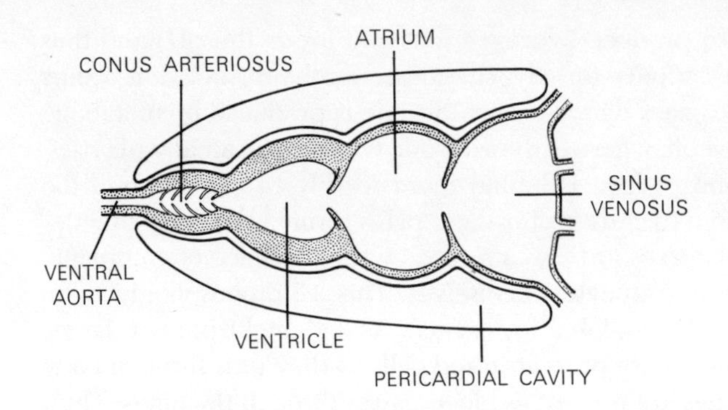

Figure 2. The probable arrangement of the heart chambers in a primitive vertebrate. As blood flows from the right to the left through the heart, its pressure is raised by action of the increasingly thick muscular walls (stippled) of the chambers. Back flow is prevented by valves between the chambers.
[After Theodore W. Torrey, _Morphogenesis of the Vertebrates,_ Second Edition, Wiley, 1967.]

ancestors of these fishes apparently evolved a hemoglobin lacking the Bohr effect, so that today's representatives can exploit a way of life, an ecological niche, not open to more lethargic relatives.

The hemoglobin of vertebrates is, of course, normally contained within erythrocytes. This is not, as long assumed, because hemoglobin solutions are particularly viscous. Among the possible reasons for the retention of the respiratory pigment within red blood cells are: (1) the possibility that free hemoglobin would excessively increase the osmotic pressure of the blood, thereby necessitating a much higher blood pressure to counteract the tendency for fluid to enter the blood; (2) the fact that cells help to cause turbulence in flowing blood, thus increasing the efficiency of gas exchange; and (3) the desirability of keeping certain enzymes or small molecules (as 2,3-diphosphoglycerate) in close proximity to hemoglobin. Besides these features of red blood cells, a crucial property is their deformability. If erythrocytes are rigidified in shape with chemical fixatives, they cannot pass through the narrow capillary beds in some parts of the body.

As the blood pigment changed in the course of time, so did the vascular system and its controls. It seems likely that, in the earliest vertebrates, blood was originally propelled by rhythmic contractions along the main anterior ventral blood vessel. As these organisms increased in size, the pumping function apparently localized in a discrete region that was to become the heart. The walls of the blood vessel in that region acquired more and more cardiac muscle cells and thickened as a result. Such a heart pump works most efficiently if its cavity is expanded somewhat beyond the resting volume before each contraction occurs. The four chambers of the primitive tubular vertebrate heart, such as that in some fish, appear to be stages for accomplishing this purpose by raising the blood pressure. When venous blood returns to the heart of a fish, it is under such low pressure that it cannot expand the thick-walled anterior pump, the ventricle. Therefore, the blood first enters a thin-walled chamber, the sinus venosus, where it is propelled forward into a somewhat thicker-walled space, the atrium, which finally raises blood pressure enough to expand the ventricle. This simplistic interpretation must be modified somewhat, because it does not take into account the elastic recoil of contracting cardiac muscle, a property that also tends to expand the chambers.

The intricate workings of the modern heart are recounted in the first article in this Section, "The Heart," by Carl J. Wiggers. In an earlier *Scientific American* article (Offprint 1067, "The Heart's Pacemaker"), E. F. Adolph pointed out that the various chambers each beat at a characteristic rate during the embryonic development of a vertebrate; the anteriormost chamber, which develops first, beats at the slowest rate, the next most posterior chamber develops next and sets the pace with a higher beat, and lastly the most posterior chamber forms and sets the final fast pace. The derivative of this chamber—the sinoatrial node—is the pacemaker in higher vertebrate adult hearts.

If we look at primitive hagfish adults (which are jawless fishes of the order Cyclostomata), we see that the heart beat is determined by the sinus venosus; the beat is wholly myogenic (intrinsic to the muscle cells) in origin and no nerves accelerate or slow the beat. Interestingly, the heart muscle does not respond to direct application of acetylcholine, the agent liberated by the vagus nerve of higher vertebrates to decrease cardiac output. In higher bony fish and sharks, however, vagal fibers do innervate the heart and their discharge of acetylcholine slows the beat rate and decreases the amplitude of contraction; consequently, less blood is pumped. Finally, in terrestrial vertebrates, we find that the vagus functions as it does in fish, and, in addition, sympathetic nerve fibers discharge noradrenaline, which accelerates the beat and increases its amplitude. Thus, we see three stages of control: myogenic, myogenic plus inhibition, and myogenic plus inhibition and stimulation. Clearly, more precise regulation is possible with such increasingly complex control circuitry. In the embryos of today's birds and mammals, the basic myogenic pattern described above develops before nerve fibers reach the heart and during that period the cardiac muscle tissue does not respond to application of acetylcholine or noradrenaline. In this way, the developmental sequence within a single embryo reflects the events that occurred during vertebrate evolution.

Although the control of heart beat is the core of vertebrate circulatory regulation, the actual cardiac output (volume of blood pumped) in a mammal is determined to a great degree by the volume of blood returning to the heart through the veins. This quantity is governed by the degree of vasoconstriction of peripheral blood vessels (arterioles, capillaries, veins); the passage of blood through these vessels is described in "The Microcirculation of the Blood" by Benjamin Zweifach). Control of venous return is still not completely understood. For example, Edwin Wood ("The Venous System," *Scientific American* Offprint 1093) has pointed out that when a human stands, return of blood upward through the veins of the leg is difficult because of gravity. Valves spaced along the veins prevent backflow and hold the blood in a series of stages. Nevertheless, a larger percent of total body blood accumulates in the legs. As this occurs, veins elsewhere in the body periphery contract and reduce their volume; this tends to maintain a constant volume of blood returning to the heart and so compensate for the temporary accumulation in the legs. In addition, the recent experiences of humans in the prolonged, gravity-free conditions of space craft, where minimal use of leg musculature occurs, suggests the importance of actual compression of the veins by the muscles in aiding venous return and maintaining general health of the blood vessels.

Exercise puts special demands upon the blood vascular system, since adequate blood flow to the brain must be maintained even if skeletal muscles require greatly increased blood supply to meet oxygen demands. Carleton B. Chapman and Jere H. Mitchell have shown in "The Physiology of Exercise" (Offprint 1011) how the flow of blood to various organs is readjusted as different levels of exercise are undertaken (see Figure 3). It can be seen that the brain's supply is maintained even though the blood flowing to the viscera and elsewhere is decreased in order to divert flow to the skeletal muscles.

Still another special demand is put on the blood vascular system in organisms such as the giraffe. This is because the brain is held high above the heart most of the time. In "The Physiology of the Giraffe," James V. Warren explains how high blood pressure and a modification in the usual breathing pattern combine to make life with such a long, erect neck feasible. Incidentally, the same problem would have required solution in some of the large, very long-necked dinosaurs, such as those in the genus *Brachiosaurus*. Calculations suggest that the blood pressure required to lift fresh blood to the brain of *Brachiosaurus* when the neck was fully erect might have required ventricular pressures of about 500 millimeters Hg (in comparison, pressures of about 125 millimeters Hg have been measured at the level of the jaw in a quietly standing giraffe). To cope with such extreme pressure, a dinosaur's heart probably would have had to be huge, perhaps weighing 1.5 tons and with extraordinarily thick walls (the heart of a whale of similar size weighs only about 200 kilograms). These considerations lead some biologists to conclude that such large dinosaurs lived in shallow aquatic habitats, where the pressure of water around their body would act like the water immersing a whale, thereby buoying the body by

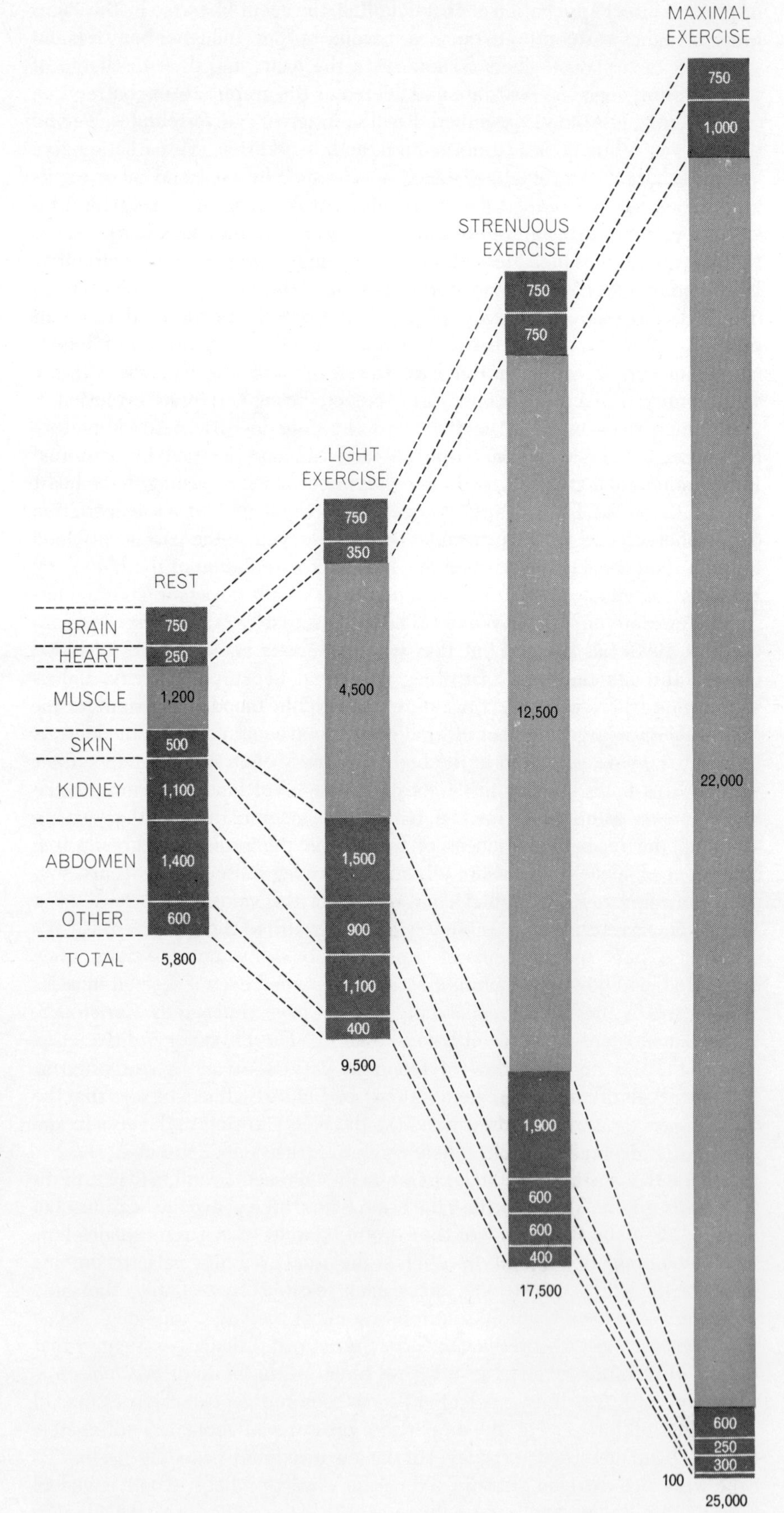

Figure 3. **Diversion of blood as a function of exercise. Numbers show blood flow in milliliters per minute.**
[From "The Physiology of Exercise" by Carleton B. Chapman and Jere H. Mitchell. Copyright © 1965 by Scientific American, Inc. All rights reserved.]

applying pressure to the sides. This would allow a much smaller heart. These kinds of factors are ignored by some paleontologists who offer fanciful descriptions of organisms such as *Brachiosaurus* galloping over the Mesozoic landscape with heads erect (and so, with immense hearts pumping ferociously)!

A specialized portion of the circulatory system is discussed in "The Lymphatic System" by H. S. Mayerson. Lymph vessels are found in all terrestrial vertebrates (and not just in mammals, as is implied by Mayerson). The higher blood pressures and rates of flow of the partially separate pulmonary circulation of amphibians and reptiles apparently produced leakage of plasma through the walls of blood capillaries. A collecting system evolved to drain the tissue spaces and to return the lost substances to the regular circulatory system. These lymphatic vessels are present in more complex form in birds and mammals.

Another type of specialization in the circulatory system will be encountered in later sections of this book, namely the rete mirabile countercurrent-flow apparatus (see "The Eland and the Oryx" in Section III and "Fishes with Warm Bodies" in Section IV). These rete beds are used in many different ways by vertebrates. Heat, gasses, and some metabolites may be passed between the tiny rete vessels in which fluid flows in opposite directions. M. A. Baker, in "The Brain Cooling System of Mammals," describes how a rete system is employed by many different mammals to keep a cool head under conditions of potential heat stress.

A particularly informative way to study the function of a rete mirabile is to examine the swim bladders of modern bony fishes. Recall that such bladders are used as hydrostatic organs that permit a trout or a sea bass to maintain neutral buoyancy at a given depth. The rete in a swim bladder is responsible for the efficient exchange of gas (oxygen) from the capillary that carries blood away from the lumen of the bladder to the capillary that brings blood to the bladder. As can be seen in Figure 4, blood passes through the rete, into an artery, through the capillary bed beneath the gas-secreting epithelium, back into a vein, and once more through the rete capillaries. This diagram makes it clear that the business of the rete is *exchange*—exchange of commodities such as oxygen or organic molecules. The separate capillary bed, which is not a countercurrent exchanger, is the *source* of gas to be passed into the swim bladder.

How can oxygen be maintained or actually secreted into the swim bladder when a fish is at great depths and the bladder already contains nearly pure oxygen at hundreds of atmospheres pressure? One factor may be lactic acid, which is apparently produced by the secretory epithelial cells of the gas gland. This lactic acid enters the blood, is exchanged across the rete to inflowing blood, and therefore increases dramatically in concentration. Hydrogen ions apparently are produced and follow the same path. Both substances serve to lower the blood pH as they accumulate in the "closed" circuit. Because of the Bohr effect, hemoglobin in the blood gives up its oxygen more easily as this point is approached. The lactic acid is also thought to cause "salting out" of oxygen from the blood fluid. In other words, as the concentration of the acid

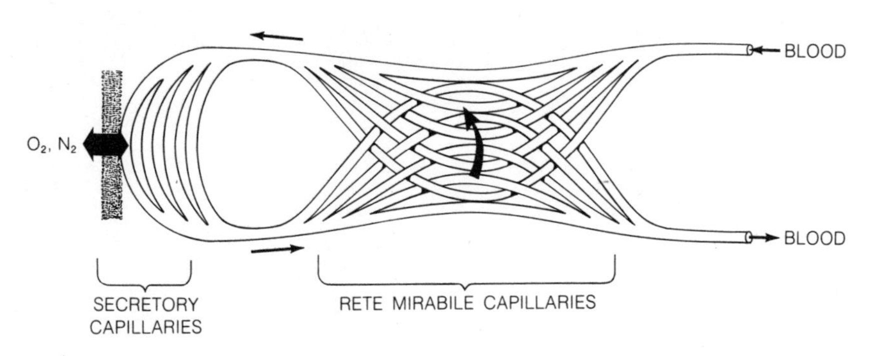

***Figure 4.* An idealized diagram demonstrating the physical separation of the secretory capillaries located near the epithelium of the gas gland in a swim bladder and the rete capillaries where countercurrent exchange takes place.**

Table 1. **Relationship between the depth at which given species of fish live in the ocean and the length of the capillaries in the rete mirabile of their swim bladders.**
[After N. B. Marshall, *Explorations in the Life of Fishes,* Harvard University Press, 1971.]

Depth (in meters)	Length of Rete Capillaries (in millimeters)
150–500	1–2.0
500–1000	2.5–6.0
750–1500	10–15

rises, oxygen solubility decreases and transfer from blood fluid to swim bladder lumen is favored. In addition to these effects of lactic acid, the hemoglobin of many fishes with swim bladders shows a "Root effect," whereby the actual quantity of oxygen that can be bound at any one partial pressure of oxygen is decreased as pH falls; the oxygen dissociation curve does not move to the right as it does in the Bohr effect, but rather the upper segment of the curve is lowered, as shown in Figure 5. Thus, hemoglobin can unload oxygen, even though hundreds of atmospheres of oxygen may be present in the swim bladder.

Interestingly, the length of the capillary vessels in the rete mirabile is a direct function of the normal depth at which a fish lives. Obviously, such anatomical modifications must have evolved for the bony fishes to attain the ability to invade great depths while maintaining buoyancy.

A survey of vertebrate vascular systems and hemoglobins provides good illustrations of the versatility of vertebrates in taking advantage of special features of the environment. For instance, Johan H. Ruud ("The Ice Fish," *Scientific American* Offprint 1025) has found that the "ice" fishes produce no hemoglobin. Instead, they seem to have an extraordinarily large blood fluid volume that meets the demands of gas transport. Low rates of metabolism in very cold ocean waters demand far less oxygen and produce less carbon dioxide than would metabolism in warmer waters. Cold water also contains more dissolved oxygen than does warm water. This is probably important to ice fish who, because no hemoglobin passes through their gill capillaries, have no pigment to bind oxygen actively and to remove substantial quantities of it from the water. Instead, simple diffusion of oxygen into the blood fluid probably occurs. Consequently, the larger quantity of dissolved oxygen in sea water, the greater the amounts that can diffuse per unit of time into the flowing blood.

Another group of vertebrates has taken advantage of the high oxygen content of cold water in quite another way. In fast-running mountain streams, several types of lungless salamanders are found. These creatures have such extensive capillary nets in their skin that all gas exchange can occur there and no lungs are formed during embryonic development. A peripheral advantage of this modification is that the air-filled lungs are not present to act as buoyancy organs; therefore specific gravity of the body is greater and there is thus less chance of the animal being swept away in the rapidly flowing water currents. But unlike the ice fish, lungless salamanders do utilize hemoglobin and erythrocytes to transport oxygen.

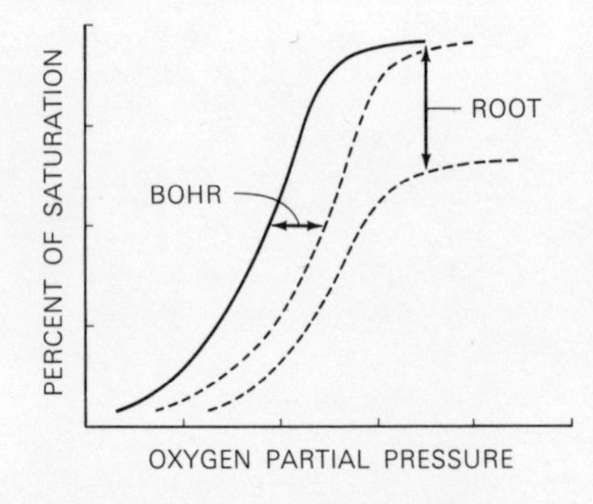

Figure 5. **Oxygen dissociation curves showing the Bohr and Root effects. In the Bohr effect the affinity of hemoglobin for oxygen decreases (the curve is displaced to the right). In the Root effect, the maximal quantity of oxygen that can combine with the respiratory pigment and be transported is decreased.**

The Heart 1

by Carl J. Wiggers
May 1957

It pumps five quarts of blood in a minute, 75 gallons in an hour, 70 barrels in a day and 18 million barrels in 70 years. It does this by means of the most intricately woven muscle in the body

The blood bathes the tissues with fluid and preserves their slight alkalinity; it supplies them with food and oxygen; it conveys the building stones for their growth and repair; it distributes heat generated by the cells and equalizes body temperature; it carries hormones that stimulate and coordinate the activities of the various organs; it conveys antibodies and cells that fight infections—and of course it carries drugs administered for therapeutic purposes. No wonder that William Harvey, the discoverer of the circulation, ardently defended the ancient belief that the blood is the seat of the soul.

The blood cannot support life unless it is kept circulating. If the blood flow to the brain is cut off, within three to five seconds the individual loses consciousness; after 15 to 20 seconds the body begins to twitch convulsively; and if the interruption of the circulation lasts more than nine minutes, the mental powers of the brain are irrevocably destroyed. Similarly the muscles of the heart cannot survive total deprivation of blood flow for longer than 30 minutes. These facts emphasize the vital importance of the heart as a pump.

The work done by this pump is out of all proportion to its size. Let us look at some figures. Even while we are asleep the heart pumps about two ounces of blood with each beat, a teacupful with every three beats, nearly five quarts per minute, 75 gallons per hour. In other words, it pumps enough blood to fill an average gasoline tank almost four times every hour just to keep the machinery of the body idling. When the body is moderately active, the heart doubles this output. During strenuous muscular efforts, such as running to catch a train or playing a game of tennis, the cardiac output may go up to 14 barrels per hour. Over the 24 hours of an average day, involving not too vigorous work, it amounts to some 70 barrels, and in a lifetime of 70 years the heart pumps nearly 18 million barrels!

The Design

Let us look at the design of this remarkable organ. The heart is a double pump, composed of two halves. Each side consists of an antechamber, formerly called the auricle but now more commonly called the atrium, and a ventricle. The capacities of these chambers vary considerably during life. In the human heart the average volume of each ventricle is about four ounces, and of each atrium about five ounces. The used blood that has circulated through the body—low in oxygen, high in carbon dioxide, and dark red (not blue) in color—first enters the right half of the heart, principally by two large veins (the superior and inferior venae cavae). The right heart pumps it via the pulmonary artery to the lungs, where the blood discharges some of its carbon dioxide and takes up oxygen. It then travels through the pulmonary veins to the left heart, which pumps the refreshed blood out through the aorta and to all regions of the body [*see diagram on following page*].

The thick muscular walls of the ventricles are mainly responsible for the pumping action. The wall of the left ventricle is much thicker than that of the right. The two pumps are welded together by an even thicker dividing wall (the septum). Around the right and left ventricles is a common envelope consisting of several layers of spiral and circular muscle [*see diagrams on page 25*]. This arrangement has a number of mechanical virtues. The blood is not merely pushed out of the ventricles but is virtually wrung out of them by the squeeze of the spiral muscle bands. Moreover, it is pumped from both ventricles almost simultaneously, which insures the ejection of equal volumes by the two chambers—a necessity if one or the other side of the heart is not to become congested or depleted. The effectiveness of the pumping action is further enhanced by the fact that the septum between the ventricles becomes rigid just before contraction of the muscle bands, so that it serves as a fixed fulcrum at their ends.

The ventricles fill up with blood from the antechambers (atria). Until the beginning of the present century it was thought that this was accomplished primarily by the contractions of the atria, *i.e.*, that the atria also functioned as pumps. This idea was based partly on inferences from anatomical studies and partly on observation of the exposed hearts of frogs. But it is now known that in mammals the atrium serves mainly as a reservoir. The ventricles fill fairly completely by their elastic recoil from contraction before the atria contract. The contraction of the latter merely completes the transfer of the small amount of blood they have left. Indeed, it has been found that the filling of the ventricles is not significantly impaired when

ANATOMY OF THE HEART and its relationship to the circulatory system is schematically depicted on the following page. The arterial blood is represented in a bright red; venous blood, in a somewhat paler red. The capillaries of the lungs are represented at left and right; the capillaries of the rest of the body, at top and bottom. The term atrium is now used in preference to auricle.

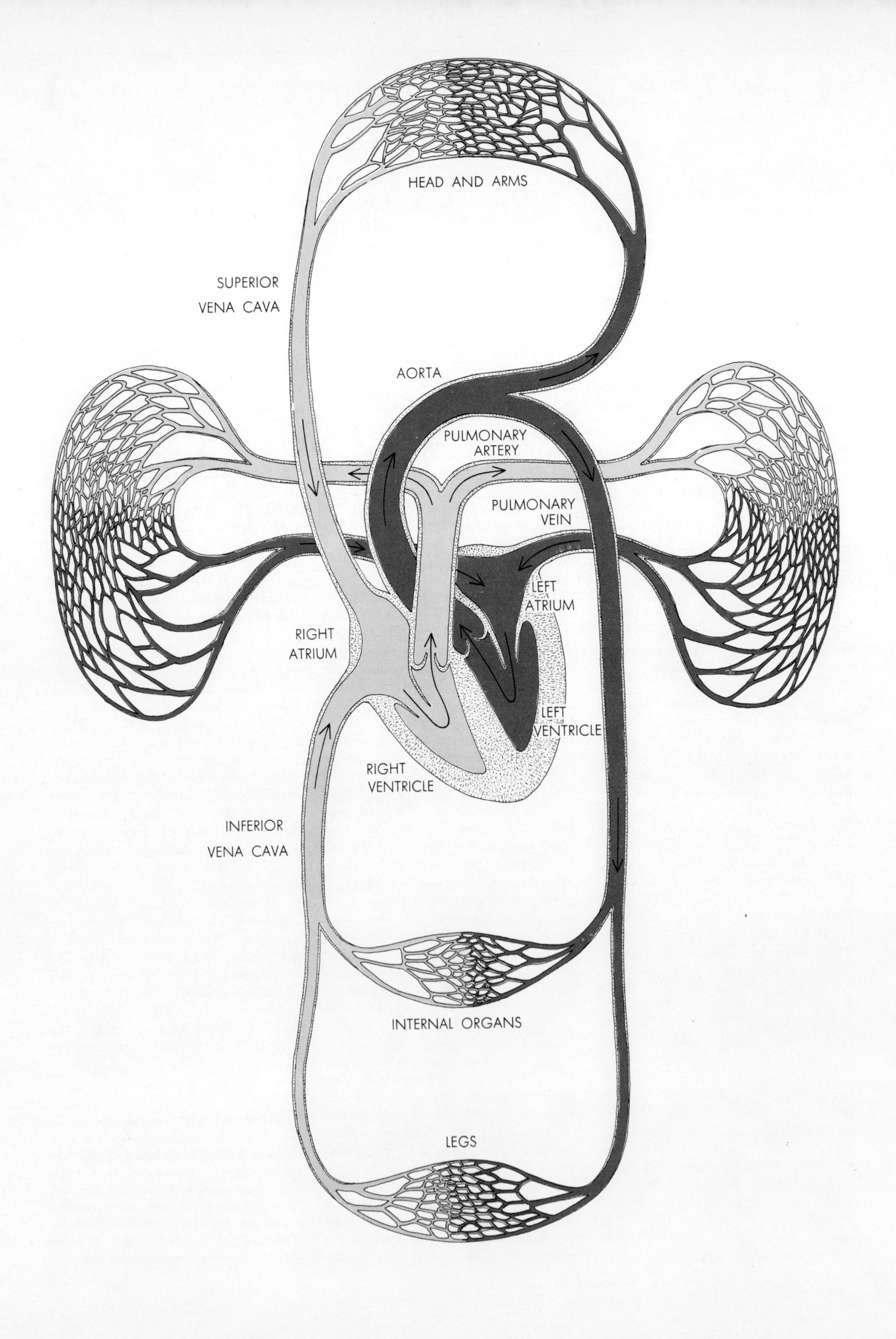
HEAD AND ARMS
SUPERIOR
VENA CAVA
AORTA
PULMONARY
ARTERY
PULMONARY
VEIN
LEFT
ATRIUM
RIGHT
ATRIUM
LEFT
VENTRICLE
RIGHT
VENTRICLE
INFERIOR
VENA CAVA
INTERNAL ORGANS
LEGS

disease destroys the ability of the atria to contract.

Factors of Safety

Since most of us believe that every biological mechanism must have some purpose, the question arises: Why complicate the cardiac pump with contractions of the atria if the ventricles alone suffice? The answer is that they provide what engineers call a "factor of safety." While the atrial contractions make only a minor contribution to filling the ventricles under normal circumstances, they assume an important role when disease narrows the valve openings between the atria and the ventricles. Their pumping action then is needed to drive blood through the narrowed orifices.

The ventricular pumps also have their factors of safety. The left ventricle can continue to function as an efficient pump even when more than half of its muscle mass is dead. Recently the astounding discovery was made that the right ventricle can be dispensed with altogether and blood will still flow through the lungs to the left heart! An efficient circulation can be maintained when the walls of the right ventricle are nearly completely destroyed or when blood is made to by-pass the right heart. Obviously the heart is equipped with large factors of safety to meet the strains of everyday life.

This applies also to the heart valves. Like any efficient pump, the ventricle is furnished with inlet and outlet valves; it opens the inlet and closes the outlet while it is filling, and closes the inlet when it is ready to discharge. The pressure produced by contraction of the heart muscles mechanically closes the inlet valve between the atrium and ventricle: shortly afterward the outlet valve opens to let the ventricle discharge its blood—into the pulmonary artery in the case of the right ventricle and into the aorta in the case of the left. Then as the muscles relax and pressure in the chambers falls, the outlet valves close and shortly thereafter the inlet valves open. The relaxation that allows the ventricles to fill is called diastole; the contraction that expels the blood is called systole.

While it might seem that competent valves are indispensable for the forward movement of blood, they are in fact not absolutely necessary. The laws of hydraulics play some peculiar tricks. As every farmhand knew in the days of hand well-pumps, if the valves of the pump were worn and leaky, one could

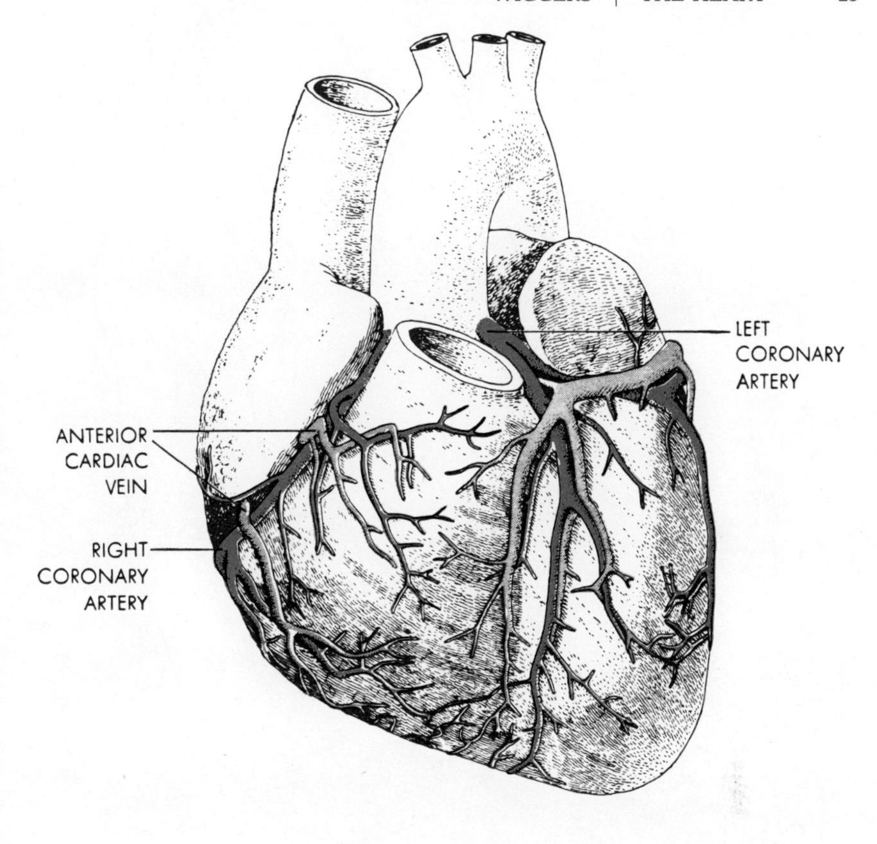

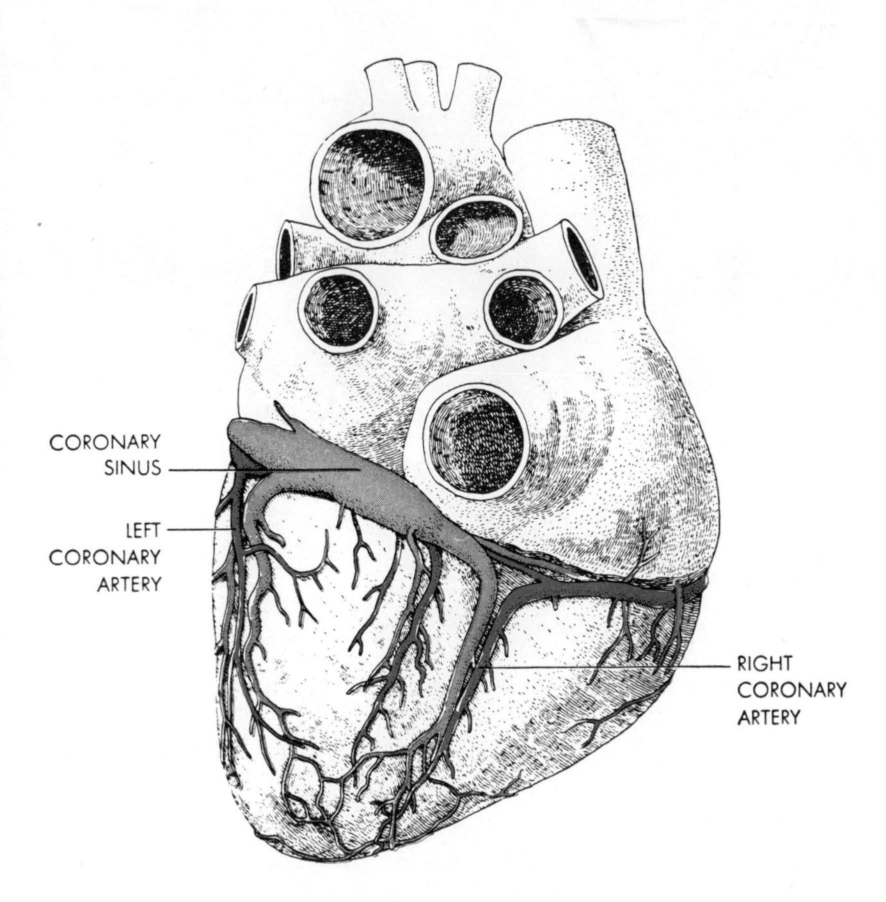

ARTERIES AND VEINS which carry blood to and from the muscles of the heart are shown from the front (*top*) and back (*bottom*). The arteries are bright red; the veins, pale red.

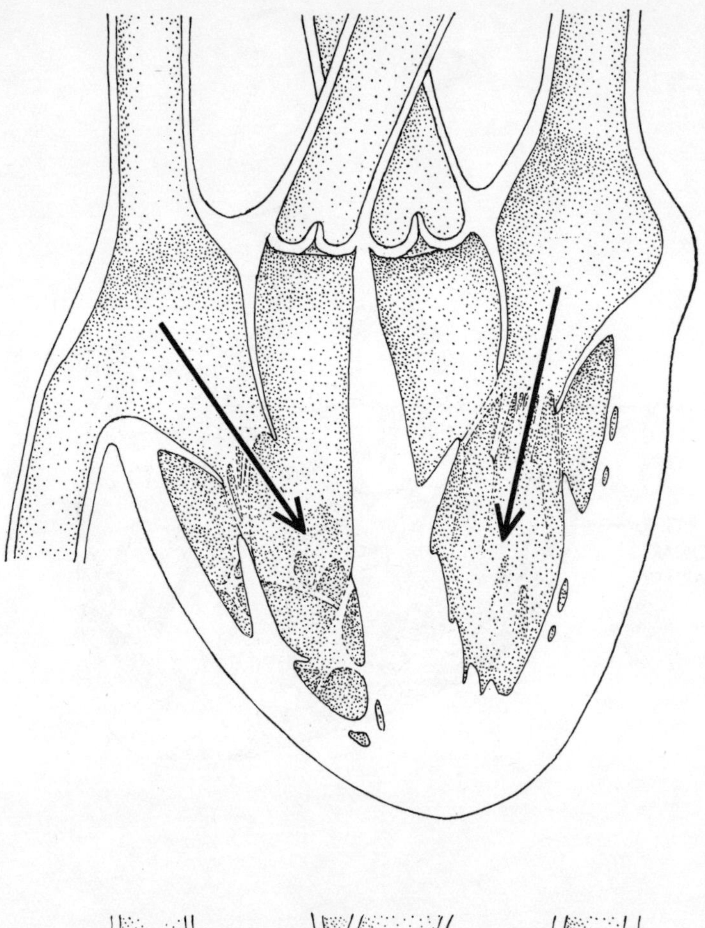

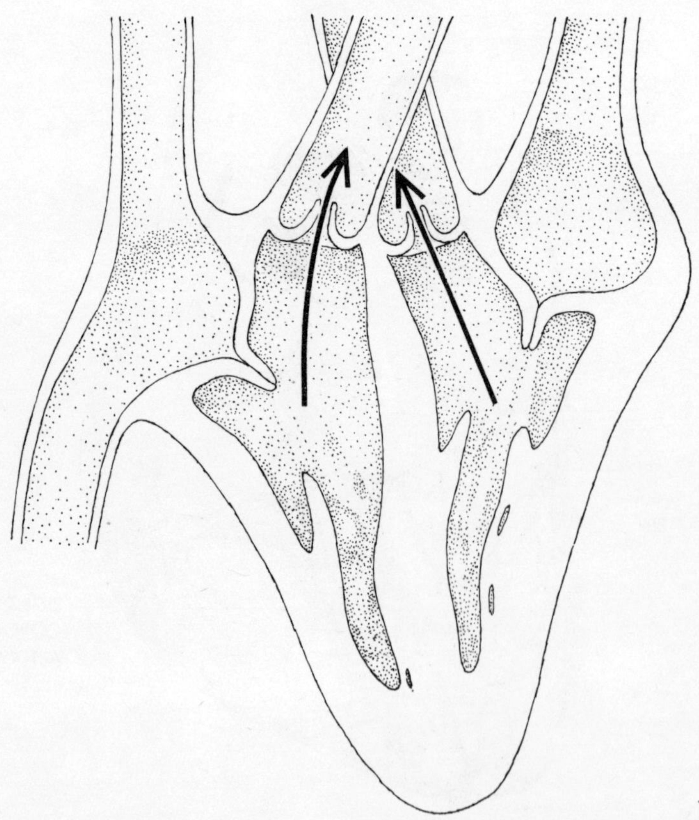

SYSTOLE AND DIASTOLE is the pumping rhythm of the heart. At the top is diastole, in which the ventricles relax and blood flows into them from the atria. The inlet valves of the ventricles are open; the outlet valves are closed. At the bottom is systole, in which the ventricles contract, closing the inlet valves and forcing blood through the outlet valves.

still draw water from the well by pumping harder. Similarly doctors have long been aware that patients can maintain a good circulation despite serious leaks in the heart valves. The factors of safety concerned are partly physical and partly physiological. The physical factor is more vigorous contraction of the heart muscles, aided by a structural arrangement of the deep muscle bands which tends to direct the blood flow forward rather than backward through the leaky valve. The physiological factor of safety is the mechanism known as "Starling's Law." In brief the rule is that, the more a cardiac muscle is stretched, the more vigorously it responds, of course within limits. The result is that the more blood the ventricles contain at the end of diastole, the more they expel. Of course they will fill with an excess of blood when either the inlet or the outlet valves leak. By pumping an extra volume of blood with each beat, the ventricles compensate for the backward loss through the atrial valves. In addition, the sympathetic nerves or hormones carried in the blood may spur the contractile power of the muscles. Under certain circumstances unfavorable influences come into play that depress the contractile force. Fortunately drugs such as digitalis can heighten the contractile force and thus again restore the balance of the circulation even though the valves leak.

Like any sharp closing of a door, the abrupt closings of the heart valves produce sounds, which can be heard at the chest wall. And just as we can gauge the vigor with which a door is slammed by the loudness of the sound, so a physician can assess the forces concerned in the closing of the individual heart valves. When a valve leaks, he hears not only the bang of the valves but also a "murmur" like the sigh of a gust of wind leaking through a broken window pane. The quality and timing of the murmur and its spread over the surface of the chest offer a trained ear considerable additional information. Sometimes a murmur means that the inlet and outlet orifices of the ventricles have been narrowed by calcification of the valves. In that case there is a characteristic sound, just as the water issuing from a hose nozzle makes a hissing sound when the nozzle is closed down.

Blood Supply

In one outstanding respect the heart has no great margin of safety: namely, its oxygen supply. In contrast to many other tissues of the body, which use as

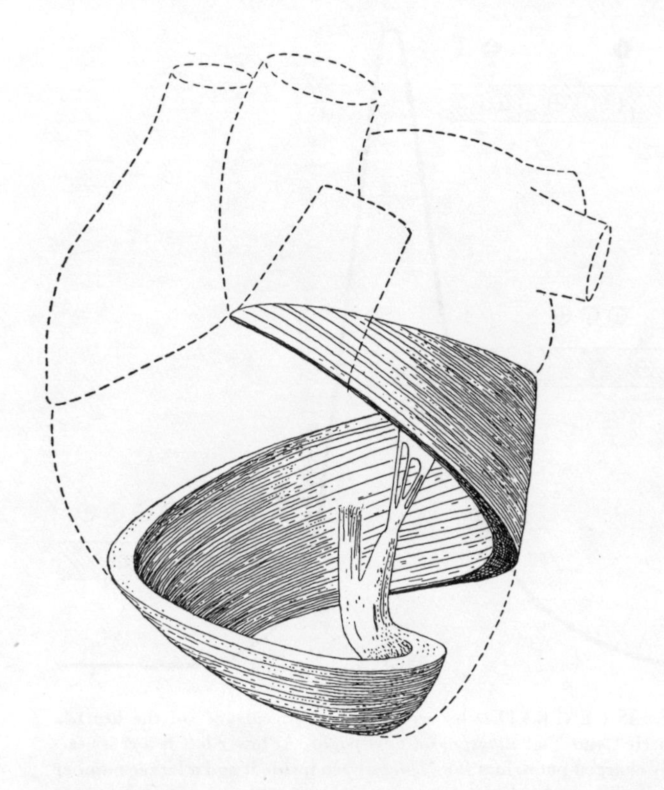

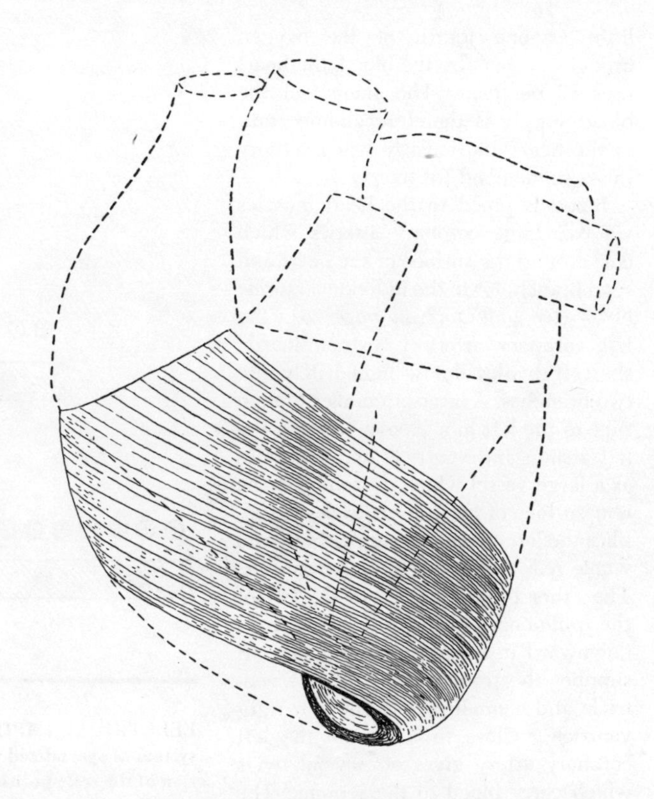

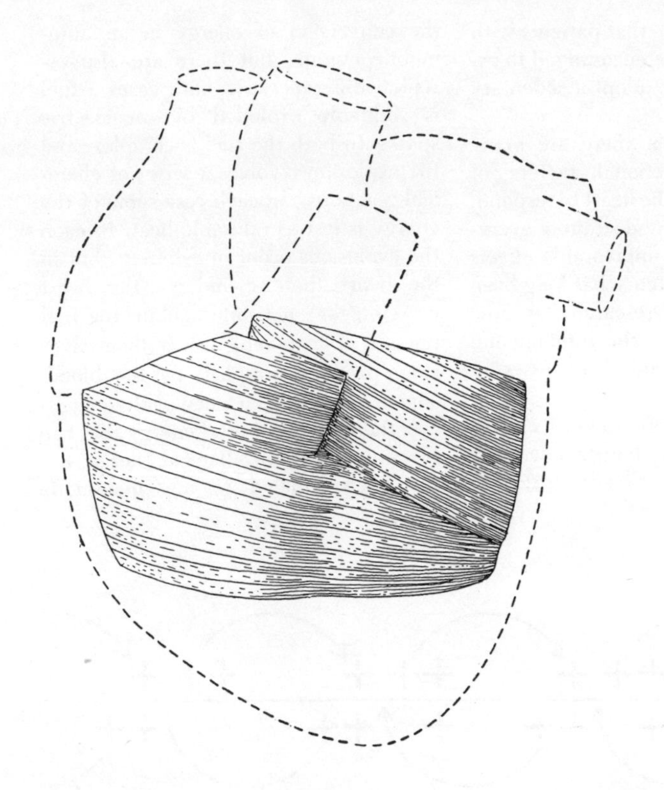

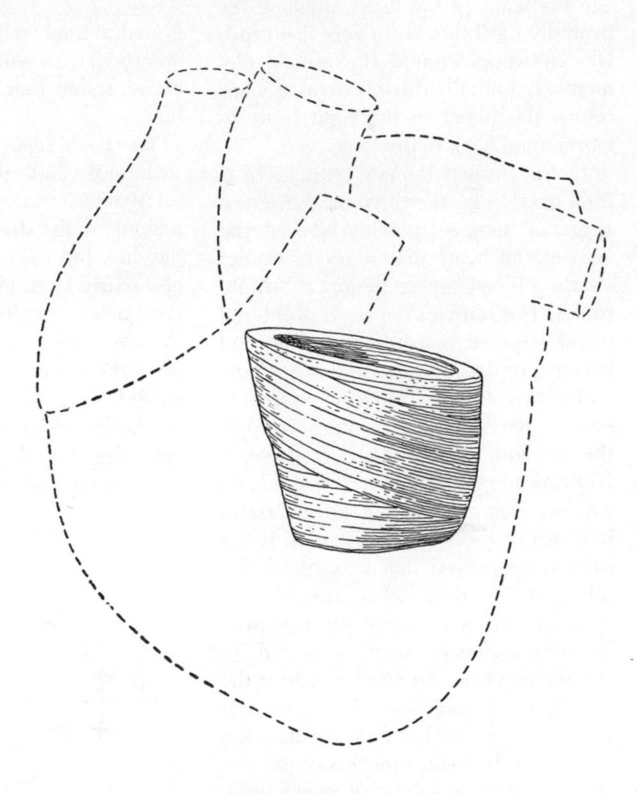

MUSCLE FIBERS of the ventricles are divided into four groups, one of which is shown in each of these four drawings. Two groups of fibers (*two drawings at top*) wind around the outside of both ventricles. Beneath these fibers a third group (*drawing at lower left*) also winds around both ventricles. Beneath these fibers, in turn, a fourth group (*drawing at lower right*) winds only around the left ventricle. The contraction of all these spiral fibers virtually wrings, rather than presses, blood out of the ventricles.

little as one fourth of the oxygen brought to them by the blood, the heart uses 80 per cent. The amount of the blood supply is therefore all-important to the heart, particularly when activity raises its demand for oxygen.

Blood is piped to the heart muscles via two large coronary arteries which curl around the surface of the heart and send branchings to the individual muscle fibers [*see diagrams on page 23*]. The left coronary artery is extraordinarily short. It divides almost immediately into two branches. A large circumflex branch runs to the left in a groove between the left atrium and ventricle and continues as a large vessel which descends on the rear surface of the left ventricle. It supplies the left atrium, the upper front and whole rear portion of the left ventricle. The other branch circles to the left of the pulmonary artery and then runs downward in a furrow to the apex. It supplies the front wall of the left ventricle and a small part of the rear right ventricle. Close to its origin the left coronary artery gives off several twigs which carry blood to the septum. The right coronary artery, embedded in fat, runs to the right in a groove between the right atrium and ventricle. It carries blood to both of them.

From the surface branches vessels run into the walls of the heart, dividing repeatedly until they form very fine capillary networks around the muscle elements. Eventually three systems of veins return the blood to the right heart to be pumped back to the lungs.

In the normal human heart there is little overlap by the three main arteries. If one of them is suddenly blocked, the area of the heart that it serves cannot obtain a blood supply by any substitute route. The muscles deprived of arterial blood soon cease contracting, die and become replaced by scar tissue. Now while this is the ordinary course of events, particularly in young persons, the amazing discovery was made some 20 years ago that the blocking of a main coronary artery does not always result in death of the muscles it serves. It has since been proved that new blood vessels grow in, from other arteries, if a main branch is progressively narrowed by atherosclerosis over a period of months or years. In other words, if the closing of a coronary artery proceeds slowly, a collateral circulation may develop. This biological process constitutes another factor of safety. Recent experiments on dogs indeed indicate that exercise will accelerate the development of collaterals when a major coronary artery is constricted. If this indication is confirmed, it may well be that patients with atherosclerosis will be encouraged to exercise, rather than to adopt a sedentary life.

We have seen that there are many structural and functional factors of safety which enable the heart to respond, not only to the stress and strain of everyday life, but also to unfavorable effects of disease. Their existence has long been recognized; modern research has now thrown some light on the fundamental physiological and chemical processes involved.

The heart's transformation of chemical energy into the mechanical energy of contraction has certain similarities to the conversion of energy in an automobile engine; but there are also essential differences. In both cases a fuel is suddenly exploded by an electric spark. In both the fuel is complex, and the explosion involves a series of chemical reactions. In each case some of the energy is lost as unusable heat. In each the explosions occur in cylinders, but in the heart these cylinders (the heart muscle cells) not only contain the fuel but are able to replenish it themselves from products supplied by the blood. The mechanical efficiency of these cells, *i.e.*, the fraction of total energy that can be converted to mechanical energy, has not been equaled by any man-made

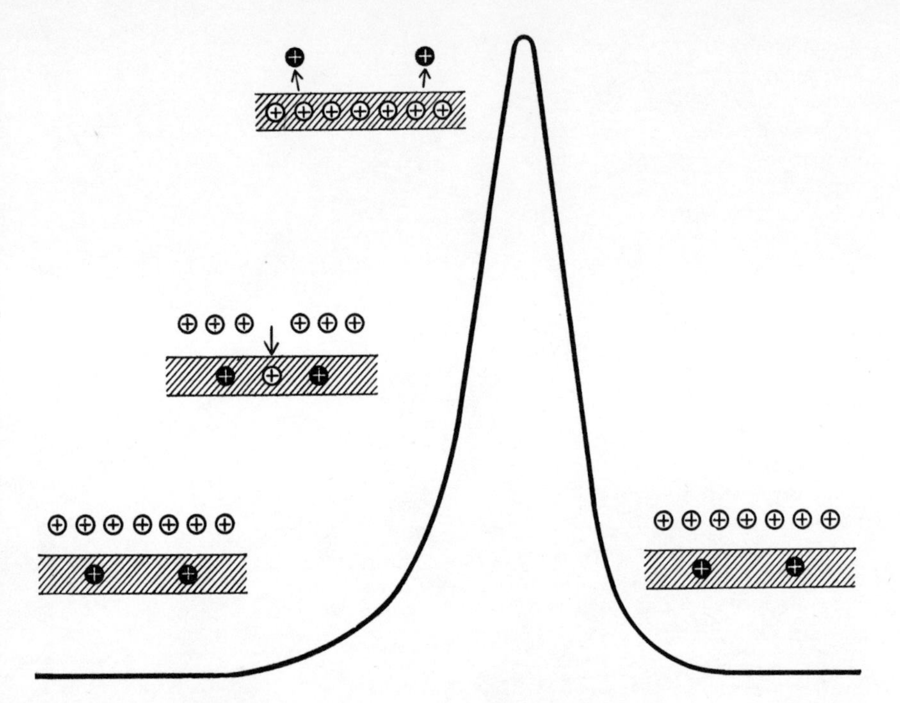

ELECTRICAL IMPULSE IS GENERATED by a cell in the "pacemaker" of the heart, a system of specialized muscle tissue (*see diagram on next page*). At lower left is a cross section of the cell; positively charged potassium ions (*black*) are inside it and a larger number of positively charged sodium ions (*white*) are outside. Because there are more positive charges outside the cell than inside, the inside of the cell is negatively charged with respect to the outside. When the cell is stimulated (*second cross section*), a sodium ion leaks across the membrane. Then many sodium ions rush across the membrane and potassium ions rush out (*third cross section*); this reverses the polarity of the cell and gives rise to an action potential (*peak in curve*). The original situation is then restored (*fourth cross section*)

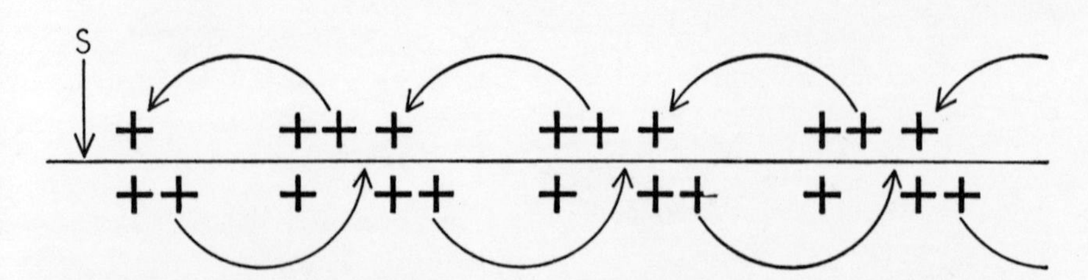

ELECTRICAL IMPULSE IS TRANSMITTED through the pacemaker system not as an electric current but as an electrical chain reaction. When a pacemaker cell is stimulated (S), it discharges and generates a local current which causes the depolarization and discharge of adjacent cells. In effect a wave of positive charge passes through the system (*curved arrows*).

machine designed in the pre-atomic age. The mechanism responsible for this efficiency is unique and very complex.

Under the microscope we can see that cardiac muscle consists of long, narrow networks of fibers, with connective tissue and tiny blood vessels filling the spaces between. Each muscle fiber is made up of innumerable fibrils embedded in a matrix. It has been demonstrated that these fibrils are responsible for the contraction of the muscle as a whole. By special and clever techniques the fibrils can be washed free of the matrix, and it has been shown that when brought into contact with the energy-rich substance ATP, the fibrils shorten.

Examinations with the polarizing and electron microscopes and with X-rays have produced a fairly good picture of the ultimate design of these microscopic fibers. Each fibril is composed of many smaller filaments, or "protofibrils," just distinguishable under the highest microscopic magnification. The fibrils of a single muscle fiber may contain a total of some 10 million such filaments. The filaments are the smallest units known to stiffen and shorten. It has been possible to extract the actomyosin of which they are composed and to reconstitute filaments by squirting the extracted protein into a salt solution. These synthetic filaments can be made to contract.

The filaments themselves remain straight during contraction; therefore the kinking or coiling necessary for their contraction must take place at a still lower level—the level of molecules. Here the picture is clouded. We know from X-ray diffraction analyses that the molecules composing myosin filaments are arranged as miniature stretched spiral springs or stretched rubber bands. But just how they effect the filaments' contraction can only be guessed. Regardless of the mechanism, there is no doubt that the stiffness and shortening which are the features of contraction are mediated by changes in the molecular arrangement. This rearrangement requires energy. The consensus is that a tiny electric spark delivered to each individual cell causes the explosion of ATP. Not all the energy released is used for shortening of the actomyosin filaments. Some of it is converted to heat and some is used to initiate a series of complex chemical reactions which replenishes the fuel by reconstituting ATP. The explosion of ATP differs from that of gasoline in that no oxygen is required. But oxygen is indispensable for the rebuilding of ATP.

The millions of cardiac cylinders, as

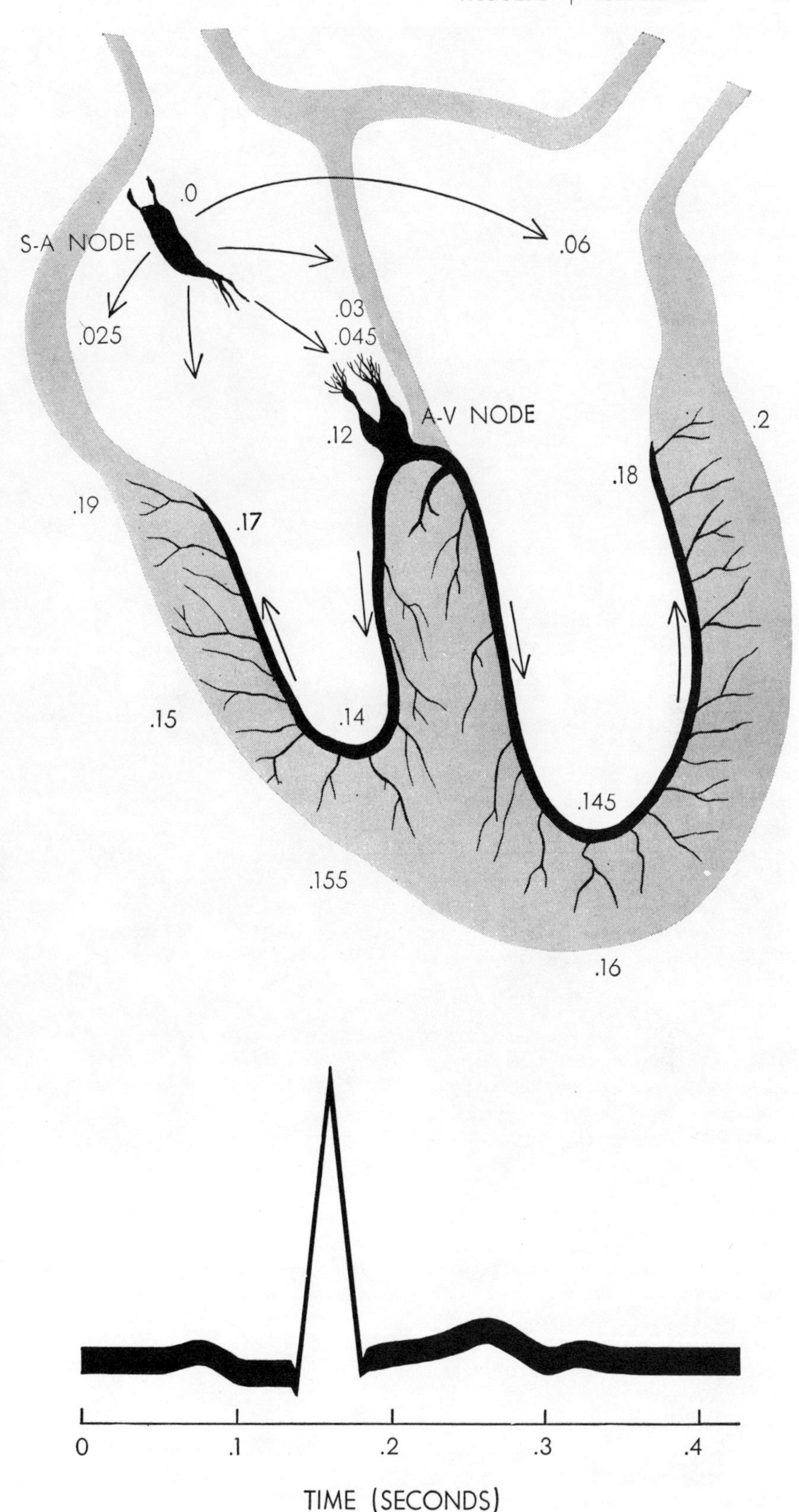

PACEMAKER SYSTEM generates and transmits the impulses which cause the contraction of the heart muscles. The impulse is generated in the sino-atrial (S-A) node and spreads across the atria, causing their contraction and stimulating the atrio ventricular (A-V) node. This in turn stimulates the rest of the system, causing the rest of the heart to contract. The numbers indicate the time (in fractions of a second) it takes an impulse to travel from the S-A node to that point. The electrocardiogram curve at bottom indicates the change in electrical potential that occurs during the spread of one impulse through the system.

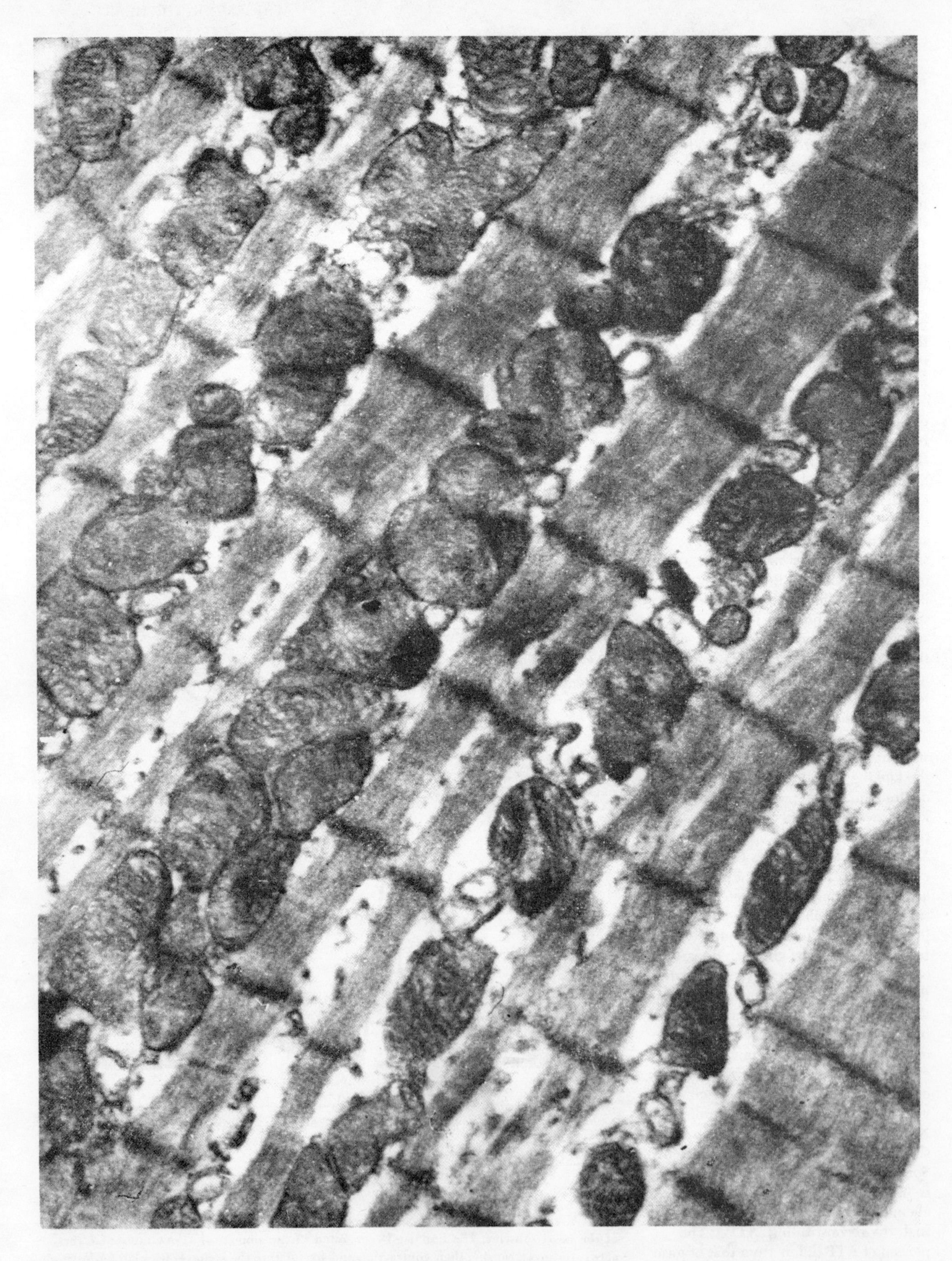

MUSCLE FIBRILS of the heart, which make up its muscle fibers, are revealed in this electron micrograph made by Bruno Kisch of the American College of Cardiology in New York. The fibrils, which are from the heart of a guinea pig, are the long bands running diagonally across the micrograph. The round bodies between the fibrils are sarcosomes, which are found in large numbers in heart muscle and which appear to supply the enzymes that make possible its tireless contractions. The fibrils themselves are made up of protofibrils, which may barely be seen in the micrograph. The micrograph enlarges these structures some 50,000 diameters.

in an automobile engine, must fire in proper sequence to contract the muscle effectively. When they fire haphazardly, there is a great liberation of energy but no coordinated action. This chaotic condition is called fibrillation—*i.e.*, independent and uncoordinated activity of the individual fibrils.

The Beat

What causes the heart to maintain its rhythmic beat? The ancients, performing sacrificial rites, must have noticed that the heart of an animal continues to beat for some time after it has been removed from the body. That the beat must originate in the heart itself was apparently clear to the Alexandrian anatomist Erasistratus in the third century B.C. But anatomists ignored this evidence for the next 20 centuries because they were convinced that the nerves to the heart must generate the heartbeat. In 1890, however, Henry Newell Martin at the Johns Hopkins University demonstrated that the heart of a mammal could be kept beating though it was completely separated from the nerves, provided it was supplied with blood. And many years before that Ernst Heinrich Weber of Germany had made the eventful discovery that stimulation of the vagus nerve to the heart does not excite it but on the contrary stops the heart. In short, it was established that the beat is indeed generated within the heart, and that the nerves have only a regulating influence.

The nature and location of the heart's "pacemaker" remained enigmatic until comparatively recent times. Within the span of my own memory there was considerable evidence for the view that the pacemaking impulses were generated by nerve cells in the right atrium and transmitted by nerve fibers to the heart-muscle cells. At present the evidence is overwhelming that the impulses are actually generated and distributed by a system of specialized muscle tissue consisting of cells placed end to end. Seventy-two times per minute—more or less—a brief electric spark of low intensity is liberated from a barely visible knot of tissue in the rear wall of the right atrium, called the sino-atrial or S-A node. The electric impulse spreads over the sheet of tissue comprising the two atria and, in so doing, excites a succession of muscle fibers which together produce the contraction of the atria. The impulse also reaches another small knot of specialized muscle known as the atrioventricular or A-V node, situated between the atria and ventricles. Here the impulse is delayed for about seven hundredths of a second, apparently to allow the atria to complete their contractions; then from the A-V node the impulse travels rapidly throughout the ventricles by way of a branching transmission system, reaching every muscle fiber of the two ventricles within six hundredths of a second. Thus the tiny spark produces fairly simultaneous explosions in all the cells, and the two ventricles contract in a concerted manner.

If the heart originates its own impulses, of what use are the two sets of nerves that anatomists have traced to the heart? A brief answer would be that they act like spurs and reins on a horse which has an intrinsic tendency to set its own pace. The vagus nerves continually check the innate tempo of the S-A node; the sympathetic nerves accelerate it during excitement and exercise.

Normally, as I have said, the S-A node generates the spark, but here, too, nature has provided a factor of safety. When the S-A node is depressed or destroyed by disease, the A-V node becomes the generator of impulses. It is not as effective a generator (its maximum rate is only 40 or 50 impulses per minute, and its output excites the atria and ventricles simultaneously), but it suffices to keep the heart going. Patients

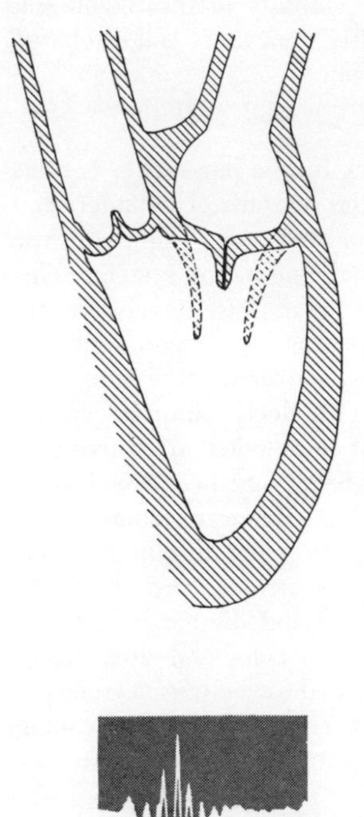
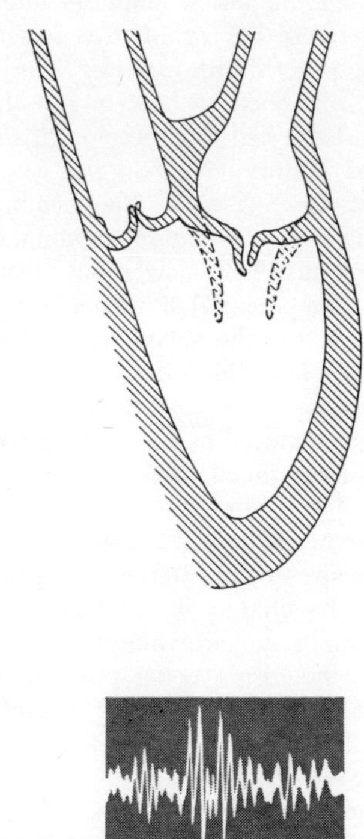
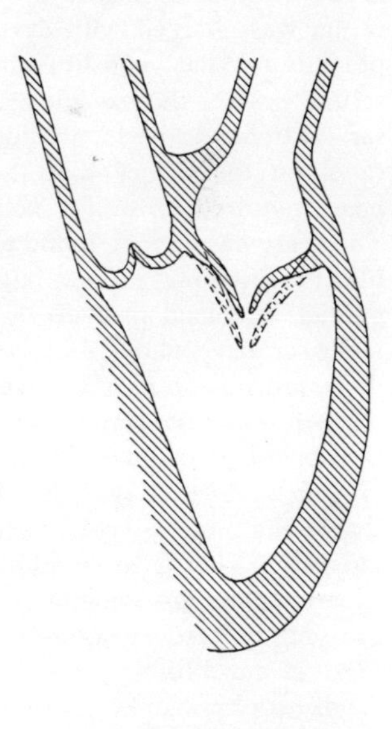
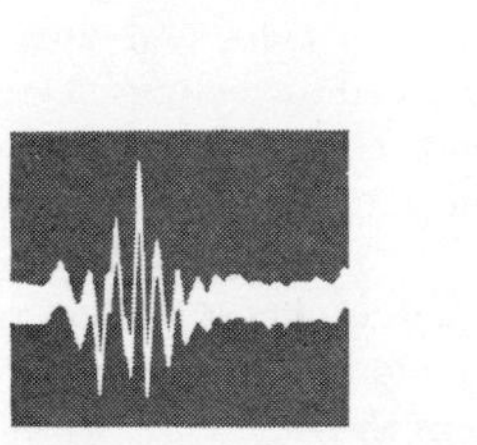
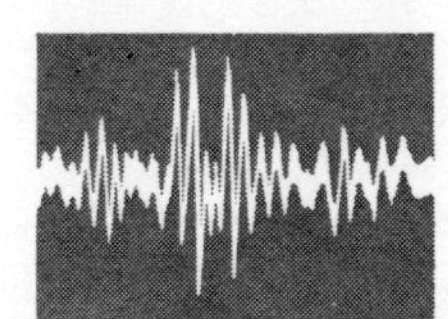
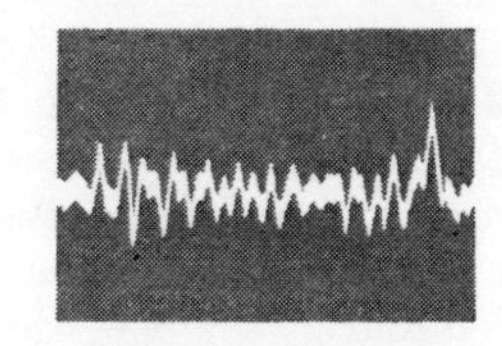

HEART SOUNDS indicate the normal and abnormal functioning of the heart valves. The three drawings at the top show the left side of the heart in cross section. The aortic, or outlet, valve is at upper left in each drawing; the mitral, or inlet, valve is at upper right. The first drawing shows the normal closing of the mitral valve; the trace below it records the sound made by this closing. The second drawing shows the partial closing of a leaky mitral valve; the trace below it records the murmur of blood continuing to flow through the valve. The third drawing shows the partial closing of a mitral valve with stiff leaves; the trace below it records a fainter murmur.

have survived up to 20 years with the A-V pacemaker substituting for a damaged S-A node.

There are still lower pacemakers which can maintain a slow heartbeat when the higher ones fail. When all the pacemakers are so weakened that, like an old battery, they are barely able to emit impulses, an anesthetic administered during an operation may stop the heart. In that case the beat can often be restored by rhythmic electric shocks —a system now incorporated in an apparatus for revival of the heart.

The Spark

What sort of mechanism exists in nodal tissues that is able to emit electric sparks with clockwork regularity 104,000 times a day? We must look first to the blood. The fact that an excised heart does not long continue to beat unless supplied artificially with blood shows that the blood must supply something essential for preservation of its beat. In the 19th century physiologists began to experiment on isolated hearts, first of frogs and turtles and later of rabbits and cats, to determine what constituents of blood could be spared without halting the rhythmic heart contractions. They found that the serum (the blood fluid without cells) could maintain the beat of a mammalian heart, provided the serum was charged with oxygen under pressure. What constituents of the serum, besides the oxygen, were necessary? Attention first focused on the proteins, on the theory that the beating heart required them for nourishment. The heart was, in fact, found to be capable of maintaining its beat fairly well on a "diet" of blood proteins or even egg white or oxygenated milk whey. But the nourishment idea received a blow when it was discovered that the heartbeat could be maintained on a solution of gum arabic! It was then suggested that serum proteins act by virtue of their viscosity. This is an example of how experimenters are sometimes led astray in trying to uncover nature's secrets.

An eventful discovery in 1882 by the English physiologist Sydney Ringer changed the direction of thinking. He showed that a solution containing salts of sodium, potassium and calcium and a little alkali, in the concentrations found in the blood, would sustain the beat of a frog's heart. It was but a step to show that Ringer's solution, when oxygenated, also keeps the mammalian heart beating for a short time. Later it was found that the addition of a biological fuel—glucose or, better yet, lactic acid—would extend the heart's performance.

Summing up the evidence, it was known at the beginning of the present century that the beat of the mammalian heart, and obviously also the generation of the spark, depends primarily on a balanced proportion of sodium, potassium and calcium plus a supply of oxygen and an energy-yielding substance such as glucose.

During the present century the scientific minds have sought to learn how these inorganic elements are involved in the initiation and spread of impulses. In order to understand the intricate mechanisms we must recall what most of us learned in high school: *viz.*, the theory that, when a salt is dissolved in water, the elements are dissociated and become ions charged with positive and negative electricity.

The delicate enclosing membrane of all cells is differentially permeable: that is, ordinarily (at rest) it allows potassium ions to enter the cell but excludes sodium ions. We may say that the potassium ions have admission tickets, while the sodium ions do not. Since sodium ions predominate in the body fluids, the positively charged potassium ions within a cell are greatly outnumbered by positively charged sodium ions around the outside of the cell; the net result is that the outside is more positive than the inside, and the interior can therefore be regarded as negative with respect to the exterior. The potential difference is about one tenth of a volt. Each cell thus becomes a small charged battery. Now in the case of cells of the S-A node, the membrane leaks slightly, allowing some sodium ions to sneak in. This slowly but steadily reduces the potential difference between the inside and the outside of the membrane. When the difference has diminished by a critical amount (usually about six hundredths of a volt), the tiny pores of the membrane abruptly open. A crowd of sodium ions then rushes in, while some of the imprisoned potassium ions escape to the exterior. As a result the relative charges on the two sides of the membrane are momentarily reversed, the inside being positive with respect to the outside. The action potential thus created is the release of the electric spark.

As soon as activity is over, the membrane repolarizes, *i.e.*, reconstitutes a charged battery. How this is accomplished is little understood, beyond the fact that oxidation of glucose or its equivalent is required. The mechanism is pictured as a kind of metabolic pump which ejects the sodium ions that have gained illegal admittance, allows potassium ions to re-enter and closes the pores again. Then the cells are ready to be discharged again.

A little reflection should make it evident that the frequency with which such cells discharge depends on at least two things: (1) the rate at which sodium ions leak into the cell, and (2) the degree to which the potential across the membrane must be reduced in order to discharge it completely. The rate of sodium entry is known to be increased by warming and decreased by cooling, which accounts in part for the more rapid firing of the pacemakers in a patient who has a fever. The magnitude of the potential difference required to discharge the cell depends on the characteristics of the membrane. In this the concentration of calcium ions plays a basic role. Calcium favors stability of the membrane: if its concentration falls below a certain critical value, the rate of discharge by the cells increases; if calcium ions are too abundant, the rate is slowed. The rate of discharge is also affected by other factors. The vagus nerves tend to reduce it, the sympathetic nerves to increase it. The blood's content of oxygen and carbon dioxide, its degree of alkalinity, hormones and drugs—these and other influences can change the stability of the membranes and thus alter the rate of cell discharge.

Transmission of Impulses

The spark from a pacemaker is transmitted to the myriads of cylinders constituting the ventricular pumps by way of the special conducting system. When we say that the impulse travels over this system, this does not mean that electricity flows, as over wires to automobile cylinders. The electric impulse spreads by a kind of chain reaction involving the successive firing of the special transmitting cells. When pacemaker cells discharge, they generate a highly localized current which in turn causes the depolarization and discharge of an adjacent group of cells, and thus the impulse is relayed to the muscle cells concerned with contraction. An advantage of this mechanism is that the strength of the very minute current reaching the contracting fibers is not reduced.

Such a mode of transmission is not unknown in the inanimate world. There is a classic experiment in chemistry which illustrates an analogous process. An iron wire is coated with a microscopic film of iron oxide and suspended in a cylinder of strong nitric acid. Protected by this coating, the iron does not dissolve. But if the coat is breached (by a scratch or by an electric current) at a spot at one end of the immersed wire, a brown bubble immediately forms at this spot and a succession of brown bubbles then traverses the whole length of the wire. An electrical recorder connected to a number of points along the wire shows that a succession of local electric currents is generated down the wire as it bubbles. At each spot the current breaks the iron oxide film and the ensuing chemical reaction generates a new action potential. The contact between bare iron and nitric acid is only momentary, because the break in the film is quickly repaired.

Summarizing, the passage of electric impulses over the conduction system of the heart represents a series of local bioelectric currents, relaying the impulses step by step over special tissue to the contracting cells. On arrival at these cells the electric charges trigger the breakdown of ATP and so release the chemical energy needed for contraction.

Diagnosis

Considering the complexity of the cardiac machinery, it is remarkable that the heartbeat does not go wrong more often. Like a repairman for an automobile or a television set, a physician sometimes has to make an extensive hunt for the source of the disorder. It seems appropriate to close this article with a list of the points at which the machinery is apt to break down.

1. The main (S-A) pacemaker, or in rare instances all the pacemakers, may fail.

2. There may be too many pacemakers. The secondary pacemakers occasionally spring into action and work at cross purposes with the normal one, producing too rapid, too slow, or ill-timed beats.

3. The system for conduction of the pacemaker impulses may break down, leading to "heart block."

4. The heart muscle cylinders may respond with little power because of poor fuel, lack of enough oxygen for building fuel, fatigue or lack of adequate vitamins, hormones or other substances in the blood.

5. Some of the cylinders may be put out of commission by blockage of a coronary artery.

6. The heart valves may leak, and in its gallant effort to compensate, the heart may be overworked to failure.

2

The Microcirculation of the Blood

by Benjamin W. Zweifach
January 1959

The primary purpose of the circulatory system is served by the microscopic vessels in which the blood flows from the arteries to the veins and thereby nourishes all the tissues of the body

When we think of the circulatory system, the words that first occur to us are heart, artery and vein. We tend to forget the microscopic vessels in which the blood flows from the arteries to the veins. Yet it is the microcirculation which serves the primary purpose of the circulatory system: to convey to the cells of the body the substances needed for their metabolism and regulation, to carry away their products—in short, to maintain the environment in which the cells can exist and perform their interrelated tasks. From this point of view the heart and the larger blood vessels are merely secondary plumbing to convey blood to the microcirculation.

To be sure, the entire circulatory system is centered on the heart. The two chambers of the right side of the heart pump blood to the lungs, where it is oxygenated and returned to the chambers of the left side of the heart. Thence the blood is pumped into the aorta, which branches like a tree into smaller and smaller arteries. The smallest twigs of the arterial system are the arterioles, which are too small to be seen with the unaided eye. It is here that the microcirculation begins. The arterioles in turn branch into the capillaries, which are still smaller. From the capillaries the blood flows into the microscopic tributaries of the venous system: the venules. Then it departs from the microcirculation and is returned by the tree of the venous system to the chambers in the right side of the heart.

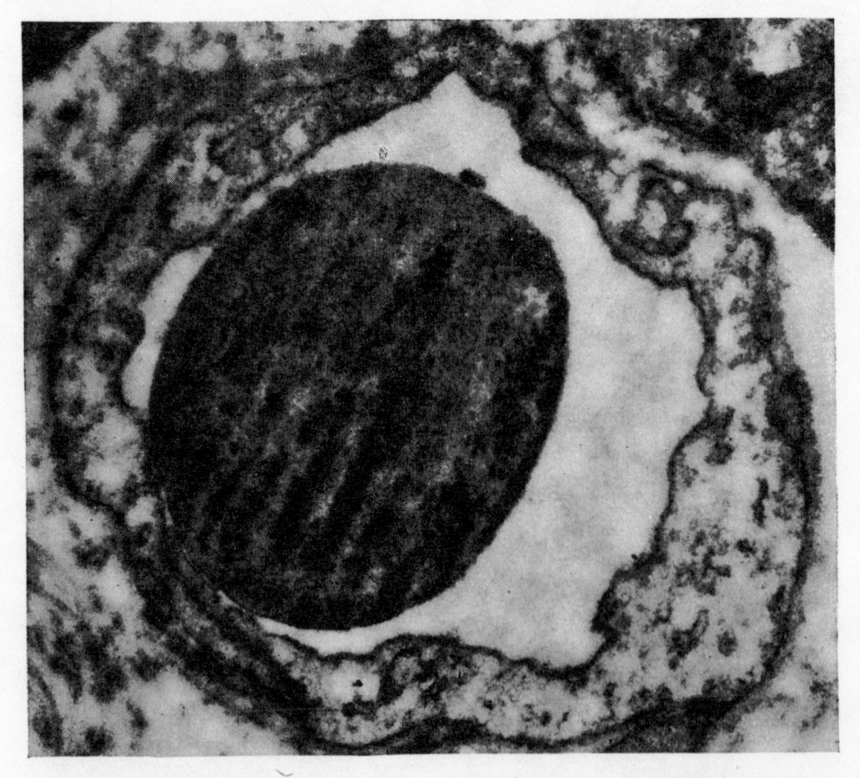

CAPILLARY from a cat's leg muscle is shown in cross section by this electron micrograph, which enlarges the structure some 20,000 diameters. The band running around the picture is the wall of the capillary. The large, dark object in the center is a single red blood cell. The micrograph was made by George D. Pappas and M. H. Ross of Columbia University.

The vessels of the microcirculation permeate every tissue of the body; they are never more than .005 inch from any cell. The capillaries themselves are about .0007 inch in diameter. To give the reader an idea of what this dimension means, it would take one cubic centimeter of blood (about 14 drops) from five to seven hours to pass through a capillary. Yet so large is the number of capillaries in the human body that the heart can pump all the blood in the body (about 5,000 cubic centimeters in an adult) through them in a few minutes. The total length of the capillaries in the body is almost 60,000 miles. Taken together, the capillaries comprise the body's largest organ; their total bulk is more than twice that of the liver.

If all the capillaries were open at one time, they would contain all of the blood in the body. Obviously this does not happen under normal circumstances, whereby hangs the principal theme of this article. How is it that the flow of blood through the capillaries can be regulated so as to meet the varying needs of all the tissues, and yet not interfere with the efficiency of the circulatory system as a whole?

It was William Harvey, physician to Charles I of England, who first demonstrated that the blood flows continuously from the arterial system to the venous. In 1661, 33 years after Harvey had published his famous work *De Motu Cordis* (*Concerning the Motion of the Heart*), the Italian anatomist Marcello Malpighi

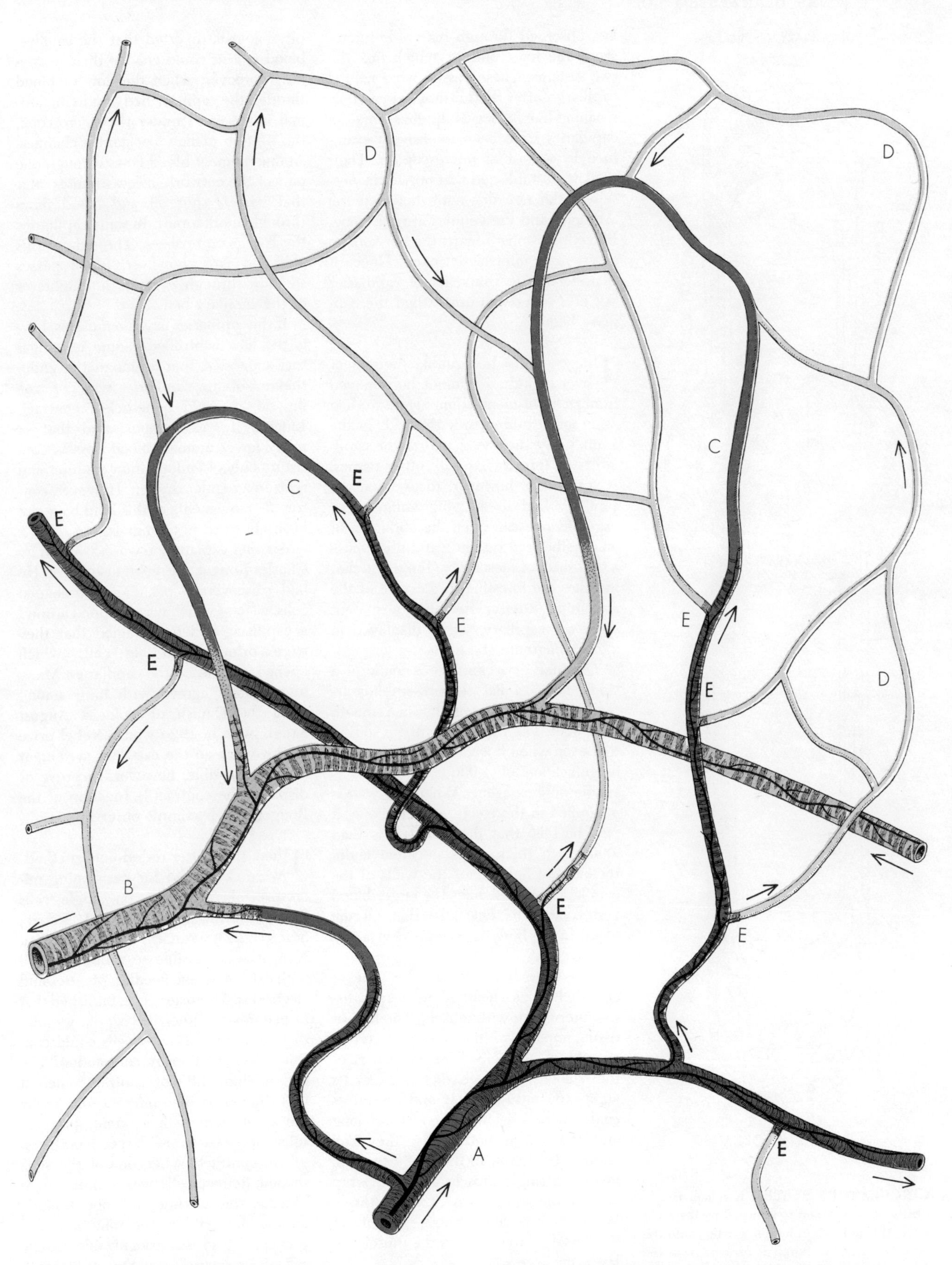

TYPICAL CAPILLARY BED is depicted in this drawing. The blood flows into the bed through an arteriole (A) and out of it through a venule (B). Between the arteriole and the venule the blood passes through thoroughfare channels (C). From these channels it passes into the capillaries proper (D), which then return it to the channels. The arteriole and venule are wrapped with muscle cells; in the thoroughfare channels the muscle cells thin out. The capillaries proper have no muscle cells at all. The flow of blood from a thoroughfare channel into a capillary is regulated by a ring of muscle called a precapillary sphincter (E). The black lines on the surface of the arteriole, venule and thoroughfare channels are nerve fibers leading to muscle cells. At lower left, between the arteriole and venule, is a channel which in many tissues shunts blood directly from the arterial system to the venous when necessary.

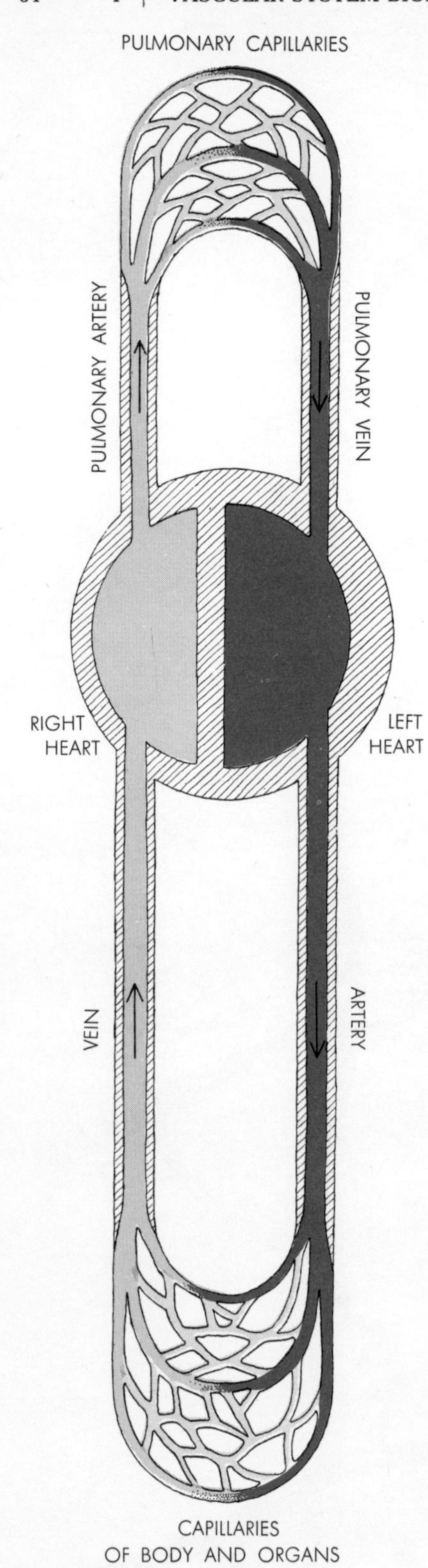

CIRCULATORY SYSTEM is schematically outlined. The blood is pumped by the right heart through the pulmonary artery into the capillaries of the lungs. It returns from the lungs through the pulmonary vein to the left heart, which pumps it through the arteries to the capillaries of the internal organs and of the rest of the body. It finally returns to the right heart through the veins.

first observed through his crude microscope the fine conduits which link the two systems. These vessels were named capillaries after the Latin word *capillus*, meaning hair. Since Malpighi's time the capillaries have been intensively examined by a host of microscopists. Their work has established that not all the vessels in the network lying between the arterioles and the venules are the same. Indeed, we must regard the network as a system of interrelated parts. Hence it is preferable to think not of capillaries but of a functional unit called the capillary bed.

The capillary bed, unlike muscle or liver or kidney, cannot be removed from an experimental animal and studied as an intact unit outside the body of the animal. By their very nature the capillaries are interwoven with other tissues. It is possible, however, to examine the capillary bed in a living animal. For example, one can open the abdomen of an anesthetized rat and carefully expose a thin sheet of mesentery: the tissue that attaches the intestine to the wall of the abdominal cavity. In this transparent sheet the capillary bed is displayed in almost diagrammatic form.

The tube of a capillary is made of a single layer of flat cells resembling irregular stones fitted together in a smooth pavement. The wall of the tube is so thin that even when it is viewed edge-on at a magnification of 1,000 diameters it is visible only as a line. When the wall is magnified in the electron microscope, it may be seen that the wall is less than .0001 inch thick. This so-called endothelium not only forms the walls of the capillaries but also lines the larger blood vessels and the heart, so that all the blood in the body is contained in a single envelope.

In a large blood vessel the tube of endothelium is sheathed in fibrous tissue interwoven with muscle. The fibrous tissue imparts to the vessel a certain amount of elasticity. The muscle is of the "smooth" type, characterized by its ability to contract slowly and sustain its contraction. The muscle cells are long and tapered at both ends; they coil around the vessel. In the tiny arterioles, in fact, a single muscle cell may wrap around the vessel two or three times. When the muscle contracts, the bore of the vessel narrows; when the muscle relaxes, the bore widens.

The muscular sheath of the larger blood vessels does not continue into the capillary bed. Yet as early as the latter part of the 19th-century experimental physiologists reported that the smallest blood vessels could change their diameter. Moreover, when the flow of blood through the capillary bed of a living animal is observed under the microscope, the pattern of flow constantly changes. At one moment blood flows through one part of the network; a few minutes later that part is shut off and blood flows through another part. In some capillaries the flow even reverses. Throughout this ebb and flow, however, blood passes steadily through certain thoroughfares of the capillary bed.

If the capillaries have no muscles, how is the flow controlled? Some investigators suggested that although the endothelium of the capillaries was not true muscle, it could nonetheless contract. Indeed, it was demonstrated that in many lower animals blood vessels consisting only of endothelium contract and relax in a regular rhythm. However, contractile movements of this kind have not been observed in mammals.

Another explanation was advanced by Charles Rouget, a French histologist. He had discovered peculiar star-shaped cells, each of which was wrapped around a capillary, and he assumed that they were primitive muscle cells which opened and closed the capillaries. Many investigators agreed with him, among them the Danish physiologist August Krogh, who in 1920 won a Nobel prize for his work on the capillary system. It was not possible, however, to prove or disprove the contractile function of the Rouget cells by simple observation.

There the matter rested until methods were developed for performing microsurgical operations on single cells [see "Microsurgery," by M. J. Kopac; SCIENTIFIC AMERICAN, October, 1950]. Now it was possible to probe the cell with extremely fine needles, pipettes and electrodes. Microsurgery established that in mammals neither the capillary endothelium nor the Rouget cells could control the circulation by contraction. The endothelium did not contract when it was stimulated by a microneedle, or by the application with a micropipette of substances that cause larger blood vessels to contract. When one of the star-shaped Rouget cells was stimulated, it became thicker but did not occlude the capillary. When the same stimulus was applied to the recognizable muscle cell of an arteriole, on the other hand, the cell contracted and the arteriole was narrowed.

The microsurgical experiments established an even more significant fact: not

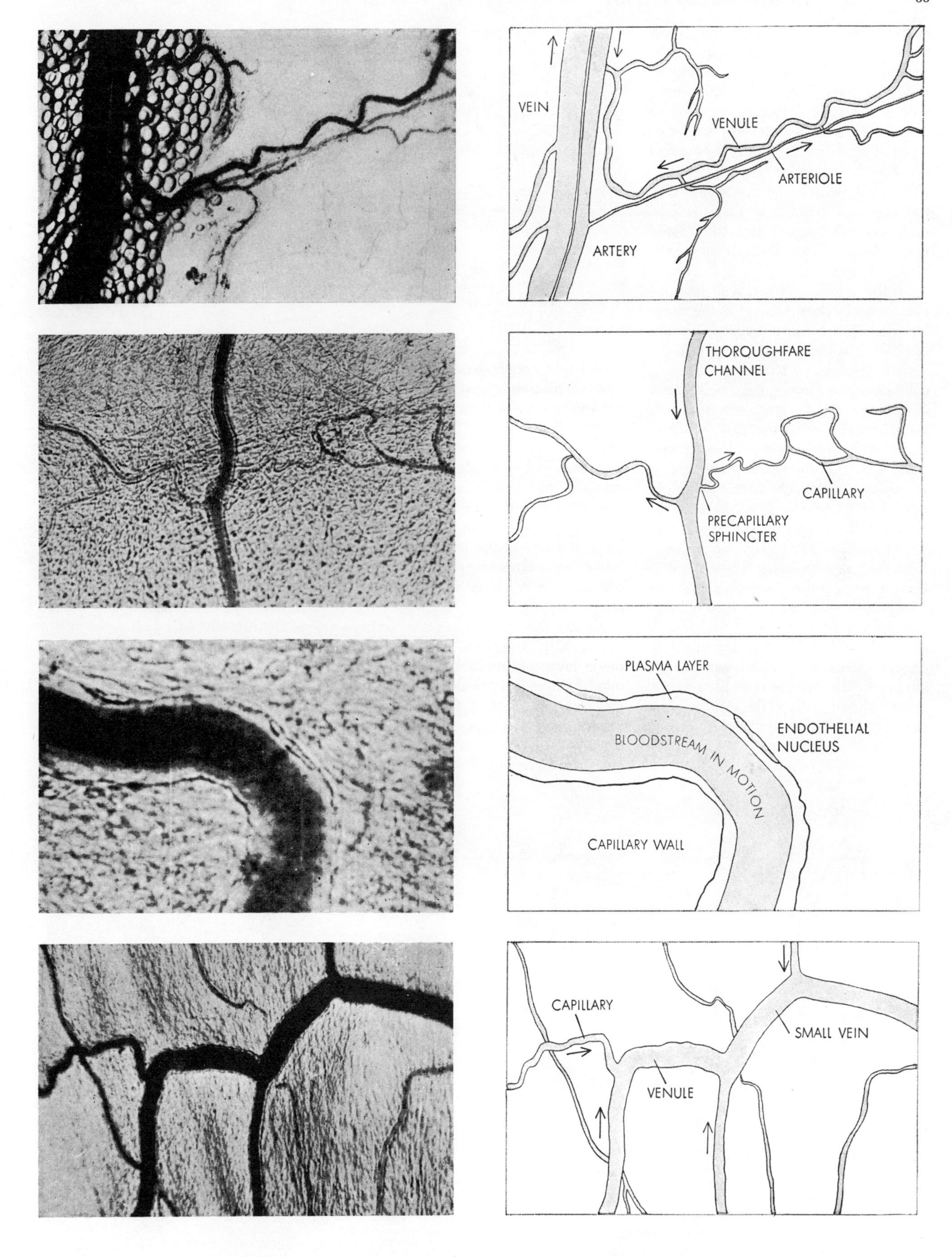

MESENTERY of a rat is photographed at various magnifications to show the characteristic structures of the microcirculation. The drawings at right label the structures. The magnification of the photomicrograph at top is 100 diameters; of the photomicrograph second from top, 200 diameters; of the third photomicrograph, 1,000 diameters; of the photomicrograph at bottom, 200 diameters.

all the vessels in the capillary bed entirely lack muscle. For example, if epinephrine, which causes larger blood vessels to contract, is injected into the capillary bed with a micropipette, some of the vessels in the bed become narrower. Even when no stimulating substances are added, the same vessels open and close with the ebb and flow of blood in the capillary bed. It is these vessels, moreover, through which the blood flows steadily from the arterial to the venous system.

So the arterial system, with its muscular vessels, does not end at the capillary bed. The blood is continuously under muscular control as it flows into the venous system. To be sure, the muscle cells along the thoroughfare are sparsely distributed. As the arterial tree branches into the tissues the muscular sheath of the endothelium becomes thinner and thinner until in the smallest arterioles it is only one cell thick. In the thoroughfare channel of the capillary bed the muscle cells are spaced so far apart that the channel is almost indistinguishable from the true capillaries. The major portion of the capillary network arises as abrupt side branches of the thoroughfare channels, and at the point where each of the branches leaves a thoroughfare channel there is a prominent muscle structure: the muscle cells form a ring around the entrance to the capillary. It is this ring, or precapillary sphincter, which acts as a floodgate to control the flow of blood into the capillary network from the thoroughfare channel.

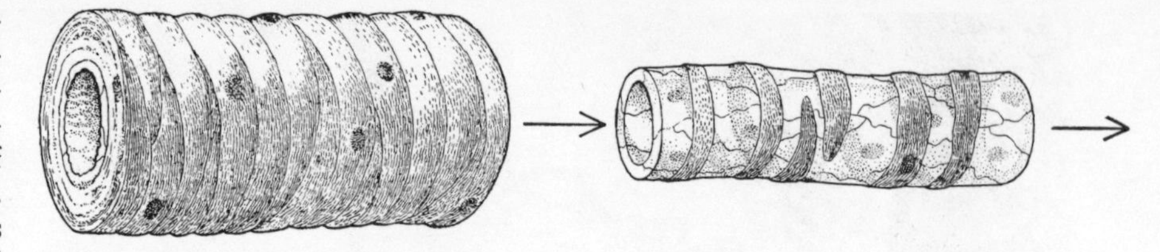

WALLS OF BLOOD VESSELS of various kinds reflect their various functions. The wall of an artery consists of a single layer of endothelial cells sheathed in several layers of muscle cells interwoven with fibrous tissue. The wall of an arteriole consists of a single layer of

The muscular specialization of the circulatory system is illuminated by its embryonic development. In the early embryo the circulatory system is a network of endothelial tubes through which the primitive blood cells flow in an erratic fashion. The tubes are at first just large enough to pass the blood cells in single file. Attached to the outer wall of the tubes are numerous star-shaped cells which have wandered in from the surrounding tissue.

As the development of the embryo proceeds, those tubes through which the blood flows most rapidly are transformed into heavy-walled arteries and veins. In the process the star-shaped cells evolve through several stages into typical muscle cells. The outer reaches of the adult circulatory system possess a graded series of muscle-cell types, which are a direct representation of this developmental process. Thus the star-shaped cells of the capillary bed—the Rouget cells—are primitive muscle elements which have no contractile function.

From this point of view the capillary bed can be considered the immature part of the circulatory system. Like embryonic tissue, it has the capacity for growth, which it exhibits in response to injury. It also ages to some extent, and ultimately becomes less capable of dealing with the diversified demands of the tissue cells.

When we put these various facts together, we see the capillary bed not as a simple web of vessels between the arterial and venous systems, but as a

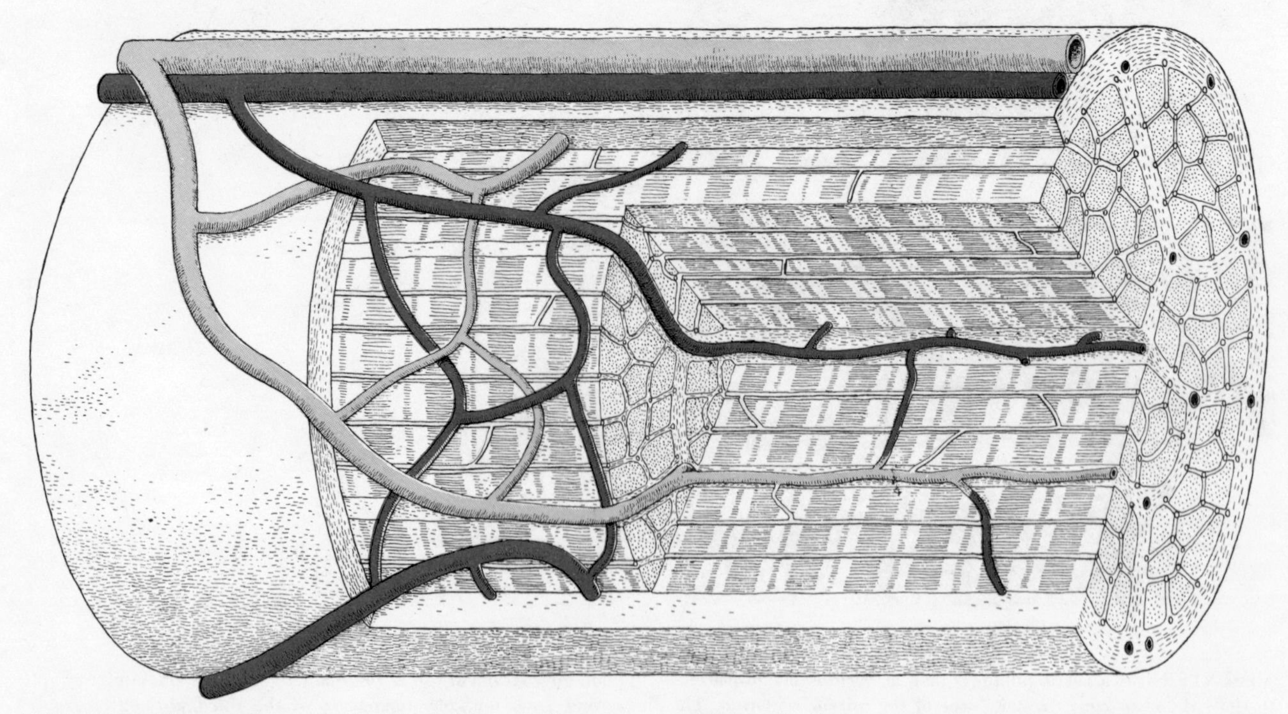

MUSCLE FIBER is richly supplied with capillaries. Lying atop this dissected muscle fiber are two blood vessels, the smaller of which is an artery and the larger a vein. Most of the capillaries run parallel to the fibrils which make up the fiber. The vessels which cut across two or more capillaries are thoroughfare channels. The system is shown in cross section at the right end of the drawing.

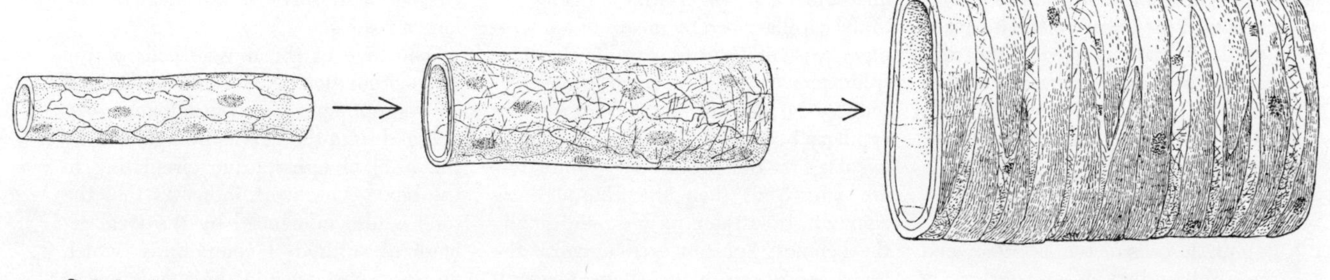

endothelial cells sheathed in a single layer of muscle cells. The wall of a capillary consists only of a single layer of endothelial cells. The wall of a venule consists of endothelial cells sheathed in fibrous tissue. The wall of a vein consists of endothelial cells sheathed in fibrous tissue and a thin layer of muscle cells. Thus a layer of endothelial cells lines the entire circulatory system.

physiological unit with two specialized components. One component is the thoroughfare channel, into which blood flows from the arteriole. The other is the true capillaries, which form a secondary network connected to the thoroughfare channel. The precapillary sphincters along the channel open and close periodically, irrigating first one part of the capillary network, then another part. When the sphincters are closed, the blood is restricted to the thoroughfare channel in its movement toward the venous system.

The structure of the physiological unit varies from one tissue to another in accordance with the characteristic needs of the tissues. For example, striated muscle, which unlike the smooth muscle of the blood vessels and other organs contracts rapidly and is under voluntary control, requires over 10 times more blood when it is active than when it is at rest. To meet this wide range of needs each thoroughfare channel in striated muscle gives rise to as many as 20 or 30 true capillaries. Glandular tissues, on the other hand, require only a steady trickle of blood, and each of their thoroughfare channels may give rise to as few as one or two capillaries. In the skin, which shields the body from its outer environment, there are special shunts through which blood can pass directly from the arteries to the veins with minimum loss of heat. Still other tissues require specialized capillary beds. The capillary beds of all the tissues, however, have the same basic feature: a central channel whose muscle cells control the flow of blood into the true capillaries.

But what controls the muscle cells? To answer this question we must draw a distinction between the control of the larger blood vessels and the control of the microcirculation. The muscle cells of the arteries and veins are made to contract and relax by two agencies: (1) the nervous system and (2) chemical "messengers" in the blood. These influences not only cause the vessels to constrict and dilate but also keep the muscle cells in a state of partial contraction. This muscle "tone" maintains the elasticity of the vessels, which assists the heart in maintaining the blood pressure. The operation of the system as a whole is supervised by special regulatory centers in the brain, working in collaboration with sensory monitoring stations strategically located in important vessels.

In the capillary bed, on the other hand, the role of the nervous system is much less significant. Most of the muscle cells in the capillary bed have no direct nerve connections at all. A further circumstance sets the response of the microscopic vessels apart from that of the larger vessels. Whereas the muscle cells of the large vessels are isolated from the surrounding tissues in the thick walls of the vessels, the muscle cells of the arterioles and the thoroughfare channels are immersed in the environment of the very tissues which they supply with blood. This feature introduces another chemical regulatory mechanism: the continuous presence of substances liberated locally by the tissue cells. As a consequence the contraction and relaxation of muscle cells in the microcirculation are under the joint control of messenger substances in the blood and specific chemical products of tissue metabolism.

The chemical substances that influence the function of the blood-vessel muscle cells comprise a subtly orchestrated system which is still imperfectly understood. Among the more important messengers are those released into the bloodstream by the cortex of the adrenal gland. These corticosteroids are essential to all cells in the body, notably maintaining the cells' internal balance of water and salts. (They have also been used with spectacular results in the treatment of degenerative diseases such as arthritis.) When the corticosteroids are deficient or absent, the muscles of the blood vessels lose their tone and the circulation collapses.

Another substance of profound importance to the circulatory system is epinephrine, which is secreted by the core of the adrenal gland (as distinct from its cortex). Epinephrine is one of two principal members of a family of substances called amines; the other principal member is norepinephrine, which is released both by the adrenal gland and by the endings of nerves in the muscles. All the amines cause the contraction of the muscle cells of the blood vessels, with the exception of certain vessels such as the coronary arteries of the heart. Also liberated at the nerve endings is acetylcholine, the effect of which is directly opposite that of the amines: it causes muscle cells to relax.

Many workers have suggested that it is norepinephrine and acetylcholine which control the flow of blood through the small vessels. Our own work at the New York University–Bellevue Medical Center leads us to conclude that such an explanation is too simple. The mechanism could not by itself account for the behavior of the small vessels.

In our view the function of the muscle cells of the small vessels is regulated not only by substances that directly cause them to contract and relax, but also by other substances that simply modify the capacity of the cells to react to stimuli and do work. It is known that a wide variety of substances extracted from tissues cause the small vessels to dilate. We postulate that when the metabolism of tissue cells is accelerated, the cells produce substances of this sort. When such substances accumulate in the vicinity of a precapillary sphincter, they depress the capacity of its muscle cells to respond to stimuli. As a result the sphincter relaxes, and blood flows from the thoroughfare channel into the capillary which nourishes the tissue.

The reaction limits itself, because the blood flow increases to the point where it is sufficient to meet the nutritional requirements of the tissue cells. This leads to a gradual disappearance of the substances liberated by accelerated metabolism, and to a gradual lessening of the inhibition of the precapillary sphincter. As the muscle cells regain their tone, the sphincter shuts off the capillary.

The muscle cells of the arterioles and the capillary bed are extraordinarily sensitive to chemical stimuli, so sensitive that they respond to as little as a hundredth of the amount of substance required to constrict or dilate a large blood vessel. This sensitivity is dramatically demonstrated by microsurgical experiments on the capillary bed of a living rat. As little as .000000001 gram (.001 microgram) of epinephrine, injected into the capillary bed by means of a micropipette, is sufficient to close its capillary sphincters completely. Such substances reduce the flow of blood through the capillary bed by an orderly sequence of events: first the precapillary sphincters are narrowed, then the thoroughfare channels, then the arterioles, and finally the venules. Substances that cause the blood vessels to dilate, such as acetylcholine, set in motion a similar sequence: first the precapillary sphincters are opened, then the thoroughfare channels, and so on. The sensitivity of the arterioles and the capillary bed to such stimuli contributes to their independent behavior. An amount of substance sufficient to cause dramatic changes in the microcirculation simply has no effect on the larger vessels.

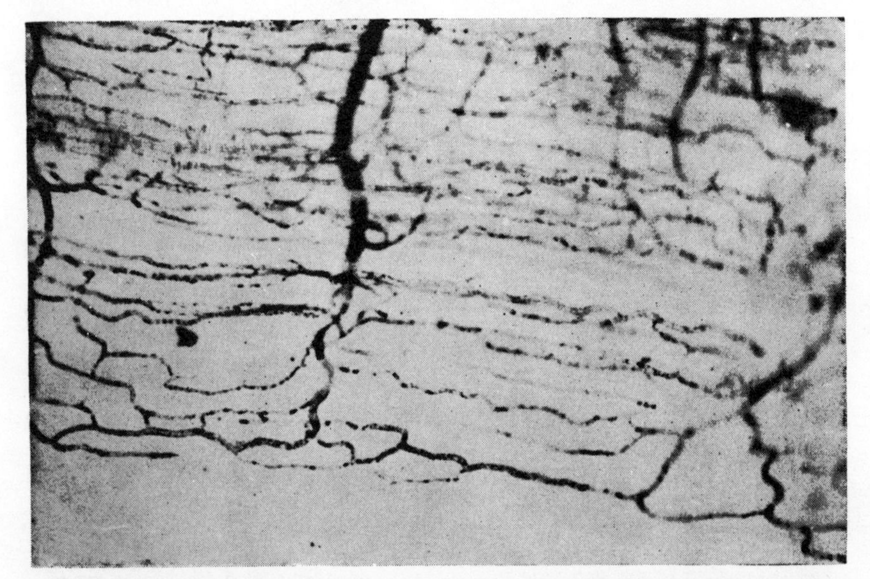

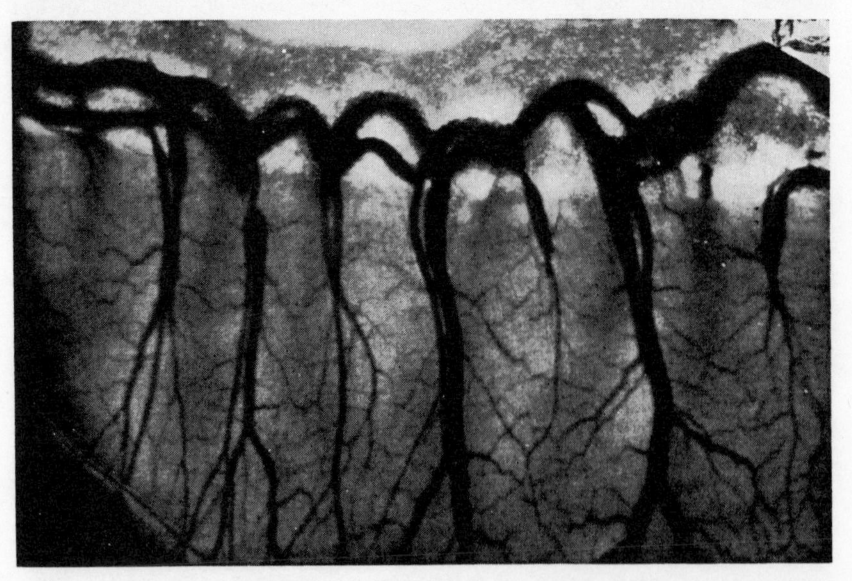

CAPILLARY BEDS OF TWO TISSUES in a living rat are enlarged 200 diameters in these photomicrographs. At top is the capillary bed of a striated muscle; the capillaries run parallel to underlying muscle fibers. At bottom are capillaries in the surface of intestine.

The tone of the muscle cells of the microcirculation may well be maintained by norepinephrine continuously discharged from the nerve endings, and by the level of epinephrine circulating in the blood. Our work indicates that the tone is also influenced by the local release of sulfhydryl compounds, which are key substances in the regulation of the oxidations conducted by cells. Now epinephrine and norepinephrine lose their activity when they are oxidized. Thus the actual level of these substances in the vicinity of muscle cells is not only dependent on their formation but also on their removal or destruction. Sulfhydryl compounds have been found to reduce the rate at which epinephrine and norepinephrine are oxidized. In this way the local release of such compounds could regulate the tone of smooth muscle.

Recently it has been suggested that a role in the local control of the microcirculation is played by the so-called mast cells, large numbers of which adjoin the small blood vessels. Various investigators have shown that the mast cells release at least three substances that strongly affect blood vessels: histamine, serotonin and heparin. It has been proposed that these substances, working alone or in certain combinations, are local regulatory factors.

It must be borne in mind that, even though the control of the microcirculation is largely independent of the rest of the circulatory system, the small blood vessels depend upon the nervous controls of the larger blood vessels for the shifting of blood from one organ to another as it is needed. Obviously the nervous controls of the larger vessels and the chemical controls of the microcirculation must be linked in some fashion. Under normal circumstances tissues that are inactive are perfused with a minimal amount of blood to allow the flow to be diverted to the tissues that need it most. During shock and acute infections, on the other hand, the demands of the tissues may be so great that the circulatory system cannot meet them, and the circulation collapses. In such conditions the effect of substances released locally to relax the muscle cells of the capillary bed has superseded the efforts of the nervous system to restrict the blood flow by the release of substances such as norepinephrine. It is ironic that this primitive response, in striving to insure the survival of individual cells, frequently overtaxes the circulation and brings about the death of the organism.

The Lymphatic System

3

by H. S. Mayerson
June 1963

This second circulation plays an essential role in maintaining the body's steady state, draining from the spaces between cells fluid, protein and other substances that leak out of the blood

Living tissue is for the most part a collection of cells bathed in a fluid medium. This interstitial fluid constitutes what the French physiologist Claude Bernard named the *milieu intérieur:* the internal environment of the organism that is the true environment of its cells. The interstitial fluid brings nutrients to the cells and carries away waste products; its composition varies in space and time under the control of the co-ordinated physiological processes that maintain homeostasis, the remarkably steady state that characterizes the internal environment of a healthy organism. In the maintenance of the homeostasis of the interstitial fluid the circulation of the blood is obviously of fundamental importance. In the higher vertebrates there is a second circulation that is equally essential: the lymphatic system. Its primary function is to recirculate the interstitial fluid to the bloodstream, thereby helping to create a proper cellular environment and to maintain the constancy of the blood itself. It also serves as a transport system, conducting specialized substances from the cells that make them into the bloodstream. In recent years physiologists, biochemists, physicians and surgeons have been studying the lymphatic system intensively, in health and in disease. Their investigations are providing much new information on how the body functions, explaining some heretofore poorly understood clinical observations and even suggesting new forms of treatment.

The fact that the lymphatic system is an evolutionary newcomer encountered only in the higher vertebrates is significant. In lower animals there is no separation between the internal and external environments; all the cells of a jellyfish, for example, are bathed in sea water. With progression up the evolutionary scale the cells become separated from the external environment, "inside" is no longer identical with "outside" and rudimentary blood circulatory systems make their appearance to conduct the exchange of nutrients and waste products. As the organism becomes more complex the blood system becomes more specialized. The system develops increasing hydrostatic pressure until, in mammals, there is a closed, high-pressure system with conduits of diminishing thickness carrying blood to an extensive, branching bed of tiny capillaries.

At this point in evolution a snag was encountered: the high pressures made the capillaries leaky, with the result that fluid and other substances seeped out of the bloodstream. A drainage system was required and lymphatic vessels evolved (from the veins, judging by embryological evidence) to meet this need.

In man the lymphatic system is an extensive network of distensible vessels resembling the veins. It arises from a fine mesh of small, thin-walled lymph capillaries that branch through most of the soft tissue of the body. Through the walls of these blind-end capillaries the interstitial fluid diffuses to become lymph, a colorless or pale yellow liquid very similar in composition to the interstitial fluid and to plasma, the liquid component of the blood. The lymphatic capillaries converge to form larger vessels that receive tributaries along their length and join to become terminal ducts emptying into large veins in the lower part of the neck. The largest of these great lymphatics, the thoracic duct, drains the lower extremities and all the organs except the heart, the lungs and the upper part of the diaphragm; these are drained by the right lymphatic duct. Smaller cervical ducts collect fluid from each side of the head and neck. All but the largest lymph vessels are fragile and difficult to trace, following different courses in different individuals and even, over a period of time, in the same individual. The larger lymphatics, like large veins, are equipped with valves to prevent backflow.

Along the larger lymphatics are numerous lymph nodes, which are of fundamental importance in protecting the body against disease and the invasion of foreign matter. The lymph nodes serve, first of all, as filtering beds that remove particulate matter from the lymph before it enters the bloodstream; they contain white cells that can ingest and destroy foreign particles, bacteria and dead tissue cells. The nodes are, moreover, centers for the proliferation and storage of lymphocytes and other antibody-manufacturing cells produced in the thymus gland; when bacteria, viruses or antigenic molecules arrive at a lymph node, they stimulate such cells to make antibodies [see "The Thymus Gland," by Sir Macfarlane Burnet, SCIENTIFIC AMERICAN, Offprint 138].

Starling's Hypothesis

The present view of the lymphatic circulation as a partner of the blood system in maintaining the fluid dynamics of the body stems from the investigations early in this century by the British physiologist Ernest H. Starling. "Starling's hypothesis" stated that the exchange of fluid between the capillaries and the interstitial space is governed by the relation between hydrostatic pressure and osmotic pressure. Blood at the arterial end of a capillary is still under a driving pressure equivalent to some 40 millimeters of mercury; this constitutes a "filtration pressure" that tends to make plasma seep out of the capillary. Starling visualized the wall of the capillary as being freely permeable to plasma and all its constituents except the plasma proteins albumin, globulin and fibrinogen, which could leak through only in very small amounts. The proteins remaining in the capillary exert an osmotic pressure that tends to keep fluid in the capillary, countering the filtration pressure. Similar forces are operative in the tissue spaces outside the capillary. At the arterial end of the capillary the resultant of all these forces is ordinarily a positive filtration pressure: water and salts leave the capillary. At the venous end, however, the blood pressure is decreased, energy having been dissipated in pushing the blood through the capillary. Now the osmotic force exerted by the proteins is dominant. The pressure gradient is reversed: fluid, salts and the waste products of cell metabolism flow into the bloodstream [*see top illustration on page 43*].

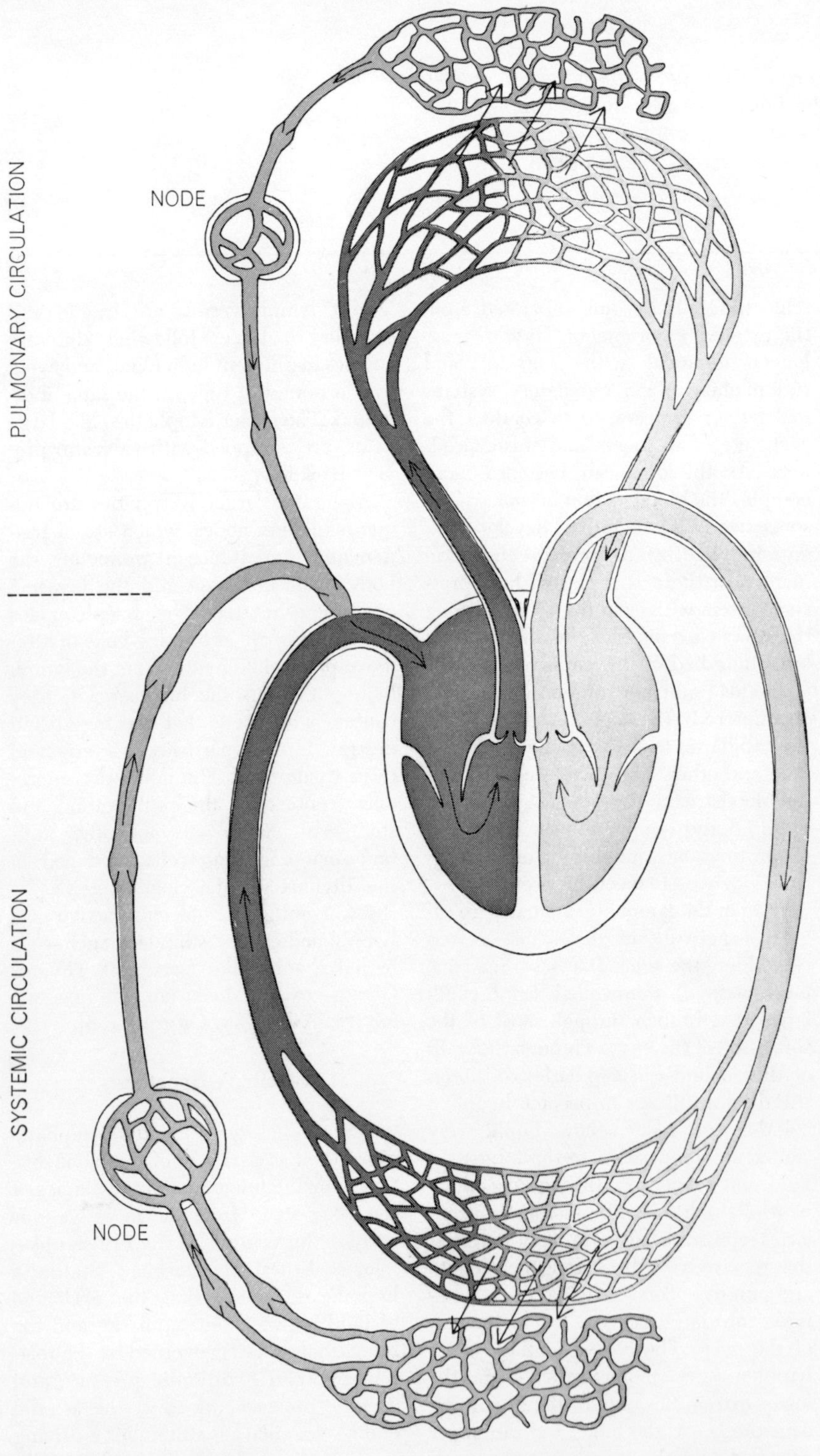

TWO CIRCULATORY SYSTEMS, the blood and the lymphatic (*color*), are related in this schematic diagram. Oxygenated blood (*light gray*) is pumped by the heart through a network of capillaries, bringing oxygen and nutrients to the tissue cells. Venous blood (*dark gray*) returns to the heart and is oxygenated in the course of the pulmonary (lung) circulation. Fluid and other substances seep out of the blood capillaries into the tissue spaces and are returned to the bloodstream by the lymph capillaries and larger lymphatic vessels.

It follows, Starling observed, that if the concentration of plasma proteins is decreased (as it would be in starvation), the return of fluid to the bloodstream will be diminished and edema, an excessive accumulation of fluid in the tissue spaces, will result. Similarly, if the capillaries become too permeable to protein, the osmotic pressure of the plasma decreases and that of the tissue fluid increases, again causing edema. Capillary poisons such as snake venoms have this effect. Abnormally high venous pressures also promote edema, by making it difficult for fluid to return to the capillaries; this is often one of the factors operating in congestive heart disease.

A fundamental tenet of Starling's hypothesis was that not much protein leaves the blood capillary. In the 1930's the late Cecil K. Drinker of the Harvard Medical School challenged this idea. Numerous experiments led him to conclude "that the capillaries practically universally leak protein; that this protein does not re-enter the blood vessels unless delivered by the lymphatic system; that the filtrate from the blood capillaries to the tissue spaces contains water, salts and sugars in concentrations found in blood, together with serum globulin, serum albumin and fibrinogen in low concentrations, lower probably than that of tissue fluid or lymph; that water and salts are reabsorbed by blood vessels and protein enters the lymphatics together with water and salts in the concentrations existing in the tissue fluid at the moment of lymphatic entrance." In other words, Drinker believed that protein is continuously filtering out of the blood; the plasma-protein level is maintained only because the lymphatic system picks up protein and returns it to the bloodstream.

Unfortunately Drinker had no definitive method by which to prove that the protein in lymph had leaked out of the blood and was not somehow originating in the cells. Perhaps for this reason his conclusions were not generally accepted. Teachers and the writers of textbooks continued to maintain that "healthy" blood capillaries did not leak protein. It was in an effort to clarify this point that I undertook an investigation of lymph

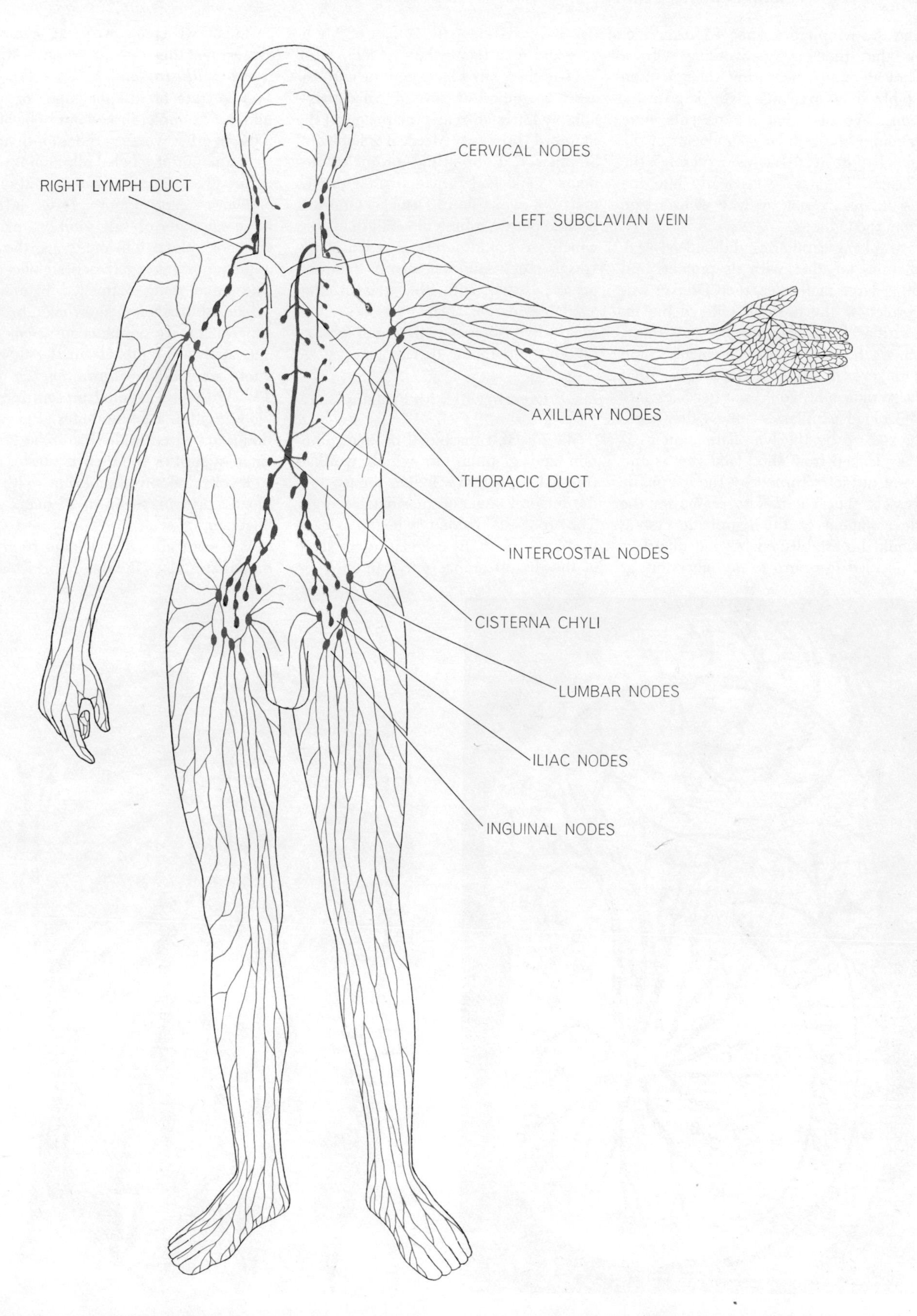

LYMPHATIC VESSELS drain the entire body, penetrating most of the tissues and carrying back to the bloodstream excess fluid from the intercellular spaces. This diagram shows only some of the larger superficial vessels (*light color*), which run near the surface of the body, and deep vessels (*dark color*), which drain the interior of the body and collect from the superficial vessels. The thoracic duct, which arises at the cisterna chyli in the abdomen, drains most of the body and empties into the left subclavian vein. The right lymph duct drains the heart, lungs, part of the diaphragm, the right upper part of the body and the right side of the head and neck, emptying into the right subclavian vein. Lymph nodes interspersed along the vessels trap foreign matter, including bacteria.

and the lymphatics some 15 years ago. At that time I was working with a clinical group measuring the retention of blood by patients given large infusions. We saw that the patients were retaining the cellular components of the blood quite well but were "losing" the plasma. The loss was clearly into the tissue spaces, not by way of excretion from the kidneys.

If blood capillaries did indeed leak plasma, together with its proteins and other large molecules, then Drinker was correct. If the proteins entered the interstitial fluid, they would stay there, since Starling's measurements and Drinker's findings made it clear that large molecules could not get back into the blood capillaries—unless they were picked up by the lymphatic system. If they leaked from the blood vessels and were in fact returned by the lymphatic vessels, the evolutionary reason for the development of the lymphatic system would be established beyond question. I decided to return to my laboratory at the Tulane University School of Medicine and investigate the problem.

Over the years I have had the enthusiastic assistance of several colleagues—notably Karlman Wasserman, now at the Stanford University Medical School, and Stephen J. LeBrie—and of many students. Time had provided us with two tools not available to Drinker. One was flexible plastic tubing of small diameter, which we could insert into lymphatic vessels much more effectively than had been possible with the glass tubing available earlier. And we now had radioactive isotopes with which to label proteins and follow their course.

Experiments with Proteins

We injected the blood proteins albumin and globulin, to which we had coupled radioactive iodine atoms, into the femoral veins of anesthetized dogs. The proteins immediately began to leave the bloodstream. By calculating the slope of the disappearance curve in each experiment we could arrive at a number expressing the rate of disappearance [*see top illustration on page 44*]. The average rate of disappearance of albumin, for example, turned out to be about .001; in other words, a thousandth of the total amount of labeled albumin present at any given time was leaking out of the capillaries each minute. If we infused large amounts of salt solution, plasma or whole blood into our dogs, the disappearance rate increased significantly. The same thing happened in animals subjected to severe hemorrhage. In some experiments we simultaneously collected and analyzed lymph from the thoracic duct [*see bottom illustration on page 44*]. As before, labeled protein left the blood; within a few minutes after injection it appeared in the lymph, at first in small quantities and then at a faster rate. It leveled off, in equilibrium with the blood's protein, seven to 13 hours after injection.

We were able to calculate from our data that in dogs the thoracic duct alone

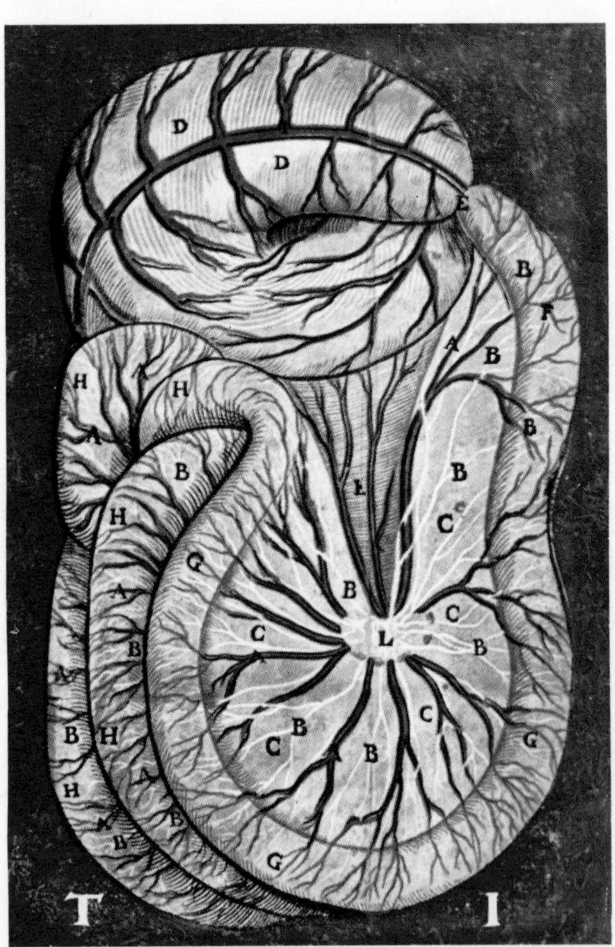

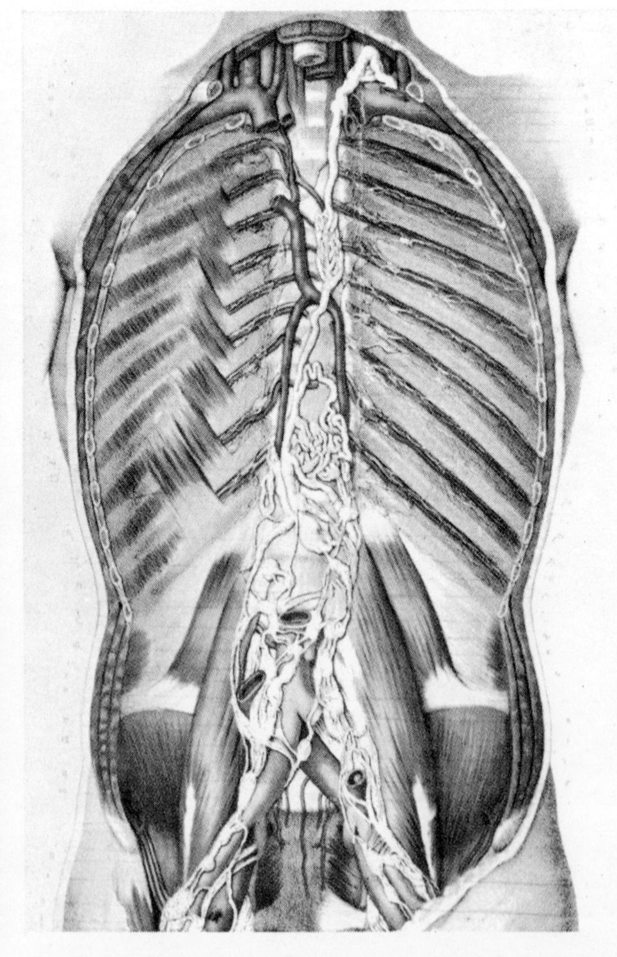

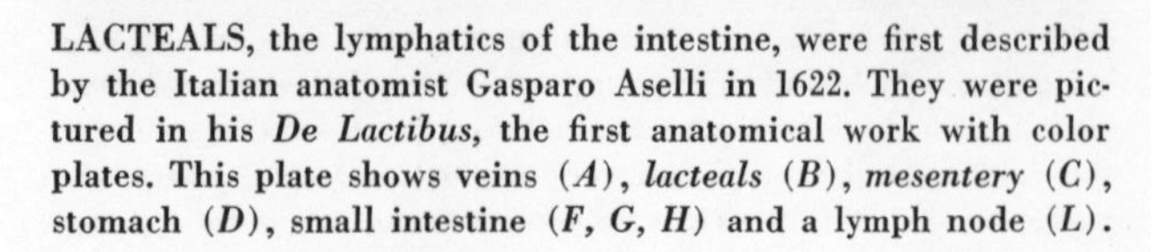

LACTEALS, the lymphatics of the intestine, were first described by the Italian anatomist Gasparo Aselli in 1622. They were pictured in his *De Lactibus*, the first anatomical work with color plates. This plate shows veins (*A*), *lacteals* (*B*), *mesentery* (*C*), stomach (*D*), small intestine (*F, G, H*) and a lymph node (*L*).

THORACIC DUCT and the major lymph vessels of the lower extremities and trunk that contribute to it are seen in this plate from a French book of 1847, *Atlas d'Anatomie Descriptive du Corps Humain*. The duct arises from a plexus of abdominal vessels and arches up into the lower neck before entering the subclavian vein.

returned about 65 per cent of the protein that leaked out of the capillaries. Extension of this kind of experiment to man showed similar rates of leakage. In the course of a day 50 per cent or more of the total amount of protein circulating in the blood is lost from the capillaries and is returned to the bloodstream by the lymphatic system.

The importance of lymphatic drainage of protein becomes clear if one considers its role in lung function, which was elucidated by Drinker. The pulmonary circulation, in contrast to the general circulation, is a low-pressure system. The pulmonary capillary pressure is about a quarter as high as the systemic capillary pressure and the filtration pressure in the pulmonary capillaries is therefore considerably below the osmotic pressure of the blood proteins. As a result fluid is retained in the bloodstream and the lung tissue remains properly "dry."

When pulmonary capillary pressure rises significantly, there is increased fluid and protein leakage and therefore increased lymph flow. For a time the lymph drainage is adequate and the lungs remain relatively dry. But when the leakage exceeds the capacity of the lymphatics to drain away excess fluid and protein, the insidious condition pulmonary edema develops. The excessive accumulation of fluid makes it more difficult for the blood to take up oxygen. The lack of oxygen increases the permeability of the pulmonary capillaries, and this leads to greater loss of protein in a vicious circle. Some recent findings by John J. Sampson and his colleagues at the San Francisco Medical Center of the University of California support this concept. They found that a gradual increase in lymphatic drainage occurs in dogs in which high pulmonary blood pressure is produced and maintained experimentally. This suggests that the lymphatic system attempts to cope with the abnormal situation by proliferating, much as blood capillaries do when coronary circulation is impaired.

Leakage from blood capillaries and recirculation by the lymphatic system is, as I indicated earlier, not limited to protein. Any large molecule can leak out of the capillaries, and it cannot get back to the bloodstream except via the lymphatics. All the plasma lipids, or fatty substances, have been identified in thoracic-duct lymph. Even chylomicrons, particles of emulsified fat as large as a micron (a thousandth of a millimeter) in diameter that are found in blood during the digestion of fat, leak out of the bloodstream and are picked

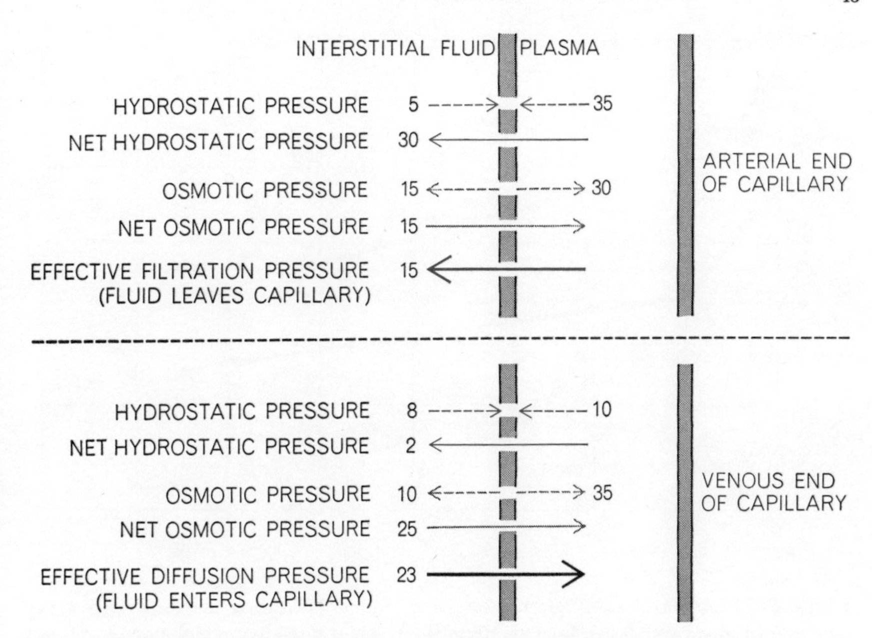

STARLING'S HYPOTHESIS explained the exchange of fluid through the capillary wall. At the arterial end of the capillary the hydrostatic pressure (*given here in centimeters of water*) delivered by the heart is dominant, and fluid leaves the capillary. At the venous end the osmotic pressure of the proteins in the plasma dominates; fluid enters the capillary.

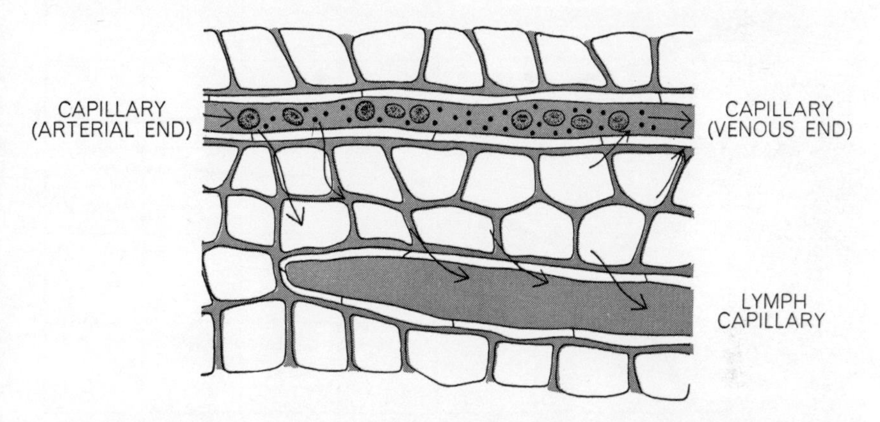

FLUID EXCHANGE is diagramed as postulated by Starling. He believed that fluid and salts (*arrows*) left the blood capillaries, mixed with the interstitial fluid and for the most part were reabsorbed by the capillaries. Excess fluid was drained by the lymph vessels. He took it for granted that most of the protein in the blood stayed inside the blood capillaries.

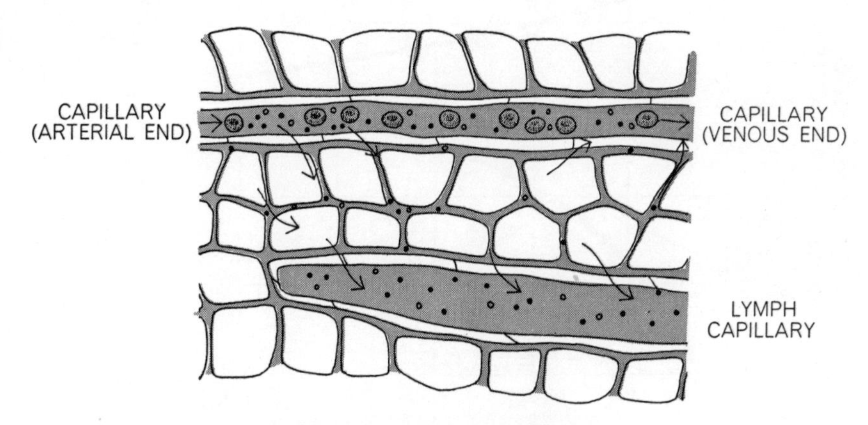

PRESENT VIEW of fluid exchange is diagramed. It appears that such large molecules as proteins (*black dots*) and lipids (*open circles*) leave the blood capillaries along with the fluid and salts. Some of the fluid and salts are reabsorbed; the excess, along with large molecules that cannot re-enter the blood capillaries, is returned via the lymphatic system.

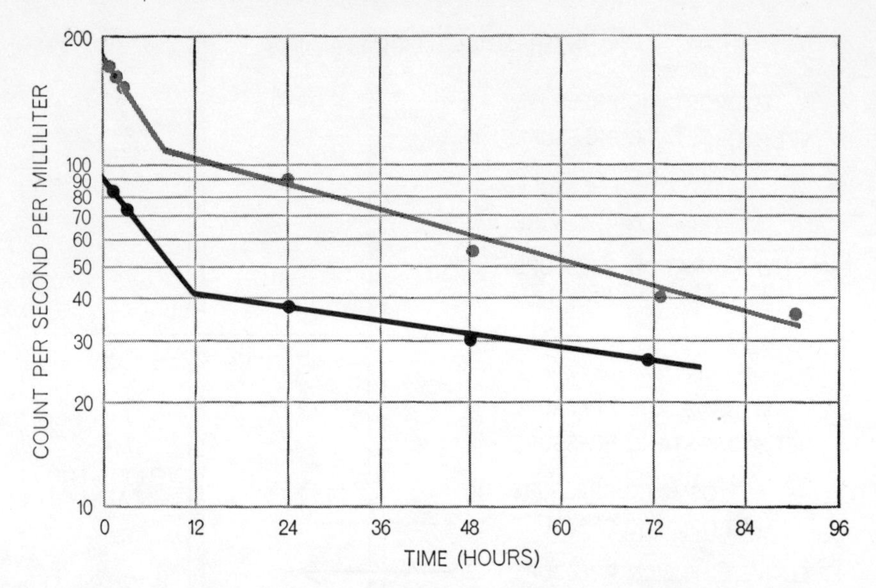

LABELED PROTEINS were injected into dogs' veins. Measuring the radioactivity per milliliter of blood withdrawn from an artery showed how quickly the globulin (*gray*) and albumin (*black*) disappeared. The steep slopes represent disappearance, the shallow slopes subsequent metabolism of the proteins. The radioactivity is plotted on a logarithmic scale.

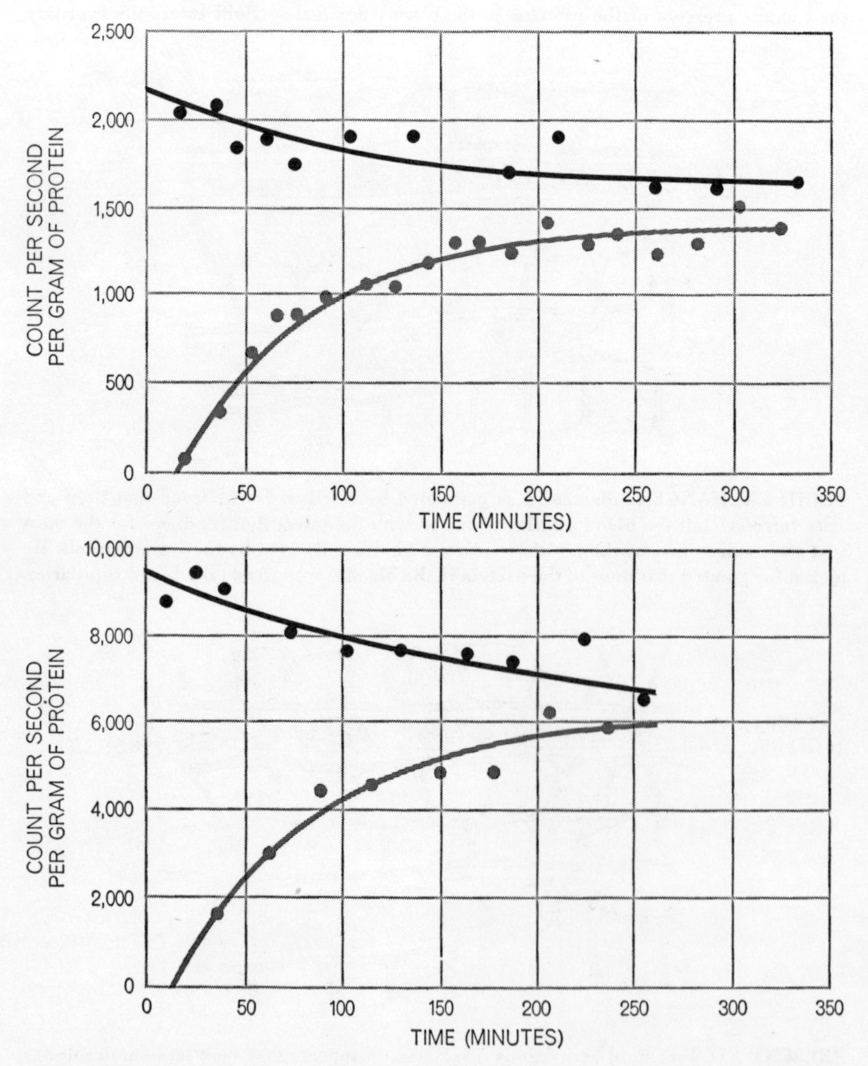

PROTEIN appeared in thoracic-duct lymph (*gray curves*) and increased in the lymph as it disappeared from the blood (*black curves*). The upper graph is for albumin, the lower one for globulin. In time the labeled protein in blood and lymph reached equilibrium.

up and recirculated to the vascular system by the lymphatics. As a matter of fact, there is evidence that they may leak out even faster than proteins. The significance of these findings remains to be explained. Aaron Kellner of the Cornell University Medical College has suggested that atherosclerosis, a form of hardening of the arteries in which there is infiltration of the walls of the arteries by lipids, may have its origin in the fact that under normal conditions there is a constant flow of fluid containing lipids and proteins across the blood-vessel lining into the vessel wall. Ordinarily this fluid is removed by the small blood vessels of the wall itself and by the lymphatics. It is conceivable that something may interfere with the removal of lipids and cause them to accumulate in the blood-vessel wall. It is even conceivable that the high capillary filtration that accompanies hypertension may increase the leakage of lipids from capillaries to a level exceeding their rate of removal from the interstitial fluid, which would then bathe even the outer surfaces of the arteries in lipids.

In addition to demonstrating that the lymph returns large molecules from the tissue spaces to the bloodstream, recent investigation has confirmed the importance of lymphatic drainage of excess fluid filtered out of the capillaries but not reabsorbed. Experiments with heavy water show that blood is unquestionably the chief source of the water of lymph. In dogs the amount of lymph returned to the bloodstream via the thoracic duct alone in 24 hours is roughly equivalent to the volume of the blood plasma. Most of this fluid apparently comes from the blood. In some of our experiments we drained the thoracic-duct lymph outside the dog's body and found that the plasma volume dropped about 20 per cent in eight hours and the plasma-protein level some 16 per cent. Translated to a 24-hour basis, the loss would be equivalent to about 60 per cent of the plasma volume and almost half of the total plasma proteins circulating in the blood. Thus the return of lymph plays an essential role in maintaining the blood volume.

One situation in which this function can be observed is the "lymphagogue" effect: the tendency of large infusions into the vascular system to increase the flow of lymph. As we increased the size of infusions in dogs, lymph flow increased proportionately; with large infusions (2,000 milliliters, about the normal blood volume of a large dog) the thoracic-duct lymph flow reached a peak value about 14 times greater than that

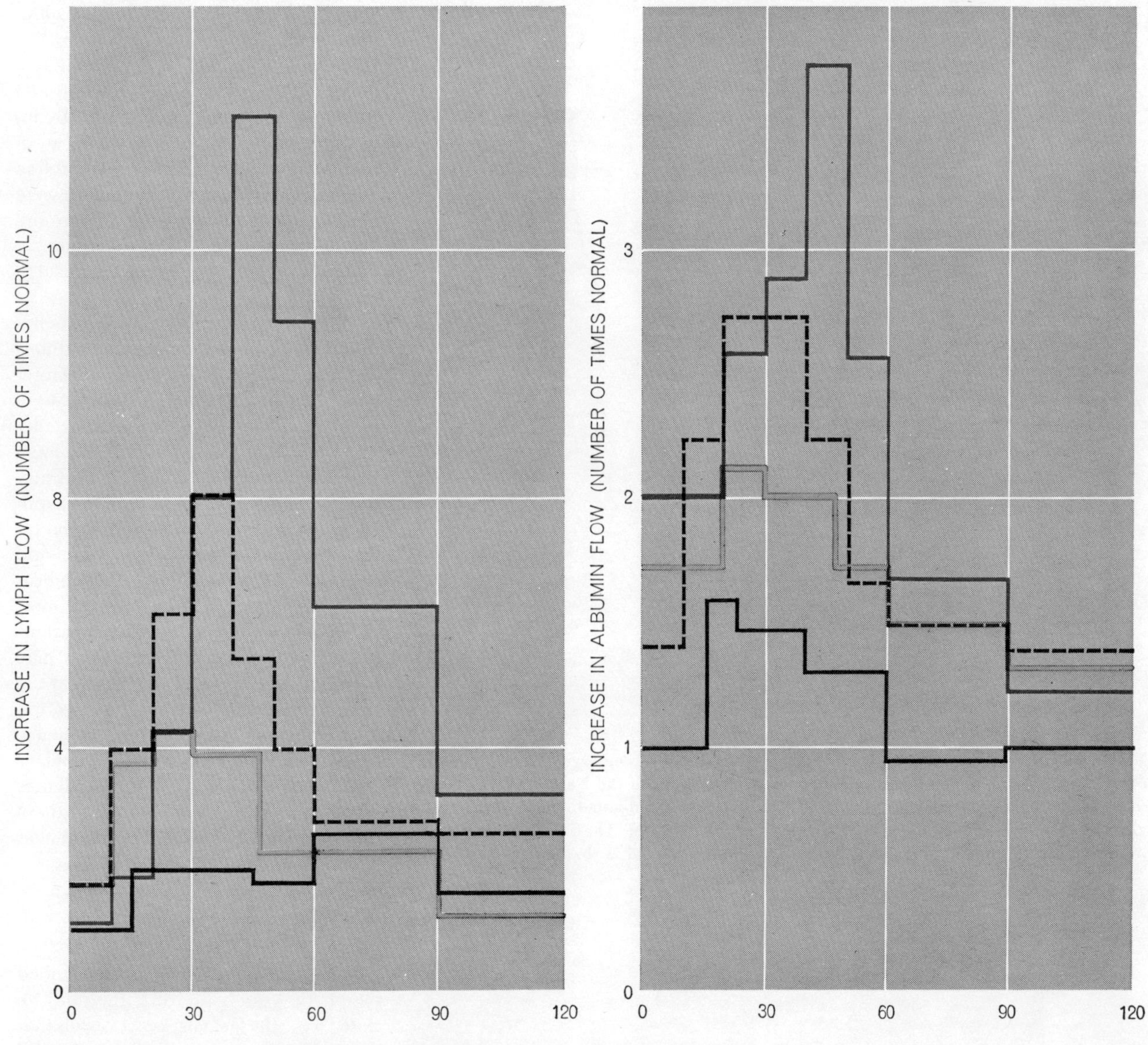

EFFECT OF INFUSIONS on lymph flow (*graph at left*) and on albumin flow in lymph (*graph at right*) in dogs is illustrated. Each curve shows the ratio of the flow after an infusion of a given size to the flow before the infusion. The curves are for infusions of 250 (*solid black*), 500 (*gray*), 1,000 (*broken black*) and 2,000 (*colored*) milliliters. The larger the infusion, the more leakage.

of the preinfusion level [*see illustration above*]. Most of the excess fluid is excreted by the kidneys in increased urine flow. But the displacement of fluid from the blood circulation into the lymph "saves" some of the fluid. In other words, it can be considered as being a fine adjustment of the blood volume so that not all the fluid is irrevocably lost from the body. Large infusions also increase protein leakage, but again the fact that the protein goes to the lymph and slowly returns to the bloodstream minimizes changes in total circulating protein and the loss of its osmotic effect.

The Mechanism of Filtration

The exact processes or sites of the filtration of large molecules through the capillary wall, and through cell membranes in general, are still unclear. As Arthur K. Solomon has pointed out in these pages [see "Pores in the Cell Membrane," by Arthur K. Solomon; SCIENTIFIC AMERICAN Offprint 76], "some materials pass directly through the fabric of the membrane, either by dissolving in the membrane or by interacting chemically with its substance. But it seems equally certain that a large part of the traffic travels via holes in the wall. These are not necessarily fixed canals; as the living membrane responds to changing conditions inside or outside the cell, some pores may open and others may seal up."

This last point appears to explain our results with infusions in dogs. The massive infusions overfill the closed blood system, raise filtration pressure in the capillaries and result in increased leakage through the capillary walls; lymph flow is copious and the lymph contains more large molecules. Small infusions do not do this. The reason, then, that patients did not do as well as expected after receiving large infusions or transfusions was that the plasma and proteins leaked out through stretched capillary pores. A similar effect accounted for the case of animals subjected to severe hemorrhage: there was not enough blood to oxygenate the capillary walls adequately, the walls became more permeable and the protein molecules passed through.

We found that the rate of leakage for any molecule depends on its size. Globulin, which has a molecular weight of

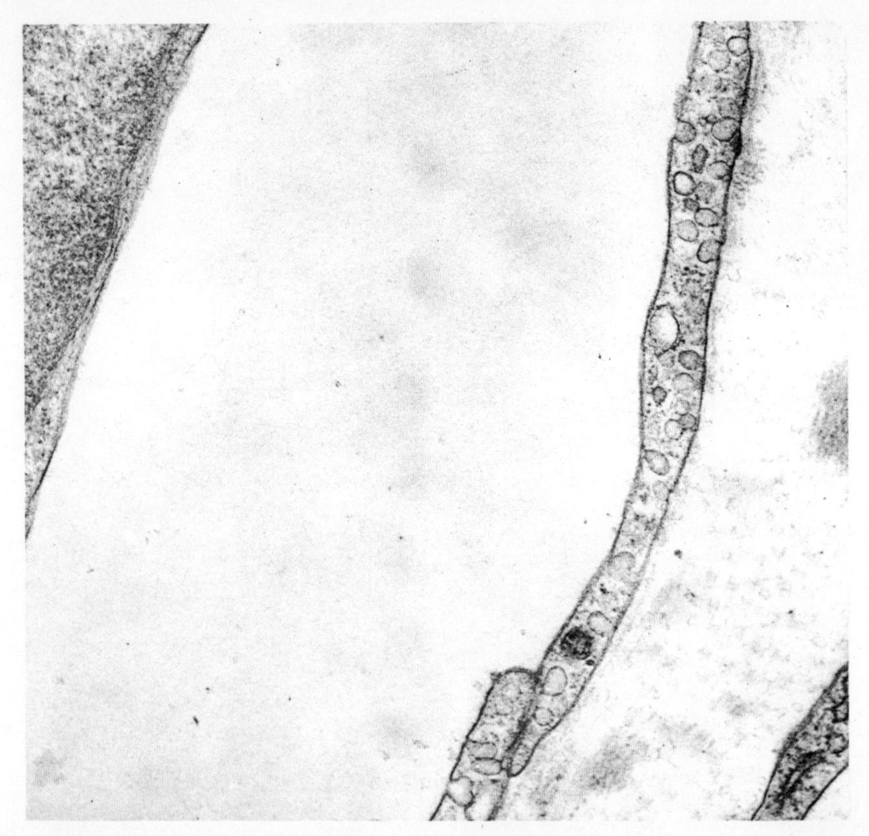

LYMPH CAPILLARY (*lightest area*) of mouse arterial tissue is enlarged 30,000 diameters in this electron micrograph made by Johannes A. G. Rhodin of the New York University School of Medicine. At upper left is part of the nucleus of an endothelial cell of the vessel wall. The thin ends of two other cells overlap near the bottom. The faint shadow in the connective tissue outside the wall is a slight indication of a basement membrane.

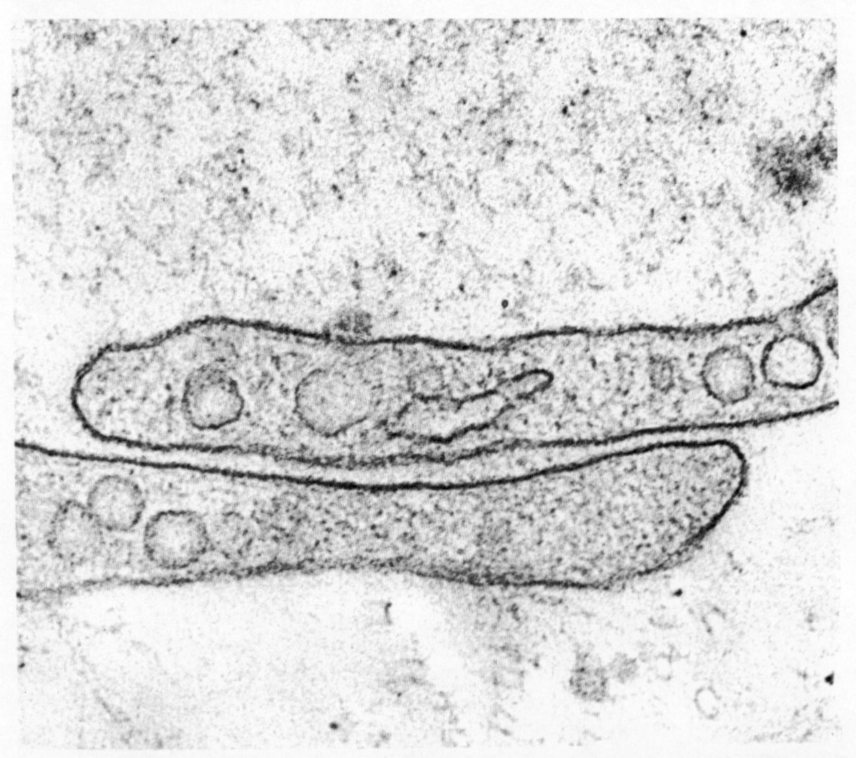

"LOOSE JUNCTION" between the overlapping ends of two endothelial cells of a vessel wall is seen in this electron micrograph, also made by Rhodin, of a lymphatic in mouse intestinal tissue. The lymph vessel is at the top, connective tissue at the bottom. In this case there is no sign of a basement membrane outside the wall. The enlargement is 90,000 diameters.

250,000, leaked more slowly than albumin, which has a weight of 70,000. The third plasma protein, fibrinogen, has a molecular weight of about 450,000 and leaves the blood still more slowly. By introducing into the blood various carbohydrate molecules, which unlike protein molecules do not carry a charge, we were able to demonstrate that it is size and not electrical charge that determines a molecule's rate of movement through the wall.

When, after infusing the carbohydrates, we collected lymph from different parts of the body, we found larger molecules in intestinal lymph than in leg lymph, and still larger molecules in lymph from the liver. This indicated that capillaries in the liver have substantially larger openings than leg capillaries do, and that the vessels of the intestine probably have both large and small openings. There are other indications that liver capillaries are the most permeable of all; for example, apparently both red cells and lymphocytes pass between the blood and the lymph in the liver. Recent studies with the electron microscope confirm the indirect evidence for variations in the size of capillary openings; the structure of the capillaries seems to vary with the organ, and these differences may be related to differences in function.

Transport by the Lymph

When Gasparo Aselli first described lymphatic vessels in 1622, the ones he noted were the lacteals: small vessels that drain the intestinal wall. The dog Aselli was dissecting had eaten recently; the lacteals had absorbed fat from the intestine, which gave them a milky-white appearance and made them far more visible than other lymphatics. The transport of certain fats from the intestine to the bloodstream by way of the thoracic duct is one of the lymphatic system's major functions. Studies in which fatty acids have been labeled with radioactive carbon show that the blood capillaries of the intestine absorb short-chain fatty acid molecules directly, together with most other digested substances, and pass them on to the liver for metabolism. But the lacteal vessels absorb the long-chain fats, such as stearic and palmitic acids, and carry them to the bloodstream via the thoracic duct. The lymphatic system is also the main route by which cholesterol, the principal steroid found in tissues, makes its way into the blood.

Since the lymphatics are interposed between tissue cells and the blood sys-

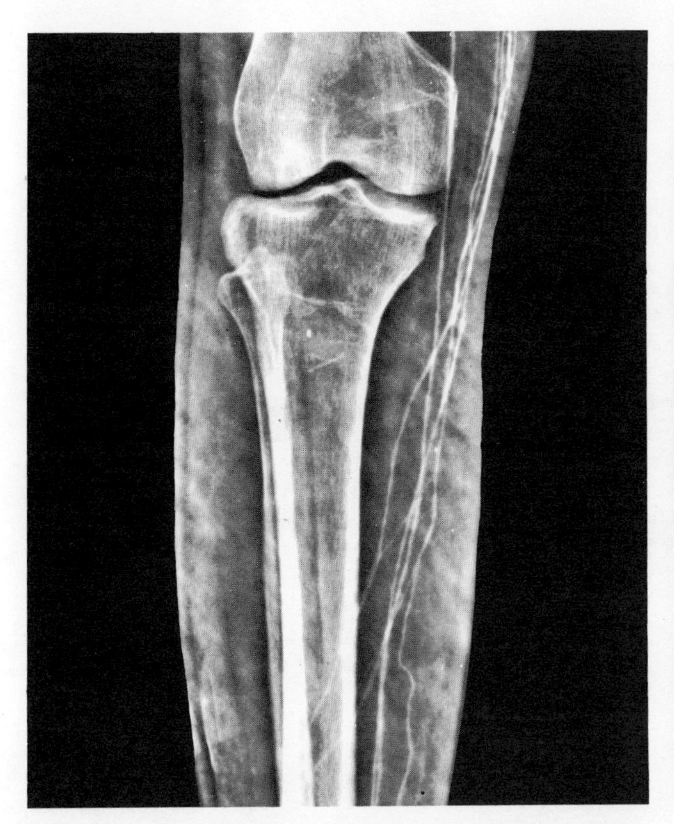

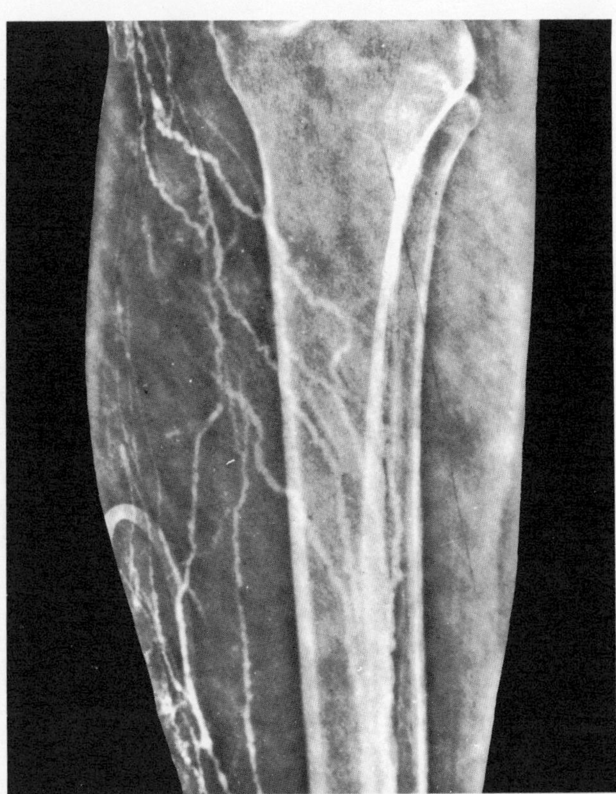

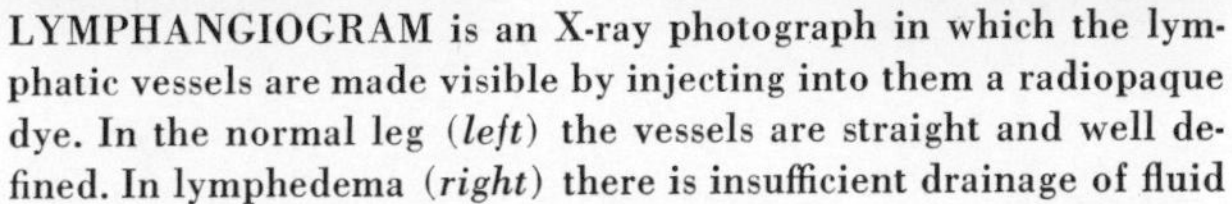

LYMPHANGIOGRAM is an X-ray photograph in which the lymphatic vessels are made visible by injecting into them a radiopaque dye. In the normal leg (*left*) the vessels are straight and well defined. In lymphedema (*right*) there is insufficient drainage of fluid and proteins, in this case because there were too few vessels in the thigh. The extra pressure on the lower-leg vessels increased their number and made them tortuous. These pictures were made by Carl A. Smith of the New York University School of Medicine.

tem it is not surprising to find that they serve as the channel for transport to the bloodstream of substances that originate in tissue cells. The lymph is probably the route by which at least some hormones, many of which are very large molecules, are carried to the blood from the endocrine glands where they are synthesized. Some enzymes, found in lymph in small concentrations, may merely have leaked out of the capillaries. But others are apparently picked up from their cells of origin and carried to the blood by the lymph. Certain enzymes, including histaminase and renin, are present in greater concentrations in lymph than in the blood. The finding on renin, reported by A. F. Lever and W. S. Peart of St. Mary's Hospital Medical School in London, is of particular interest to investigators working on the problem of hypertension. One concept ascribes high blood pressure in some individuals to the production of renin by a kidney suffering from inadequate blood circulation; the renin is thought to combine with a globulin in the plasma to form hypertensin, an enzyme that narrows the arterioles and results in high blood pressure. It has been difficult to establish this concept because no one has been able consistently to demonstrate the presence of renin in the blood of hypertensive patients. Now that it has been discovered in lymph coming from the kidney it is clear that renin is indeed being formed in these patients, although in amounts so small that it usually escapes detection after being diluted in the blood.

Recently Samuel N. Kolmen of the University of Texas Medical Branch in Galveston has provided what may be a confirmation of the renin concept. He produced hypertension in dogs by removing one kidney and partially constricting the artery supplying the other one. When he shunted the thoracic-duct flow into the dog's gullet or allowed it to escape, the hypertension diminished. The implication is that renin was being kept out of the bloodstream. When Kolmen stopped lymph flow in the shunt, presumably inhibiting the diversion of the renin, the hypertension returned.

The Lymphatic Circulation

To the investigators of the 19th century the lymphatic system was "open-mouthed": its capillaries were assumed to be open to the tissue spaces. More recent evidence has shown that the lymphatics form a closed system, that fluid enters not through the open ends of vessels but through their walls. The walls of the terminal lymph capillaries, like those of blood capillaries, consist of a single layer of platelike endothelial cells. This layer continues into the larger vessels as a lining but acquires outer layers of connective tissue, elastic fibers and muscle. Although the capillaries of the blood and lymphatic systems are structurally very similar, recent electron micrographs show differences in detail that may help to explain the ease with which the lymph vessels take up large molecules. Sir Howard Florey of the University of Oxford and J. R. Casley-Smith of the University of Adelaide in Australia believe that the most important difference is the poor development or absence in lymph capillaries of "adhesion plates," structures that hold together the endothelial cells of the blood capillaries. They suggest that as a result there are open junctions between adjacent cells in lymph capillaries that allow large molecules to pass through the walls. Johannes A. G. Rhodin of the New York University School of Medicine puts more emphasis on the apparent absence or poor development in lymph capillaries of the "basement membrane" that surrounds blood capillaries.

Certainly it is clear that very large particles do enter the lymphatic vessels: proteins, chylomicrons, lymphocytes and red cells—the last of which can be as much as nine microns in diameter. Bacteria, plastic spheres, graphite particles and other objects have been shown to penetrate the lymphatics with no apparent difficulty. Yet we have found that when we introduce substances directly into the lymphatic system, anything with a molecular weight greater than 2,000 is retained almost completely within the lymphatics, reaching the blood only by way of the thoracic duct. If large particles can get into the lymphatic vessels, why do substances with a molecular weight of 2,000 not get out by the same channels?

I have spent many hours trying to formulate an answer to this question without arriving at a sophisticated concept, and have had to be content with a simple explanation that is at least consistent with the current evidence. Assume that the smallest terminal lymphatics are freely permeable to small and large molecules and particles moving in either direction through intercellular gaps. Compression of these vessels in any way would tend to force their contents in all directions. At least some of the contents would be forced along into the larger lymph vessels, where the presence of valves would prevent backflow. And once the lymph reaches a larger vessel it can no longer lose its large particles through the thick and relatively impermeable wall of the lymphatic.

One can argue that this seems to be a rather inefficient and even casual way of getting the job done. Indeed it is, and this physiological casualness is a characteristic of the lymphatic system as a whole. There is no heart to push the lymph, and although lymphatic vessels do contract and dilate like veins and arteries this activity does not seem to be an important factor in lymph movement. The flow of lymph depends almost entirely on forces external to the system: rhythmic contraction of the intestines, changes in pressure in the chest in the course of breathing and particularly the mechanical squeezing of the lymphatics by contraction of the muscles through which they course.

Lymphatic Malfunction

In spite of the casualness of the lymphatic system, its development, as I have tried to show, was an absolute necessity for highly organized animals. Its importance is most visibly demonstrated in various forms of lymphedema, a swelling of one or more of the extremities due to the lack of lymphatic vessels or to their malfunction. In some individuals the lymphatic system fails to develop normally at birth, causing gradual swelling of the affected part. Lymphangiographic studies, in which the vessels are injected with radiopaque dyes to make them visible in X-ray photographs, show that the lymphatics are scarce, malformed or dilated. Insufficient drainage causes water and protein to accumulate in the tissues and accounts for the severe and often disabling edema. There is evidence that genetic factors may play a role in this condition. Surgical procedures that destroy lymph vessels may have a similar effect in a local area. Elephantiasis is a specific form of lymphedema resulting from the obstruction of the vessels. It can be caused by infection of the lymphatics or by infestation with a parasitic worm that invades and blocks the vessels.

The lymphedemas have been recognized as such for many years. Recently the view that the lymphatic system is essential to homeostasis—which is to say "good health"—has led to a number of investigations of its role in conditions in which no lymphatic involvement was previously suspected. Our group at Tulane has found that lymphatic drainage of the kidneys is essential in order to maintain the precise osmotic relations

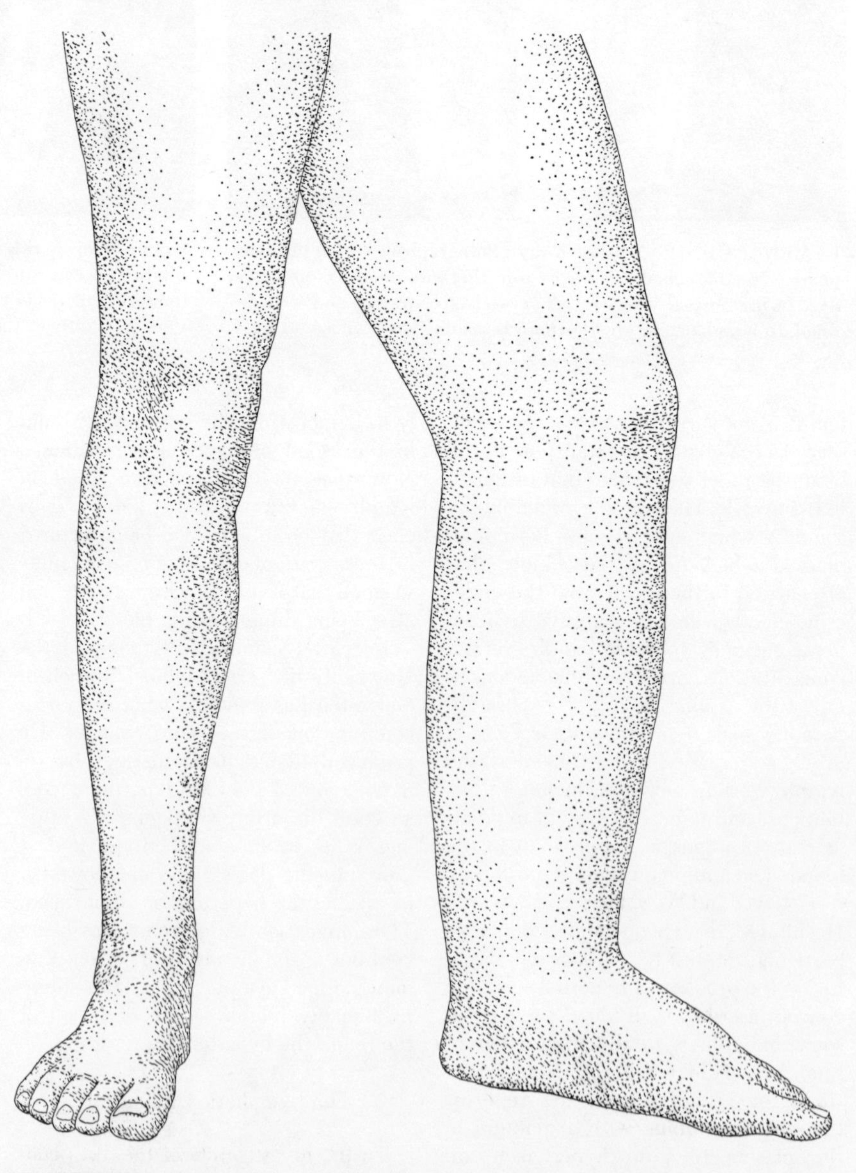

LYMPHEDEMA can cause gross deformity of a limb and even disability. The drawing is based on a photograph of an 11-year-old girl whose leg began to swell at the age of seven, probably because of an insufficiency of lymphatic vessels. The patient's condition was greatly improved by an operation in which the tissue between skin and muscles was removed.

on which proper kidney function depends. This may explain the dilution of the urine observed in some patients after kidney operations: the lymph vessels may have been damaged, decreasing their capacity for draining proteins and interfering with the reabsorption of water by the kidney tubules.

At the New York University School of Medicine, John H. Mulholland and Allan E. Dumont have been investigating the relation between thoracic-duct flow and cirrhosis of the liver. Their results suggest that the cirrhosis may be associated with increased lymph flow in the liver and that the inability of the thoracic duct to handle the flow may bring on the accumulation of fluid, local high blood pressure and venous bleeding that are frequently seen in cirrhosis patients; drainage of the duct outside the body temporarily relieves the symptoms.

These and other clinical observations are consistent with my feeling that the lymphatic system does a capable job when all is going well but that its capacity for dealing with disturbances is limited. As a phylogenetic late-comer it may simply not have evolved to the point of being able to cope with abnormal stresses and strains. The role of the second circulation in disease states is currently under intensive investigation. As more and more is learned about its primary functions and its reactions to stress, the new knowledge should be helpful in diagnosis and perhaps eventually in the treatment of patients.

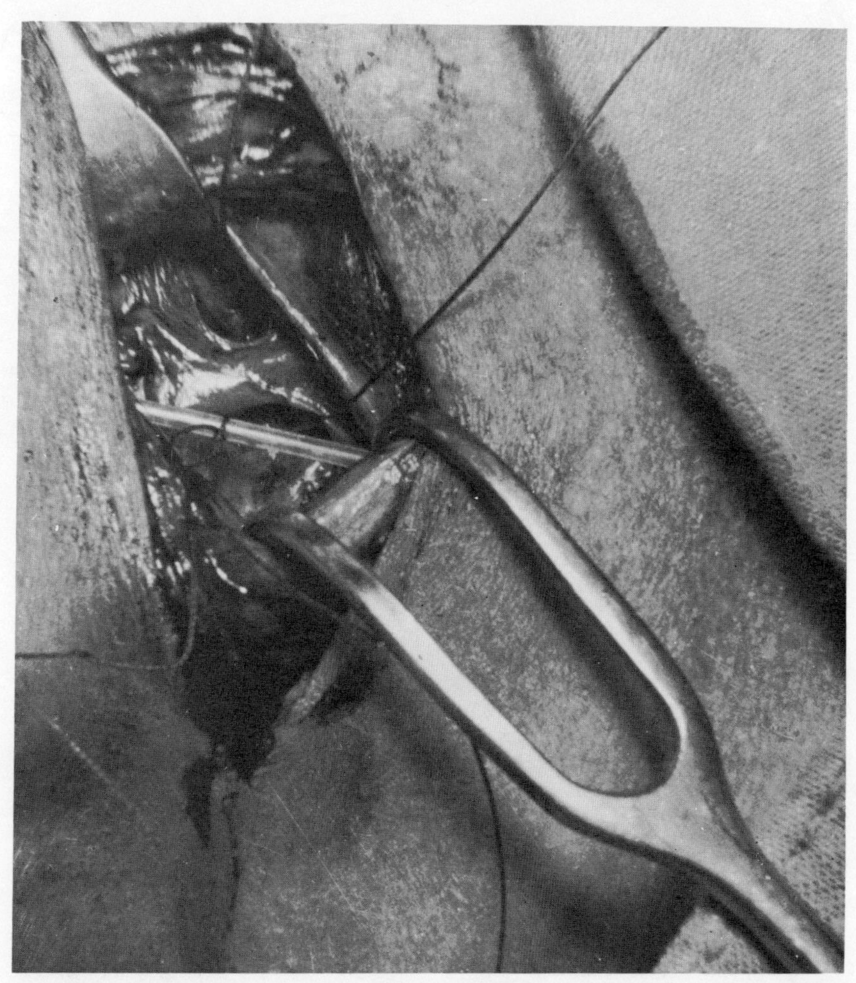

DILATED THORACIC DUCT of a patient with cirrhosis is seen in this photograph made by Allen E. Dumont and John H. Mulholland of the New York University School of Medicine and reprinted from *Annals of Surgery*. The plastic tube just below the duct is a tenth of an inch in diameter. A normal duct would be smaller than the tube.

4 A Brain-cooling System in Mammals

by Mary Ann Baker
May 1979

Carnivorous mammals and some of their mammalian prey endure lethal extremes of heat and exertion because they have a rete, or heat-exchange network, that keeps the brain from getting too hot

On a hot day a dog can keep a rabbit on the run until the rabbit dies. How is this possible? The answer is that although running raises the temperature of both animals, the dog's brain has a cooling system and the rabbit's brain does not. If the rabbit cannot find a hiding place where it can cool off, the temperature of its brain soon reaches a lethal level.

Specifically what the dog has that the rabbit does not is a countercurrent heat-exchange plexus at the base of its brain, an anatomical feature found in some mammals but not in all. This plexus, consisting of a network of small blood vessels branching from the carotid arteries, was first recognized more than two millenniums ago by the Greek physician Herophilus. As he was dissecting the head of an animal (probably a sheep) in about 300 B.C. he discovered a prominent network of blood vessels at the base of the brain.

Herophilus described the structure, but it was the anatomist Galen who made it famous nearly half a millennium later. Working in Rome in the middle of the second century A.D., he ascribed functions to the arterial network that appear to be responsible for the term that was soon applied to it: the *rete mirabile,* or wonderful net. Galen was not able to obtain human cadavers for dissection, and so his descriptive anatomy came mainly from the study of domesticated animals. In his view the carotid *rete* was a key anatomical structure: it acted to transform the "vital spirit," which was carried through the arteries, into the "psychic spirit." That spirit was then transported through the body by the nervous system, which Galen believed consisted of hollow tubes.

Galen had a profound influence on later anatomy and medicine. From his day until the time of the Renaissance, anatomists who had human heads to dissect were expected to find the wonderful net because Galen had described it, even though it is not present in man and other primates. The great Renaissance anatomist Vesalius, writing in about 1538, describes the dilemma: "I who so much labored in my love for Galen . . . never undertook to dissect a human head in public without that of a lamb or ox at hand, so as to supply what I could in no way find in that of man, and to impress it on the spectators, lest I be charged with failure to find that plexus so universally familiar by name."

Countercurrent heat-exchange networks are not limited to the brain of certain mammals; they are found in other parts of the body throughout the animal kingdom. They are responsible for the ability of wading birds to stand for long periods in cold water, for the ability of some sea mammals and fishes to live in polar seas and for the ability of such tropical mammals as sloths and anteaters to conserve body heat at night. Here, however, we shall be concerned only with the first wonderful net to be discovered, the carotid *rete,* and the role it plays in keeping the brain cool.

No carotid *rete* is found among the monotremes (the order that includes the platypus), the marsupials, the perissodactyls (the order that includes the horse, the rhinoceros and the tapir), the rodents, the lagomorphs (the order that includes rabbits and hares) and, as we have seen, the primates. For example, in man blood is supplied to the brain through the right and left internal carotid arteries and the right and left vertebral arteries. At the base of the brain the four arteries are connected in the circle of Willis, named for the English anatomist who described it in the 17th century. It is from this circle that all the major arteries supplying blood to the brain arise.

Among the artiodactyls (the order that includes cattle and the other hoofed mammals with even-numbered toes) and many carnivores the pattern of the blood supply to the brain is quite different. Instead of receiving blood from the two pairs of cranial arteries the circle of Willis in these animals is supplied mostly or entirely by the carotid *rete.* The *rete* in turn receives blood either from branches of the left and right internal carotid arteries or from branches of the external carotid arteries or from both. The right and left vertebral arteries are present, but they are often very small. Thus the *rete* is interposed between the two common carotid arteries and the circle of Willis, and almost all the arterial blood that flows into the brain passes through it.

The size and shape of the carotid *rete* varies among the species of mammals that have it, although in some of these species the *rete* has not been examined in detail. For example, among the carnivores all members of the cat family have a well-developed *rete:* a network of vessels 200 to 300 microns in diameter. These are medium-size vessels, much smaller than the arteries that enter and leave the *rete* but much larger than the capillaries. Among dogs, wolves and hyenas and among some seals and sea lions the *rete* is small and consists of no more than a few twisting medium-size arteries.

All mammals, those with a carotid *rete* and those without it, show the same anatomical arrangement at the base of the brain. There, where the arteries that supply the brain with blood enter the cranial cavity, large reservoirs of venous blood are found. Known as the venous sinuses, the reservoirs receive blood from veins both outside and inside the skull. One of the reservoirs is known as the cavernous sinus; in mammals that have no *rete* the internal carotid arteries run through the cavernous sinus on their way to the circle of Willis. In animals that have a *rete* the arterial plexus either lies inside the cavernous sinus or is associated with a similar plexus of venous blood vessels that is connected to the cavernous sinus.

In both arrangements the arteries are bathed by venous blood but the arterial blood and the venous blood do not mix. That is not surprising with respect to the internal carotid arteries, which have normally thick walls. The walls of the *rete* arteries, however, are unusually thin, as I have determined in studies

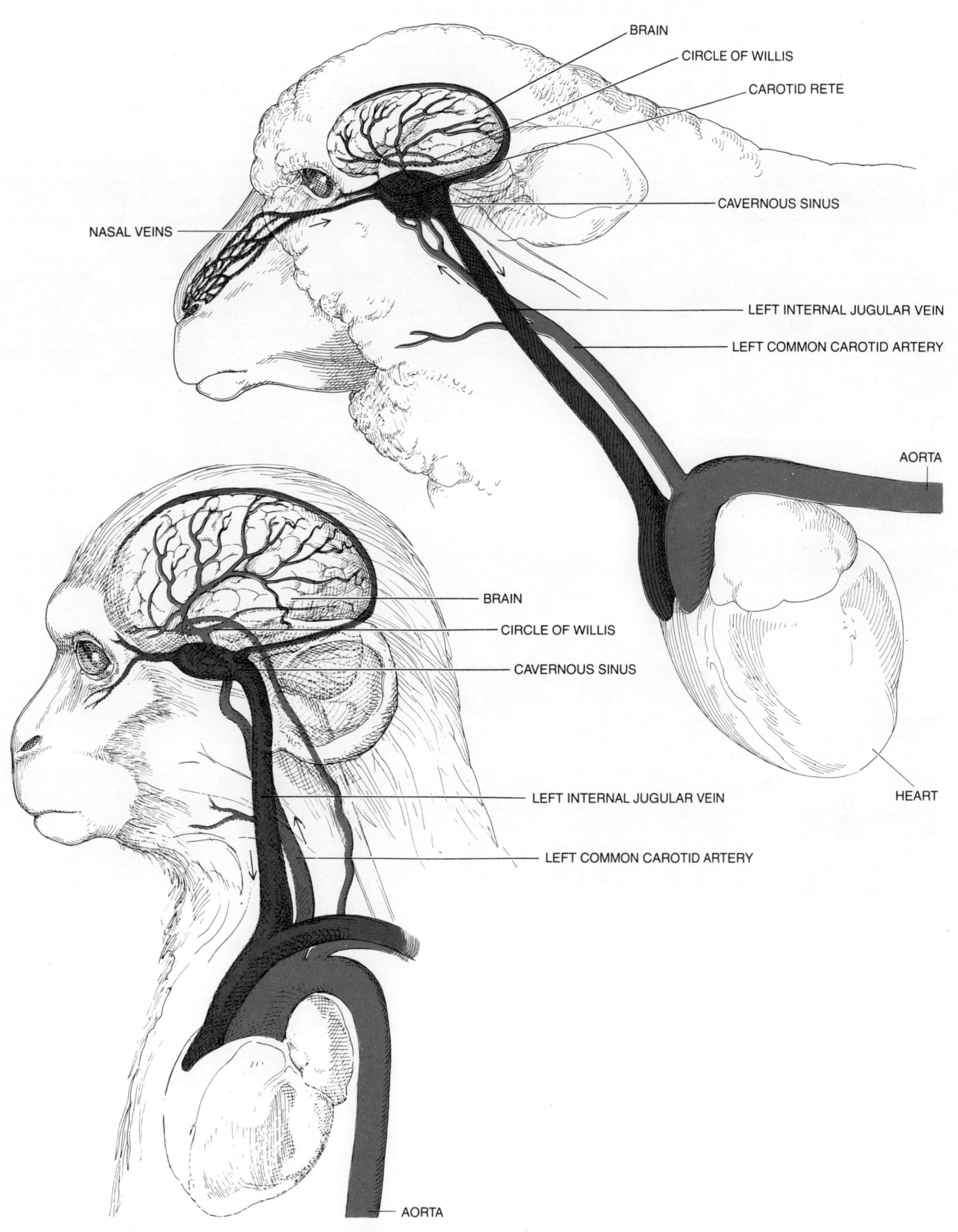

SUPPLY OF BLOOD to the brain of a mammal moves from the heart in one of two basic ways, depending on whether or not the mammal has a carotid *rete*. As illustrated schematically here, the sheep (*top*) has a *rete* but the monkey (*bottom*) does not. In the sheep arterial blood is distributed to the brain from the network of vessels known as the circle of Willis. Some of the blood arrives at the circle via the vertebral arteries but most arrives via the common carotid arteries and their branches, detouring through the network of vessels that forms the carotid *rete*. This network, located within a pool of venous blood in a cavity known as the cavernous sinus, acts as a heat exchanger: the warmer arterial blood loses heat to the cooler venous blood. As a result most of the arterial blood that reaches the sheep's brain is at a lower temperature than it was when it was pumped from the heart. In the monkey the arterial blood travels from the heart via the vertebral and carotid arteries and their branches directly to the circle of Willis. The internal carotid arteries pass through the cavernous sinus with a minimum of heat lost to the pool of venous blood there, and so the blood is supplied to the brain at heart temperature.

conducted in collaboration with Wendelin J. Paule and Sol Bernick at the University of Southern California School of Medicine. Why is the barrier between the arterial blood and the venous blood so insubstantial? The answer to this question helps to explain the major function of the carotid *rete*.

Since the supply of blood to the brain is of major physiological and clinical importance, the anatomy and physiology of the system have been studied in much detail. The metabolic processes of the brain operate at a high rate, and the organ must be supplied with blood at a rate of flow high enough to supply oxygen and nutrients and carry away waste products. If the supply is interrupted, the brain cells can be irreversibly damaged; an interruption may be damaging within a matter of minutes or in no more than a few seconds, depending on the circumstances. It is not surprising that the carotid *rete,* intimately associated with the supply of blood to the brain, has intrigued investigators for hundreds of years.

The question of whether the *rete* does or does not play some role in regulating the rate of blood flow to the brain still remains unanswered. What is now understood, however, is its function in the quite different role of a heat exchanger for cooling the arterial blood destined for the brain. The work leading to an understanding of this role began for me some 10 years ago in the laboratory of James N. Hayward at the University of California at Los Angeles. It was known that in vertebrate animals generally the temperature of the brain was important for the regulation of body temperature. As a result we were interested in finding out how brain temperature was controlled.

Our first studies arose from the observation that rather large changes in brain temperature, as much as one or two degrees Celsius, take place when mammals go from one behavioral state to another. For example, when a relaxed rabbit is suddenly alerted, its brain temperature rises, and when the rabbit relaxes again or goes to sleep, its brain temperature falls. Some investigators had suggested that such temperature

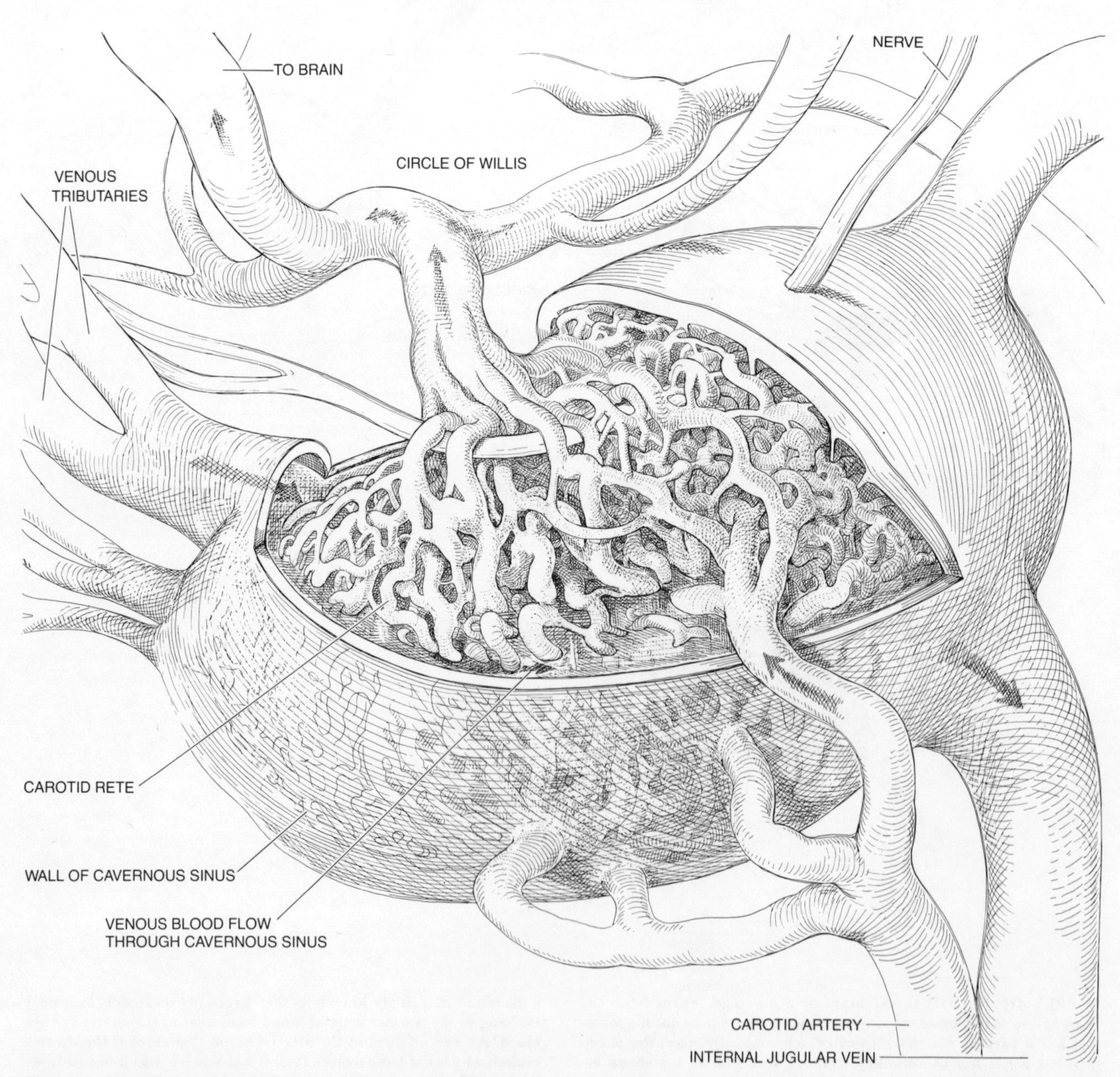

CAROTID RETE, located within a hollow known as the cavernous sinus at the base of a sheep's brain, is exposed in this cutaway drawing. Venous blood, draining from the nose and mouth, enters the sinus from the left, bathes the small arteries of the *rete* and leaves via veins that enter the internal jugular vein at the right. At the same time arterial blood from the heart enters the *rete* via branches of the external carotid artery (*right*) and travels on to the circle of Willis (*top*) after losing some of its heat to the pool of cooler venous blood.

changes were caused by increases or decreases in the metabolic rate of the brain; our studies showed that this was not so.

Selecting monkeys and rabbits as our experimental animals, we implanted very small thermocouples in each subject: one thermocouple near the arteries of the circle of Willis, a second in the animal's brain and a third in a carotid artery near the aorta. The leads to the thermocouples were attached to plugs fixed to the animal's head; long wires led from the plugs to automatic temperature recorders, making it possible for us to monitor changes in temperature as we observed the animal's behavior. We found that all the changes in brain temperature were preceded by changes in the temperature of the arterial blood being supplied to the brain.

What caused the changes in the arterial blood temperature? We found that they were the result of changes in the rate at which heat was being lost from the animals' skin. When the flow of blood through the skin increases, so does the heat loss; when the flow decreases, the heat loss is reduced.

The dilation and constriction of the blood vessels of the skin are controlled by the autonomic nervous system, a system that is affected by emotion and activity. For example, when an animal is excited, the autonomic nerves cause constriction of the blood vessels in the skin; heat loss is reduced and the body-core temperature, including the temperature of the arterial blood, rises. Soon afterward so does the temperature of the brain. Conversely, when an animal relaxes, the autonomic nervous system allows the blood vessels in the skin to dilate, heat loss through the skin increases and the temperature of the body core falls. In rabbits one of the most important heat-exchange areas is the ear; the fur there is thin and the tissues are richly supplied with blood. In monkeys the hand and the foot are important heat-exchange areas; in human beings the hand is. Indeed, this same autonomic reaction occurs in human beings. When we are excited or upset, our hands tend to get cold. When we relax, they warm up.

While we were studying the brain and blood temperatures of rabbits and monkeys we found that the temperature of the arterial blood in the carotid artery and of the arterial blood in the circle of Willis were the same. We wondered whether the same pattern would be found in other species. Our next experimental animals were cats and sheep, species that unlike rabbits and monkeys have a carotid *rete*. We implanted three thermocouples in the new subjects. When either animal was in a relaxed state, we found that, as in rabbits and monkeys, the temperature of the brain was higher than the temperature of the arterial blood at the circle of Willis. When the animals were active, the temperatures rose, as they also had in the rabbits and monkeys. We found, however, that during activity the arterial blood at the circle of Willis remained cooler than the blood at the third thermocouple, which was emplaced near the aorta. Somehow in cats and sheep the arterial blood was being cooled as it flowed from the heart to the brain. This was quite different from our findings in rabbits and monkeys.

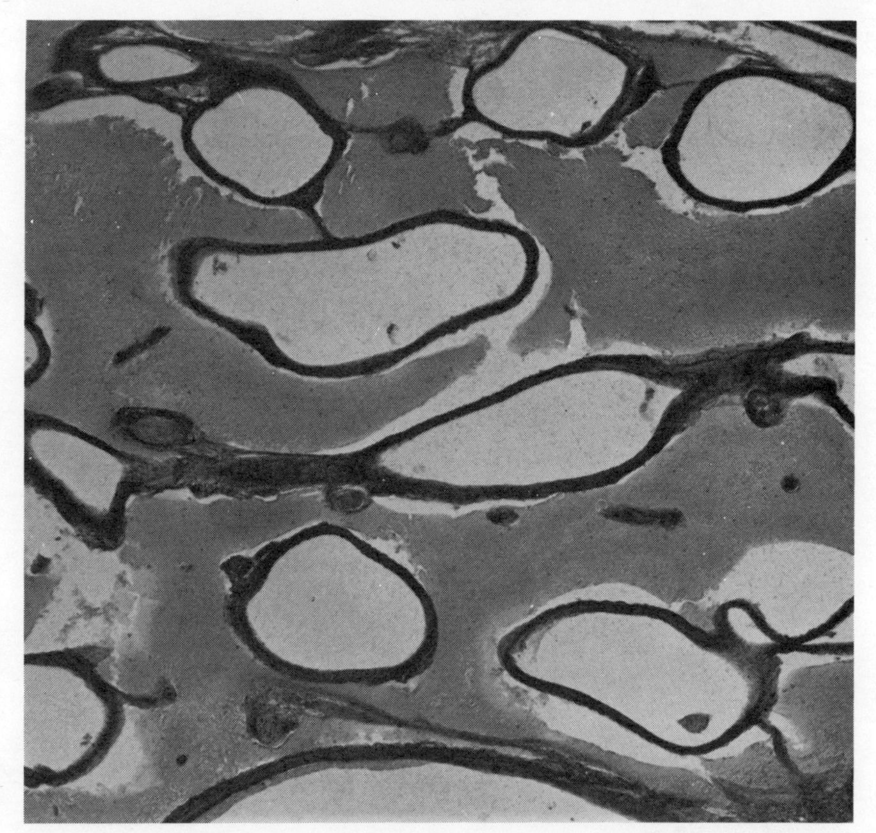

CAROTID RETE of a cat is seen in section in this micrograph. Varied in size, the thin-walled arteries lie in a lake of cool venous blood, secured in place by strands of connective tissue.

The only major difference between the two sets of experimental animals as far as the blood supply to the brain was concerned was the absence of a carotid *rete* in monkeys and rabbits and the presence of the *rete* in cats and sheep. This fact suggested that in our new subjects the *rete* was functioning as a countercurrent heat exchanger. To test this hypothesis we implanted an additional thermocouple in the animals' cavernous sinus, where the venous blood surrounds the arteries of the *rete*. We found that the venous blood also underwent temperature changes: it was cooler when the animals were relaxed or sleeping and warmer when they were alert or excited. Regardless of the animals' state, however, the venous blood was cooler than the arterial blood entering the *rete*, and so a countercurrent heat exchange was taking place.

This led us to two further questions. What is the source of the venous blood in the cavernous sinus? Why does the temperature of the venous blood change when the animal's activity changes? As to the first question, venous blood arrives at the cavernous sinus from several sources. Some of it comes from the base of the brain; we would expect this blood to be warmer than arterial blood because the brain has added heat to it. Much of the venous blood, however, comes from areas outside the cranial cavity. For example, anatomists have known for a long time that the venous blood draining from the nose and parts of the mouth can flow into the intracranial sinuses.

James H. Magilton and Curran S. Swift of Iowa State University have studied the venous drainage of the nose, with dogs as their subjects. They found that changes in temperature within a dog's nose will change its brain temperature by changing the temperature of the nasal venous blood that flows into the cranium. In an anatomical study of our own we traced a similar flow in sheep. Working with preserved sheep heads, we injected colored latex into the veins of the nose; we found that the latex filled the cavernous sinus at the base of the brain where the sheep's carotid *rete* is located.

Such studies enabled us to answer the first of our questions: the venous blood that was affecting the temperature of the arterial *rete* came from the animals' nasal passages. As for the question of changes in temperature, these changes reflect the condition of the nasal mucosa: the moist lining of the nasal passages. The blood vessels of the nasal mucosa, like the blood vessels of the skin, are under the control of the autonomic nervous system. When an animal, including the human animal, is relaxed, the blood vessels of the nasal passages are dilated and the venous blood is cooled by the evaporation of moisture from the mucosa. When the animal is excited, the blood vessels of the nasal mucosa are constricted by the autonomic nerves. The action decreases the amount of blood flowing through the nasal passages and also the amount of venous blood flowing from the nose to the cavernous sinus in the cranium. The decrease in the supply of cool venous blood diminishes the heat exchange be-

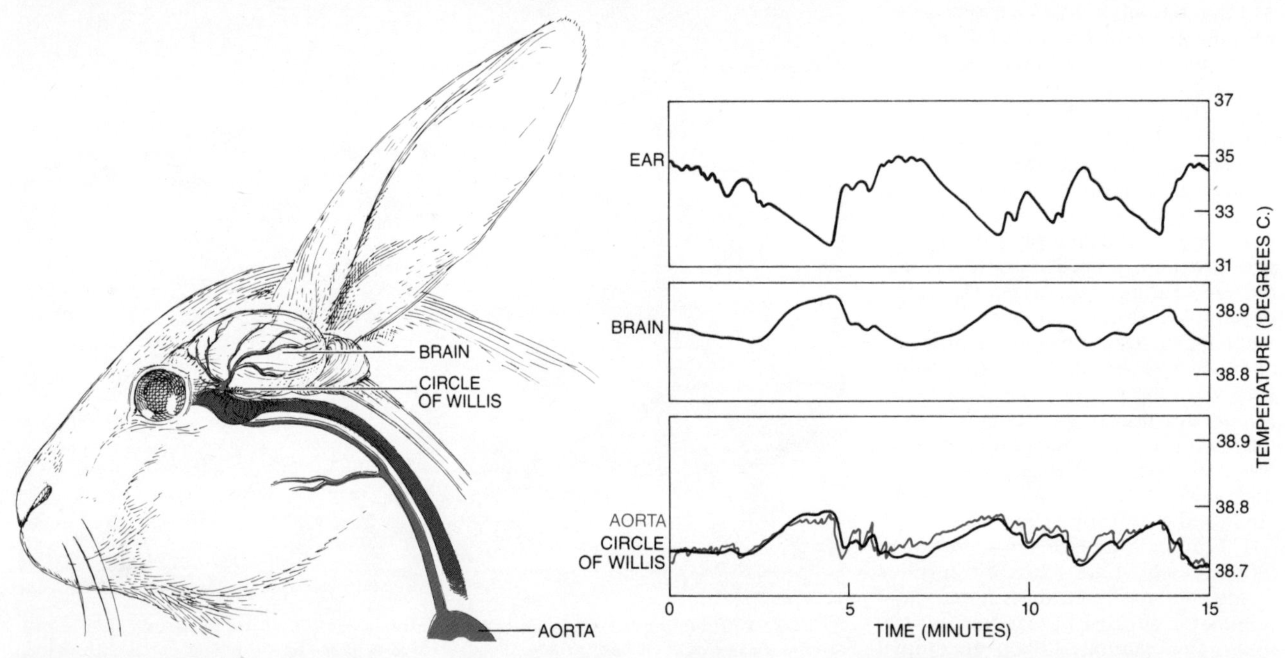

BLOOD-TEMPERATURE CHANGES in a rabbit proved to be correlated with changes in the heat loss from the ear. The dilation and constriction of the blood vessels in the ear of the rabbit, a major heat-exchange surface, are controlled by the autonomic nervous system. When the blood vessels of the ear are unconstricted and the air temperature is 25 degrees Celsius, the temperature of the ear is generally above 34 degrees. When the autonomic nerves constrict the blood vessels of the ear, the temperature of the ear drops. The drop is followed by a rise in the temperature of the arterial blood flowing from the heart through the aorta to the circle of Willis. The temperature of the brain, which is normally somewhat higher than the temperature of the blood at the circle of Willis, rises and falls similarly.

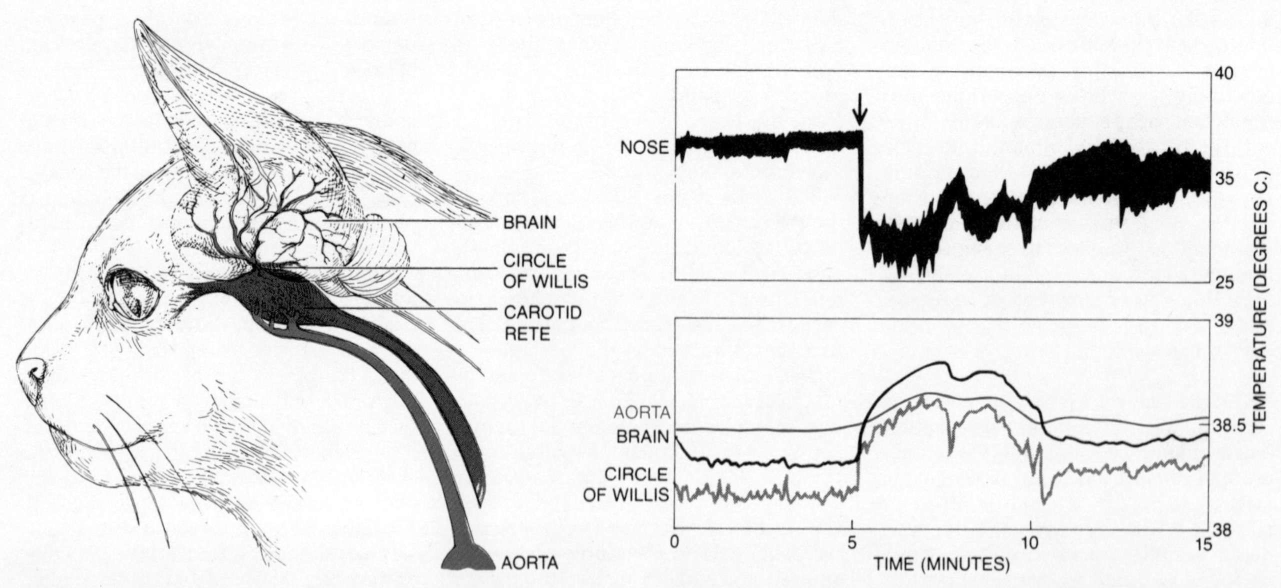

SIMILAR TEMPERATURE CHANGES in a sleeping cat that was wakened once during a 15-minute period are shown. The major heat-exchange surface, the mucosa of the cat's nose and mouth, is similarly under the control of the autonomic system. While the cat sleeps the heat loss through the nose is high and evaporation cools the venous blood draining into the cavernous sinus. There the warmer arterial blood moving from the aorta to the circle of Willis is cooled by the venous blood as it flows through the carotid *rete*. When the cat wakes (*arrow*), the autonomic nerves constrict the mucosal circulation, the nasal heat loss is reduced, as is the heat exchanged through the *rete*. The temperature of the arterial blood at the circle of Willis rises to almost match the temperature of the arterial blood from the aorta, whereas the temperature of the brain, normally lower than that of the aortic blood, briefly exceeds it. When the cat returns to sleep, the dilation of the mucosal blood vessels enhances heat exchange in the *rete*, and the other temperatures decline accordingly.

tween the venous blood and the arterial blood passing through the carotid *rete*. This results in a rise in the temperature of the arterial blood reaching the brain until this blood approaches a temperature equal to that of blood in the body core.

We confirmed the validity of this sequence of events by a series of experiments measuring the temperature of the nasal mucosa with thermocouples implanted in the nostrils of both cats and sheep. When the blood flow through the nasal mucosa increased, the temperature of the mucosa rose; when the blood flow decreased, so did the temperature of the mucosa. This exactly paralleled the changes in skin temperature we had observed in association with the constriction and dilation of the skin blood vessels. (It should be mentioned that the ambient air temperature and the animals' respiration rate remained constant during these measurements.)

Exactly how does evaporation cool the nasal mucosa? The mucosa is always moist. With each breath the animal draws relatively dry air across the moist surface, the moisture evaporates and the cooled mucosa cools the blood flowing through it. Therefore one could expect that an increase in the rate of evaporation would further decrease the temperature of the venous blood flowing from the nose and produce a consequent decrease in the temperature of the arterial blood in the carotid *rete*.

That is exactly what happened when we conducted a further experiment. We anesthetized cats and sheep and pumped air into their noses. The arterial blood supplied to the head by the heart remained constant in temperature, but the increased evaporation made the temperature of both the *rete* and the brain drop. When we repeated the experiment with a rabbit as the subject, there was no such rapid cooling. Instead the first drop in temperature took place in the carotid artery; it was followed by a lowering of the brain temperature. The sequence indicates that in the absence of a carotid *rete* the cooled venous blood from the nose must return to the heart and there cool the arterial blood moving from the heart to the brain before the temperature of the brain is affected.

Would an increase in respiratory evaporation have the same effect on animals in normal circumstances that it had on anesthetized animals? To find out we placed cats and sheep in a hot room; the flow of air through their nasal passages would naturally increase when they began to pant in response to overheating. By the time the animals were panting at a rate of 250 to 300 respirations per minute the rate of cooling of the arterial blood passing through the *rete* had also increased. Both the carotid arterial blood flowing to the input side of the *rete*

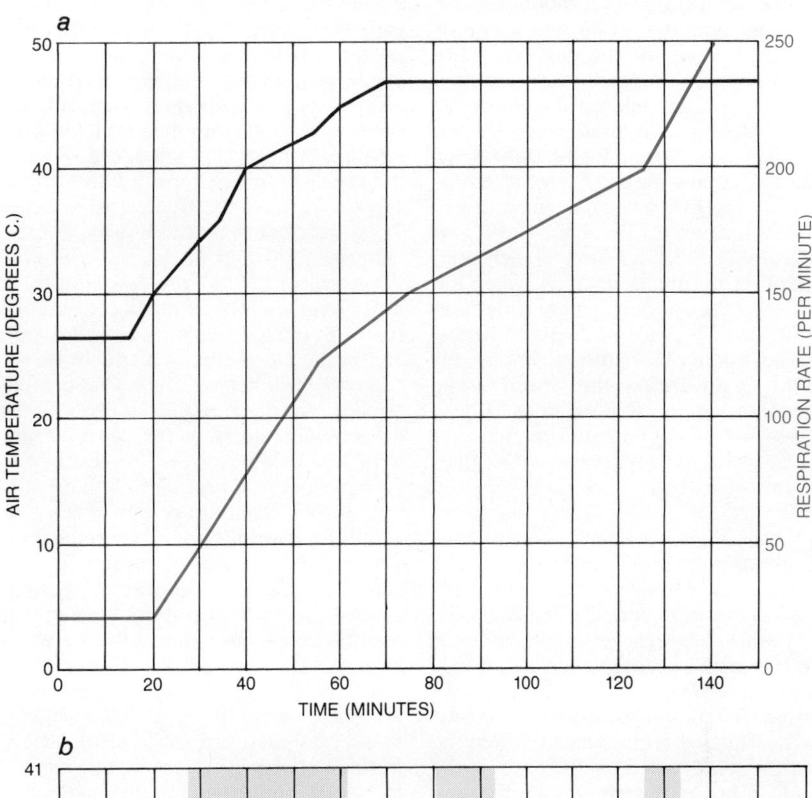

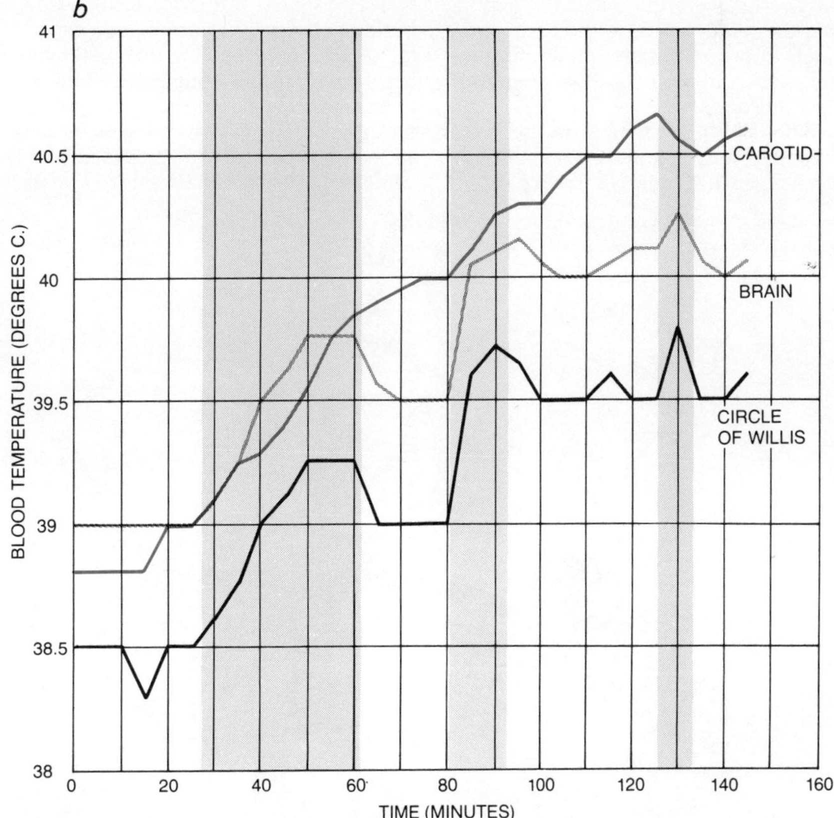

RESPONSE TO HOT ENVIRONMENT by a *rete*-protected mammal is shown in this pair of graphs. An unrestrained sheep was placed in a test chamber at an air temperature of 26 degrees C. Over a period of an hour the air was gradually heated to 47 degrees C. (117 degrees F.) and kept at that temperature for more than an hour (*a*); the respiration rate of the sheep, only 20 breaths per minute at the start, increased rapidly and continued to rise after the air temperature leveled off. The temperature of the blood in the sheep's carotid artery also rose steadily (*b*), but the animal's panting cooled the venous blood surrounding the *rete* sufficiently to keep the blood temperature in the circle of Willis below that of the blood in the carotid artery. As a result the sheep's brain, warmer than carotid blood to start, remained at or below the carotid temperature for most of the experiment. When the sheep stood up (*colored bands*) instead of lying down, both the brain and the blood at the circle of Willis became temporarily warmer.

and the cerebral arterial blood flowing from the output side showed a rise in temperature, but the rise was less for the blood that had passed through the *rete*.

As a result the difference in temperature between the arterial blood on the input side and that on the output side of the *rete* became greater. When the animals were kept at a normal temperature, the temperature of the input blood had been about .25 degree C. higher than that of the output blood. Now, when the animals were panting in a hot room, the input blood was a full degree warmer. The hot-room observations demonstrated the significance of the carotid *rete* as a heat exchanger. In conditions of heat stress the brain of a panting animal that has a carotid *rete* will remain cooler than the rest of its body.

The brain is particularly sensitive to abnormal temperatures. A rise of only four or five degrees C. above normal begins to disturb brain functions. For example, high fevers in children are sometimes accompanied by convulsions; these are manifestations of the abnormal functioning of the nerve cells of the overheated brain. Indeed, it may be that the temperature of the brain is the single most important factor limiting the survival of man and other animals in hot environments.

When the air temperature is higher than the body temperature, mammals would overheat rapidly if it were not for evaporative cooling. Some mammals cool themselves by both panting and sweating. Others, man and the rest of the primates included, depend exclusively on sweating. Still others rely entirely on the evaporation from the nasal and oral cavities that panting enhances. Of the mammals in this last group those with a carotid *rete* have the greatest advantage. This, we now realized, is because the venous blood that drains the evaporative surfaces of the nose and mouth almost immediately comes in close contact with the arteries of the carotid *rete*. Hence the blood supplying the brain receives the full benefit of the evaporative cooling. At the same time, when this venous blood returns to the heart to mix with venous blood from the rest of the animal's body, it provides further cooling for the circulatory system in general.

In most panting animals both the nose and the mouth are anatomically well suited to efficient evaporative cooling. The surface areas are large and moist, and the numerous connections between arteries and veins allow a high rate of blood flow. Considering the nose first, the complex infolding of the turbinate bones presents a very large surface area; in some mammals, such as dogs, the total surface of the nasal turbinates has been calculated as being larger than the rest of the body surface. The nasal interior not only is covered with mucosa richly supplied with interconnected veins and arteries but also houses an organ unique to panting mammals: the lateral nasal gland. The gland secretes fluid onto the mucosa, providing a constant supply of water for evaporation. In dogs the rate of secretion of the lateral nasal gland, as studies by Charles Blatt, C. R. Taylor and M. B. Habal of Harvard University have shown, is directly proportional to the air temperature.

As for the other evaporative region, the mouth, water for cooling the mucosa that lines the oral cavity is provided by the salivary glands. These glands too increase their secretion when the ambient temperature increases. For the evaporative process to reach maximum efficiency the animal's rate of ventilation must increase at the same time the glandular secretions and the rate of blood flow through the mucosa increase; the act of panting produces the required increase in ventilation.

Does exposure to high air temperatures represent the most severe thermal stress a mammal can encounter? For most mammals the answer is no. For example, in mammals the size of cats, dogs and even sheep, Taylor has found that it is exercise rather than heated air that presents the most serious threat of overheating. In these medium-size mammals the very high rate of metabolism during heavy exercise causes an explosive rise in body temperature. For the purpose of comparison, consider our finding that when cats and sheep are at rest under conditions of abnormal

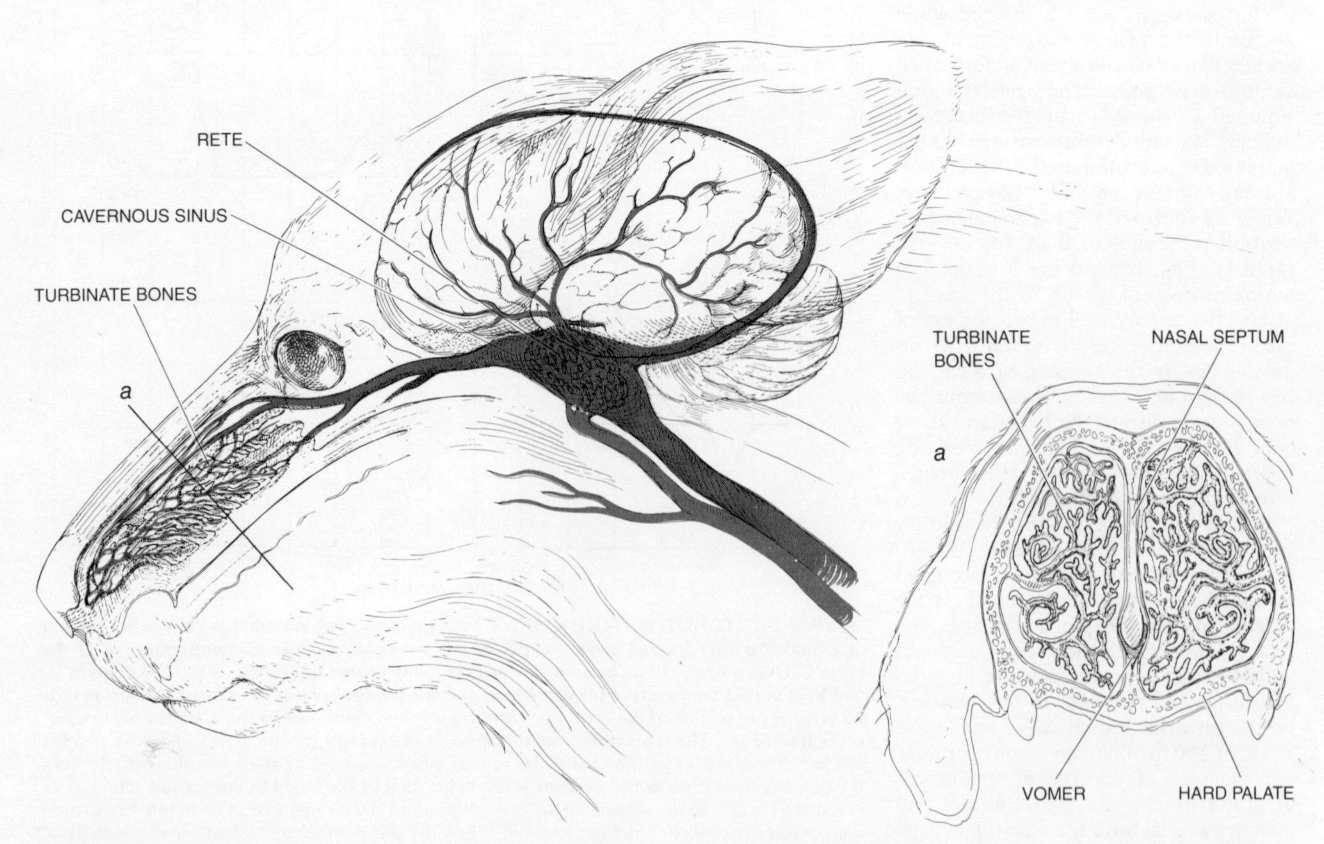

DOG'S NOSE, shown in side and front view, contains an array of complexly infolded turbinate bones that support the nasal mucosa, a membrane that is cooled by evaporation. Cool venous blood, draining from the mucosa into the cavernous sinus, cools the dog's *rete*.

heat, they display a brain temperature one degree C. lower than their body temperature. Taylor and his colleague Charles P. Lyman made similar measurements of brain and body temperature in Thompson's gazelle (an African antelope with a carotid *rete*) when the animals were running at full speed. They found that the brain temperature dropped to almost three degrees C. below the body temperature.

This unexpectedly large cooling effect led me and my colleagues at the University of California at Riverside to suspect that the carotid *rete* might not reach its maximum efficiency as a heat exchanger except during exercise. Recent studies we have conducted with exercising dogs as subjects have confirmed the conjecture, even though dogs do not have a carotid *rete* as well developed as that of antelopes, sheep or cats.

When the dogs were put in hot rooms, they reacted scarcely at all to the abnormal temperature. Both the body and the brain rose in temperature, and the brain remained only slightly cooler than the body. We found this puzzling, because dogs are known to be among the most heat-tolerant of mammals and to have a marked capacity for enduring long periods of exertion in extreme desert conditions.

We set up a treadmill in a warm, rather than hot, room and measured the dogs' brain and body temperatures as they ran on the treadmill. From the moment the dogs began to run their body temperature rose rapidly. Within the first few minutes of exercise, however, the brain temperature fell; it remained about 1.3 degrees C. below the body temperature throughout the run. That is almost three times the cooling we had observed when the dogs were at rest in a hot room.

Two of the physiological factors responsible for this high rate of brain cooling have recently been measured. The first is a large increase in the rate of ventilation when dogs exercise. R. Flandrois and his colleagues at the University of Lyons have found that a dog at rest breathes five to six liters of air per minute. In the first few seconds of hard exercise (about the same degree of exertion that is demanded of the dogs in our laboratory) Flandrois's dogs increased their rate of ventilation to 30 liters per minute. The rate thereafter rose less dramatically and reached 40 liters per minute after 15 minutes of exercise. When exercise was terminated, the rate of ventilation dropped abruptly. The primary purpose of the increase in ventilation, of course, is to deliver more oxygen and flush out more carbon dioxide in order to meet the physiological demands associated with exercise. At the same time the increase has the secondary effect of accelerating evaporation in the dogs' nasal and oral pas-

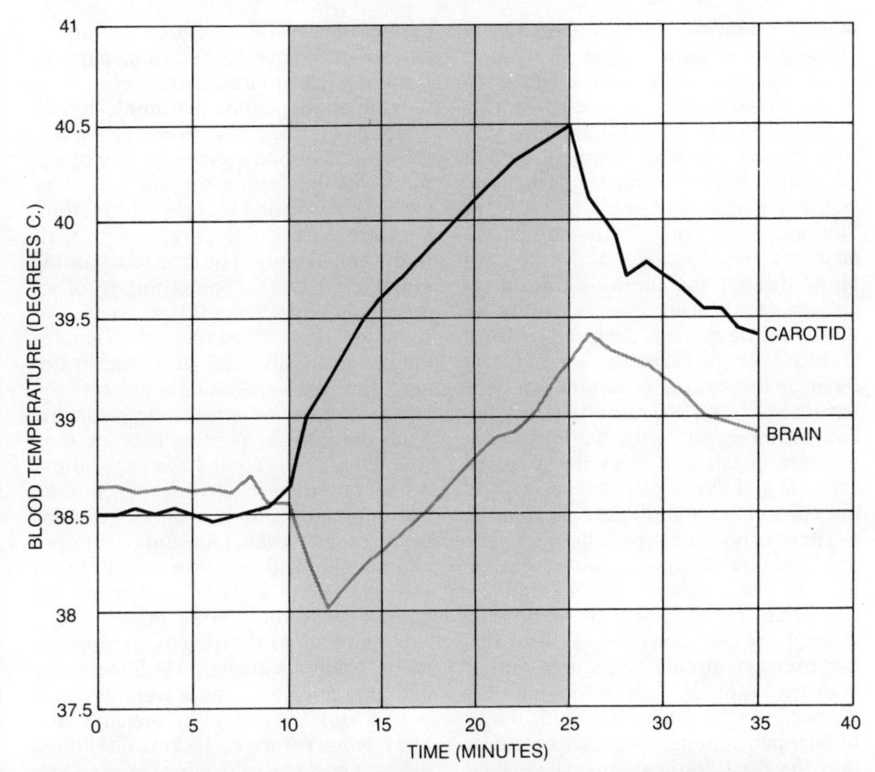

EXERCISING DOGS maintained a brain temperature significantly lower than body temperature during a 15-minute trot at 7.2 kilometers per hour (*colored band*) on a treadmill with a slope of 13 degrees. Warmer than the body while the dogs were at rest, the brain rapidly grew cooler at the onset of exercise and remained more than one degree C. cooler throughout the exercise period. The two traces show the mean values recorded for two dogs in four trials each.

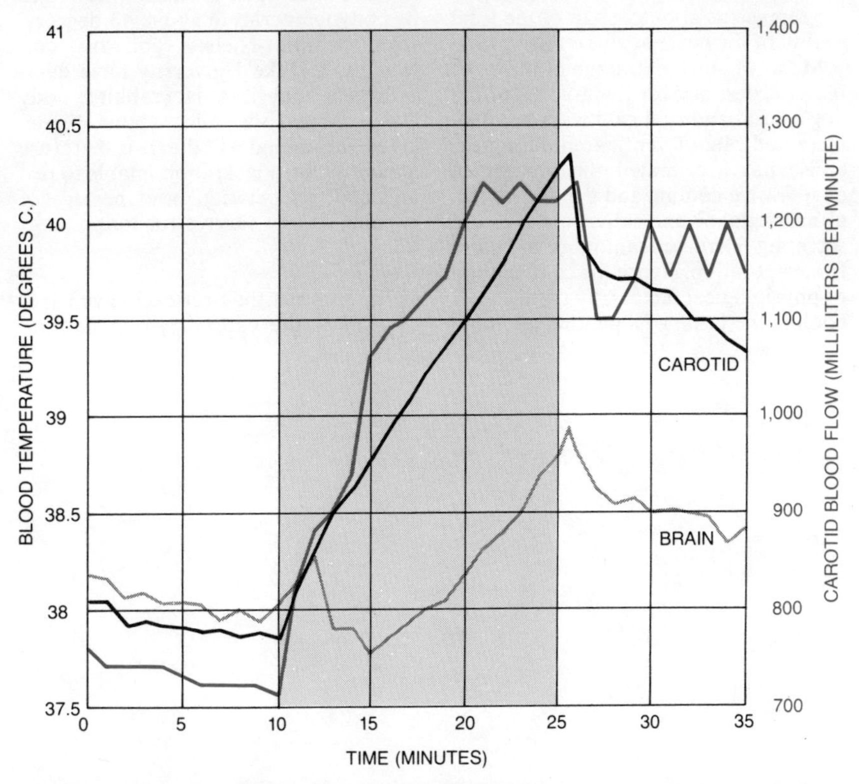

INCREASED CIRCULATION during exercise is traced in another exercising dog, this one trotting on a treadmill with a slope of 18 degrees. At rest the blood flowed through the dog's left common carotid artery (*color*) at a rate of less than 800 milliliters per minute. During the 15-minute exercise period (*colored band*) the rate of blood flow increased to between 1.2 and 1.25 liters and remained at 1.1 liters or above for several minutes after exercise ended. The enhanced flow of blood to the nose and mouth caused an increase in the heat exchange in the carotid *rete*, as is evidenced by the consistently lower temperature of the animal's brain.

sages, thereby lowering the temperature of the venous blood.

The second brain-cooling factor to be measured is the increase in the rate of blood flow to the nasal and oral mucosa during exercise. Working in collaboration with Roland D. Rader and William Kirtland of the University of Southern California School of Medicine, we have measured the flow of blood through the common carotid artery in dogs during exercise, using an ultrasonic flow probe designed by Rader. In a large dog at rest the flow of blood through the common carotid artery is between 600 and 700 milliliters per minute. In our experiments we found that the rate began to rise at the onset of exercise and eventually reached a level in excess of 1,200 milliliters per minute.

There is no reason to believe the rate of blood flow to the brain increases during exercise. Accordingly we can assume that the additional blood flowing through the common carotid artery during exercise circulates to areas other than the cranium. Subtracting the flow destined for the brain (roughly 250 milliliters per minute), we can calculate that the flow destined for extracranial areas is about 400 milliliters per minute when the dog is resting and a full liter per minute when the dog is exercising. Since the arterial circulation passes through both common carotid arteries, the combined flow of two liters per minute represents about a sixth of the total output of the heart of an exercising dog.

Most of this extracranial flow of blood is destined for the mucosa of the dog's nasal and oral cavities, where the increased rate of ventilation during exercise has accelerated the process of evaporative cooling and the dog's lateral nasal glands and salivary glands are secreting increased amounts of water for evaporation. Hence the combination of physiological factors, by significantly cooling the brain of a panting mammal that has a carotid *rete*, provides an increased tolerance to exertion, particularly in a hot environment.

What about panting mammals that do not have a carotid *rete?* We have studied the brain and body temperatures of domestic rabbits while the animals were running. We found that their brain temperature increased in parallel with their body temperature. The animals' panting nonetheless can do something to offset the rise in brain temperature, as M. Caputa and his colleagues at Nicholas Copernicus University of Toruń in Poland have demonstrated. Measuring the temperature of venous blood returning from the nose of panting rabbits, they found that it was cooled by evaporation. As a consequence cooling by conduction lowered the temperature of regions of the rabbits' brain adjacent to the cool venous blood; these parts of the brain proved to be .5 degree C. lower in temperature than the rabbits' body.

In nature this offsetting factor appears to be trivial. Vaughan H. Shoemaker and his colleagues at Riverside have used radio telemetry to measure the body temperature of jackrabbits living unrestrained in the Mojave Desert. On a hot summer day a jackrabbit at rest has a body temperature of about 41 degrees C. When the animal is flushed from its resting place and chased, it shows a rapid rise in body temperature. Between five and 10 minutes of exertion can raise its body temperature above 43 degrees. Knut Schmidt-Nielsen and his colleagues at Duke University have demonstrated that if a jackrabbit's body temperature rises much above 44 degrees, the animal will die. It is therefore apparent that a jackrabbit unable to rest and cool off between brief periods of running cannot survive for long.

The study of the heat-exchange function of the carotid *rete* in panting mammals has raised intriguing questions about mammalian temperature regulation in general. It has long been known that there is a heat-sensitive region in the brain stem of mammals and that by sensing the temperature of the arterial blood flowing from the core of the body to the brain this region probably acts as a thermostat controlling the body-core temperature. This seems to be a valid assumption for mammals that lack the carotid *rete;* in those mammals the arterial blood does not change temperature as it passes from the body to the brain. What, however, does the brain-stem thermostat sense in a mammal that has a carotid *rete?*

What it senses is at the very least equivocal. For example, during exertion the temperature sensed by the thermostat is quite different from the actual temperature of the body core; the brain temperature and the body temperature change in opposite directions at the onset of exercise and remain at different levels during most of the exercise period. How, then, does the brain thermostat know what the body temperature is? In mammals that have a carotid *rete* are the temperatures of the brain and the body regulated independently?

We can speculate that in such animals temperature-sensitive nerve cells in parts of the body other than the brain are important in the regulation of temperature during exertion. For example, such sensors exist in the spinal cord and in the abdominal cavity and have been shown to play a thermoregulatory role in some mammals. In mammals with a carotid *rete* the input from these extracranial sensors to the region of the brain that controls body temperature may be more significant during a period of exertion than during a period of rest. What the right answers are to such questions will be learned from further study of thermoregulation in a wide variety of mammals.

The Physiology of the Giraffe

5

by James V. Warren
November 1974

The head of the animal is so far above the heart and the lungs that the task of supplying it with oxygenated blood calls for a remarkably high blood pressure and unusually deep breathing

The giraffe, the spectacular mammalian adaptation to browsing on tree leaves, has been an object of curiosity since prehistoric times. There is archaeological evidence that it was then found not only in Africa but also in southern Europe and the Near East. Although the species later receded to Africa as its sole habitat, the animal continued to be known in the ancient world around the Mediterranean as one of the wonders of nature and was imported for exhibition. Historical accounts record that Julius Caesar paraded giraffes in the Colosseum, and that other heads of state presented the animal as a unique gift to a distinguished visitor (as the panda has on occasion been presented in recent years).

We are no less intrigued by the giraffe in our own time, but it is a curious fact that not until comparatively recent decades was any question raised about the physiological puzzles presented by this improbable animal. The first physiologist to ask such a question, as far as the record shows, was the late August Krogh of Denmark, winner of the Nobel prize in physiology and medicine in 1920 for his studies of capillary blood vessels. In the Silliman Lecture at Yale University in 1929 Krogh wondered about the great pressure the capillaries in the legs of the giraffe must bear because of the high column of blood weighing on them. Pointing out that the giraffe's heart is eight feet or more above its feet, he observed that the effect of gravity, added to the heart's pumping action, should be expected to make the blood pressure in the legs so high that it would force fluid out of the capillaries.

"It would be extremely interesting," Krogh remarked, "to know just how the giraffe avoids the development of filtration edema in its long legs. Unfortunately, we have not found it possible to obtain giraffe blood for determination of the osmotic pressure." Nor was Krogh ever able to obtain a giraffe for study.

Krogh's question did not arouse wide interest at the time, and it was not revived until nearly two decades later. This time the query was put in another form: How does the giraffe manage to pump blood up its long neck to its brain? In a giraffe standing upright the head is between seven and 10 feet above the heart, whereas in man the vertical distance from the heart to the brain is only a foot or so, and in most other animals, including large ones, it is little more.

The performance of the giraffe's heart and circulatory system in driving the blood against gravity to such a height above the heart was quite remarkable when one thought about it. The phenomenon attracted the attention of physicians interested in problems of aviation medicine during World War II. They calculated that the gravitational force a giraffe must overcome to supply blood to its brain corresponds roughly to the force of acceleration, amounting to several *g*, that fliers encounter in maneuvering high-speed military aircraft. In order to prevent blacking out because of the great change of blood pressure in the brain under those circumstances a "*g* suit" that buffered the acceleration impact had been developed for fliers. Medical investigators thought it might be worth while to look into the natural mechanisms with which the giraffe handles its severe circulatory challenges. Here was a living experiment of nature presenting highly unusual capabilities. Study of the giraffe's cardiovascular system might be expected to yield new and perhaps useful information.

Interest in the problem persisted after World War II. When John L. Patterson, Jr., Otto H. Gauer, Joseph T. Doyle and other investigators in the U.S. learned of a similar interest on the part of Robert H. Goetz and his colleagues in South Africa, they suggested setting up an international study group. In the late 1950's the American group visited South Africa and subsequently did studies in the U.S., using cows as the experimental animals, to get comparative data on circulatory and respiratory functions. More recently an American team that included Robert L. Van Citters, Dean L. Franklin, Stephen F. Vatner, Thomas E. Patrick and me undertook further field investigations in Kenya. By the use of telemetry we obtained direct measurements of blood pressure and blood flow in giraffes roaming free in the field.

The height of a large adult giraffe is commonly in the range from 15 to 18 feet. The animal's heart lies approximately midway between its head and its feet, that is, usually seven feet or more below its brain. For analysis of the operation of the giraffe's circulatory system the comparative blood pressures at two significant levels were needed: at the heart and at the brain. Our group arranged to obtain measurements of these pressures, and of the blood flow as well, under a variety of natural circumstances: with the animal lying flat on the ground, standing quietly erect and actively moving about.

For direct and accurate measurement of the blood pressure we used a sensitive pressure gauge implanted in a carotid artery. The amount of pressure, sensed by a diaphragm in the device, was transduced into a radio-frequency signal that could be picked up by a receiver as much as a third of a mile away. The animal was followed during the period of measurement, and a record of the blood-pressure variations was made on magnetic tape and charted later.

The subjects were wild giraffes cap-

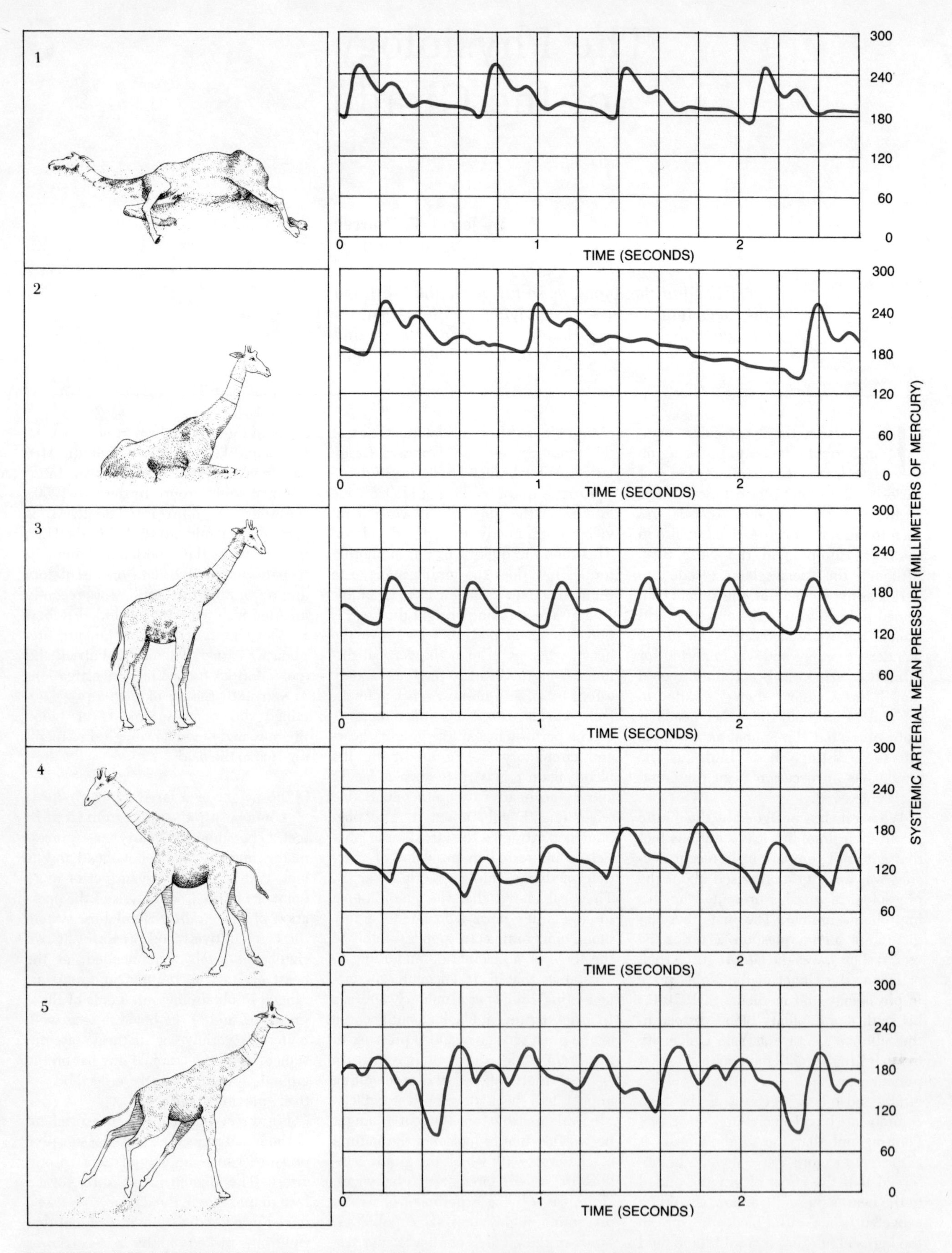

PRESSURE AND HEARTBEAT vary in accordance with the activity of the giraffe. When the animal is lying down (*1*), with its heart and its brain at the same level, the blood pressure in the carotid artery ranges from 180 to 240 mm. Hg. (millimeters of mercury). The heart rate, with four major peaks in 2.5 seconds, is about 96 beats per minute. As the animal raises its head (*2*) the blood pressure stays much the same but the heart rate slows temporarily. Standing (*3*) and walking (*4*) raise the heart rate to some 150 beats per minute, while arterial blood pressure falls to a range of from 90 to 150 mm. Hg. Galloping (*5*) brings the heart rate to maximum: 170 beats per minute. Highest blood pressures in galloping are 220 mm. Hg.; deep drops in pressure coincide with front-hoof beats.

tured by lasso from a fast-moving vehicle on the Kenya plains. After the giraffe was captured and tied down flat on the ground the pressure gauge was implanted in the animal's neck through a small slit made under local anesthesia in the carotid artery. In a giraffe the carotid is usually more than six feet long, is relatively straight and has few branches. The gauge was placed in the upper part of the neck near the head. At the same time a small ultrasonic flowmeter, measuring the velocity of blood flow by means of the backscattering of high-frequency sound waves from the red cells, was implanted. The apparatus, including a small transmitter and mercury batteries, was taped onto the giraffe's neck. At the end of the period of observation the animal was recaptured, the instruments were removed, the incisions were repaired and the giraffe was released unharmed to join its fellows in the bush.

The initial blood-pressure measurement, made while the giraffe was still lying on the ground, could be considered to represent the pressure generated at the animal's heart level, even though the gauge was near the head. Obviously in an animal stretched out on the ground the blood circulation is not much influenced by gravity; therefore the blood pressure in the giraffe's carotid artery should be about the same near the heart as near the head. (Experiments with the gauge implanted in the carotid at the base of the neck have confirmed this.)

Not surprisingly, the giraffe's blood pressure at the heart level proved to be very high—higher, indeed, than that recorded in any other animal. The systolic pressure at the heart level ranged from 200 to 300 millimeters of mercury and the diastolic pressure from 100 to 170; the average was 260/160. Compared with man, the giraffe is "hypertensive": even at 200/100 millimeters of mercury its blood pressure is far above the 120/80 average in a man at rest. The giraffe's hypertension, however, is not hypertensive vascular disease but a necessary condition for supplying the brain with blood at sufficient pressure when the animal stands erect. Other factors may also be involved in the circulation to the brain, as we shall see.

When the giraffe was freed, it usually raised its head and remained in a squatting position with its legs under its body for a few seconds before rising to its feet. Conceivably this pause may be a behavioral adaptation to allow time for adjustment of the circulation. It is well known that in man an individual with low blood pressure sometimes suffers dizziness or even blacks out after standing up abruptly from a reclining position. At any rate, the blood pressure in the giraffe's carotid was substantially less after it stood up than it was when the measurement was recorded at heart level. Taking into account that the instrument was implanted in the neck 14 inches below the brain, calculations based on recordings when the giraffe was standing quietly indicated that the pressure of the blood entering the brain itself averaged approximately 120/75. That is about the pressure at which blood perfuses the brain in man and in most other animals. Apparently the barostat, or pressure control, is set at about the same reading in most mammals, thereby providing that the brain will generally be bathed with blood at a pressure of about 120/80.

Blood pressure, of course, is by no means constant; it fluctuates rather widely even in "normal" individuals not affected by hypertensive illness. The pressure is influenced by many kinds of physical or mental stress. After a giraffe was released and had run off to rejoin its herd, its blood-pressure records generally reflected the nature of its activities. While it was standing still or walking, the carotid pressure ranged between 140/90 and 180/120. During a hard run, as when it was being chased for recapture, the pressure rose to about 220/150. That was still considerably lower than the pressure recorded when the animal lay prone.

What happens when a giraffe bends its neck down to drink? The pressure within the blood vessels of the brain and the eyes should then be higher than that at the level of the heart. Why does the high pressure not rupture those delicate vessels or at least force leakage from

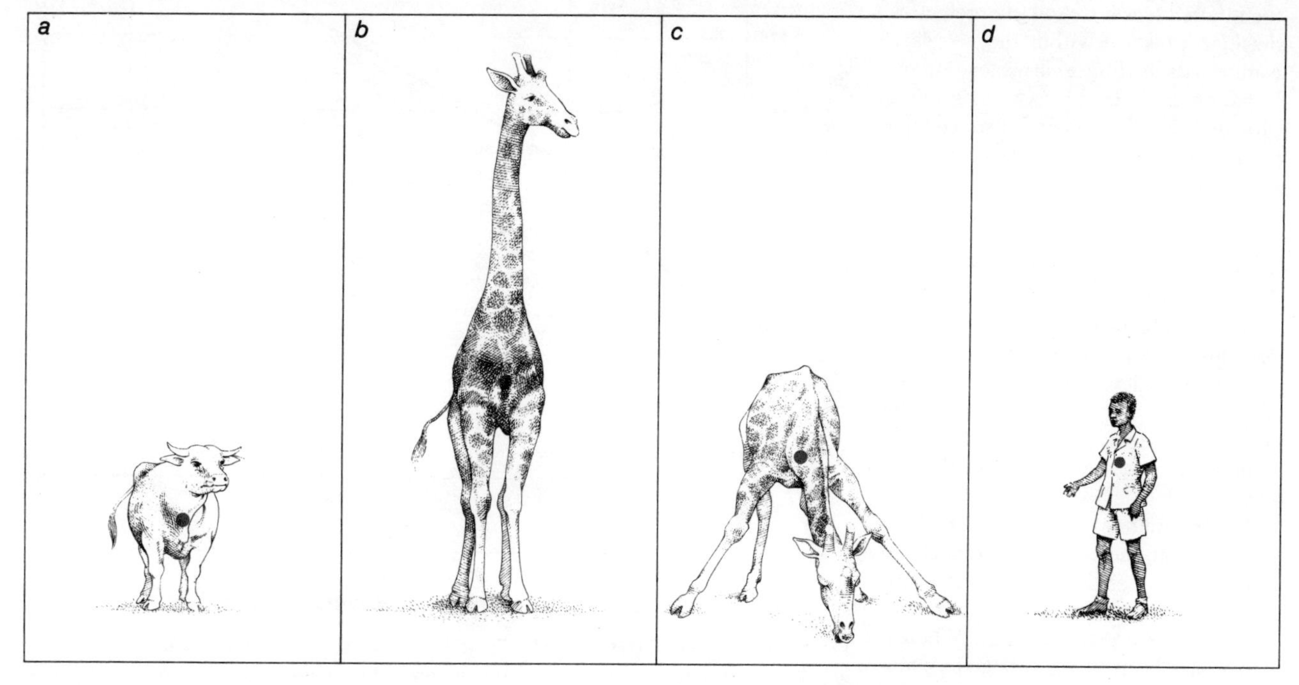

HEART OF A GIRAFFE is located from seven to 10 feet below its brain when the animal is standing (*b*). In an average quadruped, a cow in this example (*a*), and in man (*d*) the distance between the heart and the brain is far shorter. In a giraffe the brain's supply of blood is ensured by "abnormally" high blood pressure. This might be a severe handicap when the animal is drinking (*c*), except that its posture reduces the distance between the heart and the brain. The back pressure of the cerebrospinal fluid is added protection.

them? This brings us to the question originally asked by Krogh: How does the giraffe avoid the development of edema in its long legs? On the basis of established facts we must consider what mechanisms may control the blood circulation and pressure in the giraffe.

Analysis of giraffe blood has shown that its chemistry is not grossly different from that of human blood, and so it is unlikely that its composition makes it unusually resistant to passing through the blood-vessel walls. In the giraffe the walls of the arteries (and of the heart itself) are thicker than they are in other mammals. This would not, however, prevent leakage through the walls of the finer blood vessels in the event of a large difference in pressure between one side of the wall and the other. We are therefore led to the reasonable hypothesis that in the giraffe a heightened pressure within the blood vessels is counterbalanced in some way by an increase in the pressure outside the vessels. In the brain such a function may be served by the cerebrospinal fluid that bathes the brain and spinal cord. In effect this fluid, in which the blood vessels are immersed, serves as a *g* suit buffering the impact of the downward rush of blood when the giraffe bends down to drink. The animal also employs a behavioral device that minimizes the gravity problem: when it bends to drink, it spreads its front legs wide so that its chest is lowered, bringing the heart closer to the ground.

It seems likely that the capillaries in the giraffe's legs are similarly protected; the high pressure within the vessels is counterbalanced by a corresponding external pressure. In this case the *g* suit is provided by the general extracellular fluid bathing the body tissues. The giraffe has a thick, tight skin that easily sustains these high pressures.

In short, part of the answer both to Krogh's query and to the bending-to-drink problem is rather simple: The principal factor preventing edema below the heart level seems to be high extravascular pressure counteracting the high pressure within the blood vessels. Quite clearly, however, that is only part of the story.

In the giraffe as in other animals, including man, the pumping action of the muscles plays an important part in assisting blood circulation and thus relieving the pressure in the legs. When a person has been sitting or standing motionless for some time, it is not uncommon to find that his ankles have begun to swell; this slight edema results from the sluggishness of the circulation in the

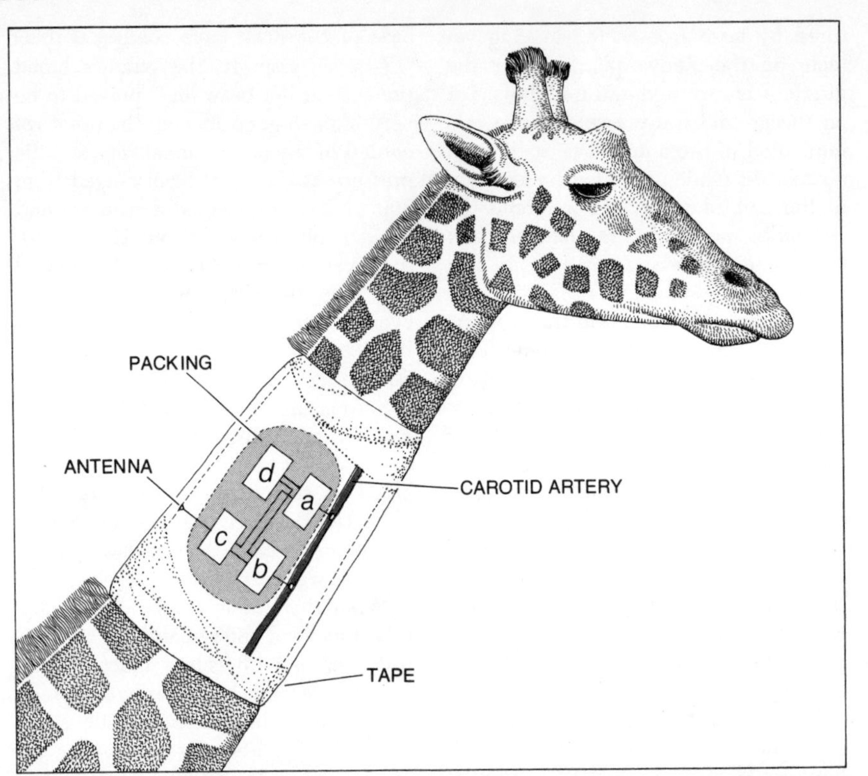

TELEMETRY PACKAGE included a pressure gauge (*a*) containing a sensing diaphragm that transduced pressure variations into radio-frequency signals, a flowmeter (*b*) that sensed blood velocity by means of the backscattering of high-frequency sound waves from red blood cells and signaled a radio transmitter (*c*) accordingly, and the batteries (*d*) that powered the transmissions. Auxiliary components were protected from shock with foam-rubber padding. The gauges were implanted through separate incisions in the carotid artery.

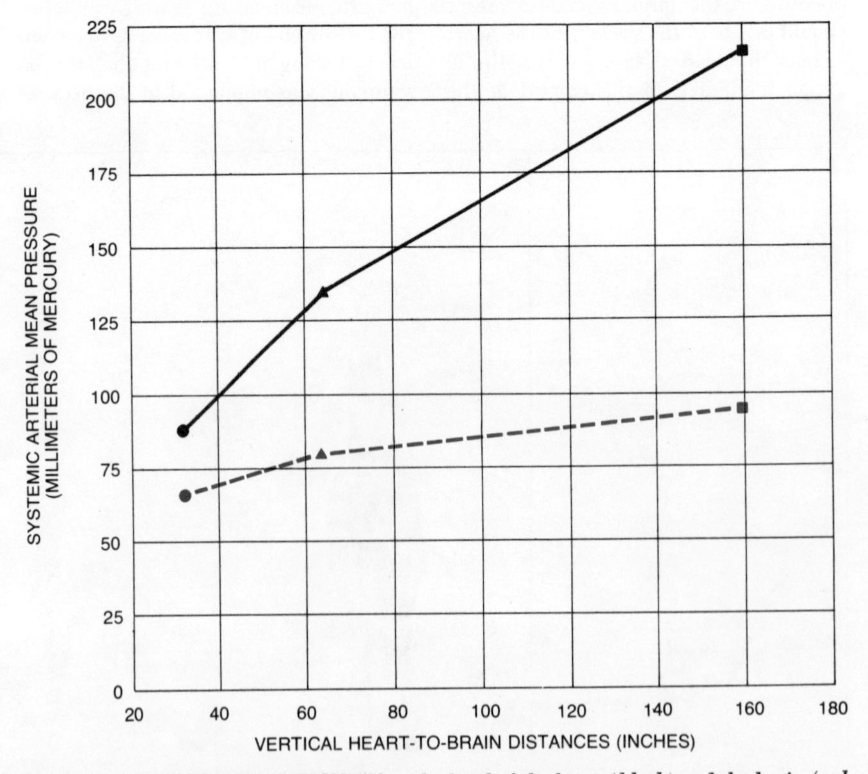

COMPARATIVE BLOOD PRESSURES at the level of the heart (*black*) and the brain (*color*) show a consistent increase in relation to heart-brain distance. In man, with an assumed heart-brain distance of 34 centimeters, a mean arterial pressure equal to 90 millimeters of mercury at the level of the heart suffices to deliver blood to the brain at a mean pressure of about 65 mm. Hg. In the giraffe the equivalent values are 215 mm. Hg. at the level of the heart and 90 mm. Hg. at brain level. The equivalent values in the cow are closer to man's.

inactive legs. Movement of the legs or exercise of the body as a whole is indeed a major factor in the pumping of blood from the legs back to the heart. (In individuals with certain types of varicose veins the mechanisms responsible for aiding the circulation may operate inadequately.) The force required to raise the blood from the legs to the heart is not great, either in man or in the giraffe, because the artery-vein system from the heart to the legs and back again simulates a *U*-tube. That is to say, on the same principle as the operation of a sump pump, little pressure is needed to bring the blood back up the return arm to the same level as the source of the pumping.

As for the question of how the giraffe drives blood from the heart up to the brain, it appears that this remarkable performance can be accounted for fully by the high pressure at which the blood is discharged into the arteries by the heart. Conceivably the artery-vein system in the long neck might make a contribution to the circulation by acting as an inverted *U*-tube, producing a siphonlike effect. There is no evidence, however, that such assistance is needed. The arterial pressure itself is sufficient to deliver the blood to the brain at a pressure adequate to nourish it.

It has been suggested that the giraffe's carotid arteries possess an interesting special feature. Before entering the brain each of these arteries apparently divides into a network of fine vessels that resembles a system found in certain other animals: the rete mirabile, or "wonderful net." In some animals this network serves as a countercurrent heat exchanger that enables species such as wading birds to maintain a normal body temperature even though their extremities become very cold; blood returning from the extremities is warmed before it reenters the general circulation. Similar networks have been found to perform other exchange functions, such as transferring oxygen or electrolytes. It may be that the rete mirabile at the entrance to the giraffe's brain acts in some way to regulate the blood pressure in the brain. This is a question that has not yet been satisfactorily answered.

In addition to circulatory problems the giraffe is burdened with certain other challenges. One of the most interesting has to do with respiration. The trachea in the giraffe's long neck may be more than five feet long and two or more inches in diameter, so that its total capacity may amount to 2.5 quarts or more. This is "dead space" in the sense that, although each breath of air must fill the windpipe as well as the lungs, the air in the pipe serves no useful purpose for gas exchange. If man had to deal with such a large dead space, he would almost suffocate. His inhalation of a breath of fresh air would hardly suffice to fill his windpipe, hence none would reach the lungs; conversely, the carbon-dioxide-laden exhalation from the lungs would be trapped in the windpipe and come back into the lungs in the next inhalation. The giraffe solves this problem by hyperventilation, breathing more deeply and more rapidly than man. Giraffes even at rest have been found to have a respiration rate of more than 20 breaths per minute, whereas the corresponding rate in man is 12 to 15. In contrast to man and other mammals, the giraffe spends more time in inhalation than in exhalation. Nevertheless, its arterial blood carries a marginally low content of oxygen. It has been found that giraffes show distress in situations where the air is low in oxygen, as at high altitudes. They are much less adaptable than man is to such conditions.

The giraffe is a fascinating anomaly of nature. Investigators have learned much from unusual kidneys in primitive fish, from the urine-concentrating ability of desert rats for water conservation, from the electrocardiograms of whales and from heart failure in cattle at high altitudes. The giraffe too, as odd an animal as any in man's eyes, probably has a great deal to tell us about how we are put together.

CAPTIVE GIRAFFE lies outstretched in a corral in Kenya while experimenters implant a pair of gauges in a neck artery. The animal is one of several giraffes that carried telemetering apparatus so that observers could record natural variations in blood pressure.

II

GAS EXCHANGE AND THE LUNGS

It is universally admitted that a very few other creatures in the sea (other than whales) also breathe, those whose internal organs include a lung, since it is thought that no animal is able to breathe without one. Those who hold this opinion believe that the fishes possessing gills do not alternatively expire and inspire air. . . . Nor do I pretend that I do not myself immediately accept this view of theirs, since it is possible that animals may also possess other respiratory organs in place of lungs, if nature so wills, just as also many possess another fluid instead of blood. . . . Undoubtedly to my mind there are additional facts that make me believe that in fact all creatures in the water breath . . . in the first place a sort of panting that has often been noticed in fishes during summer heat, and another form of gasping, so to speak, in calm weather.

Pliny
NATURAL HISTORY, IX, vi

II

GAS EXCHANGE AND THE LUNGS

INTRODUCTION

Almost all vertebrates use gills or lungs as respiratory organs. Except for rare cases (see the first article in this Section, Kjell Johansen's "Air-Breathing Fishes"), individual gill filaments of fishes cannot function adequately in air. The reason is simply that adjacent filaments tend to adhere to each other, making it impossible for air to flow between the sheets of tissue and near the blood capillaries where gas exchange occurs. Consequently, all terrestrial vertebrates and many fish that obtain oxygen from the air possess lungs or lung-like sacs in which oxygen and carbon dioxide may be exchanged.

It seems likely that the simple lungs of modern lungfish are good facsimiles of early vertebrate lungs that appeared long ago in the history of fishes. The circumstance most conducive to the appearance of lungs was probably a fresh-water habitat that was subject to heating and drying; under such conditions, the amount of dissolved oxygen would decrease, creating an absolute necessity for ability to breathe air. Johansen presents data pertinent to these considerations and also discusses specializations other than lungs that permit various fishes to breathe air.

Once vertebrates began living on land, lungs began to increase in surface area. With the appearance of the dry skin in reptiles, all respiratory exchange was restricted to the lungs, so that the necessity for more surface area was greater. The two groups of temperature-regulating animals have radically different gas-exchange structures: mammals have large numbers of minute blind sacs, alveoli; birds, tiny tubes (air capillaries with a diameter of only 10 microns), which have the same function as the alveoli.

Birds also fill their lungs in a different manner than mammals. The expandable rib cage, first seen in reptiles, is still present in birds, but additional filling and emptying is caused by movement of the abdominal muscles. Mammals have a diaphragm—a muscular sheet of tissue that closes off the lung cavity from the abdominal cavity—as well as the expanding rib cage system. Three articles in this section, "The Lung" by Julius H. Comroe, Jr., "How Birds Breathe" by Knut Schmidt-Nielsen, and "Surface Tension in the Lungs" by John A. Clements, explain in detail the function of the avian and mammalian lung systems.

Recent study of the "surfactant" responsible for controlling the surface tension of lungs indicates that the surfactant components are stored in special lamellated bodies within so-called "Type II cells" of the alveoli. Two phospholipids (phosphatidylcholine and phosphatidylglycerol) are present in the cells as well as in the surfactant film lining the alveoli. Surfactant has now been found in lungs of lungfishes, amphibians, reptiles, birds, and mammals. It may be presumed, therefore, that the use of these lipids to lower surface tension was an early adaptation associated with breathing air.

The crucial role of the lung surfactant is nowhere better illustrated than in cases of hyaline membrane disease, a potentially lethal condition found at birth in some human infants. Because of insufficient surfactant, the alveoli in the lungs of these newborn infants collapse each time air is expelled from the body. For the baby, each new breath is an exhausting fight to re-expand the alveoli. In the absence of intensive care for at least five or six days, such infants usually die. It now seems likely that the disease results from the fact that the alveolar Type II cells fail to differentiate during the latter part of gestation. In exciting experiments reported by M. E. Avery, N. S. Wang, and H. W. Taeusch (*Scientific American,* April 1973), it is shown that early injection of adrenal steroids such as cortisone into other mammals (rabbits and sheep) causes the precocious maturation of Type II cells; the result is production of surfactant. These observations have led to the first attempts in pregnant human females to accelerate fetal lung development and thus prevent hyaline membrane disease from appearing at birth.

In recent years, study of the respiratory control system has verified that respiratory chemosensitive cells are not located in the inspiratory and expiratory centers of the brain. Instead, they appear to be in the ventrolateral walls of the medulla of the brain. In these walls, the cells are sensitive to the level of hydrogen ions (H^+) in the cerebrospinal fluid. (The acid level is, of course, related to the partial pressure of carbon dioxide (CO_2) in the fluid by the reactions $CO_2 + H_2O \leftrightarrows H_2CO_3 \leftrightarrows H^+ + HCO_3^-$. Thus, elevated carbon dioxide produces hydrogen ions.) In fact, the pH of the cerebrospinal fluid is normally kept very constant by active transport regulation. However, the pH apparently does change in localized regions near the receptor cells to modify their activity. As a result the main control of breathing can occur. Complex exchange of Cl^- ions for the HCO_3^- that is a byproduct of H^+ production is an additional ingredient in the process.

The relative roles of the CO_2 sensors of the medulla and O_2 sensors of the great vessels is demonstrated by the following exercise: a student breathes into a paper bag. When the CO_2 level rises to 3.5 percent and the O_2 level falls to 17 percent, breathing increases in speed. However, if a container of hydroxide is placed in the bottom of the bag so that CO_2 is absorbed, then the O_2 level must fall to 14 percent before breathing accelerates. From these and other types of observations, it is concluded that the "fine" control of respiratory rate is carried out in the medulla and involves CO_2 measurement.

Reptiles, birds, and mammals have all given rise to lines of organisms that reentered the water to make it their prime habitat. All of these organisms retained their lungs, failed to develop new gill systems, and continued to breathe air. In part this results from the fact that there is not enough oxygen dissolved in water to permit gills to extract sufficient amounts to meet the needs of such organisms (particularly the high-temperature birds and mammals). Furthermore, as was pointed out in the Introduction to Section I, the hemoglobin of a mammal or a bird could not load and serve as a respiratory carrier under such conditions, because the partial pressure of oxygen in water is too low. Quite simply, the high metabolic rate of homeotherms can be supported only by the respiratory efficiency of lungs and air breathing. In spite of the handicap of having to come to the surface periodically, several mammals and a few birds have come to compete successfully with the fishes.

A common feature of terrestrial vertebrates that have returned to the water as a primary habitat concerns a slowing of the heart beat during diving (see P. Scholander's "The Master Switch of Life," *Scientific American,* December 1963). Suk Ki Hong and Hermann Rahn, in "The Diving Women of Korea and Japan," report the same phenomenon in women adapted to prolonged diving. Another cardiovascular adaptation essential for prolonged dives of whales or porpoises is a diversion and restriction of blood flow to the brain and heart muscle; even skeletal muscle receives less than might be anticipated, but it is adapted to withstand severe oxygen debt without cramping and becoming nonfunctional.

6 Air-breathing Fishes

by Kjell Johansen
October 1968

Lack of oxygen in water forced certain Devonian fishes to develop air-breathing organs. Some left the water and colonized the land; descendants of others are today's remarkable air-breathing species

The transition from breathing water to breathing air was perhaps the most significant single event in the evolution of vertebrate life. How did it come about? The change in respiration is commonly identified with the emergence of animals from the water to dry land. Actually the fossil and living evidence shows that air-breathing by vertebrates began long before that development. Well before the evolution of amphibians some fishes had begun to use air instead of water for respiration. The crucial steps toward the development of lungs came about not as an adaptation to living out of the water but in response to changes in the aquatic environment itself.

During the Devonian period of some 350 million years ago a respiration crisis developed for much of the vertebrate life inhabiting the freshwater basins of the earth. The oxygen content of the waters gradually declined, as a result of high temperatures and the oxygen-consuming decay of dead organic material in shallow lakes, rivers and swamps. Throughout a vast portion of the aquatic environment the oxygen supply dropped to a marginal level. For the fishes, equipped only with gills for obtaining oxygen from water, life became quite precarious. Yet they were mere inches away from the limitless reservoir of oxygen in the atmosphere above the surface of the water. This saving circumstance led to the evolution of a variety of organs that enabled fishes to obtain oxygen from the air.

We can see the adjustments that were made to the oxygen problem in many fishes today, among both modern and archaic species. Certain species living in stagnant tropical pools, for instance, spend their entire existence close to the surface, within ready reach of the thin top layer of water that contains more oxygen than the waters below. Some of these species come to the surface layer to replenish their oxygen at frequent intervals; others visit the layer less frequently but stay in it longer each time. During a recent expedition to the Amazon River area I witnessed a dramatic demonstration of the latter performance. We had placed a freshwater stingray in a tank containing only a small amount of water. We noted that as the oxygen content of the water declined the fish pumped water over its gills more and more rapidly. Suddenly, when the oxygen pressure dropped below 20 millimeters of mercury, the fish swam straight to the surface, began to draw the surface film of water into the intake openings on its upper side and continued this performance at the surface for more than 15 minutes. The whole act was beautiful, and we were excited to discover this illuminating behavior—which the Amazon stingray probably has been performing routinely for millions of years. All the close relatives of this stingray are marine or brackish-water fishes with the flattened shape of bottom-dwellers and with water-intake openings on their upper surface, where the openings cannot be clogged by debris from the bottom. In the amazing freshwater stingray the

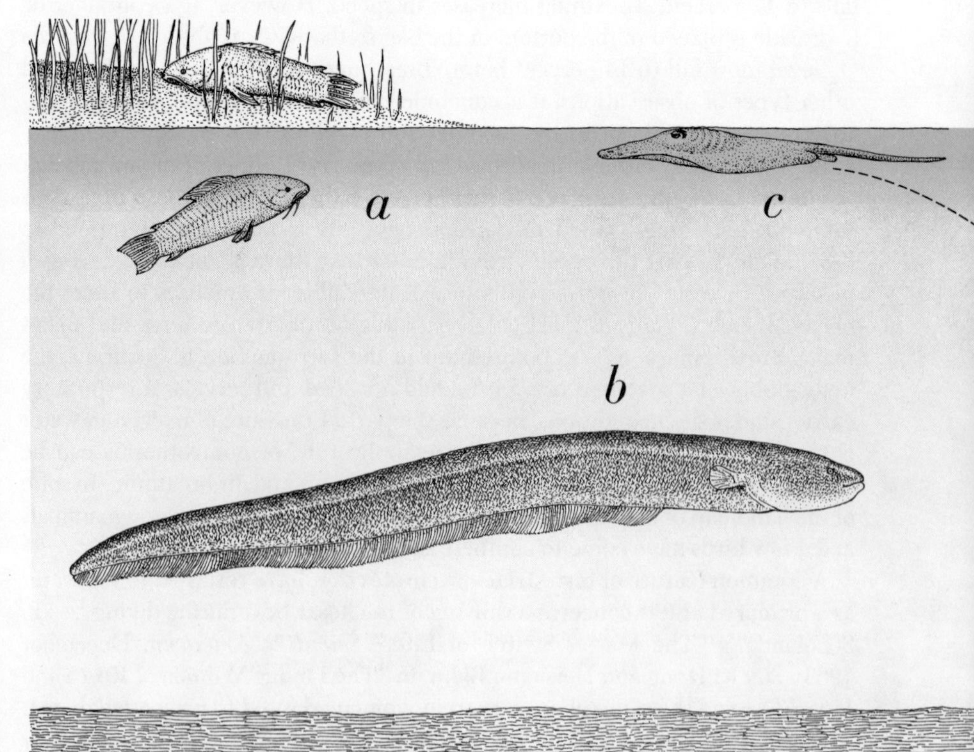

OXYGEN-DEFICIENT TROPICAL SWAMP is the habitat of air-breathing fishes and some that breathe near the surface where the oxygen supply is best. The relative oxygenation is shown by the color intensity. *Hoplosternum littorale* (*a*) is an intestinal breather that gets

same morphology exploits the opposite side of the hydrosphere: the interface with the air! It is also significant that the gills of the stingray (and of other fishes with similar habits) are large and well developed, maximizing gas exchange so that the fish can make efficient use of the limited supply of oxygen in the surface water.

The step from tapping surface water for oxygen to obtaining oxygen from the air above the surface was of course a drastic one for fishes, requiring radical modification of their respiratory structures. The typical gills of a fish, although well suited for gas exchange in the water, are totally unsuited to performance of that function in the air. The thin lamellae of the gills, with their fine blood vessels, collapse under gravity in the air. In order to breathe air fishes had to develop air-holding chambers of one kind or another. Some species became air-breathers by adapting the stomach or a segment of the intestine to this purpose; they swallowed air and expelled it again at the mouth or the cloaca. Others developed elaborate air-holding structures in the mouth or throat. The most common adaptation for breathing air was the development of the special organ now recognized in many fishes as the swim bladder. This chamber serves other purposes in water-breathing fishes: it maintains the fish's buoyancy by adjusting to the hydrostatic pressure, and it may also act as an organ for detecting and producing sounds. The fossil evidence indicates, however, that the swim bladder of fishes originally came into being as an organ for respiration, and in some species it eventually evolved into a true lung.

A surprisingly large number of modern species of teleosts (bony fishes) are capable of breathing air. Evidently conditions like those that existed in Devonian times—swampy, oxygen-deficient waters—have given rise again and again to air-breathing fishes. Apart from these latter-day examples, we have considerable direct information on the respiratory revolution that led to the liberation of vertebrates from the aquatic environment during the Devonian. The information is provided not only by fossils from that time but also by living representatives of several archaic groups of air-breathing fishes that still exist today. These include two African genera in the primitive order Chondrostei, two American genera in the order Holostei and three genera of lungfishes (Dipnoi) in Africa, Australia and South America. These descendants from Devonian times inhabit oxygen-deficient waters and are so remarkably little changed from their ancestors that we can consider them representative of the early vertebrate forms that made the first crucial step toward air-breathing. The amphibians are believed to have arisen from the archaic ancestors of the lungfishes.

The lungfishes have an air bladder that developed originally as a diverticulum, or pouch, from the gut. The acquisition of this incipient lung was of course only a first step toward effective air-breathing. It had to be accompanied by changes in the fishes' system of blood circulation that would provide efficient transportation to the body tissues of blood oxygenated in the lung. In the lungfishes we see the circulation transformed in that way. In short, the lungfishes are truly lung breathers, with a respiratory and circulatory system that forecasts the system later developed by the higher vertebrates—the birds and the mammals. Fortunately the three living genera of lungfishes depict three stages in the transition from water-breathing to air-breathing, so that we are able to trace the anatomical changes that effected this transition.

Curiously, although the lungfishes offer an inviting opportunity to investigate the evolution of air-breathing by experimental studies, it is only within the past decade that such studies have been undertaken. With the collaboration of Claude Lenfant and other colleagues I have been pursuing this inquiry with lungfishes and certain other

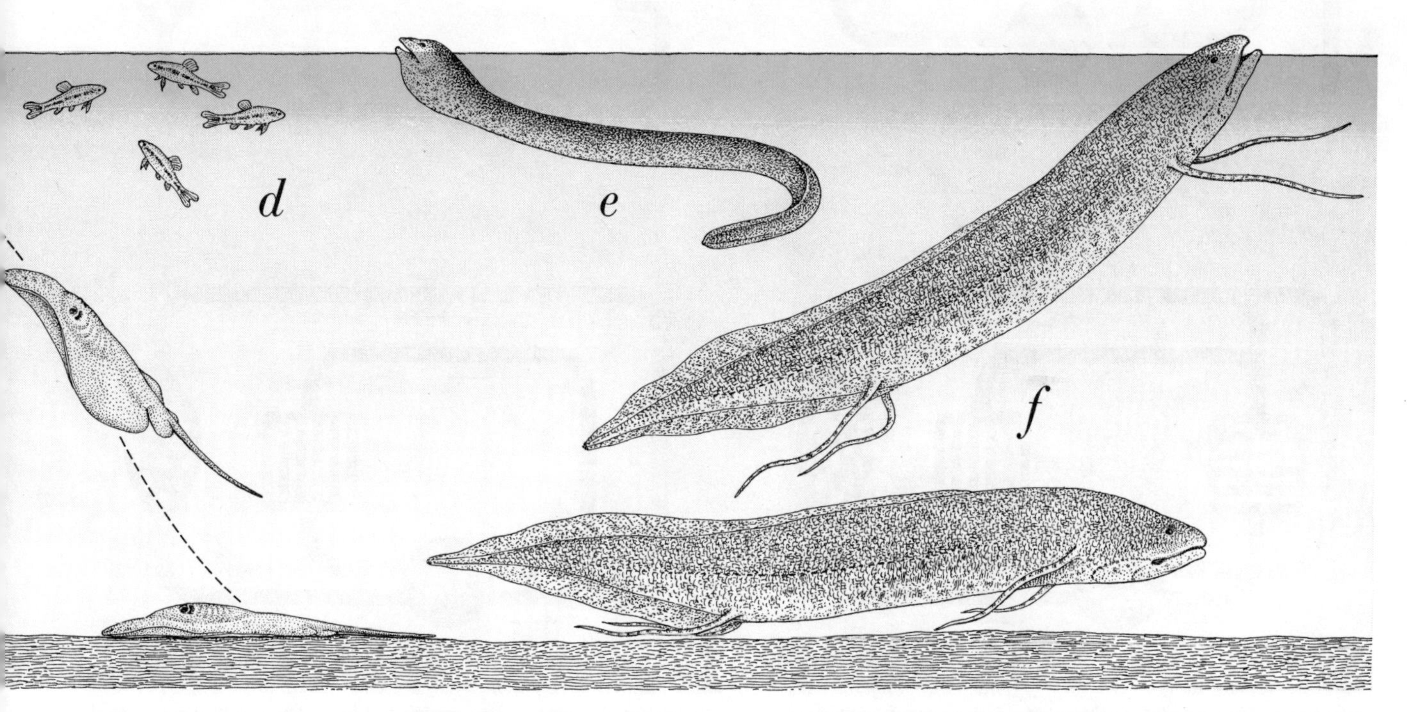

oxygen from air it swallows. *Electrophorus*, the electric eel (*b*), gulps air and obtains oxygen through the walls of the mouth. The freshwater stingray (*c*) visits the surface to breathe oxygenated water, as does *Characidium* (*d*). The gills of *Symbranchus marmoratus* (*e*) function in air as well as water. In lungfishes such as *Protopterus* (*f*) the swim bladder has become a well-developed lung.

air-breathing fishes in Africa, Australia, South America, Norway and in my laboratory at the University of Washington. The following is an account of our findings.

The lungfishes have long been known to man; they have been an important source of food for primitive peoples, in part because many of the species go into estivation (a dormant state) during dry seasons and therefore are easy to catch. Some of the lungfishes grow to considerable size; one African specimen displayed in a Nairobi museum measures seven feet in length. It was only about 130 years ago that lungfishes began to attract the attention of scientists. The discovery at that time of their remarkable ability to breathe air evoked considerable excitement and confusion in zoological societies. Reluctant to accept an air-breather as a fish, the zoologists at first classified the lungfishes as amphibians, and the surprise occasioned by the discovery is still preserved in the species name of a South American lungfish: *paradoxa*.

In the three lungfish genera we can observe a sequence of development from very little dependence on air-breathing to a stage where air-breathing became the principal means of respiration. The

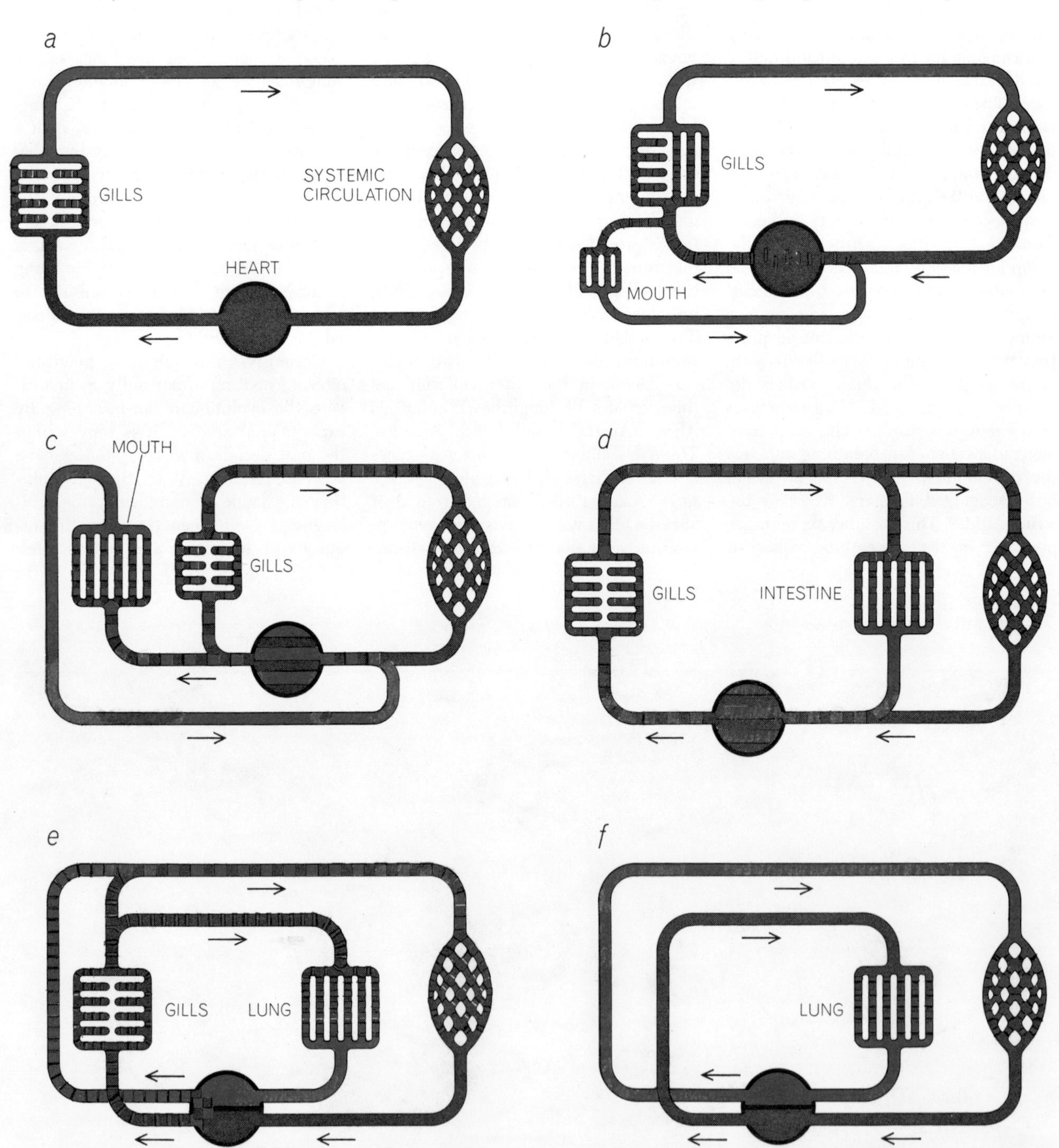

VASCULAR SYSTEM is diagrammed in relation to air- and water-breathing organs in typical fishes (*a*), birds and mammals (*f*) and various "in-between" air-breathing fishes: *Symbranchus* (*b*), *Electrophorus* (*c*), *Hoplosternum* (*d*) and the lungfishes (*e*). Water-breathing fishes have a single circulation with no mixing between oxygenated (*red*) and deoxygenated (*gray*) blood (*a*). Birds and mammals have a double circulation with no mixing (*f*). In the others there is more or less mixing of blood. The lungfish arrangement (*e*) was conducive to development toward the vertebrate condition: two circuits are arranged in parallel, with variable mixing.

Australian lungfishes (genus *Neoceratodus*), living today in rivers that are never severely deficient in oxygen, are primarily water-breathers, with well-developed and functioning gills. They do, however, possess lungs and can raise their heads out of the water to breathe air when necessary. The South American lungfishes (*Lepidosiren*) and the African (*Protopterus*) live in swamps whose waters are severely deoxygenated and periodically dried up by droughts. In dry periods these fishes entomb themselves in burrows or cocoons in the mud and go into a profound state of suspended animation; one might say their metabolic furnace is turned down all the way to the pilot light. Within minutes or hours after the swamp is flooded again by rainfall they revert to active life. Although water is their natural habitat, they depend mainly on their lungs and air-breathing for the absorption of oxygen. In the adult South American and African lungfishes large portions of the gills have degenerated to the point where they are little more than vestiges.

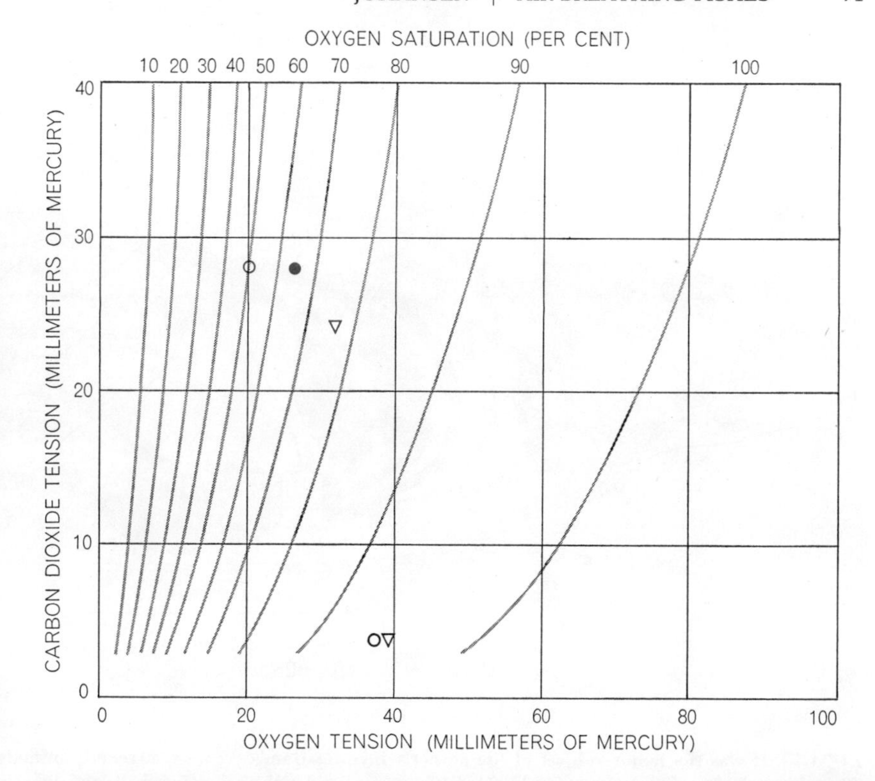

MEASUREMENTS of the oxygen and carbon dioxide content of the blood in the Australian lungfish *Neoceratodus* (*black*) and the African lungfish *Protopterus* (*color*) show that the latter depends more on its lung even in the water. In *Neoceratodus*, blood going to the lung in the pulmonary artery (*circle*) is about as well saturated with oxygen (*gray curves*) as blood coming from the lung in the pulmonary vein (*triangle*). In *Protopterus*, however, oxygen saturation is higher and carbon dioxide tension lower in blood coming from the lung (*triangle*) than in blood going to it (*open circle*); the oxygenated blood is channeled preferentially to the systemic arteries (*dot*) rather than to the lung. The lower carbon dioxide tensions in *Neoceratodus* reflect the dominance of water-breathing with efficient gills.

For examination of the functioning of the lungfishes' respiratory apparatus we have used catheters and transducers that enable us to determine blood flow and blood pressure in the vessels supplying the respiratory organs. We implant the instrumentation in the important blood vessels and in the lung itself, and after the fishes' recovery from the surgery we are able to obtain a continuous record of blood flow and pressure as the animals swim about in an aquarium, and to measure the oxygen tension in the blood and lung by analyzing samples from those organs. With this technique we have studied the lungfishes' respiratory behavior under normal conditions and under experimental variations of their environment.

The observations that are thus made possible show clearly the extent of the difference between the Australian lungfish and, say, the African lungfish with regard to breathing behavior [*see illustration on this page*]. With the fish swimming in well-oxygenated water, the blood coming from the gills of the Australian lungfish is almost fully saturated with oxygen. The lung then has little or no importance; there is no need to add to the oxygen supply by means of air in the lung. Furthermore, the carbon dioxide in the blood is kept at a very low level, because carbon dioxide has a relatively high solubility in water and the gills are remarkably efficient in exchanging this gas. In the African lungfish, on the other hand, the situation is somewhat reversed. The blood delivered to the lung from this fish's vestigial gills is not nearly as rich in oxygen, and the lung makes a large additional contribution to the oxygen tension, raising the oxygen saturation of the blood to about 80 percent. The lung also serves to eliminate some carbon dioxide. It is less efficient in this function than gills are, however, and the carbon dioxide tension in the blood is therefore considerably higher in the African lungfish than it is in the Australian lungfish.

Now let us drain the water from the tank and leave both types of fish exposed to air. The Australian lungfish frantically searches for a return to water and at the same time begins to gulp air into its lung at a rapid rate. This effort succeeds in saturating the blood flowing through the lung with oxygen, but the rate of blood flow is not sufficient to maintain the oxygen level in the arterial system. The oxygen tension in the arteries of the fish's body rapidly declines. Moreover, the lung does not eliminate carbon dioxide at the requisite rate, and in less than 30 minutes the carbon dioxide tension in this fish's arterial blood rises more than fivefold. In contrast, the African lungfish fares much better when it is left completely exposed to air. The intensification of this fish's air-breathing succeeds in keeping the oxygen tension in the arteries at a high level. The carbon dioxide concentration in the blood does rise for a time, but the elimination of that gas through the lung is such that the carbon dioxide tension soon levels off [*see top illustration on page 73*]. Thus the African lungfish clearly shows that its air-breathing system is sufficiently developed to enable it to survive out of water, as it must do during periods of drought, whereas the Australian lungfish normally never leaves the water.

Obviously the African lungfish's superior performance in air-breathing must depend on a circulatory advantage that enables it to make more effective use of the oxygen taken up through the lung. The possession of a lung, even one well supplied with blood vessels, cannot efficiently provide oxygen to an animal for its metabolic needs unless it is accom-

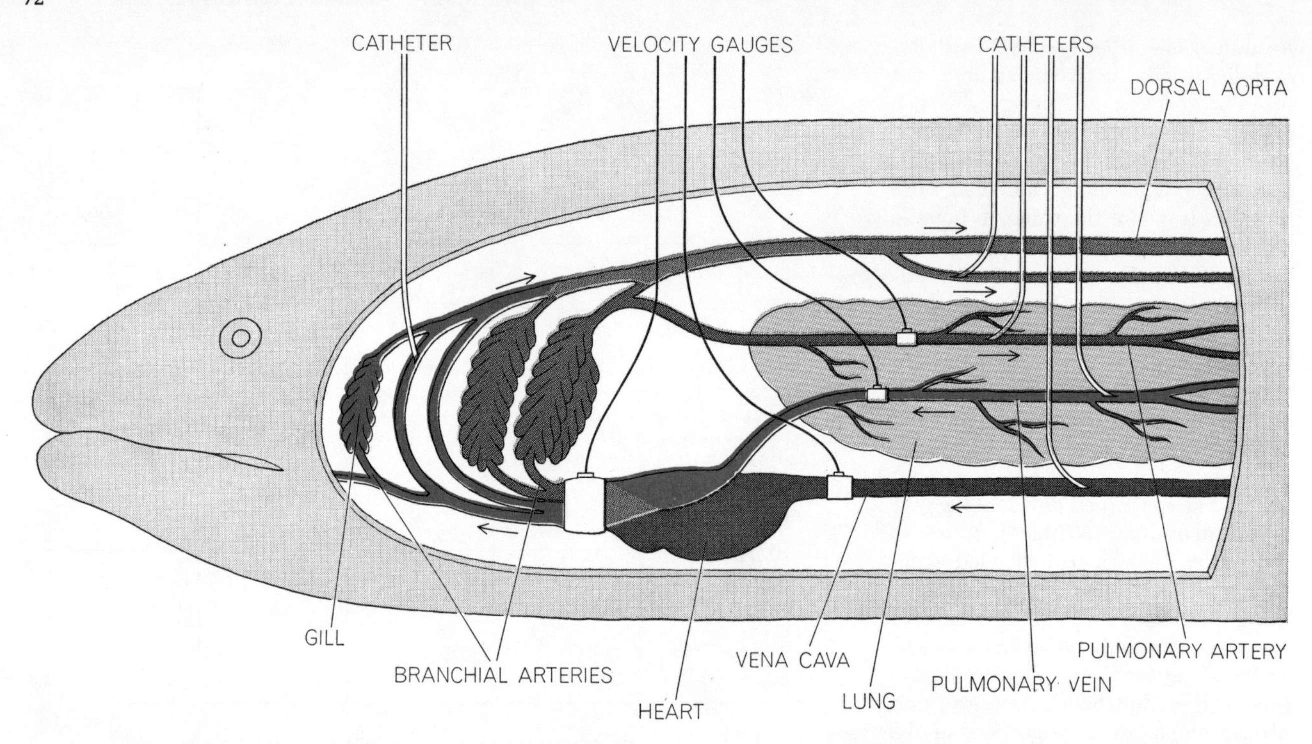

LUNGFISH was the major subject of the author's investigations. The schematic diagram shows how blood-velocity gauges and catheters for the measurement of blood pressure and blood-gas tensions were attached; animals swam freely after the instruments were implanted. In the lungfish blood is shunted either to the gills and lung for gas exchange or to the dorsal aorta for systemic circulation.

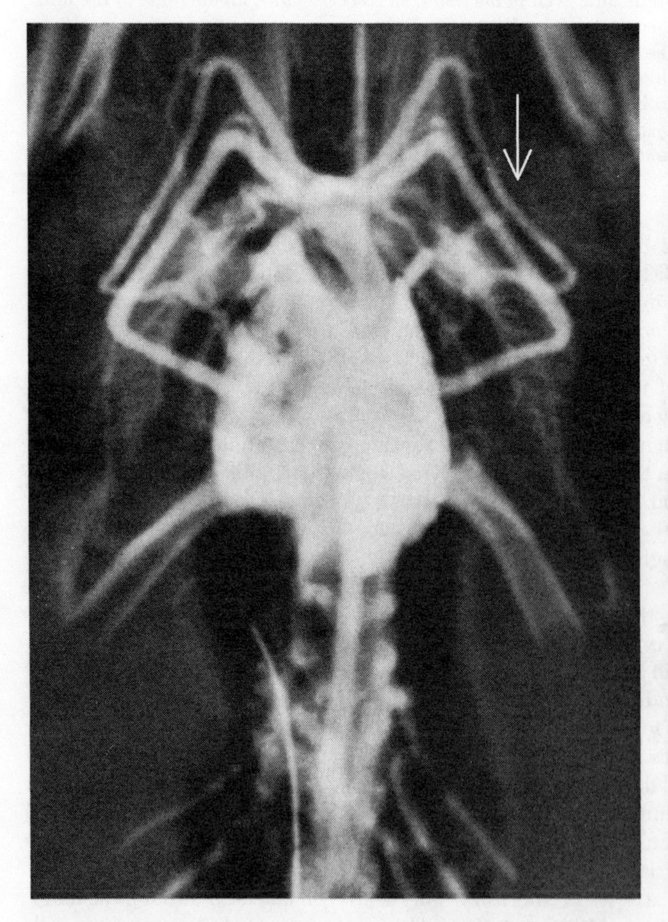

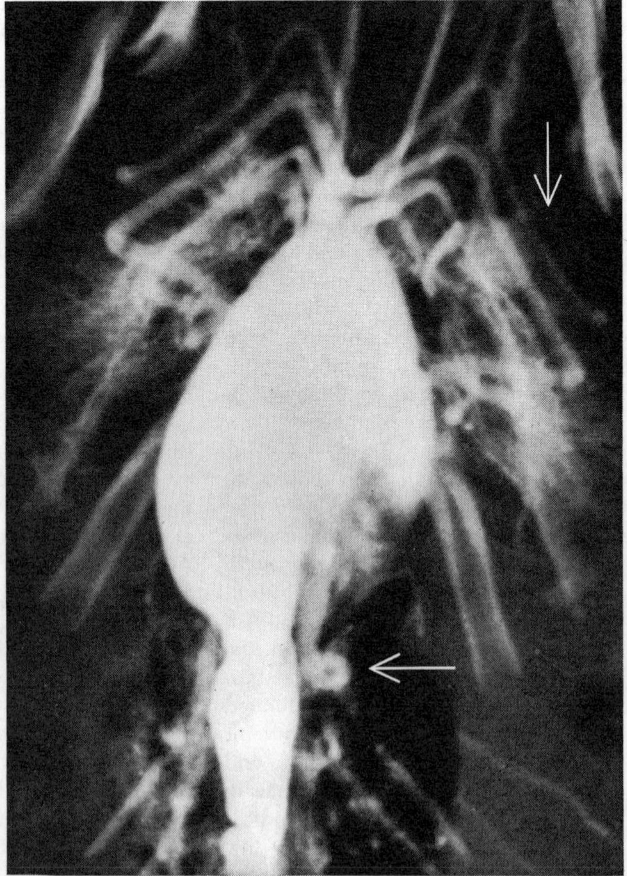

X-RAY ANGIOGRAPHY of *Protopterus* shows that blood returning to the heart from the lung is preferentially sent to the systemic circulation, with slight recirculation to the lung. When contrast medium is injected in the pulmonary vein (*left*), it appears mainly in two gill-less branchial arteries (*arrow*) and does not enter the pulmonary artery. Injected in the vena cava, however (*right*), it appears in all the branchial arteries and fine gill capillaries (*vertical arrow*) and also enters the pulmonary artery (*horizontal arrow*).

panied by an appropriate circulation for effective direct delivery of the oxygen to the metabolizing tissues. The circulatory requirement for an air-breather is fundamentally different from that for a water-breather. Hence a principal key to the transition from water-breathing to air-breathing lies in the evolution of the cardiovascular apparatus. The circulatory systems of the air-breathing fishes, differing one from another, show the steps in this evolution.

In ordinary water-breathing fishes the blood issuing from the pumping heart is not oxygenated. It flows to the gills, picks up oxygen there, emerges into a dorsal aorta and then flows through arteries to the various parts of the body, returning to the heart by way of the venous system. In short, the water-breathing fish has a single circulation, in contrast to the double circulation of a mammal, which first sends the venous blood from the right ventricle of the heart to the lungs and then, on return of the oxygenated blood to the left ventricle, pumps it out into the systemic circulation. For the fish the single circulation is perfectly adequate; the gills are highly efficient gas exchangers, and the blood flowing freely through them needs no extra input of energy in order to travel on through the arterial system.

When a fish acquires a lung and breathes air, complications begin to arise. The fish tends to retain its gills, at least at first, as an escape hatch for carbon dioxide, which is eliminated much more readily through gills than through a lung. Now, with gas exchange taking place both in the lung and in the gills, and with new avenues of blood circulation developing, the oxygenated blood from the lung becomes mixed with deoxygenated blood from the veins [*see illustration on page 70*]. Many of the air-breathing fishes still retain a single-circulation system in spite of this disadvantage. In the lungfishes, however, we see the beginnings of development of a double circulation. The two circulations are parallel, one passing through the lung, the other through the gills. They are not distinctly separated, and consequently there is some mixing of blood between them.

The Australian lungfish, as we have noted, is principally a water-breather and still has well-developed gills. It resorts to air-breathing only when the oxygen content of the water falls below normal. This fish could not survive long in severely deoxygenated water because the oxygen it gained by breathing air would be lost to the water as its oxy-

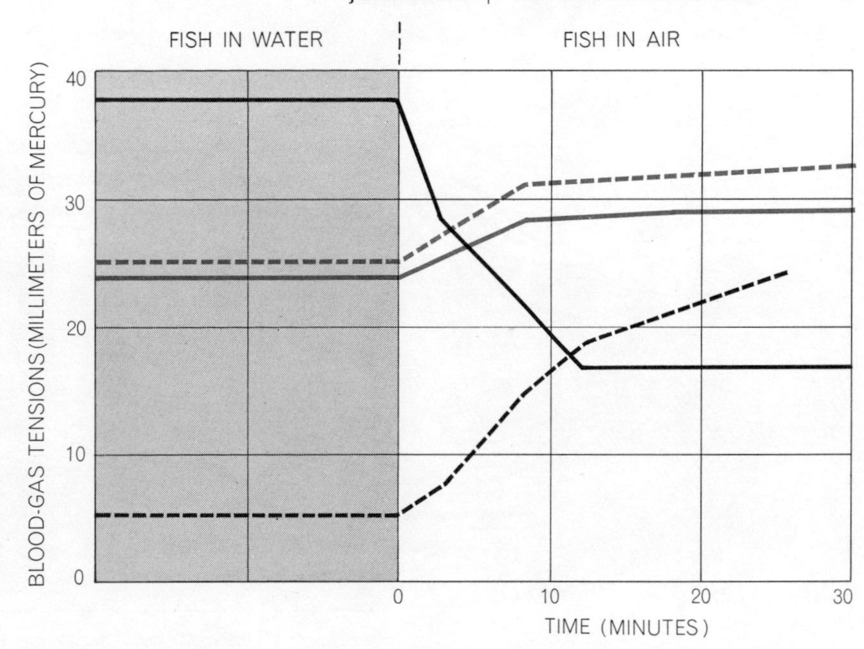

ADAPTABILITY TO AIR exposure varies in *Protopterus* (*color*) and *Neoceratodus* (*black*). When the tank is drained, both fish intensify air-breathing. *Protopterus* increases arterial oxygen tension (*solid curve*) and controls carbon dioxide (*broken curve*). *Neoceratodus* cannot get enough oxygen into its circulation or eliminate carbon dioxide efficiently.

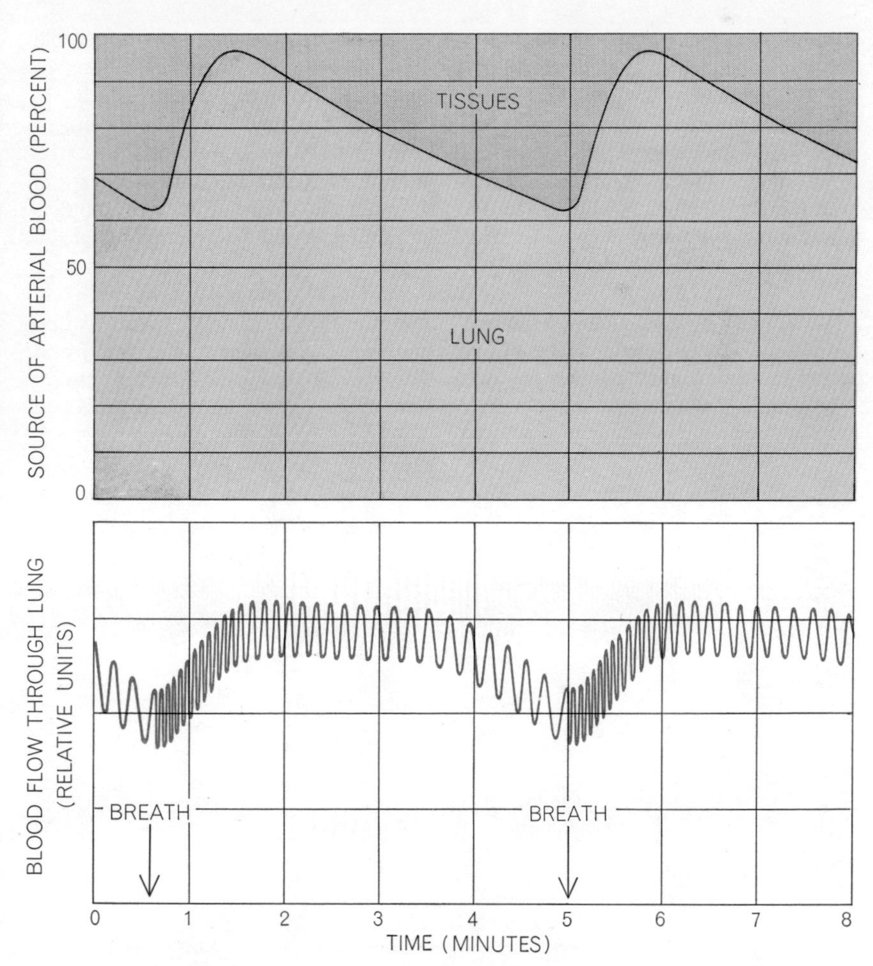

SHUNTING of oxygenated blood changes with the breathing cycle in *Protopterus*. Right after each breath almost all blood entering the systemic arteries is oxygenated blood from the lung; later in the cycle the proportion is lower. Blood flow through the lung also tends to be highest just after a breath and then to diminish until the next breath is taken.

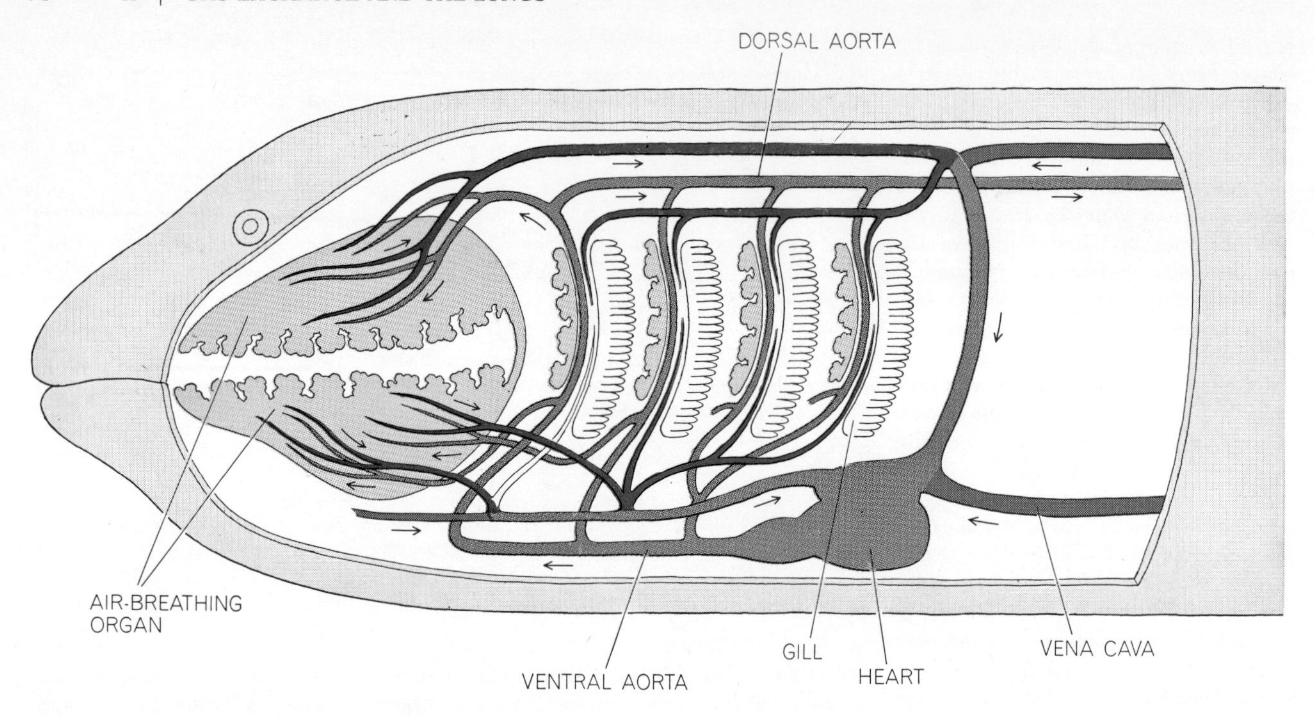

ELECTRIC EEL is a mouth breather. The mucous membrane lining its mouth is creased and wrinkled, providing a large surface richly supplied with blood vessels for gas exchange. Blood moves past the vestigial gills, which are unimportant in gas exchange.

genated blood circulated through the well-developed gills. The African lungfish, however, gets along without stress in oxygen-depleted waters partly because its gills are degenerated and even bypassed by the two largest arteries from the heart, which go directly to the dorsal aorta, and partly because its lung circulation is more fully developed.

By means of X-ray angiography, using special film changers capable of exposing several frames per second, we have examined the pattern of circulation in the African lungfish in considerable detail. We find that the channeling of venous and arterial blood is highly selective in this fish; the recirculation, or mixing, of blood is very slight in the pulmonary circuit and only moderate in the systemic circuit. The X-ray studies of blood flow and oxygen analysis of the blood also show that the blood flow through the lung is greatest immediately after the intake of an air breath (when the lung gas is richest in oxygen) and that the blood in the arteries issuing from the heart to the tissues reflects a similar cycle: it is richest in oxygen right after a breath and gives way to deoxygenated blood from the tissues in the interval before the next breath.

Evidently the African lungfish has a control system that coordinates the respiratory and circulatory mechanisms to achieve a maximal yield in gas exchange and gas transport. Interestingly enough, an interaction of the two mechanisms such as we observe in the African lungfish can also be seen in man. When part of the human lung is blocked from contact with air or is poorly ventilated for

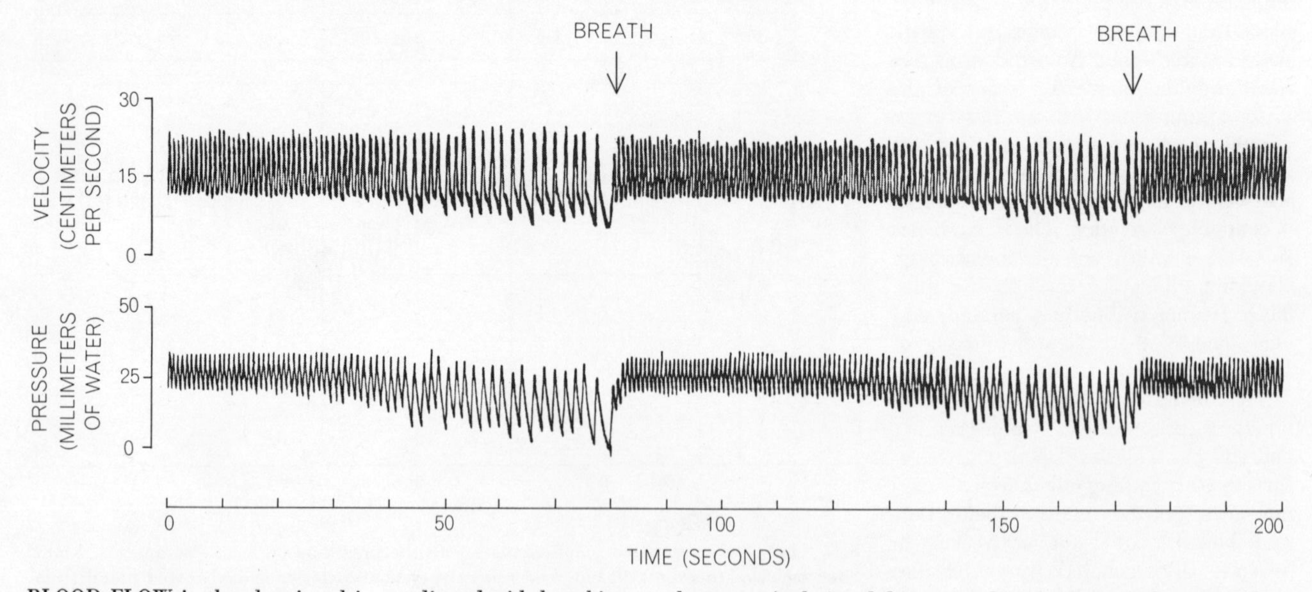

BLOOD FLOW in the electric eel is coordinated with breathing. Right after each breath (*arrows*) the cardiac output is high; the heart rate (pulse) and the average blood velocity and pressure then decline until the next breath, when they return to the high level.

some other reason, the vascular system reduces the flow of blood to that part. That is to say, the blood flow is adjusted to the availability of oxygen at the lung interface, as it is in the lungfish.

An even more basic parallel between the lungfish and human respiratory systems can be noted. Human respiration is regulated by an elaborate process of negative feedback in which the respiratory activities are controlled by changes in the oxygen and carbon dioxide tensions and the acidity of the blood. A rise in the blood's carbon dioxide tension, for instance, stimulates more rapid breathing to reduce it to the normal level. In air the African lungfish displays a similar sensitivity to specific internal cues. If it is exposed to air deficient in oxygen, the lungfish responds by breathing more rapidly than it would on taking normal atmospheric air into its lungs. Evidently its internal receptors sense hypoxia (oxygen deficiency) within the blood, and this acts as negative feedback to stimulate increased respiratory activity. In water, however, the response is different. If an adult African lungfish is placed in even severely deoxygenated water, the rate of its air-breathing, the gas tensions in its arterial blood and the normally slow rate of water-pumping across its gills all remain unchanged. The fish has been relieved of the threat of internal deoxygenation from oxygen-poor water; the gills no longer offer effective communication between the blood and the external environment and the respiratory control system monitors predominantly the activity of what is now the principal organ for oxygen extraction: the lung.

(It is easy to see that a feedback response prompting increased water-breathing in severely oxygen-depleted water would be harmful rather than advantageous to a fish that depends primarily on water-breathing. For an inhabitant of a tropical swamp it would lead only to an energy-consuming effort to obtain oxygen that is simply not available. The few truly water-breathing fishes living in such waters depend on a high tolerance for lack of oxygen in their blood and an ability to stay away from water masses most deficient in oxygen.)

What causes a lungfish to intensify its breathing efforts when it is totally exposed to air? This response is so prompt that it seems unlikely it is initiated by the operation of some internal chemical mechanism. My own surmise is that the response is probably triggered by the sudden translation of the fish from its weightless condition in water to the exposure to net gravitational force in the air. This represents a massive physical stimulus that may well jolt the animal into a burst of rapid breathing. We may have here a phylogenetic parallel to mammalian ontogeny. The newborn baby's first breath after it emerges from the amniotic pool of its mother's womb seems to be triggered by the massive impingement of physical stimuli to which it is suddenly exposed in the air.

We come now to certain fishes in which the mouth serves as a lung. One example is the famous climbing perch of India (which, contrary to its mythology, does not actually climb trees). Probably the most remarkable of the mouth breathers is the electric eel (*Electrophorus electricus*) that lives in shallow, muddy pools along creeks and rivers in tropical South America. This fish is an obligate air-breather; it could not survive if it did not come up frequently for air. At regular intervals—every minute or two—the fish rises to the surface, gulps air into its mouth and then sinks back to the bottom. It expels the air through flap-covered opercular openings.

The electric eel has markedly degenerate gills that play no significant part in its respiration. The uptake of oxygen occurs in the mouth, which is almost entirely lined with papillae (protuberances) that provide a large total surface area for gas exchange. Comparatively little carbon dioxide is released through the mouth; most of it is discharged into the water by way of the skin and the vestigial gills.

A rich network of blood vessels is embedded in the lining of the mouth for the absorption of oxygen. For the carnivorous electric eel the presence of these fragile vessels would be a serious drawback if it had to kill and chew its prey with its mouth, but it avoids that problem by stunning its victim with a powerful electric charge (up to 500 volts) and swallowing the food whole. The electric eel's respiration is handicapped, however, by a relatively inefficient circulatory system. The blood, after its oxygenation in the mouth, is not dispatched directly to the body tissues by way of the arteries but goes into veins that carry it to the heart [*see illustration on page 70*]. Consequently the oxygenated blood is mixed with the deoxygenated blood. As a result of the circulatory pattern, in which the blood is shunted through the respiratory organ between the arterial side and the venous side of the circulation, the arteries receive only part of the cardiac output, and the blood delivered to the tissues is a mixture of oxygenated and deoxygenated blood.

Our studies of blood flow in the electric eel showed that it pulses in a cyclic pattern like the pattern in the lungfish. Immediately after the eel has taken a breath the heart steps up its output; the

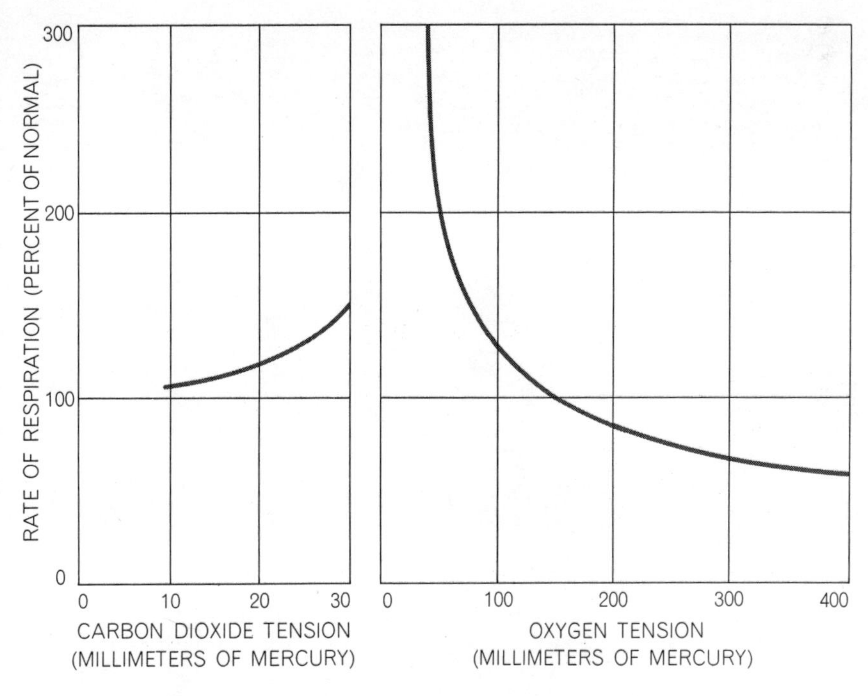

RESPONSE of electric eel to changes in the ambient atmosphere is clear and prompt. If it surfaces in an atmosphere low in oxygen (or high in carbon dioxide), its breathing rate is accelerated. If there is more oxygen present than in normal air, the breathing rate declines.

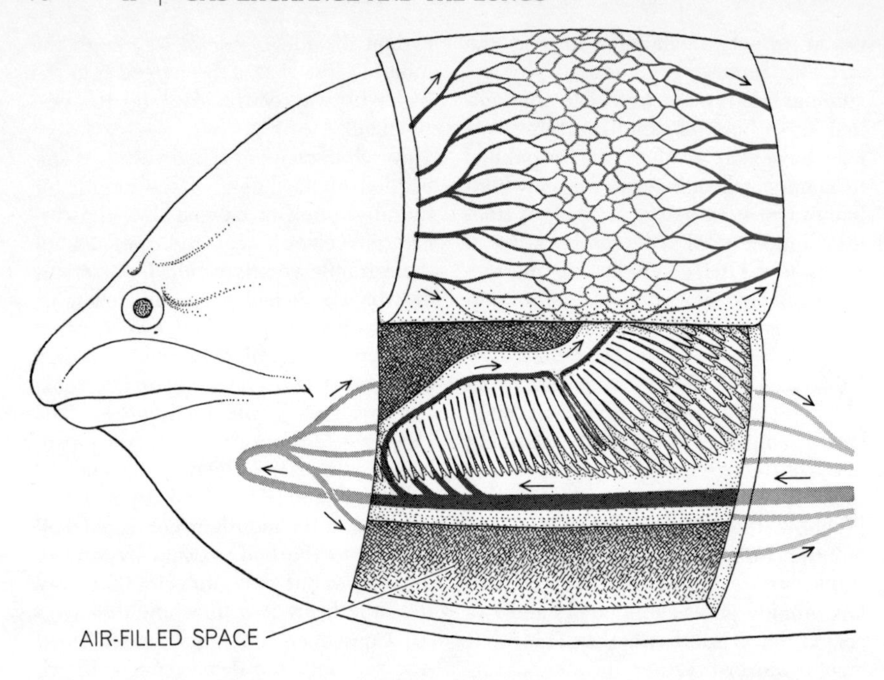

***SYMBRANCHUS* is unusual in that it can use the same organ, its gills, for air- and water-breathing, and can do both quite successfully. In addition to gills modified for air-breathing, it has a richly vascularized mouth lining (*flap*) that participates in gas exchange.**

heart rate, output and blood pressure then decline until the next air breath. During the accelerated phase of the cardiac output a high proportion of the blood flow goes to the mouth, where it can be oxygenated; then, as the oxygen in the mouth is used up, the proportion of the cardiac output delivered there declines. Thus the operation of the circulatory system partly compensates for the inefficiency of its structure by delivering more blood to the respiratory organ when the availability of oxygen in the organ rises to a peak. The increase in cardiac output evidently stems from stimulation of mechanical receptors in the mouth by the act of inhalation; we have found that it can be elicited by inflating the eel's mouth with pure oxygen or with nitrogen. Our experiments have also shown that the electric eel, like the African lungfish, is responsive to the oxygen content of the air it breathes. When it is subjected to air that is low in oxygen, the eel speeds up its breathing efforts; when the air is abnormally high in oxygen, the fish's breathing slows down.

A few remarkable fishes are able to use their gills for breathing air. An outstanding example is an eellike freshwater fish of South America named *Symbranchus marmoratus.* The gills of this fish are so organized that their fine blood vessels do not collapse under the force of gravity in the air; hence the blood flow through the gills remains normal. In *Symbranchus* the gills serve as an effective breathing organ both in the water and in the air. In air they are assisted by the lining of the mouth cavity, which also is well vascularized and provides gas exchange between the air and the blood. As long as the water contains an adequate oxygen content the fish spends all its time underwater. When the oxygen content drops, however, as commonly happens in tropical waters during the heat of the day, the fish rises to the surface, inflates its mouth and throat with air, closes its mouth and uses the trapped air for respiration. The inflation of the head makes the fish buoyant, and it may float on the surface for a considerable time, sometimes for hours, opening its mouth at intervals to expel the spent air and take a fresh breath. *Symbranchus'* air-breathing ability is so well developed that during a dry season it can survive for several months out of the water in the dormant state.

According to our measurements of the gases in this fish's blood, the oxygen tension of the arterial blood rises to higher levels when the fish is breathing air than when it breathes water. The carbon dioxide tension of the blood, however, also rises during air-breathing, both because the gills are less efficient in eliminating carbon dioxide in air than in water and because, as the fish rebreathes the air trapped in its closed mouth, waste carbon dioxide accumulates there. We found that when we removed *Symbranchus* from the water, it changed its breathing behavior so that the mouth was ventilated more or less continually, which allowed more carbon dioxide to escape. The fish kept its mouth open and made intermittent inhaling movements with its lower jaw. This change in breathing behavior on dry land may also be related to the fact that the fish no

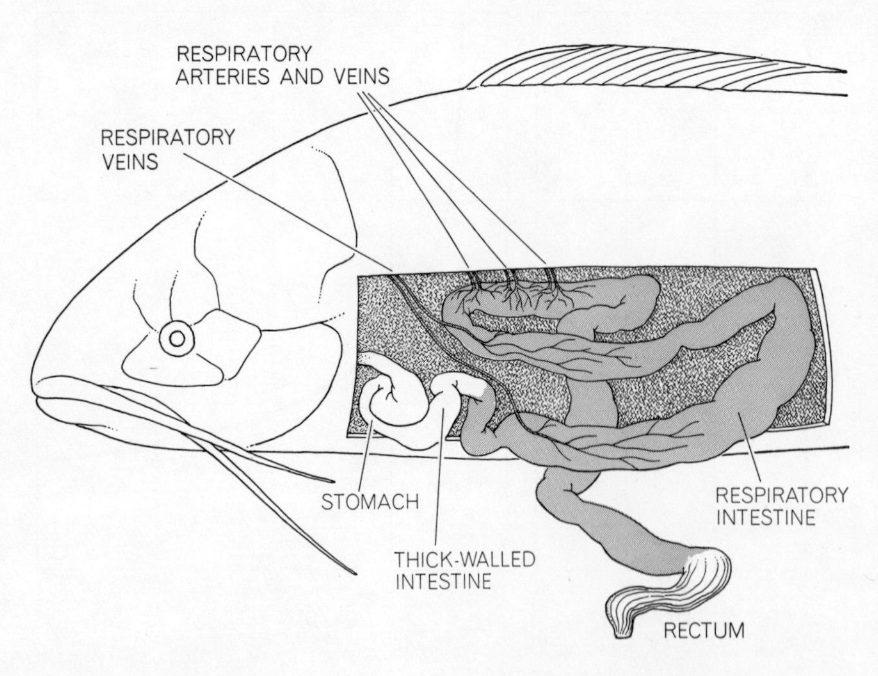

***HOPLOSTERNUM* is one of several kinds of intestinal-tract breathers. It swallows air, which fills a long, coiled portion of the intestine that is thin-walled and richly supplied with blood vessels. Only a short segment of the intestine remains to function in digestion.**

longer needs to maintain positive buoyancy.

Several species of tropical freshwater fishes use rhythmically ventilated air pockets in the gastrointestinal tract for air respiration. Air is swallowed and oxygen is delivered to the blood at the specialized, thin-walled and richly vascular portions of the gut. The circulatory system in these cases is generally inefficient for delivery of the oxygen to the body tissues because the arrangement is such that the oxygenated and deoxygenated blood is completely mixed and the arteries carry this mixture to the tissues [*see illustration on page 70*].

For the gastrointestinal-breathing species air-breathing by way of the gut is accessory to water-breathing with the gills. Nonetheless, the intestinal breathers are vitally dependent on air; most of them cannot survive if they are denied occasional access to the air, and for some species air-breathing has become the dominant method of respiration. A case in point is *Hoplosternum littorale,* a common fish of tropical South America that cannot live by water-breathing alone even if the water is saturated with oxygen, but that on the other hand is able to survive indefinitely in waters very severely deficient in oxygen by breathing air. Incidentally, this fish, like most other intestinal breathers, presents a puzzling question. Most of its intestine, a coiled tube that occupies nearly the entire body cavity, is modified for air-breathing. How can it carry out the usual intestinal functions, or, for that matter, how can it convey foodstuffs without disturbing the gas-exchange process of respiration? In our examinations we have always found the intestine full of gas and empty of solid food. It appears that the fish must feed only at long intervals and perform the digestive and absorptive functions with the short sections of the intestine that have not been modified for breathing.

The air-breathing fishes give a striking exhibition of the combination of exploratory and conservative forces that shaped the course of animal evolution. Changes in the oxygen tension of their aquatic environment forced the fishes to develop an apparatus for breathing air, yet they clung to the advantages that living primarily in water had to offer. At the same time, the crucial acquisition of the ability to breathe air opened a new world that inexorably led some fishes to develop the structural modifications that enabled vertebrates to step out of the water and walk onto dry land.

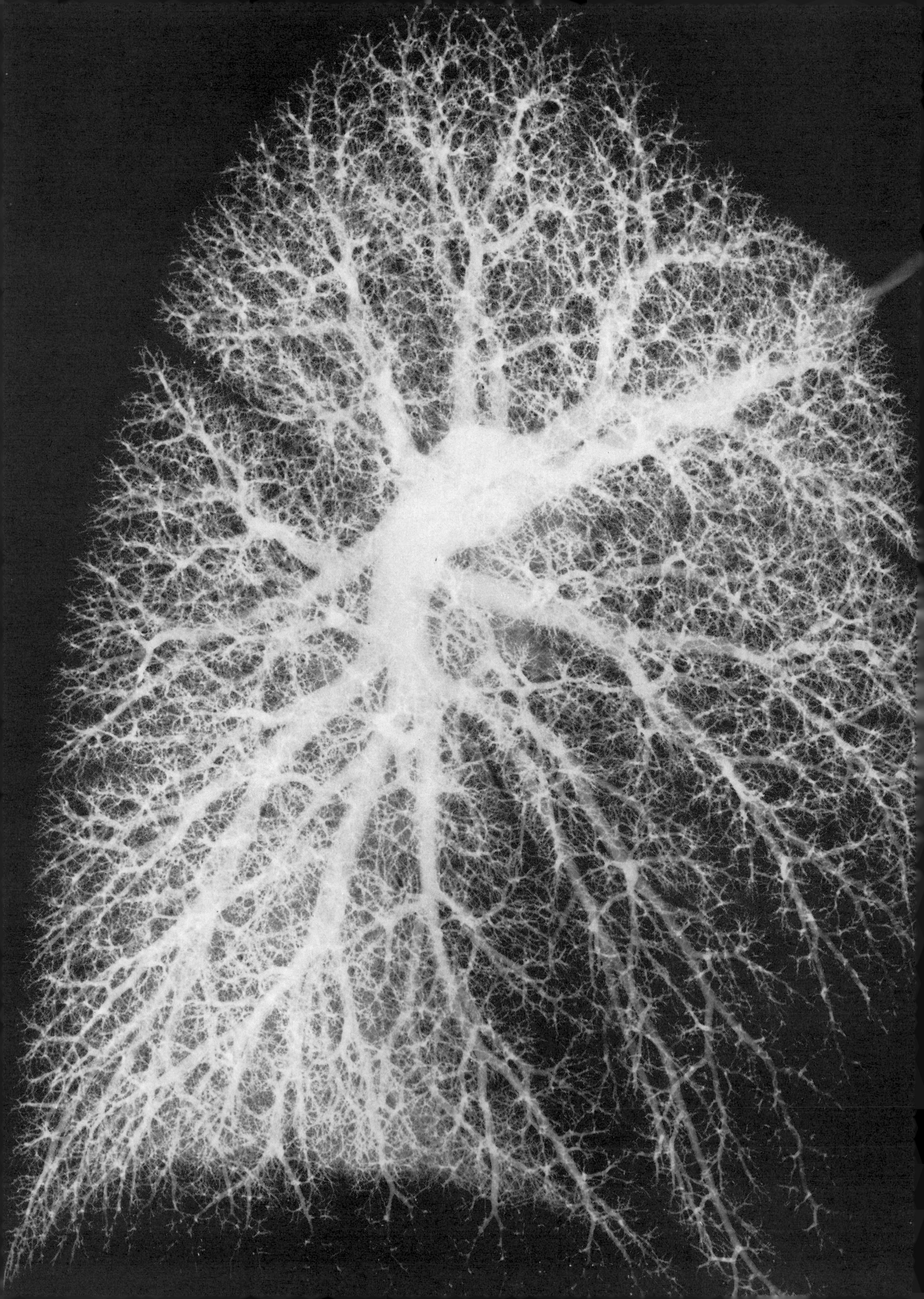

The Lung

7

by Julius H. Comroe, Jr.
February 1966

This elaborately involuted tissue of air sacs and blood vessels serves to exchange gases between the air and the blood. In man the total area of the membrane between the two systems is 70 square meters

Each year an adult human inhales and exhales between two million and five million liters of air. Each breath consists of about half a liter of air, 20 percent of which is molecular oxygen (O_2); the air swirls briefly through a maze of branching ducts leading to tiny sacs that comprises a gas-exchange apparatus in which some of the oxygen is added to the blood. The apparatus I am describing is of course the lung: the central organ in the system the larger land animals have evolved as a means of supplying each of their cells with oxygen.

Oxygen is essential to most forms of life. A one-celled organism floating in water requires no complex apparatus to extract oxygen from its surroundings; its needs are satisfied by the process of diffusion—the random movement of molecules that results in a net flow of oxygen from regions of abundance to regions of scarcity. Over very short distances, such as the radius of a cell, the diffusion of oxygen is rapid. Over larger distances diffusion is a much slower process; it cannot meet the needs of any many-celled organism in which the distance between the source of oxygen and the most remote cell is greater than half a millimeter.

The evolution of large animals has therefore required the development of various special systems that deliver oxygen from the surrounding medium to each of the animal's cells. In higher vertebrates such as man this system consists of a gas pump (the thorax) and two fluid pumps (the right and left ventricles of the heart). The fluid pumps are linked to networks of capillaries, through the walls of which the actual exchange of gas molecules takes place. The capillary network that receives its blood from the left ventricle distributes oxygen throughout the body. The network that receives its blood from the right ventricle serves a different purpose. In it the actions of the fluid pump and the gas pump are combined to obtain the oxygen needed by the body from the surrounding atmosphere.

This integrated system works in the following manner. The expansion of the thorax allows air to flow into ducts that divide into finer tubes that terminate in the sacs called alveoli. The rhythmic contraction of the right ventricle of the heart drives blood from the veins through a series of vessels that spread through the lung and branch into capillaries [*see illustration on opposite page*]. These blood-filled pulmonary capillaries surround the gas-filled alveoli; in most places the membrane separating the gas from the blood is only a thousandth of a millimeter thick. Over such a short distance the molecules of oxygen, which are more abundant in the inhaled air than in the venous blood, diffuse readily into the blood. Conversely the molecules of the body's waste product carbon dioxide, which is more abundant in the venous blood than in the inhaled air, diffuse in the opposite direction. In the human lung the surface area available for this gas exchange is huge: 70 square meters—some 40 times the surface area of the entire body. Accordingly gas can be transferred to and from the blood quickly and in large amounts.

The gas exchange that occurs in the lungs is only the first in a series of events that meet the oxygen needs of the human body's billions of cells. The second of the two liquid pumps—the left ventricle of the heart—now distributes the oxygenated blood throughout the body by means of arteries and arterioles that lead to a second gas-exchange system. This second system is in reality composed of billions of individual gas-exchangers, because each capillary is a gas-exchanger for the cells it supplies. As the arterial blood flowing through the capillary gives up its oxygen molecules to the adjacent cells and absorbs the cells' waste products, it becomes venous blood and passes into the collecting system that brings it back to the right ventricle of the heart [*see illustration on next page*].

The customary use of the term "pulmonary circulation" for that part of the circulatory system which involves the right ventricle and the lungs, and of the term "systemic circulation" for the left ventricle and the balance of the circulatory system, seems to imply that the body has two distinct blood circuits. In actuality there is only one circuit; the systemic apparatus is one arc of it and the pulmonary apparatus is the other. The systemic part of the circuit supplies arterial blood, rich in oxygen and poor in carbon dioxide, to all the capillary gas-exchangers in the body; it also collects the venous blood, poor in oxygen and rich in carbon dioxide, from these exchangers and returns it to the right ventricle. The pulmonary part of the circuit delivers the venous blood to the pulmonary gas-exchanger and then sends it on to the systemic part of the circuit.

This combination of two functions not only provides an adequate exchange of oxygen and carbon dioxide but also

ARTERIAL SYSTEM of the human lung (*opposite page*) is revealed by X-ray photography following the injection of a radio-opaque fluid. The finest visible branches actually subdivide further into capillaries from five to 10 microns in diameter that surround the lung's 300 million air sacs.

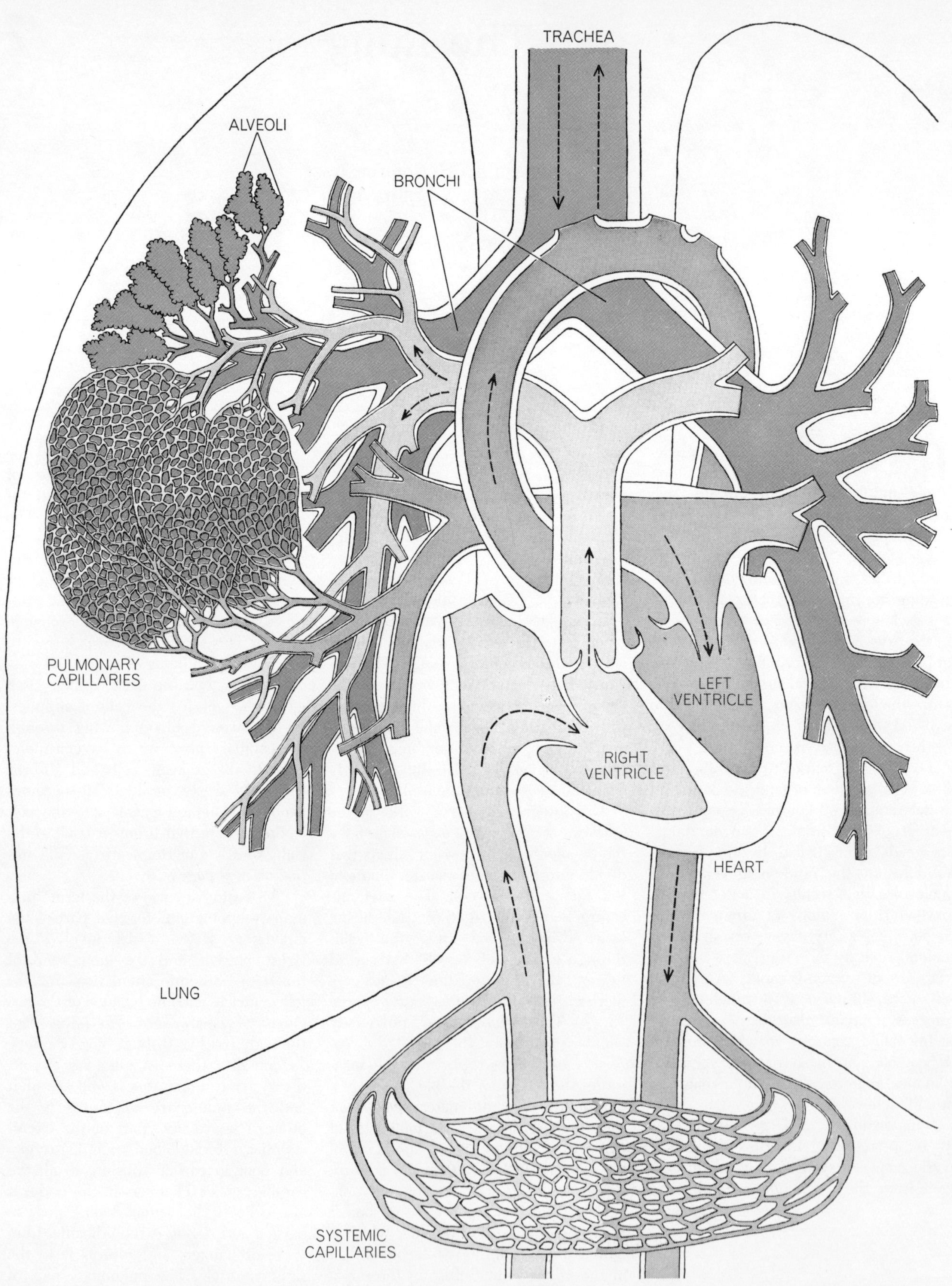

THREE PUMPS operate the lungs' gas-exchange system. One gas pump, the thorax (*see illustration on opposite page*), moves from five to seven liters of air (*color*) in and out of the lungs' air sacs every minute. At top left these alveoli are shown without their covering of pulmonary capillaries, which are shown at center left. One fluid pump, the heart's right ventricle, forces from 70 to 100 cubic centimeters of blood into the pulmonary capillaries at each contraction. This blood (*light gray*) is low in oxygen and high in carbon dioxide; oxygen, abundant in the air, is diffused into the blood while carbon dioxide is diffused from the blood into the air. Oxygenated and low in carbon dioxide, the blood (*dark gray*) then reaches the third pump, the left ventricle, which sends it on to the systemic capillaries (*example at bottom of illustration*) that deliver oxygen to and collect carbon dioxide from all the body's cells.

supplies a variety of nutrients to the tissues and removes the products of tissue metabolism, including heat. Complex regulatory mechanisms ensure enough air flow and blood flow to meet both the body's overall needs and the special needs of any particular part of the body according to its activity.

The Respiratory System

To the physiologist respiration usually means the movement of the thorax and the flow of air in and out of the lungs; to the biochemist respiration is the process within the cells that utilizes oxygen and produces carbon dioxide. Some call the first process external respiration and the second internal respiration or tissue respiration. Here I shall mainly discuss those processes that occur in the lung and that involve exchanges either between the outside air and the gas in the alveoli or between the alveolar gas and the blood in the pulmonary capillaries.

The structure of the respiratory system is sometimes shown in an oversimplified way that emphasizes only the conducting air path and the alveoli. The system is far more complex. It originates with the two tubes of the nose (the mouth can be regarded as a third tube), which join to become one: the trachea. The trachea then subdivides into two main branches, the right bronchus and the left. Each of the bronchi divides into two, each of them into two more and so on; there are from 20 to 22 bronchial subdivisions. These subdivisions give rise to more than a million tubes that end in numerous alveoli, where the gas exchange occurs. There are some 300 million alveoli in a pair of human lungs; they vary in diameter from 75 to 300 microns (thousandths of a millimeter). Before birth they are filled with fluid but thereafter the alveoli of normal lungs always contain gas. Even at the end of a complete exhalation the lungs of a healthy adult contain somewhat more than a liter of gas; this quantity is known as the residual volume. At the end of a normal exhalation the lungs contain more than two liters; this is called the functional residual capacity. When the lungs are expanded to the maximum, a state that is termed the total capacity, they contain from six to seven liters.

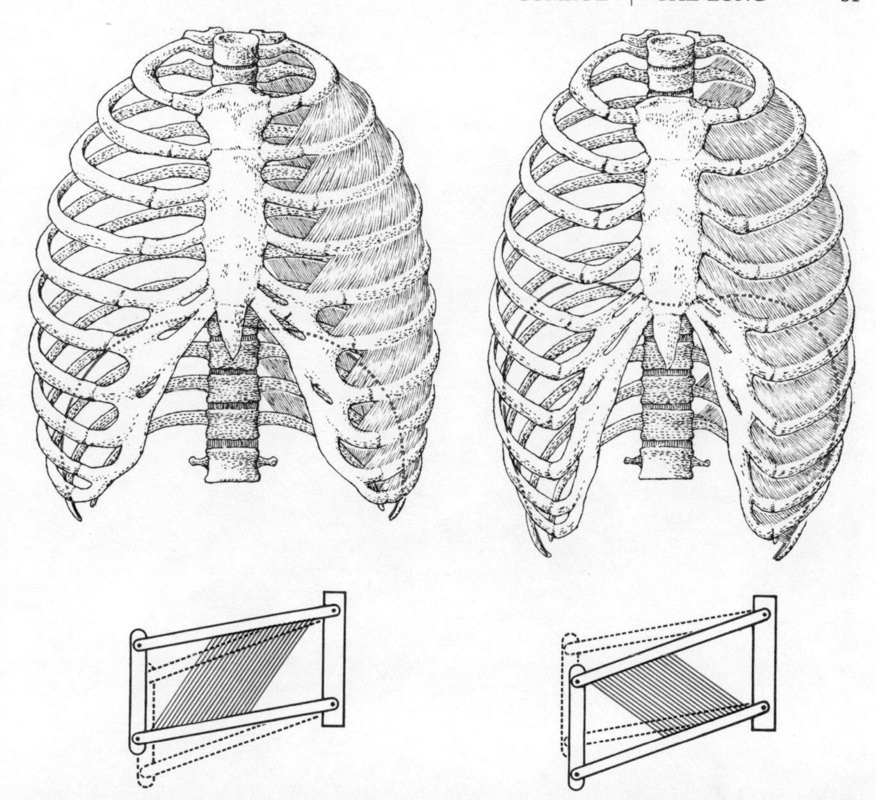

INHALATION takes place when the thorax expands, thus forcing the lungs to enlarge and bringing the pressure of the gas within them below atmospheric pressure. The expansion (*top left*) is principally the work of the diaphragm but also involves the muscles of the rib cage (*detail at bottom left*). Exhalation occurs when passive relaxation reduces the size of the thorax (*top right*), thereby raising the gas in the lungs to greater than atmospheric pressure. Only under conditions of stress do muscles (*detail at bottom right*) aid exhalation.

More important than total capacity, functional residual capacity or residual volume is the amount of air that reaches the alveoli. An adult human at rest inhales and exhales about half a liter of gas with each breath. Ideally each breath supplies his two-liter reservoir of gas—the functional residual capacity—with a volume of oxygen equal to the volume absorbed from the reservoir by the blood flowing through the pulmonary capillaries. At the same time each breath removes from the reservoir a volume of carbon dioxide equal to the volume produced by the body's cells and yielded to the lungs by the venous blood. The resting adult breathes from 10 to 14 times a minute, with the result that his ventilation—the volume of air entering and leaving the lungs—is from five to seven liters per minute. The maximum human capacity for breathing is about 30 times the resting ventilation rate, a flow of from 150 to 200 liters per minute. Even during the most intense muscular exercise, however, ventilation averages only between 80 and 120 liters per minute; man obviously has a great reserve of ventilation.

The volume of ventilation that is important in terms of gas exchange is less than the total amount of fresh air that enters the nose and mouth. Only the fresh air that reaches the alveoli and mixes with the gas already there is useful. Unlike the blood in the circulatory system, the air does not travel only in one direction. There are no valves in the bronchi or their subdivisions; incoming and outgoing air moves back and forth through the same system of tubes. Little or no gas is exchanged in these tubes. They represent dead space, and the air that is in them at the end of an inhalation is wasted because it is washed out again during an exhalation. Thus the useful ventilation obtained from any one breath consists of the total inhalation—half a liter, or 500 cubic centimeters—minus that part of the inhalation which is wasted in the dead space. In an adult male the wasted volume is about 150 cubic centimeters.

From an engineering standpoint this dead space in the air pump represents a disadvantage: more ventilation, which necessitates more work by the pump, is required to overcome the 30 percent inefficiency. The dead space may nonetheless represent a net advantage. The use of a single system of ducts for incoming and outgoing air eliminates the need for a separate set of ducts to carry each flow. Such a dual system would certainly encroach on the lung area available for gas exchange and might well result in more than a 30 percent inefficiency.

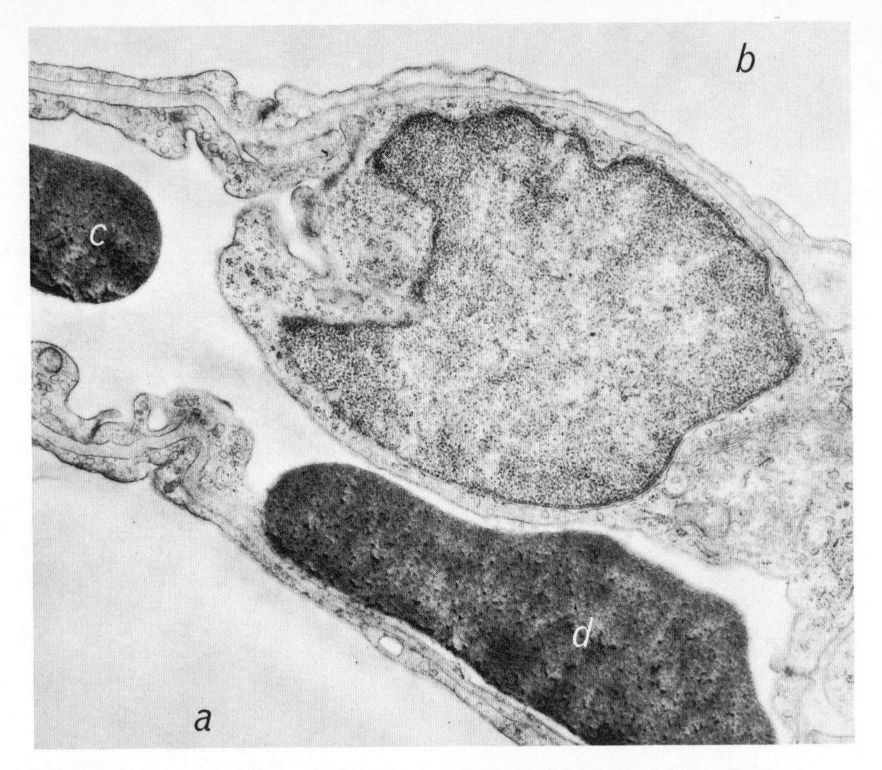

DIFFUSION PATH from gas-filled air sac to blood-filled capillary can be extremely short. In this electron micrograph of mouse lung tissue, *a* and *b* are air spaces; between these lies a capillary in which there are two red blood cells, *c* and *d*. The large light gray mass is the nucleus of an endothelial cell. The distances from air space *a* to red cell *d* and from air space *b* to red cell *c* are less than half a micron; the diffusion of gases across these gaps is swift.

Ventilation of the alveoli is not enough to ensure an adequate supply of oxygen. The incoming air must also be distributed uniformly so that each alveolus receives its share. Some alveoli are very close to the main branchings of the bronchi; others at the top and bottom of the lung are 20 or 30 centimeters away from such a branch. It is a remarkable feat of engineering to distribute the proper amount of fresh air almost simultaneously to 300 million alveoli of varying sizes through a network of a million tubes of varying lengths and diameters.

By the same token, ideal gas exchange requires that the blood be evenly distributed to all the pulmonary capillaries—that none of it should escape oxygenation by flowing through shunts in the capillary bed. Under abnormal circumstances, however, some alveoli are poorly ventilated, and venous blood flowing through the adjacent capillaries is not properly oxygenated. Conversely, there are situations when the flow of blood through certain capillaries is inadequate or nonexistent; in these cases good ventilation of the neighboring alveoli is wasted.

In spite of such deficiencies, the gas exchange can remain effective if the defects are matched (if the same regions that have poor ventilation, for example, also have poor blood flow), so that regions with increased ventilation also have increased blood flow. Two mechanisms help to achieve this kind of matching. First, a decrease in the ventilation of a group of alveoli results in a constriction of the blood vessels and a decrease in the blood flow of the affected region. This decrease is not the result of a nerve stimulus or a reflex but is a local mechanism, probably initiated by oxygen deficiency. Second, a local deficiency in blood flow results in a constriction of the pathways that conduct air to the affected alveoli. The resulting increase in airway resistance serves to direct more of the air to alveoli with a normal, or better than normal, blood supply.

The Pulmonary Circulation

The system that distributes blood to the lungs is just as remarkable as the system that distributes air. As I have noted, the right ventricle of the heart receives all the venous blood from every part of the body; the contraction of the heart propels the blood into one large tube, the pulmonary trunk. Like the trachea, this tube divides and subdivides, ultimately forming hundreds of millions of short, narrow, thin-walled capillaries. Each capillary has a diameter of from five to 10 microns, which is just wide enough to enable red blood cells to pass one at a time. The wall of the capillary is less than .1 micron thick; the capillary's length ranges from .1 to .5 millimeter.

If the pulmonary capillaries were laid end to end, their total length would be hundreds of miles, but the overall capillary network offers surprisingly little resistance to the flow of blood. In order to pump from five to 10 liters of blood per minute through the pulmonary system the right ventricle needs to provide a driving pressure of less than 10 millimeters of mercury—a tenth of the pressure required of the left ventricle for systemic circulation. With only a small increase in driving pressure the pulmonary blood flow can be increased to 30 liters per minute.

Although the total surface area provided by the pulmonary capillaries is enormous, at any instant the capillary vessels contain only from 70 to 100 cubic centimeters of blood. This volume is almost identical with the volume of blood the right ventricle ejects with each contraction. Thus with each heartbeat the reoxygenated blood in the pulmonary capillaries is pushed on toward the left ventricle and venous blood refills the capillaries. In the human body at rest each red blood cell remains in a pulmonary capillary for about three-quarters of a second; during vigorous exercise the length of its stay is reduced to about a third of a second. Even this brief interval is sufficient, under normal conditions, for gas exchange.

Electron microscopy reveals that the membrane in the lung that separates the gas-filled alveolus from the blood-filled capillary consists of three distinct layers. One is the alveolar epithelium, the second is the basement membrane and the third is the capillary endothelium. The cellular structure of these layers renders between a quarter and a third of the membrane's 70 square meters of surface too thick to be ideal for rapid gas exchange; over the rest of the area, however, the barrier through which the gas molecules must diffuse is very thin—as thin as .2 micron. Comparison of the gas-transfer rate in the body at rest with the rate during vigorous exercise provides a measure of the gas-exchange system's capacity. In the body at rest only 200 to 250 cubic centimeters of oxygen diffuse per minute; in the exercising body the system can deliver as much as 5,500 cubic centimeters per minute.

The combination of a large diffusion area and a short diffusion path is responsible for much of the lung's efficiency in gas exchange, but an even more critical factor is the remarkable ability of the red blood pigment hemoglobin to combine with oxygen. If plasma that contained no red cells or hemoglobin were substituted for normal blood, the adult human heart would have to pump 83 liters per minute through the pulmonary capillaries to meet the oxygen needs of a man at rest. (Even this assumes that 100 percent of the oxygen in the plasma is delivered to the tissues, which is never the case.) In contrast, blood with a normal amount of hemoglobin in the red cells picks up 65 times more oxygen than plasma alone does; the heart of a man at rest need pump only about five and a half liters of blood per minute, even though the tissues normally extract only from 20 to 25 percent of the oxygen carried to the cells by the red blood corpuscles.

The Mechanics of Breathing

Just as water inevitably runs downhill, so gases flow from regions of higher pressure to those of lower pressure. In the case of the lung, when the total gas pressure in the alveoli is equal to the pressure of the surrounding atmosphere, no movement of gas is possible. For inhalation to occur, the alveolar gas pressure must be less than the atmospheric pressure; for exhalation, the opposite must be the case. There are two ways in which the pressure difference required for the movement of air into the lungs can be created: either the pressure in the alveoli can be lowered below atmospheric pressure or the pressure at the nose and mouth can be raised above atmospheric pressure. In normal breathing man follows the former course; enlarging the thorax (and thus enlarging the lungs as well) enables the alveolar gas to expand until its pressure drops below that of the surrounding atmosphere. Inhalation follows automatically.

The principal muscle for enlarging the thorax is the diaphragm, the large dome-shaped sheet of tissue that is anchored around the circumference of the lower thorax and separates the thoracic cavity from the abdominal cavity. When the muscle of the diaphragm contracts, the mobile central portion of the sheet moves downward, much as a piston moves in a cylinder. In addition, skeletal muscles enlarge the bony thoracic cage by increasing its circumference [*see illustration on page 81*]. The lungs, of course, lie entirely within the thorax. They have no skeletal muscles and cannot increase their volume by their own efforts, but their covering (the visceral pleura) is closely linked to the entire inner lining of the thorax (the parietal pleura). Only a thin layer of fluid separates the two pleural surfaces; when the thorax expands, the lungs must follow suit. As the pressure of the alveolar gas drops below that of the atmosphere, the outside air flows in through the nose, mouth and tracheobronchial air paths until the pressure is equalized.

This kind of pulmonary ventilation requires work; the active contraction of the thoracic muscles provides the force necessary to overcome a series of opposing loads. These loads include the recoil of the elastic tissues of the thorax, the recoil of the elastic tissue of the lungs, the frictional resistance to the flow of air through the hundreds of thousands of ducts of the tracheobronchial tree, and the surface forces created at the fluid-gas interfaces in the alveoli [see "Surface Tension in the Lungs," by John A. Clements; SCIENTIFIC AMERICAN, February, 1966].

In contrast to inhalation, exhalation is usually a passive process. During the active contraction of muscles that causes the enlargement of the thorax, the tissues of thorax and lungs are stretched and potential energy is stored in them. The recoil of the stretched tissue and the release of the stored energy produce the exhalation. Only at very high rates of ventilation or when there is an obstruction of the tracheobronchial tree is there active contraction of muscles to assist exhalation.

Artificial ventilation can be produced either by raising external gas pressure or by lowering internal pressure. Body respirators of the "iron lung" type lower the pressure of the air surrounding the thorax in part of their cycle of operation. As a consequence the volume of the thorax increases, the alveolar pressure falls and, since the patient's nose and mouth are outside the apparatus, air at atmospheric pressure flows into his lungs. Later in the cycle the pressure within the respirator rises; the volume of the thorax decreases and the patient exhales.

Other types of artificial ventilation depend on raising the external pressure at the nose and mouth above the atmospheric level. Some mechanical respirators operate by supplying high-pressure

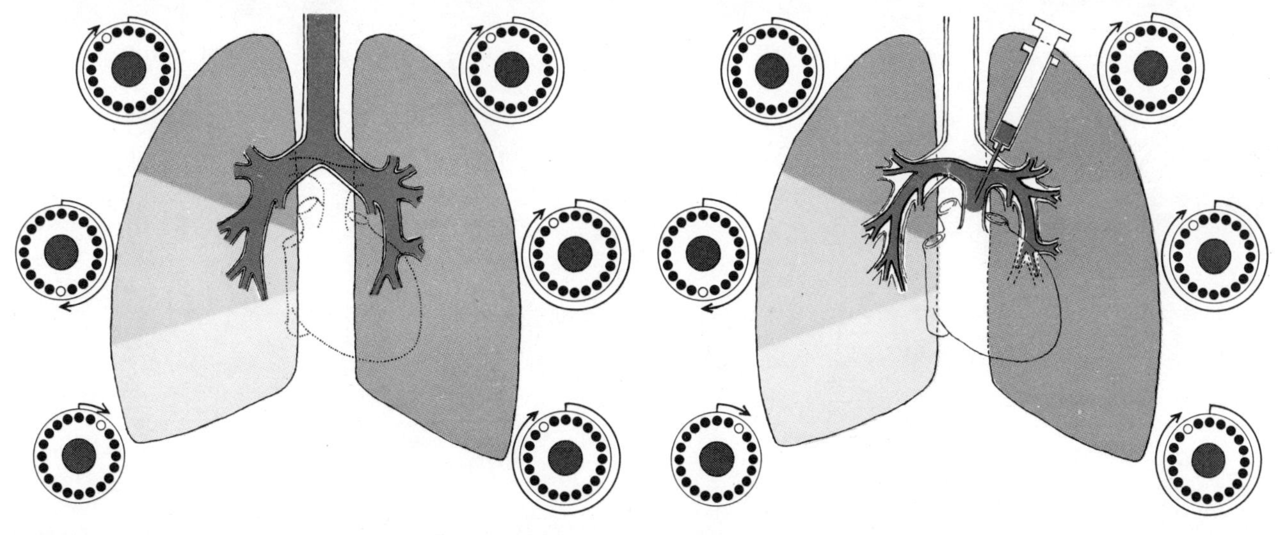

RADIOACTIVE TRACERS such as xenon 133 can assess the performance of the two components in the lungs' gas-exchange system. Inhaled as a gas (*left*), the xenon will be unevenly distributed if the air ducts are blocked; zones that are poorly ventilated will produce lower scintillation-counter readings than normal ones will. Injected in a solution into the bloodstream (*right*), the xenon will diffuse unevenly into the air sacs if the blood vessels are blocked. Such faulty blood circulation also causes low readings.

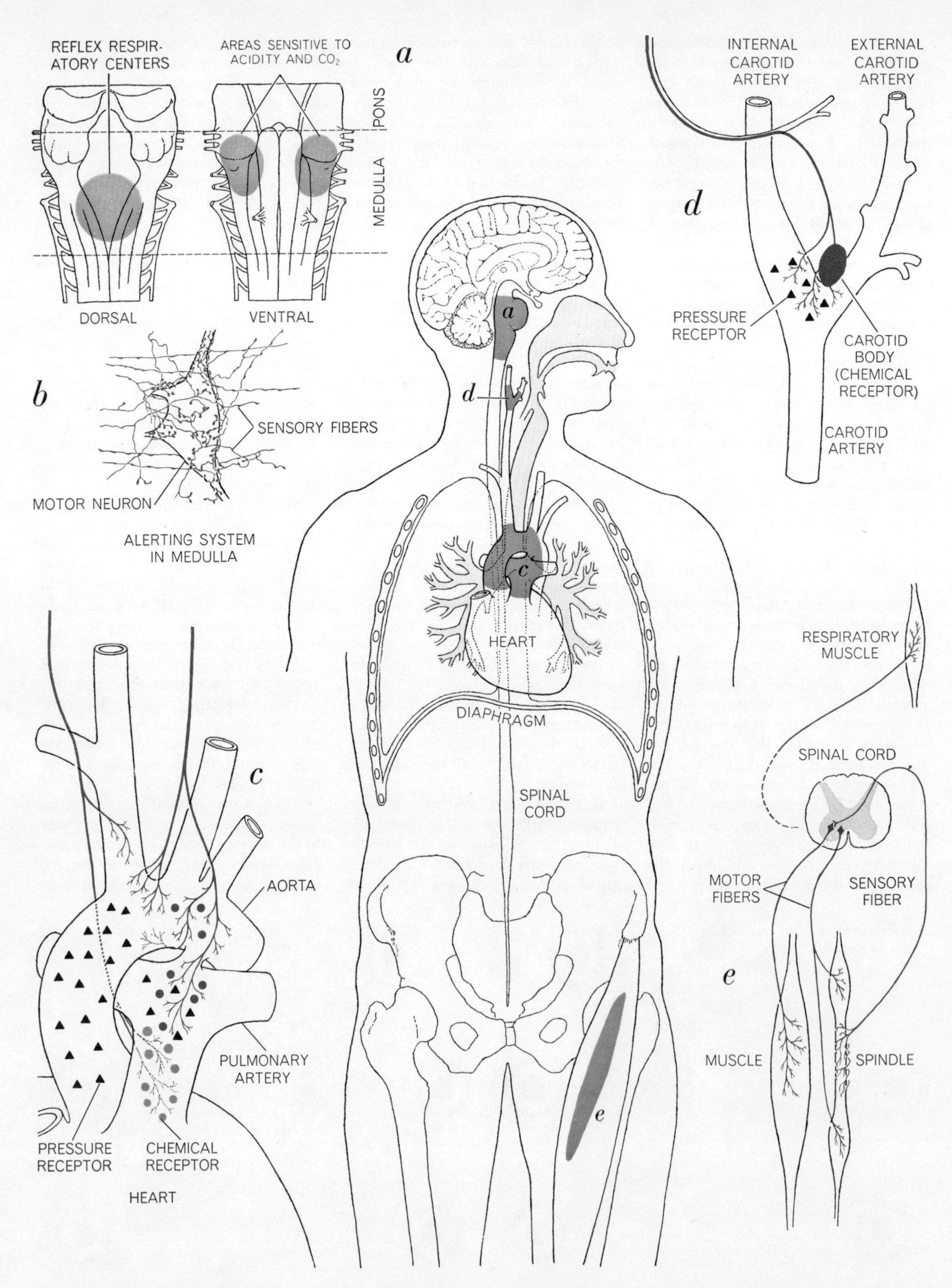

REGULATION OF BREATHING is controlled not only by the lower brain but also by a variety of other receptor and reflex centers. Portions of the pons and medulla of the lower brain (*a*) react to increases in acidity and carbon dioxide pressure. Other sensitive cells are attached to the aorta and pulmonary artery (*c*) and the carotid artery (*d*). The spindle receptors of skeletal muscles (*e*) also affect breathing, as do such external factors as temperature and vibration and such psychological ones as anxiety and wakefulness.

air at rhythmic intervals to the nose or mouth or more directly to the trachea. In mouth-to-mouth resuscitation the person who administers the air conveys it to the person who receives it by contracting his thorax and pushing air into the other person's lungs. Frogs can push air into their lungs by first filling their cheeks and then contracting their cheek muscles; some people whose respiratory muscles are paralyzed but whose head and neck muscles are not can learn to do this "frog breathing."

Still another way to increase the amount of oxygen available to human tissues is to place a hospital patient in an oxygen tent, in which the concentration of oxygen is increased, or to supply oxygen to him in a hyperbaric chamber, in which the total gas pressure is increased to two or three atmospheres. The extra oxygen that is taken up by the blood under these circumstances may be of help if the patient's clinical problem requires for its correction only more oxygen in the blood and tissues; such is the case, for example, when a patient has an infection caused by anaerobic bacteria. But supplying oxygen in greater than normal amounts does not increase ventilation and therefore cannot eliminate more carbon dioxide, nor does it increase the amount of blood being circulated. In the first instance, help in eliminating carbon dioxide is of prime importance in cases of pulmonary or respiratory disease; in the second, the tissues of patients with circulatory problems need not only more oxygen but also the added glucose, amino acids, lipids, white blood cells, blood platelets, proteins and hormones that can only be obtained from adequate blood flow.

The Measurement of Lung Function

Before the 1950's only a few tests of specific lung function had been devised. These included measurements of the amount of air in the lungs at the end of a full inhalation, the amount in the lungs at the end of a full exhalation, and the maximum amount exhaled after a full inhalation; these three volumes were known respectively as the total lung capacity, the residual volume and the vital capacity. In addition, some measurements had been made of the way in which gas was distributed throughout the tracheobronchial tree during inhalation by means of observing the dilution of a tracer gas such as helium or the exhalation of nitrogen following the inhalation of pure oxygen. Since the 1950's, however, pulmonary physiologists have developed a number of new instruments and test techniques with which to measure objectively, rapidly and accurately not only lung volume and the distribution and diffusion of gases but also the pulmonary circulation and the physical properties of the lung and its connecting air paths.

One such new instrument is the plethysmograph (also known as the body box), which can measure the volume of gas in the lungs and thorax, the resistance to air flow in the bronchial tree and even the blood flow in the pulmonary capillaries. For the first measurement the subject sits in the box (which is airtight and about the size of a telephone booth) and breathes the supply of air in the box through a mouthpiece fitted with a shutter and a pressure gauge. To measure the volume of gas in the subject's lungs at any moment an observer outside the box triggers a circuit that closes the shutter in the mouthpiece and then records the air pressure both in the subject's lungs and in the body box as the subject attempts to inhale; Boyle's law yields a precise measurement of the volume of gas in the subject's lungs.

In order to measure the blood flow in the pulmonary capillaries the subject in the body box is provided with a bag that contains a mixture of 80 percent nitrous oxide and 20 percent oxygen. At a signal the subject inhales a single breath of this mixture and holds the breath for a few seconds. Nitrous oxide dissolves readily in the blood; as its molecules diffuse from the alveoli to enter the blood flowing through the pulmonary capillaries, the total number of gas molecules in the alveoli obviously decreases. But the nitrous oxide that dissolves in the blood does not increase the volume of the blood; therefore the total gas pressure must decrease as the nitrous oxide molecules are subtracted. Knowing the total volume of gas in the lungs, the volume of nitrous oxide and the solubility of nitrous oxide in the blood, one can calculate the flow of blood through the pulmonary capillaries instant by instant. These calculations can be used both to measure the amount of blood pumped by the heart—and thus to arrive at an index of cardiac performance—and to detect unusual resistance to blood flow through the pulmonary capillaries.

To measure resistance to air flow in the bronchial tree the subject in the body box is instructed to breathe the air about him without any interruption while the observer continuously records changes in pressure in the box. One would expect that in a closed system the mere movement of 500 cubic centimeters of gas from the supply in the box into the lungs of a subject also in the box would not bring about any overall change in pressure. In actuality the pressure in the box increases. The reason is that gas can flow only from a region of higher pressure to a region of lower pressure. The subject cannot inhale unless the pressure of the gas in his lungs is lower than the pressure of the gas in the box; therefore the molecules of gas in his lungs during inhalation must occupy a greater volume than they did in the box before inhalation. The expansion of the subject's thorax, in turn, compresses—and thus is the equivalent of adding to—the rest of the gas in the body box; the effect is reversed as the thorax contracts during exhalation. An appropriately calibrated record of these pressure changes makes it possible for the pressure of the gas in the subject's alveoli to be calculated at any moment in the respiratory cycle. This test can be used to detect the increased resistance to air flow that arises in patients with bronchial asthma, for example, and to evaluate the effectiveness of antiasthmatic drugs.

Another useful instrument for the study of lung function is the nitrogen meter, developed during World War II by John C. Lilly at the University of Pennsylvania in order to detect leaks in or around aviators' oxygen masks. The instrument operates as an emission spectroscope, continuously sampling, analyzing and recording the concentration of nitrogen in a mixture of gases; its lag is less than a tenth of a second. Pulmonary physiologists use the nitrogen meter to detect uneven distribution of air within the lung. Assume that when one breathes ordinary air, the lungs contain 2,000 cubic centimeters of gas, 80 percent of which is nitrogen. If one next inhales 2,000 cubic centimeters of pure oxygen, and if this oxygen is distributed uniformly to the millions of alveoli, each alveolus should now contain a gas that is only 40 percent nitrogen instead of the former 80 percent. If, however, the 2,000 cubic centimeters of oxygen are not evenly distributed, some alveoli will receive less than their share, others will receive more, and the composition of the alveolar gas at the end of the oxygen inhalation will be decidedly nonuniform. In the alveoli that receive the most oxygen the proportion of nitrogen may be reduced to 30 percent; in those that receive little oxygen the proportion of nitrogen may remain as high as 75 percent.

It is impossible to put sampling

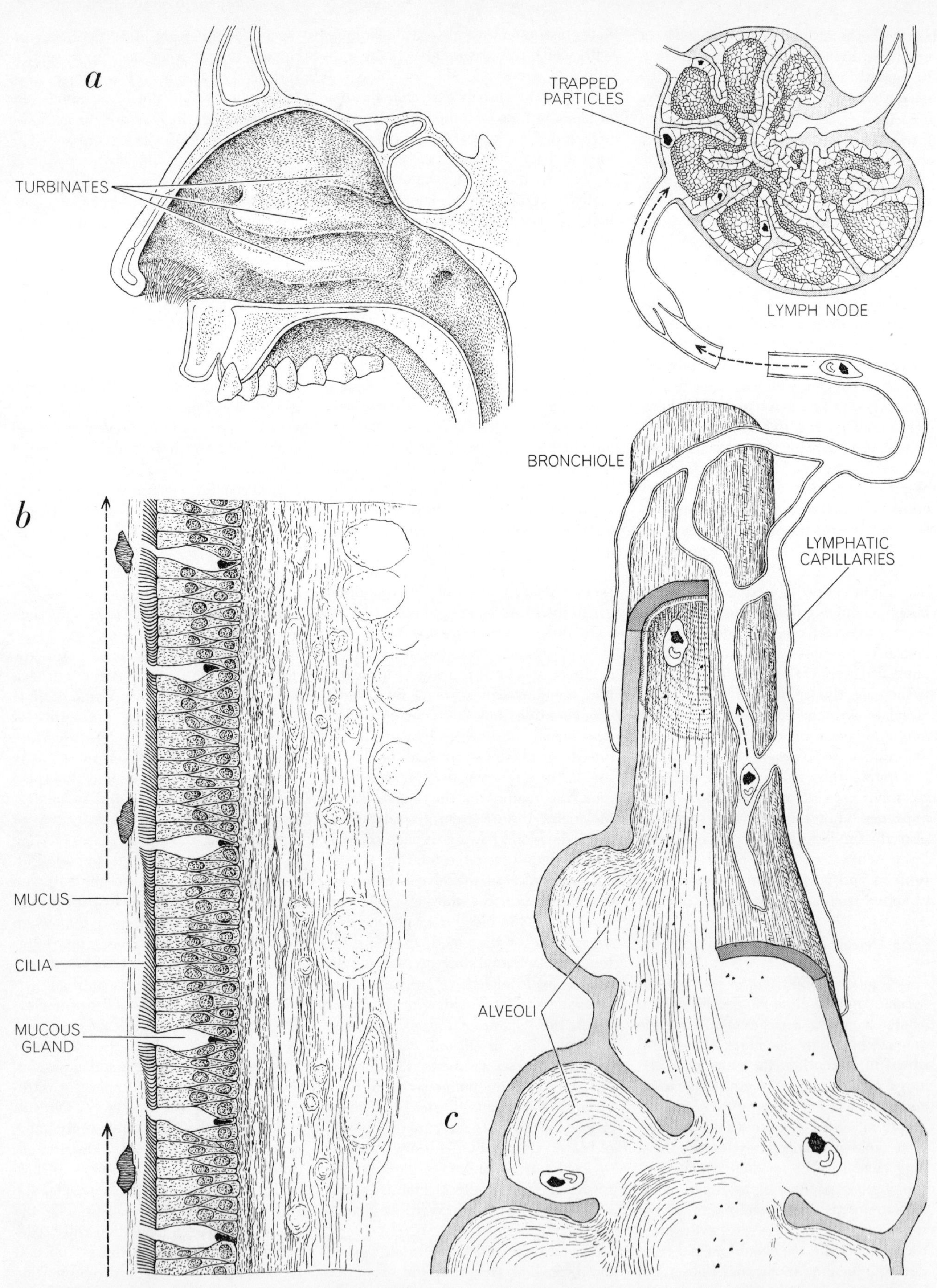

BAFFLE SYSTEM protects the lung's interior from intrusion by particles of foreign matter. Hairs in the nose and the convolutions of the turbinate bones (*a*) entrap most particles larger than 10 microns in diameter. Particles of from two to 10 microns in diameter usually settle on the walls of the trachea, the bronchi or the bronchioles, where the escalator action of mucus-covered cilia (*b*) carries them up to the pharynx for expulsion. Particles smaller than two microns in diameter reach the lung's air sacs (*c*). Some are engulfed by scavenger cells; others are carried to the nearest lymph node. Any that remain may cause fibrous tissue to form.

needles into thousands of alveoli in order to determine what mixture of gas each of them contains, but the nitrogen meter easily samples and analyzes the gas leaving the alveoli as the subject exhales. The first part of the exhaled gas comes from the outermost part of the tracheobronchial tree; it contains pure oxygen that has not traveled far enough down the conducting air path to mix with any alveolar gas. The second part of the exhalation shows a rapidly rising concentration of nitrogen; it represents alveolar gas that has washed some pure oxygen out of the conducting air path during exhalation and has mixed with it in the process. Once the conducting air path has been washed clear of oxygen, the remainder of the exhalation will be entirely alveolar gas. Analysis of the amount of nitrogen in the third part of the exhalation quickly shows whether or not the oxygen is distributed uniformly. If the distribution is uniform, the nitrogen-meter record for this part of the exhalation will be a horizontal line: from beginning to end the alveolar gas will be 40 percent nitrogen. If, on the other hand, the distribution is uneven, the nitrogen-meter record for the third part of the exhalation will rise continuously because the first part of the alveolar gas will come from well-ventilated areas of the lung and the last from poorly ventilated ones. In less than a minute the nitrogen meter can separate individuals with uneven ventilation from individuals whose ventilation is normal.

Having discovered by means of the nitrogen meter that a subject suffers from uneven ventilation somewhere in his lungs, the pulmonary physiologist can now use radioactive gases to determine exactly here the unevenness lies. The subject inhales a small amount of a relatively insoluble radioactive gas such as xenon 133 and holds his breath. A battery of three radiation counters on each side of the thorax measures the amount of radioactivity in the alveolar gas contained in the upper, middle and lower portions of each lung. Well-ventilated lung areas show a high level of radioactivity; poorly ventilated areas, a low level.

Radiation-counter readings can also be used to measure the uniformity of blood flow through the pulmonary capillaries. In order to do this the radioactive xenon is dissolved in a saline solution and the solution is administered intravenously. As it flows through the pulmonary capillaries the xenon comes out of solution and enters the alveolar gas. A high local concentration of xenon is an indication of a good flow of blood in that area; a low concentration indicates the contrary.

Another use of radioactive tracers to check on blood circulation involves the deliberate clogging of some fine pulmonary blood vessels. Radioactive albumin is treated so that it forms clumps that are about 30 microns in diameter—a size somewhat larger than the pulmonary capillaries or the vessels that lead into them. When the clumps are administered intravenously, they cannot enter blood vessels that are obstructed by disease; instead they collect in the parts of the lung with good circulation, where they block some of the fine vessels for a few hours. The whole thorax can now be scanned for radioactivity; in 10 or 20 minutes the activity of the albumin produces a clear image in which the regions of the lung with good pulmonary blood flow are clearly delineated.

Still another test, which measures the rate of gas exchange across the alveolar membrane, is made possible by the fact that hemoglobin has an extraordinary capacity for combining with carbon monoxide. The subject inhales a very low concentration of this potentially toxic gas. The carbon monoxide molecules diffuse across the capillary membranes and combine with the hemoglobin in the red blood cells. Assuming a normal amount of blood in the pulmonary capillaries, the rate at which the carbon monoxide disappears from the alveolar gas is directly proportional to its rate of diffusion. Unlike the somewhat similar test involving nitrous oxide, the carbon monoxide test measures only the rate of gas diffusion, not the rate of capillary blood flow. The affinity between the gas and the hemoglobin is so great that, even if the circulation of the blood were briefly halted, the stagnant red blood cells could still absorb carbon monoxide. A slow rate of carbon monoxide diffusion therefore indicates that the alveolar membranes have become thickened or that some abnormal fluid or tissue is separating many of the alveoli from the pulmonary capillaries.

The Regulation of Ventilation

The blood and air pumps that feed the lung's gas-exchange apparatus must be able to vary their performance to suit environments that range from sea level to high altitudes and activities that range from complete rest to violent exercise. Whatever the circumstances, exactly the right amount of oxygenated blood must be provided to meet the body's needs; to achieve this result responsive decision centers in the body, controlling both respiration and circulation, must not only be supplied with the necessary information but also possess the capacity to enforce decisions.

The first of these respiratory control centers was discovered by César Legallois of France in 1811. He found that if the cerebrum, the cerebellum and part of the upper brainstem were removed from a rabbit, the animal's breathing remained rhythmic, but that if a small region of the lower medulla was damaged or removed, breathing ceased. In the century and a half since Legallois' time physiologists have continued to accord this region of the brain —a group of interconnected nerve cells in the lower medulla—the paramount role in the control of respiration. This is not, however, the only region of the brain concerned with the regulation of breathing; there are chemically sensitive regions near the lateral surfaces of the upper medulla that call for an increase in ventilation when their carbon dioxide pressure or their acidity increases. Some other parts of the medulla, the cerebral cortex and the part of the brain called the pons can also influence respiration.

In addition to these areas in the brain a variety of respiratory receptors, interconnecting links, pathways and reflexes are found elsewhere in the body. Chemically sensitive cells in the regions of the carotid artery and the aorta initiate reflexes that increase respiration when their oxygen supply is not sufficient to maintain their metabolic needs or when the local carbon dioxide pressure or acidity increases. Stretch-sensitive receptors in the major arteries act through reflexes to increase or decrease respiration in response to low or high arterial blood pressure. Other receptors in the circulatory system, sensitive both to chemical stimuli and to mechanical deformation, can set off reflexes that slow or stop breathing. Respiration can also be regulated by the degree of inflation or deflation of the lungs, by the individual's state of wakefulness or awareness, by the concentration of certain hormones in the blood and by the discharge of the special sensory receptors known as spindles in skeletal muscles (including the respiratory muscles themselves).

We still do not know how some or all of these central and peripheral components interact to achieve the most important (and the most frequent) change in ventilation: the change that takes place when the body's metabolic activity increases. We know that during exercise both the body's oxygen consumption and its carbon dioxide production increase, and that so does the rate of ventilation. It is therefore logical to assume that ventilation is regulated by receptors somewhere in the body that are sensitive to oxygen or to carbon dioxide or to both. A puzzling fact remains: mild and even moderate exercise simply does not decrease the amount of oxygen or increase the amount of carbon dioxide in the arterial blood, yet the respiration rate rises. What causes this increase, which is enough to satisfy both the ordinary needs of the body and the extraordinary needs of exercising muscles? No one knows.

The Upper Respiratory Tract

In the course of taking from four to 10 million breaths a year each individual draws into the alveoli of his lungs air that may be hot or cold, dry or moist, possibly clean but more probably dirty. Each liter of urban air, for example, contains several million particles of foreign matter; in a day a city-dwelling adult inhales perhaps 20 billion such particles. What protects the lungs and the air ducts leading to them from air with undesirable physical characteristics or chemical composition? Sensory receptors in the air ducts and the lungs can initiate protective reflexes when they are suitably stimulated; specialized cells can also engulf foreign particles that have penetrated far into the lung. The main task of protecting the lungs, however, is left to the upper respiratory tract.

The nose, the mouth, the oropharynx, the nasopharynx, the larynx, the trachea and those bronchi that are outside the lung itself together constitute the upper respiratory tract. Although the obvious function of this series of passages is to conduct air to and from the lung, the tract is also a sophisticated air conditioner and filter. It contains built-in warning devices to signal the presence of most pollutants and is carpeted with a remarkable escalator membrane that moves foreign bodies upward and out of the tract at the rate of nearly an inch a minute. Within quite broad limits the initial state of the air a man breathes is of little consequence; thanks to the mediation of the upper respiratory tract the air will be warm, moist and almost free of particles by the time it reaches the alveoli.

The first role in the conditioning process is played by the mucous membrane of the nose, the mouth and the pharynx; this large surface has a rich blood supply that warms cold air, cools hot air and otherwise protects the alveoli under a wide range of conditions. Experimental animals have been exposed to air heated to 500 degrees centigrade and air cooled to −100 degrees C.; in both instances the trip through the respiratory tract had cooled or warmed the air almost to body temperature by the time it had reached the lower trachea.

The upper respiratory tract also filters air. The hairs in the nose block the passage of large particles; beyond these hairs the involuted contours of the nasal turbinate bones force the air to move in numerous narrow streams, so that suspended particles tend to approach either the dividing septum of the nose or the moist mucous membranes of the turbinates. Here many particles either impinge directly on the mucous membranes or settle there in response to gravity.

The filter system of the nose almost completely removes from the air particles with a diameter larger than 10 microns. Particles ranging in diameter from two to 10 microns usually settle on the walls of the trachea, the bronchi and the bronchioles. Only particles between .3 micron and two microns in diameter are likely to reach the alveolar ducts and the alveoli. Particles smaller than .3 micron, if they are not taken up by the blood, are likely to remain in suspension as aerosols and so are washed out of the lungs along with the exhaled air.

Foreign bodies that settle on the walls of the nose, the pharynx, the trachea, the bronchi and the bronchioles may be expelled by the explosive blast of air that is generated by a sneeze or a cough, but more often they are removed by the action of the cilia. These are very primitive structures that are found in many forms of life, from one-celled organisms to man. Resembling hairs, they are powered by a contractile mechanism; in action each cilium makes a fast, forceful forward stroke that is followed by a slower, less forceful return stroke that brings the cilium into starting position again. The strokes of a row of cilia are precisely coordinated so that the hairs move together as a wave. The cilia of the human respiratory tract do not beat in the open air; they operate within a protective sheet of the mucus that is secreted by glands in the trachea and the bronchi. The effect of their wavelike motion is to move the entire mucus sheet—and anything trapped on it—up the respiratory tract to the pharynx, where it can be swallowed or spat out.

The ciliary escalator is in constant operation; it provides a quiet, unobtrusive, round-the-clock mechanism for the removal of foreign matter from the upper respiratory tract. The speed of this upward movement depends on the length of the cilia and the frequency of their motion. Calculations show that a cilium that is 10 microns long and that beats 20 times per second can move the mucus sheet 320 microns per second, or 19.2 millimeters per minute. Speeds of 16 millimeters per minute have actually been observed in experiments.

In spite of all such preventive measures some inhaled particles—particularly those suspended in fluid droplets—manage to pass through the alveolar ducts and reach the alveoli. How do these deeper surfaces, which have no cilia or mucous glands, cleanse themselves? The amoeba-like lymphocytes of the bloodstream and their larger relatives the macrophages engulf and digest some particles of foreign matter. They can also surround the particles in the air ducts and then ride the mucus escalator up to the nasopharynx. Other particles may pass into lymphatic vessels and come to rest in the nearest lymph nodes. Some remain permanently attached to the lung tissue, as the darkened lungs of coal miners demonstrate. Many such intrusions are essentially harmless, but some, for example particles of silica, can result in the formation of tough fibrous tissue that causes serious pulmonary disease.

The filtration mechanism of the upper respiratory tract can thus be credited with several important achievements. It is responsible for the interception and removal of foreign particles. It can remove bacteria suspended in the air and also dispose of bacteria, viruses and even irritant or carcinogenic gases when they are adsorbed onto larger particles. Unless the filter system is overloaded, it keeps the alveoli practically sterile. This, however, is not the only protection the lungs possess. Among the reflex responses to chemical or mechanical irritation of the nose are cessation of breathing, closure of the larynx, constriction of the bronchi and even slowing of the heart. These responses are aimed at preventing potentially harmful gases from reaching the alveoli and, through the alveoli, the pulmonary circulation.

In many animals, for instance, the act of swallowing results in reflex closure of the glottis and the inhibition of respiration. Because the pharynx is a passageway both for air and for food and water, this reflex prevents food or water from entering the respiratory passages during the journey from the mouth to the esophagus. Because the reflex does not operate during unconsciousness it is dangerous to try to arouse an unconscious person by pouring liquids such as alcohol into his mouth.

When specific chemical irritants penetrate beyond the larynx, the reflex response is usually a cough combined with bronchial constriction. Like the swallowing reflex, the cough reflex is depressed or absent during unconsciousness. It also is less active in older people; this is why they are more likely to draw foreign bodies into their lungs.

Bronchial constriction is a response to irritation of the air paths that is less obvious than a cough. When the concentration of dust, smoke or irritant gas is too low to elicit the cough reflex, this constrictive increase in air-path resistance is frequently evident. Smoking a cigarette, for example, induces an immediate twofold or threefold rise in air-path resistance that continues for 10 to 30 minutes. The inhalation of cigarette smoke produces the same effect in smokers and nonsmokers alike. It does not cause shortness of breath, as asthma does; the air-path resistance must increase fourfold or fivefold to produce that effect. Nor has the reflex anything to do with nicotine; no increase in air-path resistance is caused by the inhalation of nicotine aerosols, whereas exactly the same degree of resistance is induced by smoking cigarettes with a normal (2 percent) or a minimal (.5 percent) nicotine content. The reflex is evidently triggered by the settling of particles less than a micron in diameter on the sensory receptors in the air path.

Other air pollutants—irritant gases, vapors, fumes, smokes, aerosols or small particles—may give rise to a similar bronchial constriction. It is one of the ironies of man's urban way of life that exposure to the pollutants that produce severe and repeated bronchial constriction results in excessive secretion of mucus, a reduction in ciliary activity, obstruction of the fine air paths and finally cell damage. These circumstances enable bacteria to penetrate to the alveoli and remain there long enough to initiate infectious lung disease. They are also probably a factor in the development of such tracheobronchial diseases as chronic bronchitis and lung cancer. Thus man's advances in material culture increasingly threaten the air pump that helped to make his evolutionary success possible.

8 How Birds Breathe

by Knut Schmidt-Nielsen
December 1971

The avian respiratory system is different from the mammalian one. The lungs do not simply take air in and then expel it; the air also flows through a series of large sacs and even hollow bones.

A bird in flight expends more energy, weight for weight, than a mammal walking or running on the ground. Moreover, the bird's respiratory system can deliver enough oxygen for the animal to fly at altitudes where a mammal can barely function. How do the birds do it? It turns out that the avian respiratory system is quite different from the mammalian one. The remarkable anatomical details of the avian system have been elucidated over a period of three centuries, but precisely how the system operates has been worked out only recently.

One of the first clues to the distinctive nature of the avian respiratory system was the discovery that a bird with a blocked windpipe can still breathe, provided that a connection has been made between one of its bones and the outside air. This phenomenon was demonstrated in 1758 by John Hunter, a fellow of the Royal Society, who wrote: "I next cut the wing through the *os humeri* [the wing bone] in another fowl, and tying up the trachea, as in the cock, found that the air passed to and from the lungs by the canal in this bone. The same experiment was made with the *os femoris* [the leg bone] of a young hawk, and was attended with a similar result."

The bones of birds contain air, not marrow. This is true not only of the larger bones but also often of the smaller ones and of the skull bones, particularly in birds that are good fliers. As Hunter's experiments showed, the air spaces are connected to the respiratory system.

Like mammals, birds have two lungs. They are connected to the outside by the trachea, much as in mammals, but in addition they are connected to several large, thin-walled air sacs that fill much of the chest and the abdominal cavity [*see top illustration on next page*]. The sacs are connected to the air spaces in the bones. The continuation of the air passages into large, membranous air sacs was discovered in 1653 by William Harvey, the British anatomist who became famous for discovering the circulation of blood in mammals.

The presence in birds of these large air spaces, much larger in volume than the lungs, has given rise to considerable speculation. It has often been said that the air sacs make a bird lighter and are therefore an adaptation to flight. Certainly a bone filled with air weighs less than a bone filled with marrow. The large air sacs, however, do not in themselves make a bird any lighter. As a student I heard a professor of zoology assert that the sacs did make a bird lighter and therefore better suited to flight. Somewhat undiplomatically I suggested that if I were to take a poor flier such as a chicken and pump it up with a bicycle pump, the chicken would be neither any lighter nor better able to fly. The simple logic of the argument must have convinced the professor, because we did not hear any more about the function of the air sacs.

In order to understand the function of the sacs, how air flows in them, how oxygen is taken up by the blood and carbon dioxide is given off and so on it is necessary to consider the main structural features of the system. In this context it is helpful to compare birds with mammals. Birds as a group are much more alike than mammals. In size they range from the hummingbird weighing some three grams to the ostrich weighing about 100 kilograms. In terms of weight the largest bird is roughly 30,000 times bigger than the smallest one.

All birds have two legs and two wings, although the ostrich cannot use its wings for flying and penguins have flipper-like wings modified for swimming. All birds have feathers, and all birds have a similar respiratory system, with lungs, air sacs and pneumatized bones. (Even the ostrich has the larger leg bones filled with air. Air sacs and pneumatized bones are therefore not restricted to birds that can fly.)

Mammals, on the other hand, range in size from the shrew, which weighs about as much as a hummingbird, to the 100-ton blue whale. The largest mammal is therefore 30 million times bigger than the smallest one. Mammals can be four-legged, two-legged or no-legged (whales). They can even have wings, as bats do. Most mammals have fur, but many do not.

It was once argued that birds needed a respiratory system particularly adapted for flight because of the high requirement for energy and oxygen during flight. At rest birds and mammals of similar size have similar rates of oxygen consumption, although in both birds and mammals the oxygen consumption per unit of body weight increases with decreasing size. In recent years the oxygen consumption of birds in flight has been determined in wind-tunnel experiments [see "The Energetics of Bird Flight," by Vance A. Tucker; SCIENTIFIC AMERICAN Offprint 1141]. The results show that the oxygen consumption in flight is some 10 to 15 times higher than it is in the bird at rest. This performance is not much different from that of a well-trained human athlete, who can sustain a similar increase in oxygen consumption.

Small mammals such as rats or mice, however, seem unable to increase their oxygen consumption as much as tenfold. Since birds and mammals of the same body size show similar oxygen consumption when they are at rest, is it the special design of the bird's respiratory system that allows the high rates of oxygen consumption during flight?

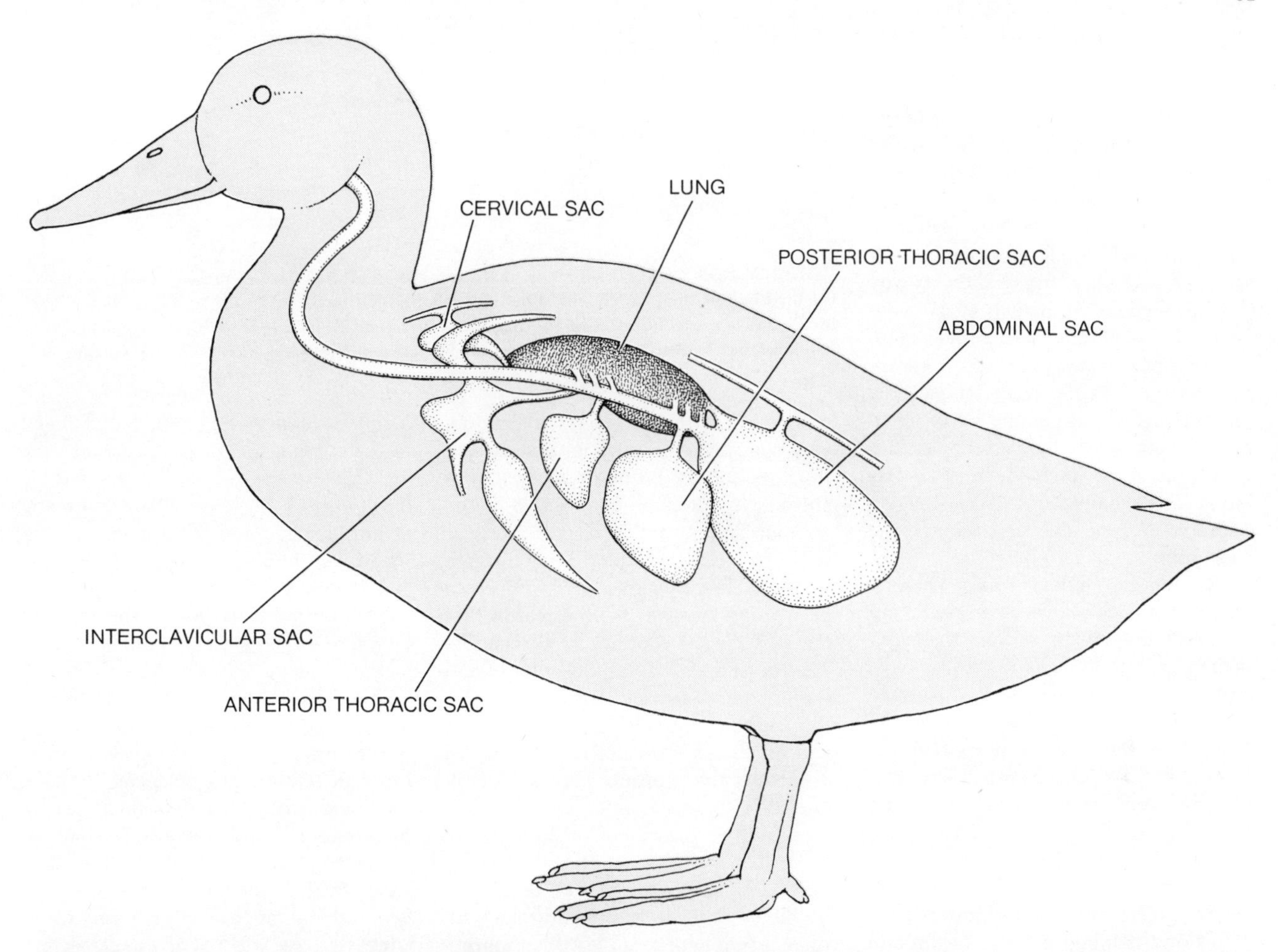

AVIAN RESPIRATORY SYSTEM, here represented by the system of a mallard duck, has as a distinctive feature a number of air sacs that are connected to the bronchial passages and the lungs. Most of the air inhaled on a given breath goes directly into the posterior sacs. As the respiratory cycle continues the air passes through the lungs and into the anterior sacs, from which it is exhaled to the outside in the next cycle. The mechanism provides a continuous flow of air through the lung and also, by means of the holding-chamber function of the anterior sacs, keeps the carbon dioxide content of the air at an appropriate physiological level.

The best argument against considering the unusual features of the avian respiratory system as being necessary for flight is provided by bats. They have typical mammalian lungs and do not have air sacs or pneumatized bones, and yet they are excellent fliers. Moreover, it has recently been shown by Steven Thomas and Roderick A. Suthers at Indiana University that bats in flight consume oxygen at a rate comparable to the rate in flying birds. Clearly an avian respiratory system is not necessary for a high rate of oxygen uptake or for flight.

One highly significant phenomenon is that many birds can fly at high altitudes where mammals suffer seriously from lack of oxygen. This fact points to what is perhaps the most important difference between the avian system and the mammalian one.

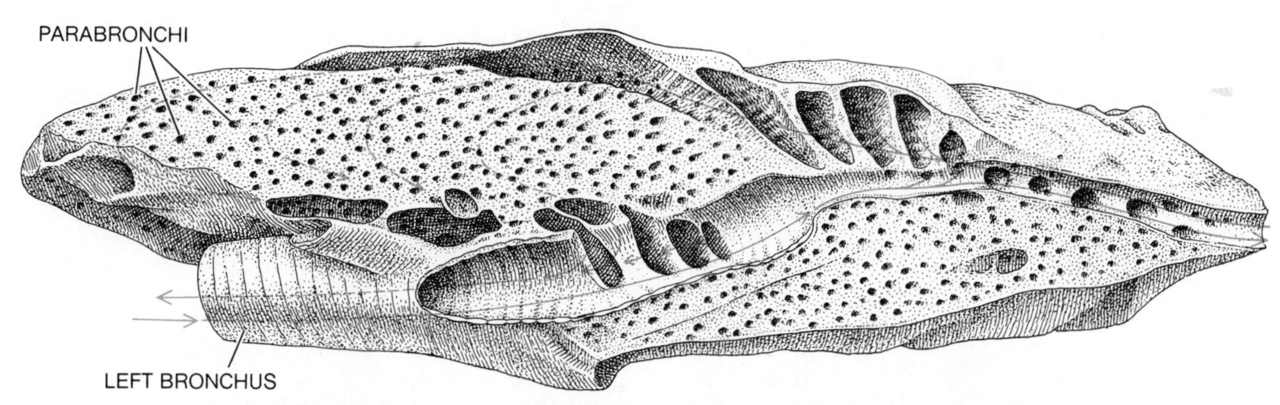

AVIAN LUNG is shown in longitudinal section from the left side of the bird. Since the bird has two lungs, one sees here half of the lung system. The orientation is as it would be with the bird's head at left. Air enters the bronchus and for the most part passes through to the posterior sacs. On its return through the lungs, assisted by a bellows-like action of the posterior sacs, it flows into the many parabronchi, where gases are exchanged with the blood. The flow has some similarity to the flow of water through a sponge.

Let us look more closely at the main features of the avian respiratory system. The trachea of a bird branches into two main bronchi, each leading to one of the lungs. So far the system is similar to the mammalian system. In birds, however, each main bronchus continues through the lung and connects with the most posterior (and usually the largest) air sacs: the abdominal air sacs. On its way through the lung the main bronchus connects to the anterior air sacs and also to the lung [*see bottom illustration on opposite page*]. In the posterior part the main bronchus has another set of openings that connect to the posterior air sacs as well as to the lung. The air sacs also have direct connections to the lung in addition to the connection through the bronchus.

The lung itself has a most peculiar characteristic: it allows air to pass completely through it. In contrast, the mammalian lung terminates in small saclike structures, the alveoli; air can flow only in and out of it. The bird lung is perforated by the finest branches of the bronchial system, which are called parabronchi. Air flows through the lung somewhat the way water can flow through a sponge.

This feature of the bird lung has led to the suggestion that the air sacs act as bellows helping to push air through the lung, which thus could be supplied with air more effectively than the mammalian lung is. Before accepting this hypothesis one must be sure that the air sacs do not have a lunglike function, that is, that they do not serve as places where oxygen is taken up by the blood. Since the air sacs are thin-walled, they could perhaps be important in the exchange of gases between the air and the blood.

The fact is that the sacs are poorly supplied with blood. Moreover, they have smooth walls, which do not provide the immensely enlarged surface that the finely subdivided lung has. A crucial experiment was performed some 80 years ago by the French investigator J. M. Soum, who admitted carbon monoxide into the air sacs of birds in which he had blocked the connections to the rest of the respiratory system. If the air sacs had played any major role in gas exchange, the birds would of course have been rapidly poisoned by the carbon monoxide. They remained completely unaffected. We can therefore conclude that the air sacs have no direct function in gas exchange. Since the volume of the sacs changes considerably during the respiratory cycle, one can accept the hypothesis that they serve as a bellows.

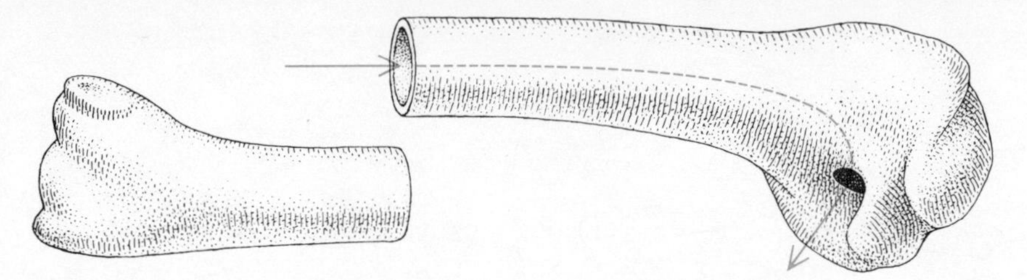

HOLLOW BONE filled with air is characteristic of bird skeleton. Such a structure makes the bird lighter than a bird would be with mammalian bones and so is an aid to flying. The bird's bones are connected to the respiratory system. A bird with a blocked trachea can still breathe if a connection has been made between the wing bone and the outside air.

A suggestion made long ago is that the large sacs could be filled with fresh air and, by alternate contraction of the anterior and posterior sets of sacs, air could be passed back and forth through the lung. The hypothesis has proved to be wrong, however, for the reason that the sacs do not contract in alternation. The pressure changes in the anterior and posterior sacs are similar: on inhalation the pressure drops in both sets of sacs, and all the sacs are filled with air; on exhalation the pressure increases simultaneously in the anterior and posterior sacs, and air passes out of both sets of sacs.

It has even been suggested that birds, by filling their air sacs, could take with them a supply of air to last them during a flight. This adventurous suggestion was supported by the speculation that the chest of a flying bird is so rigidly constrained by muscular contractions that breathing is impossible. The reasoning disregards the most elementary considerations of the amount of oxygen needed for flight.

The question of how air flows in the avian lung can be studied in a number of ways. One useful approach is to introduce a foreign gas as a marker. The flow of the gas and its time of arrival at various points in the respiratory system yield much information. Another approach is to use small probes that are sensitive to airflow and to place them in various parts of the elaborate passageways. In this way the flow directions can be determined directly during the phases of the respiratory cycle. My colleagues and I at Duke University have used both of these approaches, and we believe we now know with reasonable certainty the main features of avian respiration.

The use of a tracer gas has been quite successful in clarifying the flow of air. Our first experiments were with ostriches, which have the advantage of rather slow respiration. An ostrich breathes about six times per minute, and changes in the composition of gas in its air sacs can therefore be followed rather easily. If an ostrich is given a single breath of pure oxygen and is then returned to breathing normal air, which has an oxygen content of 21 percent, an increased concentration of oxygen in the respiratory system will indicate how the single marked inhalation is distributed.

We used an oxygen electrode to follow changes in oxygen concentration. In the posterior air sacs we picked up a rapid increase in oxygen near the conclusion of the inhalation that carried pure oxygen. In other words, the marker gas flowed directly to the posterior sacs. In contrast, in the anterior sacs the oxygen did not appear until a full cycle later; the rise was noted as the second inhalation was ending [*see illustration on page 94*]. This finding must mean that the anterior air sacs do not receive inhaled air directly from the outside and that the marker gas that arrived on the second cycle or later must meanwhile have been in some other part of the respiratory system. We concluded that the posterior sacs are filled with air coming from the outside and that air entering the anterior sacs must come from elsewhere, presumably the lungs. Outside air thus enters the anterior sacs only indirectly, through other parts of the respiratory system, and it is delayed by at least one cycle.

It would be tempting to conclude from this experiment that the posterior sacs are well ventilated and that the anterior sacs do not receive much air but contain a rather inert and stagnant mass of air. The composition of the gas in the sacs might seem to support such a conclusion. The posterior air sacs usually contain about 3 or 4 percent carbon dioxide and the anterior sacs 6 or 7 percent, which is comparable to the carbon dioxide content of an air mass that is in equilibrium with venous blood. The conclusion would be wrong.

Whether or not an air sac is well ventilated can be ascertained by introducing a marker gas directly into the sac and determining how fast the marker disappears on being washed out by other air. In the ostrich we injected 100 milliliters of pure oxygen directly into an air sac and measured the time required to reduce by half the increase in oxygen concentration thereby achieved. We found that all the air sacs in the ostrich are highly ventilated and that they wash out rapidly.

The results showed that none of the air sacs contained a stagnant or relatively inert air mass. Since the anterior sacs have about the same washout time as the posterior sacs, they must be equally well ventilated. Why, then, since the renewal rate of air is high in the anterior sacs, do they contain a high concentration of carbon dioxide? This phenomenon can best be explained by postulating that the anterior sacs receive air that has passed through the lungs, where during passage it has exchanged gases with the blood, taking up carbon dioxide and delivering oxygen.

When we had arrived at this stage, it became essential for us to obtain unequivocal information about the flow of air in the bird lung. For this purpose W. L. Bretz in our laboratory designed and built a small probe that could record the direction of airflow at strategic points in the respiratory system of ducks. The information obtained in these experiments can best be summarized by going through the events of inhalation and then through the events of exhalation.

On inhalation air flows directly to the posterior sacs, which therefore initially receive first the air that remained in the trachea from the previous exhalation and then, immediately afterward, fresh outside air. The posterior sacs thus become filled with a mixture of exhalation air and outside air. Experiments with marker gas showed just this sequence; the marker arrived in the posterior sacs as the first inhalation was ending. The flow probe did not show any flow in the connections to the anterior sacs during inhalation, which was what we had expected from the fact that marker gas never arrived directly in the anterior sacs. Since the anterior sacs do expand during inhalation, the air that fills them can come only from the lung. Another finding is that air flows in the connection from the main bronchus to the posterior part of the lung, indicating that some of the inhaled air goes directly to the lung.

During exhalation the posterior sacs decrease in volume. Since the flow probe shows little or no flow in the main bronchus, the air must flow into the lung. The anterior sacs also decrease in volume. A probe placed in their connection to the main bronchus shows a high flow, consistent with direct emptying of these sacs to the outside.

The most interesting conclusion to be drawn from these patterns of flow is that air flows continuously in the same direction through the avian lung during both inhalation and exhalation. This suggestion is not new, but once we are certain that it is correct we can better examine its consequences. The air flowing through the lungs comes mostly from the posterior sacs, where the combination of dead-space air and outside air supplies a mixture that is high in oxygen but also contains a significant amount of carbon dioxide. Here we encounter one of the most elegant features of the system. If completely fresh outside air, which contains only .03 percent carbon dioxide, were passed through the lung, the blood would lose too much carbon dioxide, with serious consequences for the acid-base regulation of the bird's body. Another consequence of excessive loss of carbon dioxide arises from the fact that breathing is regulated primarily by the concentration of carbon dioxide in the blood. An increase in carbon dioxide stimulates breathing; a decrease causes breathing to slow down or even stop for a time.

Hence we see that the avian lung is continuously supplied with a mixture of air that is high in oxygen without being too low in carbon dioxide. The anterior sacs serve as holding chambers for the air coming from the lungs. This air is later discharged to the outside on exhalation, but enough of it remains in the trachea to ensure the right concentration of carbon dioxide in the posterior sacs after the next inhalation.

A few disturbing questions remain. One is why, since the pressure in the anterior air sacs falls during inhalation, air from the outside does not enter these sacs. The system has no valves that can open and close to help direct the flow. The answer is probably that the openings from the main bronchus have an aerodynamic shape that tends to lead the air past the openings. The avian respiratory tract is a low-pressure, high-velocity system in which gas flow may be governed by the principles of fluidics without the need for anatomical values [see "Fluid Control Devices," by Stanley W. Angrist; SCIENTIFIC AMERICAN, December, 1964].

Another conceptual difficulty is why air moves from the posterior sacs to the lungs during exhalation and from the lungs to the anterior sacs during inhalation. These movements require both suitable pressure gradients and a change in the volume of the lungs. It has been said that the bird lung changes little in volume because it is much firmer and less distensible than the mammalian lung. A bird's lung removed from the body retains its shape instead of collapsing to a small fraction of its normal volume as a mammal's lung does.

Another anatomical feature that has been misinterpreted is the bird's diaphragm. Birds have no muscular diaphragm, which is a most important feature in mammalian respiration. In its place they have a thin membrane of connective tissue. The membrane is con-

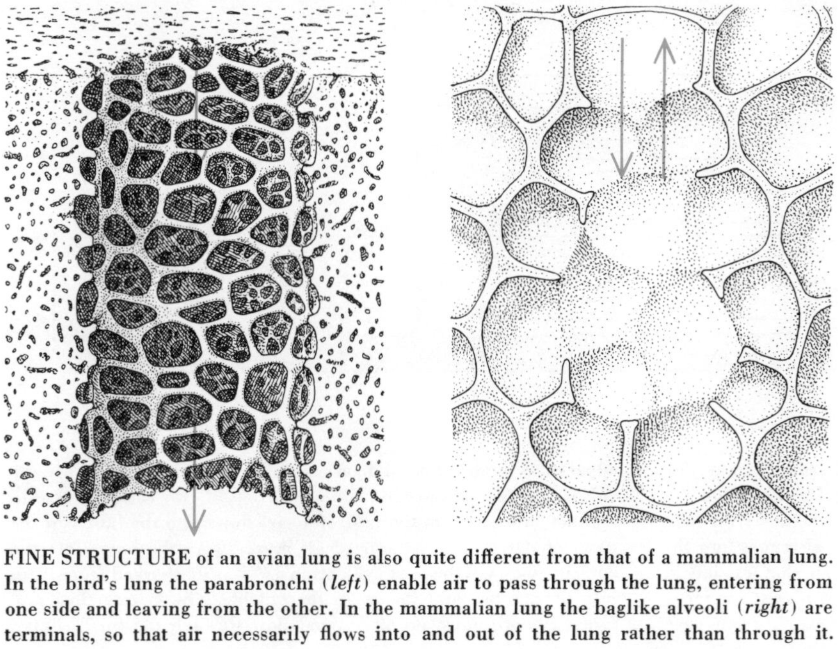

FINE STRUCTURE of an avian lung is also quite different from that of a mammalian lung. In the bird's lung the parabronchi (*left*) enable air to pass through the lung, entering from one side and leaving from the other. In the mammalian lung the baglike alveoli (*right*) are terminals, so that air necessarily flows into and out of the lung rather than through it.

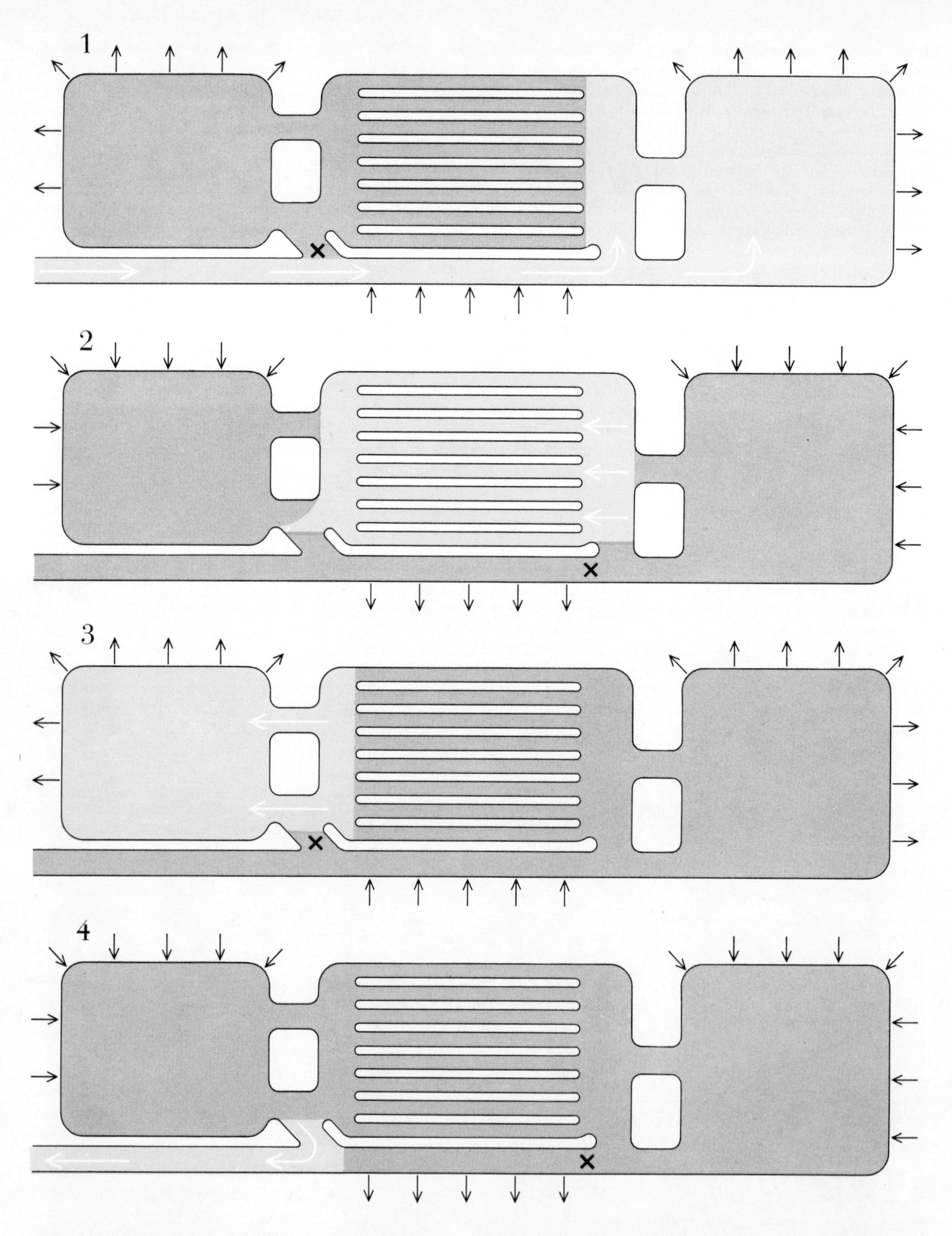

RESPIRATORY CYCLE in a bird is depicted schematically, following a single slug of air through two breaths. On inhalation (*1*) air flows through the bronchus and mainly into the posterior sacs, represented here by a single chamber. Some air also goes into the lung. The first air to reach the posterior sacs is air that was left in the trachea after the previous exhalation, so that it contains more carbon dioxide than fresh air does. The anterior sacs are bypassed, apparently under fluid-dynamical influences since there are no valves. The air sacs expand. On exhalation (*2*) the sacs decrease in volume, and air from the posterior sacs flows into the lung. On the next inhalation (*3*) the slug of air moves from the lungs into the anterior sacs. On the next exhalation (*4*) it is discharged from them into the trachea and thence to the outside. The system thus provides a continuous, unidirectional flow through the bird's lungs.

nected to muscles that are attached to the body wall. When the muscles contract, they flatten out the membranous diaphragm, thus pulling on the ventral surface of the lung in a manner that is mechanically similar to the pull of the mammalian diaphragm. The avian diaphragm, however, works on a cycle opposite to that of the mammalian diaphragm: it tends to make the lungs expand on exhalation and the volume of the lungs to diminish on inhalation.

This paradoxical cycle provides the necessary mechanism for the movement of air into the lungs. As the lungs expand on exhalation, air flows from the posterior sacs to fill the lungs. As the lungs diminish in volume on inhalation, air flows from the lungs to the anterior sacs.

Earlier in this article I remarked that the complex lung and air sac system of birds is not a prerequisite for flight, but I suggested that it confers a considerable advantage at high altitude. Man and other mammals begin to show marked symptoms of oxygen deficiency at an altitude of 3,000 to 4,000 meters (10,000 to 13,000 feet). A man moving to such an altitude from sea level finds it difficult to exert himself in physical work, although he gradually acclimatizes and is able to perform normally. At higher altitudes work and acclimatization become increasingly difficult; the limit for moderately active functioning of a man, even after long acclimatization, is about 6,000 meters.

Birds, in contrast, have been observed to move about freely and fly at altitudes above 6,000 meters. Airplanes have collided with flying birds as high as 7,000 meters. Birds might reach these altitudes by riding on strong upcurrents of wind, but this would not explain the fact that they fly actively and without apparent difficulty once they are there.

A few years ago Vance A. Tucker of Duke University simultaneously exposed house sparrows and mice to a simulated altitude of 6,100 meters, which represents somewhat less than half the atmospheric pressure at sea level and therefore less than half the partial pressure of oxygen at sea level. At this low level of oxygen the sparrows were still able to fly, but the mice were comatose. The blood of the sparrow does not have any higher affinity for oxygen than the blood of the mouse; otherwise the ability of the sparrows to take up oxygen at low pressure could be explained as a difference in the blood. What can explain the difference between birds and mammals under these conditions is the unidirectional flow of air in the bird's lungs.

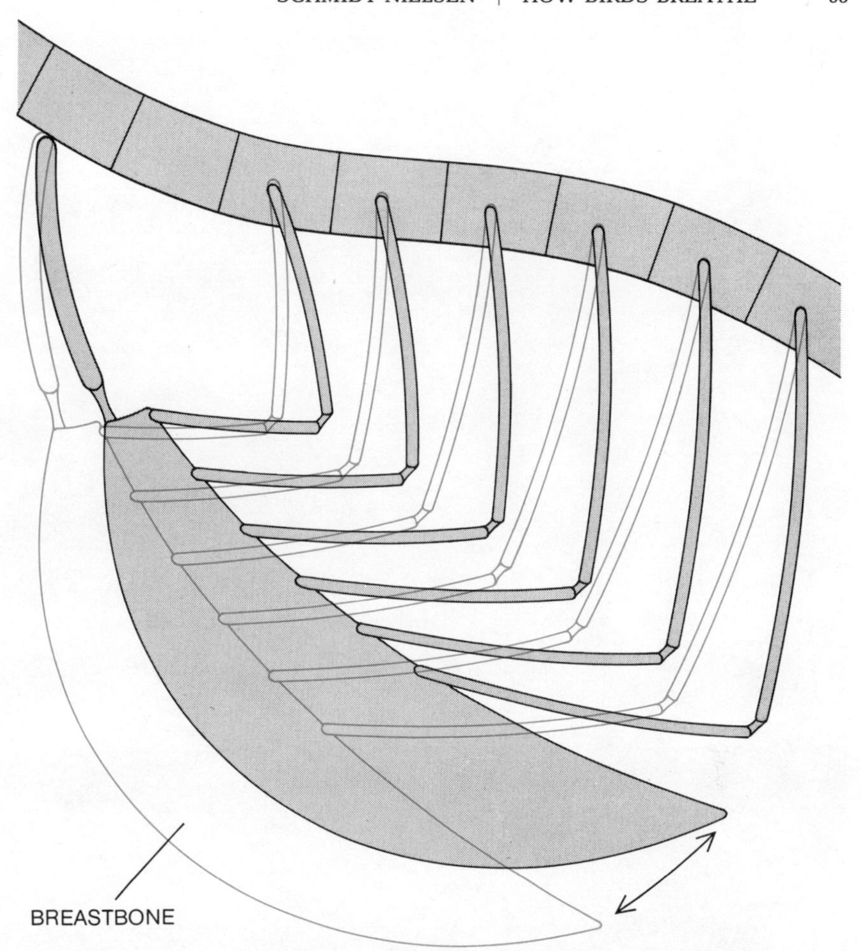

RIB STRUCTURE of a bird is related to respiration by being hinged in such a way that on inhalation the breastbone is lowered. The chest expands, as do the air sacs, but the lung diminishes in volume. On exhalation the process is reversed. Because the lungs expand on exhalation, air flows into them from the posterior sacs. Similarly, as the lungs decrease in volume on the next inhalation, air flows out of them and into the anterior-sac system.

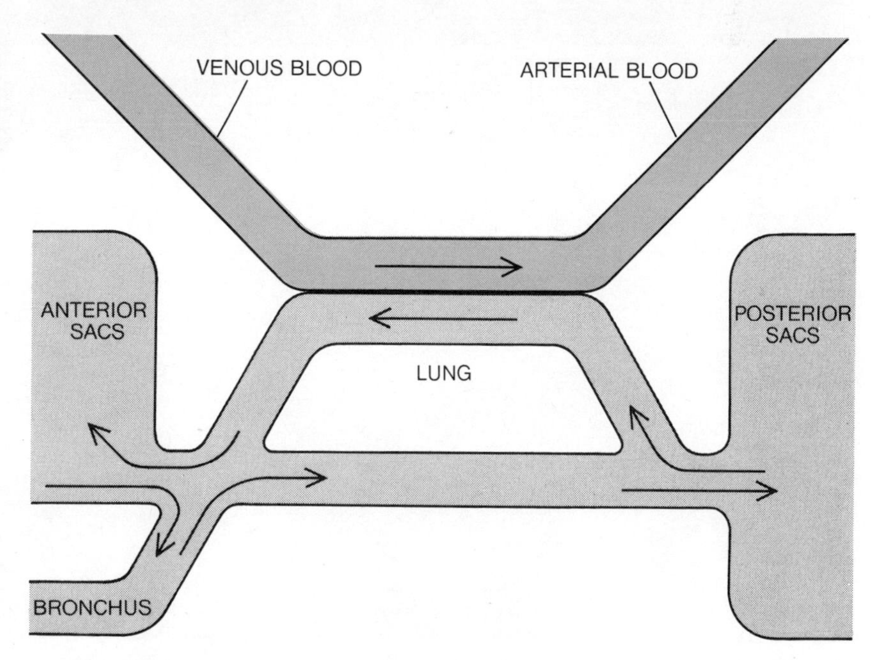

COUNTERCURRENT FLOW OF BLOOD AND AIR in the lung of the bird is the key to the bird's efficient extraction of oxygen and so to its ability to fly at high altitudes. Air flowing through the lung from the posterior sacs gives up more and more oxygen to the blood, and the blood can continuously take up more and more oxygen. Even as blood enters the lung it can take up oxygen because blood at that point has a low oxygen concentration.

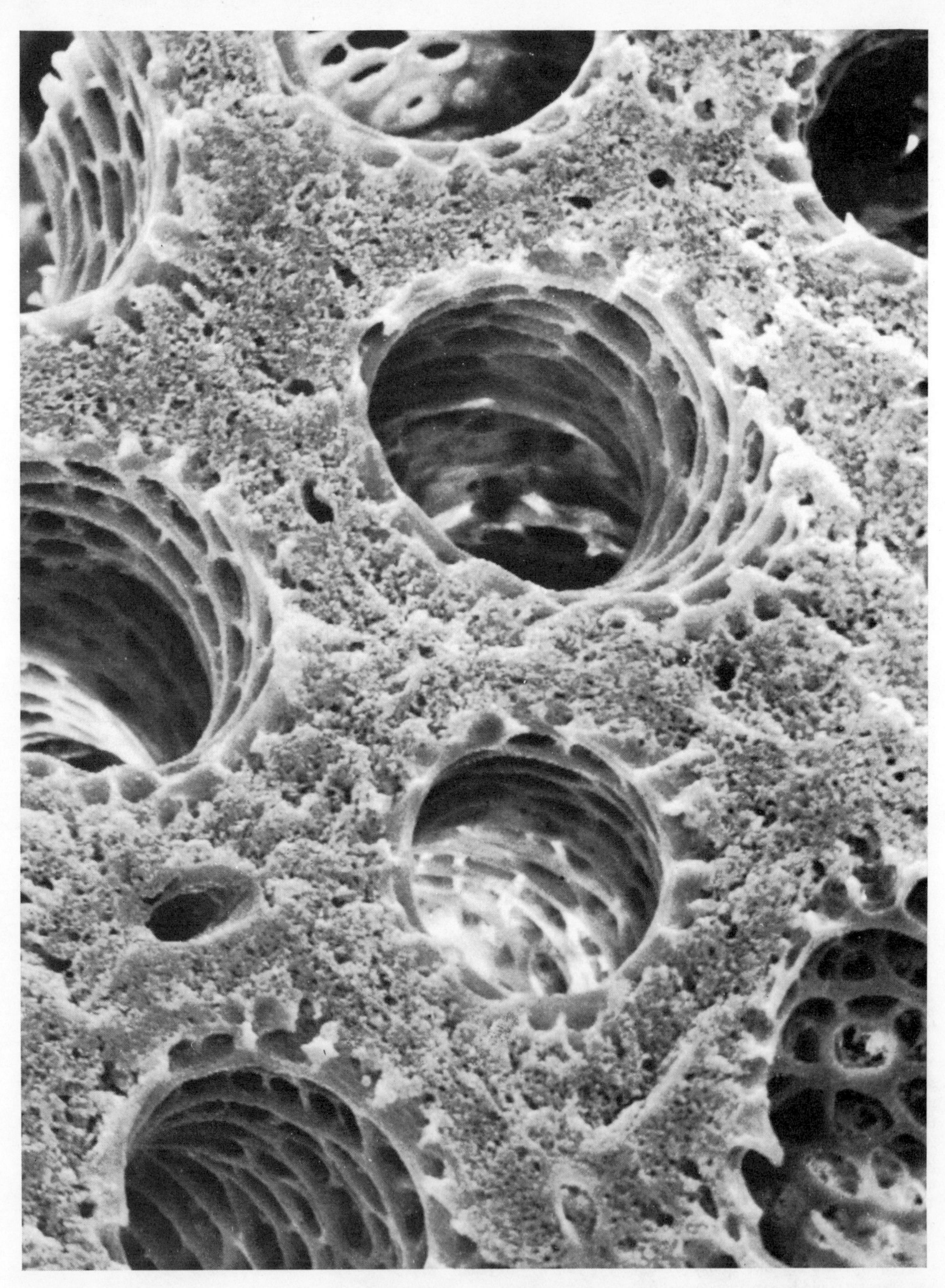

INTERIOR OF BIRD'S LUNG shows the structure that enables air to flow through the lung instead of in and out of it, as in the mammalian lung. This scanning electron micrograph was made by H. R. Duncker of Justus Liebig University at Giessen in West Germany. The circular structures are parabronchi, which are fine branches of the bronchial system, and the surrounding material is lung tissue. Equivalent structures in the mammalian lung are the saclike alveoli. The parabronchi in this micrograph, which are enlarged 180 diameters, are in the lung of a domestic fowl that was 14 days old. The micrograph shows the parabronchi in transverse section.

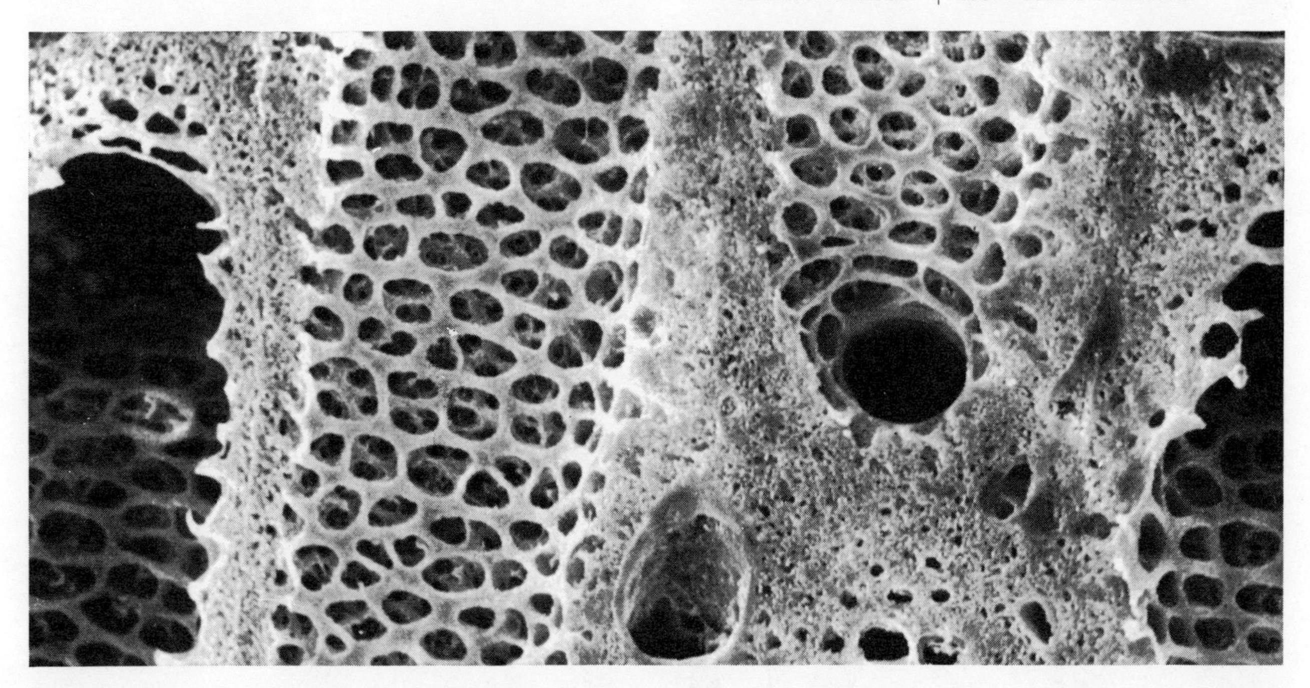

LONGITUDINAL SECTION of parabronchi in a bird's lung shows the spongy structure that enables air to flow through the lung as water flows through a sponge. This scanning electron micrograph was also made by Duncker; the enlargement is 90 diameters.

One can depict the flow of air and blood in the bird's lungs with a simple diagram [*see bottom illustration on page 95*]. In the diagram the airflow through the lungs is shown as a single stream and the flow of blood as another single stream. The salient point is that the two flows are in opposite directions.

In this way it becomes apparent that the blood, as it is about to leave the lungs, can take up oxygen from air that has the highest oxygen concentration available anywhere in the system. As the air flows through the lungs it gives up more and more oxygen to the blood before it enters the anterior sacs, where it is held until it is exhaled. It will be noted that the air, just before it leaves the lungs, encounters venous blood that is low in oxygen. This blood is therefore able to take up some oxygen, even though much of the oxygen in the air has already been removed. As the blood passes through the lungs it meets air of increasing oxygen concentration and therefore can continuously take up more oxygen until, just before it leaves the lungs, it meets the maximally oxygen-rich air coming from the posterior sacs.

The end result of this countercurrent flow is that more oxygen can be extracted from the air than would otherwise be possible. The system is similar to the flow through the gill of the fish, where the blood and the water flow in opposite directions. The blood just as it leaves the gill therefore encounters water with the highest possible oxygen content. Because of this type of flow, fish can extract from 80 to 90 percent of the oxygen in the water. The oxygen extraction normally reached in mammals under normal conditions is about 20 to 25 percent of the oxygen present in the air.

We are still trying to obtain better evidence that the flow of air and blood in the bird's lungs is as proposed in the scheme I have described, but the performance of birds at high altitude could hardly be explained in any other way. Examining the exchange of carbon dioxide rather than oxygen, we found several years ago that the air in the anterior sacs has a content of carbon dioxide that is much higher than the concentration in the arterial blood. This relation too can only be explained if the air coming from the lungs to the anterior sacs has received carbon dioxide from venous blood instead of being in equilibrium with arterialized blood as exhaled air in mammals is.

To what I have said so far, which I regard as hypotheses well supported by physical evidence, I should like to add a wild speculation. It is well known that some large birds, notably cranes and swans, have an extremely elongated trachea. This long trachea would seem to be a disadvantage, since at the end of an exhalation it would represent dead space filled with exhaled air that would have to be reinhaled at the beginning of the next breath, thus diluting the fresh outside air that follows.

The usual interpretation of the long trachea of swans and cranes is that it aids in vocalization. Such a luxury could not be allowed, however, if the large increase in dead space were physiologically detrimental. In fact, the increase in dead space may be an advantage. For aerodynamic reasons large birds have a slow wingbeat. For anatomical reasons the wingbeat and breathing in flying birds may be synchronized, since the large muscles that provide the downstroke of the wing are inserted at the keel of the breastbone and pull on it. It would therefore seem simple to attain simultaneous movements of wing and chest; indeed, it may be difficult to avoid.

The reasoning now goes as follows. If a slow wingbeat is determined by the size of the bird, and if respiration is synchronized with wingbeat, enough air can be taken in only by making each breath deeper. If each breath is deeper, and it is necessary (as I pointed out earlier) to achieve a certain level of carbon dioxide in the posterior air sacs, the amount of exhaled air reinhaled with each breath must be increased. In other words, to achieve the necessary concentration of carbon dioxide it is necessary to increase the volume of dead space.

Perhaps this speculation will have to be modified as more evidence becomes available. At present, we do not have adequate information about the synchronization of wingbeat and respiration in any of the larger birds. In fact, the respiration of birds during flight remains an interesting and almost uncharted field of physiology.

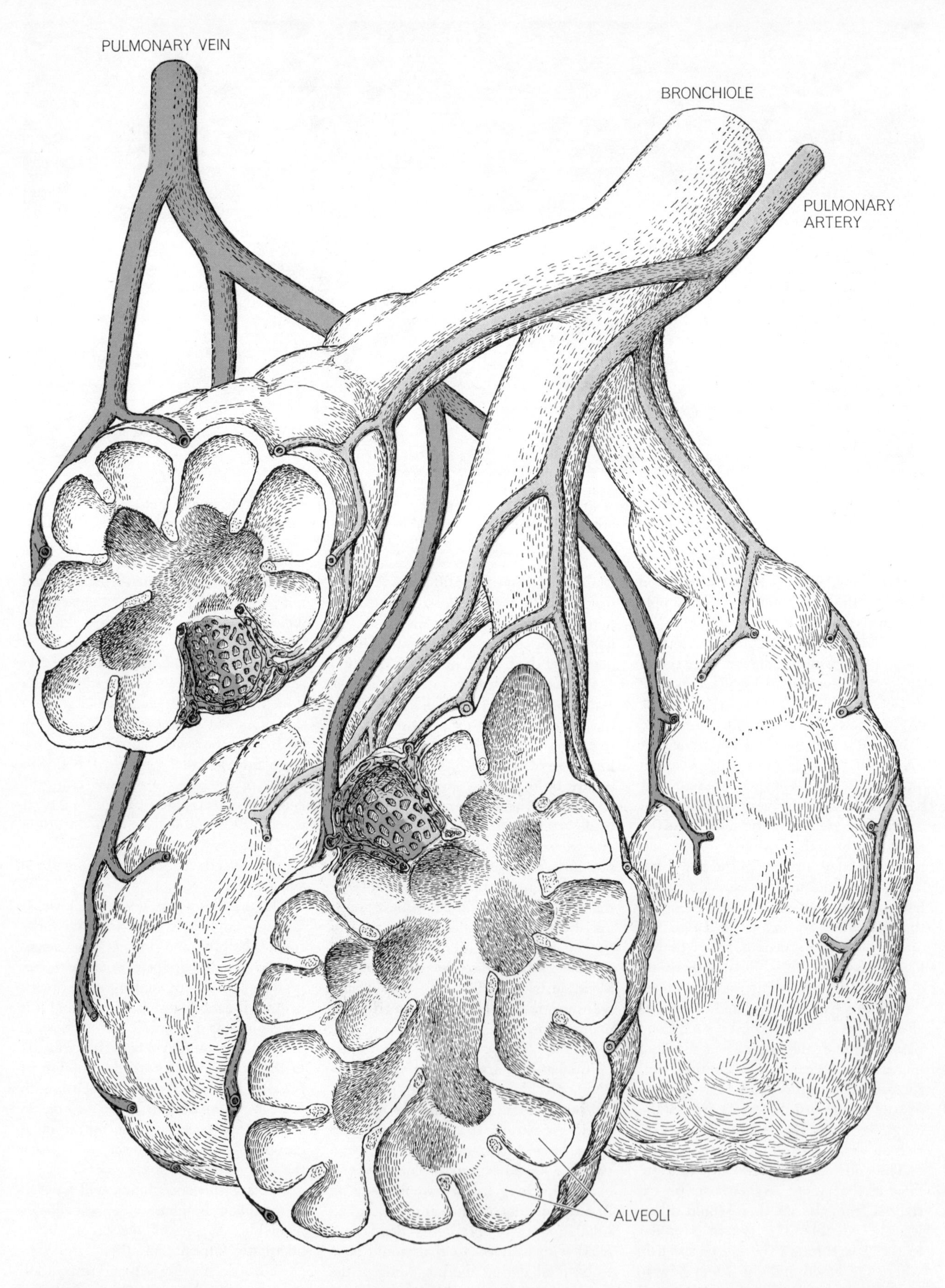

ALVEOLI are the air spaces in the lungs through which oxygen enters the blood and carbon dioxide leaves. A surface-active agent coats the moist alveoli and regulates the elasticity and tension of the lungs as a whole. In this schematic diagram nearly all the smaller blood vessels have been omitted except where capillary networks embedded in the alveolar walls show through from the backs of two alveoli seen in cross section. The average alveolus expands and contracts more than 15,000 times a day during breathing.

Surface Tension in the Lungs

by John A. Clements
December 1962

Recent investigations have shown that the air spaces of the lung are coated with a complex substance that lowers surface tension. It now appears that this substance keeps the lungs from collapsing

By far the most extensive surface of the human body in contact with the environment is the moist interior surface of the lungs. To carry on the exchange of carbon dioxide and oxygen between the circulating blood and the atmosphere in sufficient volume to sustain life processes requires approximately one square meter of lung surface for each kilogram of body weight. In the normal adult this amounts to the area of a tennis court. Such an area is encompassed in the comparatively small volume of the chest by the compartmentation of the lungs into hundreds of millions of tiny air spaces called alveoli. These air spaces are connected by confluent passages through the bronchial tree and the trachea to the atmosphere and are thus, topologically speaking, outside the body. Within the walls of the alveoli the blood is spread out in a thin sheet, separated from the air by a membrane about one micron (.001 millimeter) thick.

Since the primary function of the lungs is to present the inner surface of the alveoli to the air, it is not surprising to learn that the vital process of respiration is critically dependent on the physical properties of this surface. There is, of course, much more to the anatomy and physiology of the lungs. In recent years, however, the attention of investigators has been drawn increasingly to the role that is played by surface tension: the manifestation of the universal intermolecular forces that is observed in the surfaces of all fluids. The surface tension in the outermost single layer of molecules in the film of tissue fluid that moistens the surface of the lungs has been found to account for one-half to three-quarters of the elasticity with which the air spaces expand and contract in the course of the 15,000 breaths that are drawn into the lungs of the average individual each day.

As this knowledge suggests, it has also been found that the body has a way of regulating the surface tension of the lungs. Certain cells in the walls of the alveoli secrete a sort of detergent or wetting agent. This "surface-active" substance tends to weaken the surface tension. Its presence in the monomolecular layer on the surface of the film of moisture coating the air spaces serves to stabilize the dynamic activity of the lungs. It equalizes the tension in the air spaces as they expand and contract; it

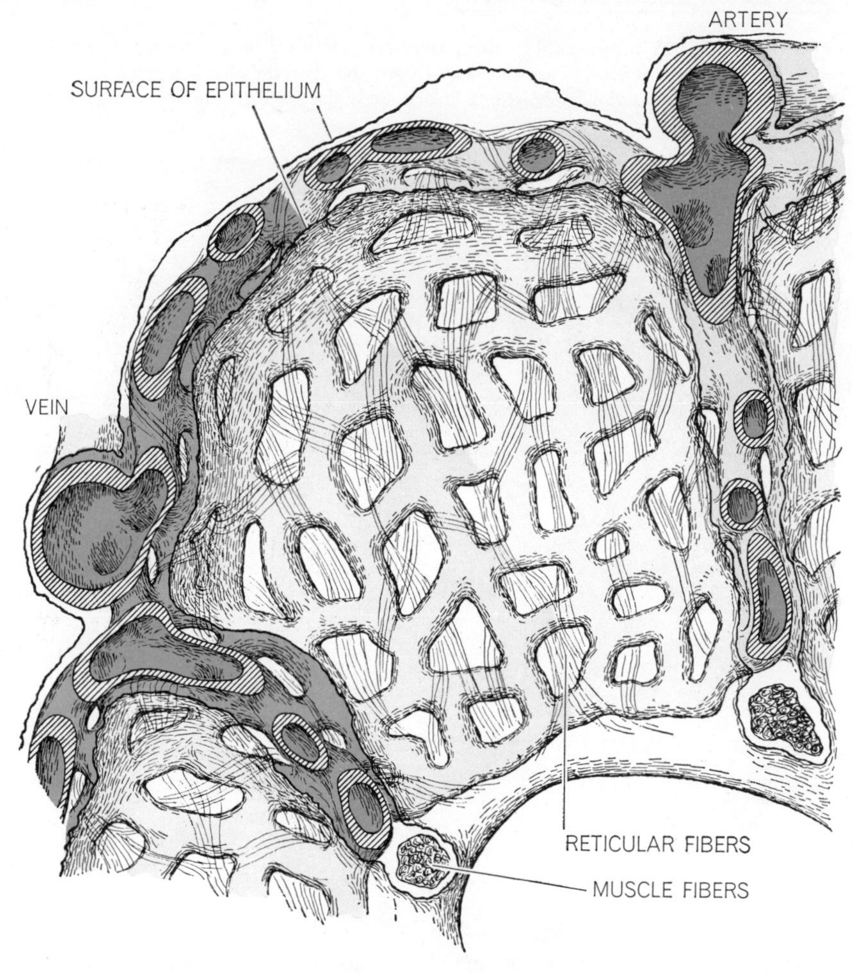

SINGLE ALVEOLUS is actually microscopic in size. The alveolar wall has been rendered transparent in this schematic cross section so that the rich network of blood capillaries and fibers that support the alveolus can be seen. The surface-active substance that plays a key role in stabilizing lung function normally coats the epithelium of every healthy alveolus.

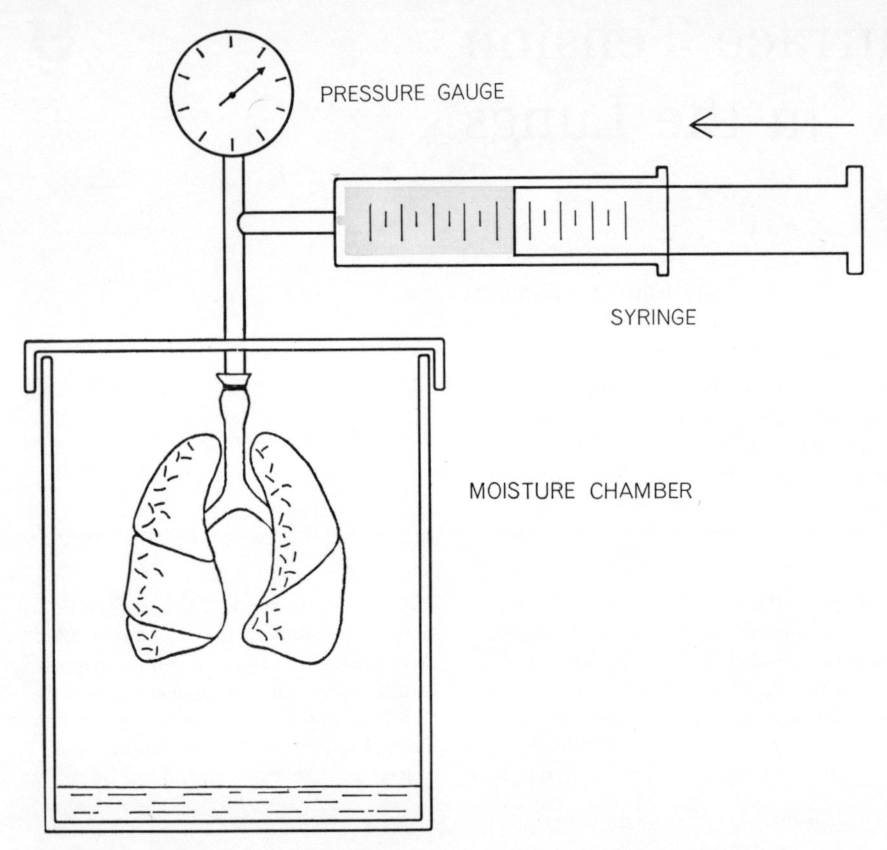

LUNG ELASTICITY is measured by using large syringe to fill lungs with a gas or a fluid (*color*). Lungs taken at autopsy are placed in a moist chamber and attached to a manometer, or pressure gauge. Karl von Neergaard of Zurich made the first such measurement in 1929.

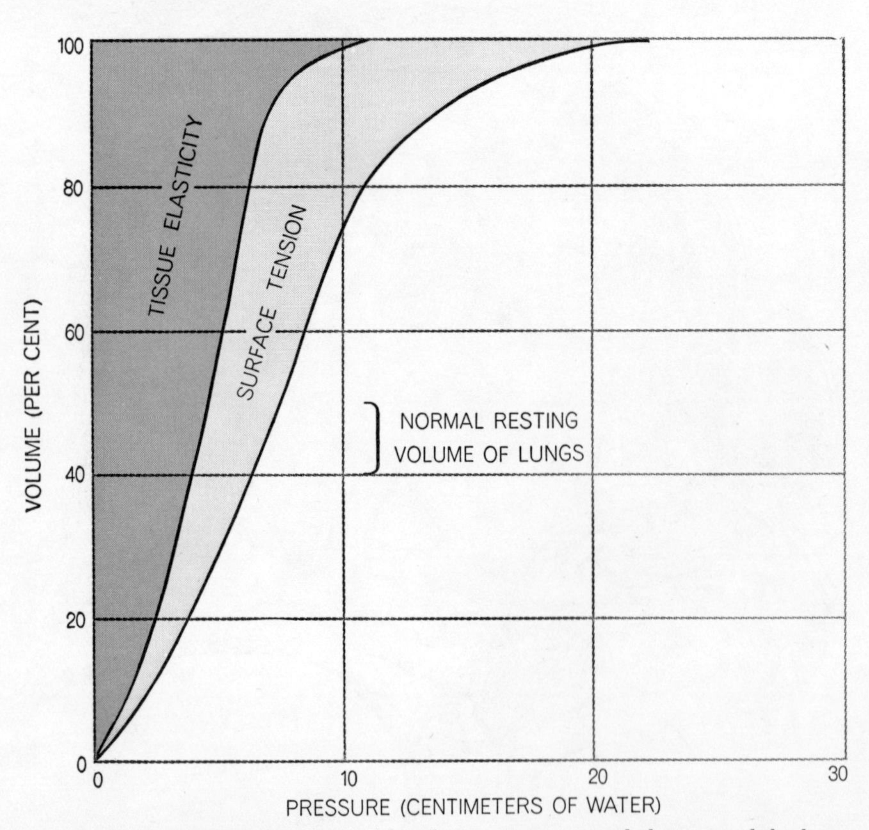

ELASTICITY CURVES are obtained by plotting pressure needed to expand the lungs against volume of lungs. The lungs show much less elasticity when filled with fluid than they do when filled with gas. Surface tension plays a significant role only when lungs are filled with gas. The amount of elasticity contributed by surface tension shows up as the difference between the curves at any particular volume of the lungs. It rises as the lungs expand.

brings about an even distribution of pressure between large and small alveoli, and by decreasing the over-all pressure it reduces the muscular effort required for respiration. This substance has the further function of assisting the osmotic forces acting across the surface of the lungs and so keeping the film of moisture on the surface from drawing fluid into the air spaces.

The critical importance of the surface-active agent becomes apparent when it is not there to do its work. Its absence explains some of the symptoms in the complex organic disease of the newborn recorded variously as fatal respiratory distress, hyaline-membrane disease or atelectasis. The collapse of the lungs and the filling of the air spaces with fluid observed in this disease is promoted by abnormally high surface tension in the alveoli. Some 25,000 newborn infants die of the disease in the U.S. each year. A similar syndrome, although it is not always fatal, has recently appeared as a complication attending heart surgery in some patients whose lungs have been temporarily disengaged from the respiratory function by diversion of the blood through a "heart-lung" machine.

The contribution of surface tension to the elasticity of the lungs was first demonstrated in 1929 by Karl von Neergaard of the University Clinic in Zurich. He distended lung preparations alternately with air and with saline solution and compared the pressures required to do so with each. This experiment, since repeated many times by other workers, showed that it takes a higher pressure to distend the lungs with air. The interpretation of this experiment calls for a more precise definition of a surface: it is an interface between two substances and it is established by the relative cohesion of their constituent molecules. Thus when the fluid on the surface of the lungs forms an interface with air, it exhibits a stronger surface tension than it does at an interface with saline solution. In fact, the tissue fluid forms essentially no interface at all with saline solution of the right concentration and the surface tension is reduced to almost zero. Distention of the lungs with saline solution can therefore be used to measure the elastic properties of the tissue alone, uncomplicated by the effects of surface tension. Since inflation with air yields a measure of both the tissue and the surface-tension components, the effect of surface tension can be derived by subtracting the pressure required to distend the lungs with saline solution from the pressure required to inflate them to the same volume with air.

The technique has recently provided conclusive evidence that surface tension is abnormally high in the fatal respiratory distress of the newborn. After autopsy the lungs of such infants can be expanded at almost normal pressures with saline solution but require three to four times the normal pressure for air inflation. Moreover, the alveoli collapse at abnormally high air pressures during deflation.

In everyday experience with surface tension there is little to suggest that it has such formidable power. It has barely measurable effects on the properties of solids, showing up for example in the measurement of the elasticity of fine-drawn wires. Its action is more prominent in the behavior of liquids, as in the shaping of raindrops or the providing of a platform for certain aquatic insects. But it seems no more than an incidental effect of the geometry that accounts for it. Whereas the molecules in the bulk of a liquid experience forces of mutual attraction that are balanced in all directions, the molecules at the surface are attracted more strongly to their neighbors below the surface and are attracted only weakly to the sparser population of molecules in the air above the surface. Because the net pull is downward, the surface particles tend to dive and the surface shrinks to the least possible area.

The resultant force of cohesion at the interface between a liquid and the air can be demonstrated with the help of a Maxwell frame, named for the 19th-century physicist James Clerk Maxwell. This is a U-shaped wire, with the open end of the U closed by a cross wire that can slide along the legs of the U. A film of liquid stretched out on the frame tends to pull the cross wire to the bottom of the U. The force necessary to resist this pull—to maintain a constant area of film—provides a measure of the surface tension. Since the film in this experiment has two surfaces, the measured force must be divided by twice the width of the U; the result is usually expressed in dynes per centimeter. (A dyne is the force required to accelerate a one-gram mass one centimeter per second.) The surface tension of pure water at body temperature is equal to about 70 dynes per centimeter; that of blood plasma and

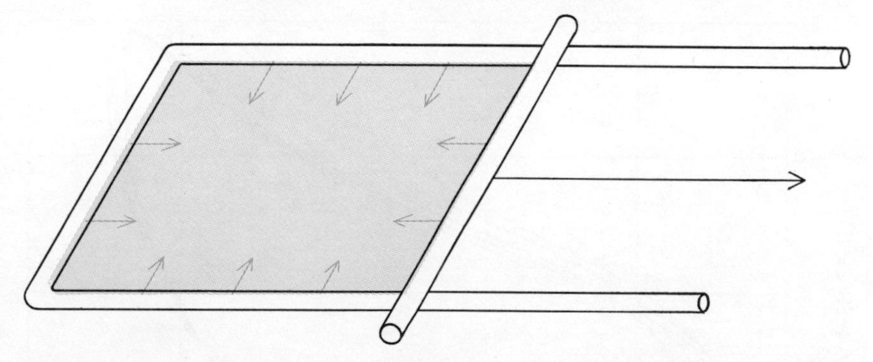

MAXWELL FRAME, used to demonstrate surface tension, consists of U-shaped wire with separate wire across open end. Liquid film (*color*) pulls at cross wire. Force needed to prevent cross wire from moving to bottom of U (*black arrow*) is proportional to the tension.

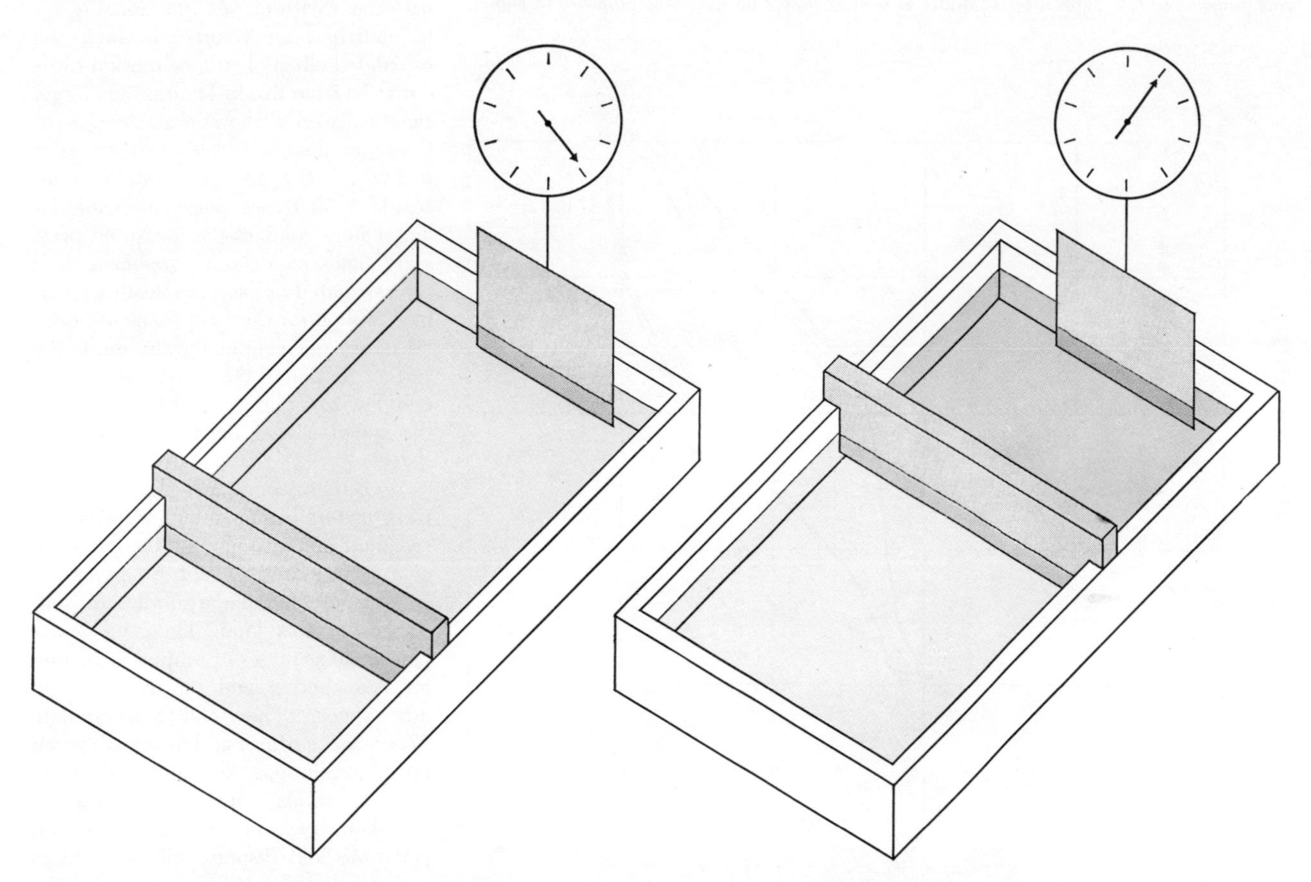

SURFACE BALANCE measures change in surface tension as area of film of surface-active agent on water increases and decreases. Surface tension pulls down on platinum strip (*attached to gauge*). Water alone produces pull of about 70 dynes per centimeter. A detergent in the water makes surface tension about 30 dynes per centimeter but the tension does not change as barrier moves slowly back and forth. Surface-active agent from lungs forms a film on the water and makes the surface tension about 40 dynes per centimeter (*left*). As barrier moves toward strip, compressing the agent, tension drops (*right*). Surface tension rises as the barrier moves back.

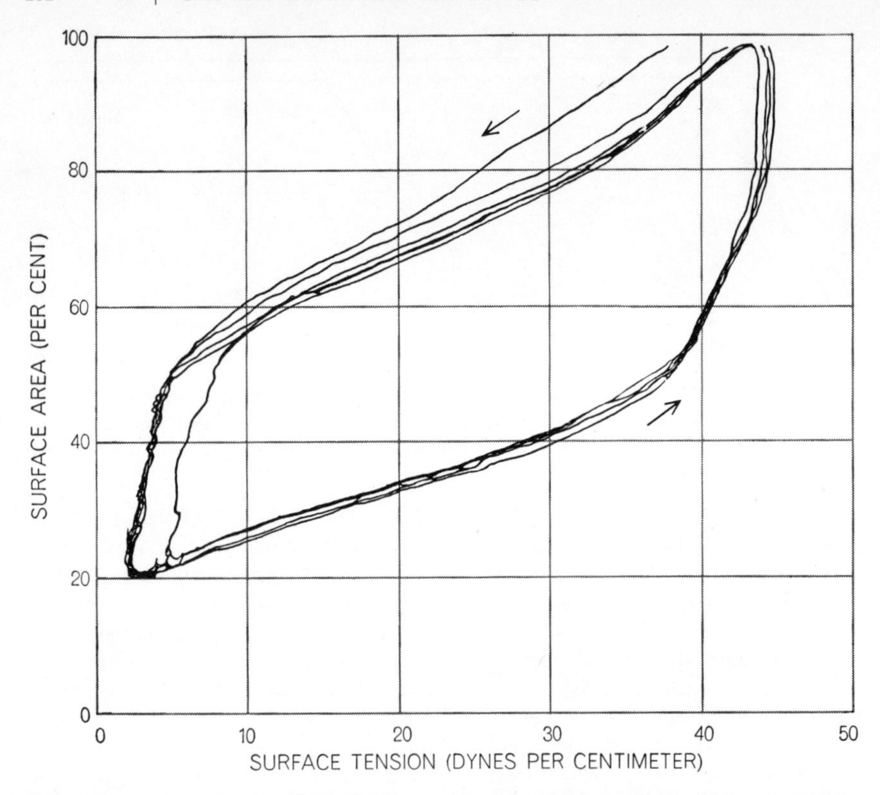

CHANGE IN SURFACE TENSION with area, as measured by surface balance, is large when surface-active agent from normal lungs covers the water. Moving barrier made several trips back and forth during test. The tension at first drops rapidly as barrier moves in (*arrow pointing to left*), and it rises rapidly as barrier moves back (*arrow pointing to right*).

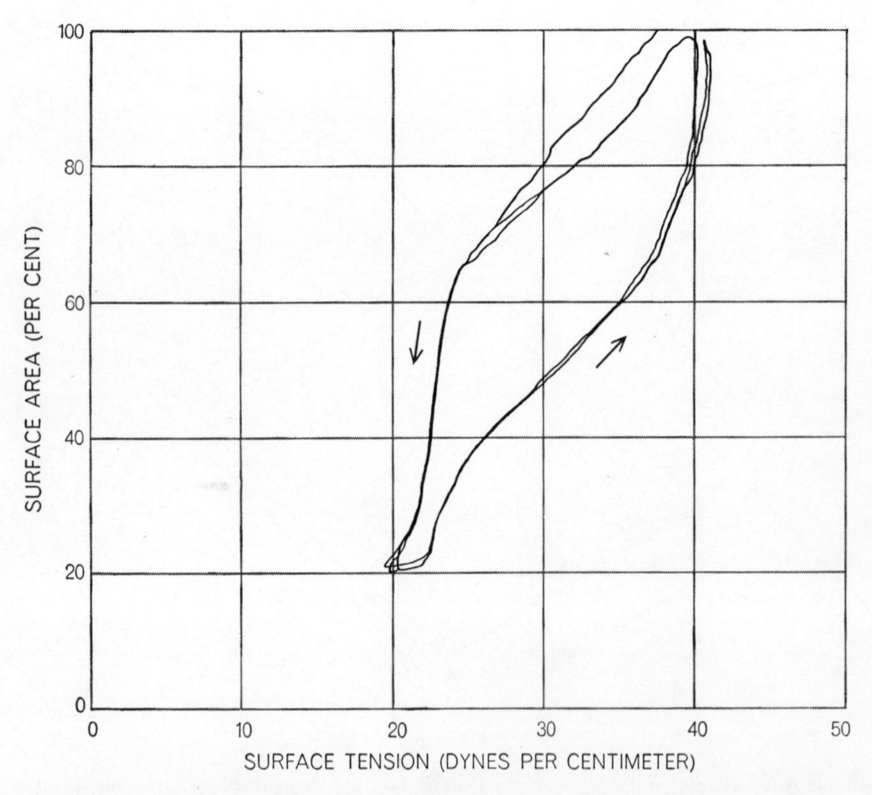

CHANGE IS SMALL when the alveolar coating comes from lungs of newborn infant who succumbed to acute respiratory distress. In such a case the surface tension is about 18 dynes per centimeter. Relative lack of surface activity plays a key role in the fatal disease.

tissue fluid, to about 50 dynes per centimeter.

The results conform with the observation that surface tension does not exert very strong forces. At this point, however, geometry enters the picture again. According to a formula of Pierre Simon de Laplace, the 18th-century astronomer and mathematician, the force exerted in a given surface is equal to twice the tension divided by the radius of the surface. In a flat surface with, so to speak, infinite radius the force is zero. Given the tiny dimensions of the average-sized alveolus, calculation shows that the surface tension of tissue fluid should exert a considerable force. At 50 dynes per centimeter in a surface with a radius of only .05 millimeter, it would produce a force of 20,000 dynes per square centimeter. Expressed as pressure this is equal to 20 centimeters of water.

This computation explains why surface tension influences the elasticity of the lungs so greatly. It does not, however, agree with the actual values for the surface tension in normal lungs obtained by comparison of the pressures required to distend the lungs with liquid and to inflate them with air. At functional or intermediate lung volume, in fact, the calculated effect of surface tension turns out to be from five to 10 times too large. In other words, the surface tension of the tissue fluid would have to be closer to five or 10 dynes per centimeter instead of 50 dynes per centimeter. At larger lung volumes the measured pressure comes into closer agreement with the calculated pressure, indicating a surface tension for the lung tissue of about 40 dynes per centimeter. In short, the surface tension of the tissue is unexpectedly low in the lungs, and it varies with the inflation and deflation of the air spaces.

With these suggestive clues in hand, investigators began to look into the tissue fluid of the lungs for the presence of a surface-active agent. Soaps or detergents are familiar examples of substances of this kind. Their molecules have weaker forces of mutual attraction for one another and for molecules of other species. They tend to accumulate in excess at surfaces and interfaces when mixed in solutions. Acting as bridges between dissimilar substances such as oil and water or water and air, they wet, penetrate and disperse oily substances and stabilize emulsions and foams. The concentration of their weaker attractive forces at an interface reduces the surface tension. At a number of laboratories the presence of a surface-active agent was soon demonstrated in the tissue fluid extracted from the lungs. The extraction can be accomplished in a number of ways: by rinsing the alveoli with saline solution via the air passages; by generating a foam in the alveoli; and by filtration from minced whole tissue. Each of these procedures yields an extract that contains a powerful surface-active agent on which accurate measurements can be made.

The laboratory technique for detecting the presence of this agent and measuring its effect on the surface tension of the tissue fluid provides a nice demonstration of its mode of operation in the alveoli. The extract is placed in a shallow tray and a .001-inch-thick platinum strip is suspended in it from the arm of a sensitive electrobalance or strain gauge. The pull of the surface on the strip provides a measure of the surface tension. A motor-driven barrier slowly sweeps the surface from the far end of the tray, reducing the area of liquid surface in which the platinum strip is hanging to 10 or 20 per cent of its initial size. Since the surface-active agent in the extract spontaneously forms a film at the surface, it is concentrated in the area in front of the barrier. As the concentration builds up, the surface tension falls to low values.

Extracts from normal lungs show a change in surface tension, when measured this way, from about 40 dynes per square centimeter to two dynes per square centimeter—in excellent agreement with the surface tension as estimated in the lung itself. In contrast, the surface tension does not fall below 18 dynes per square centimeter in extracts from the lungs of newborn infants that have succumbed to hyaline-membrane disease.

The change in surface tension with the change in surface area is the key to the action of the surface-active agent in the lung. In an expanded air space the layer of surface-active agent is attenuated, and surface tension is increased accordingly. The increase in tension is partly offset, however, by the increase in the radius of the air space, and the increase in force or counterpressure exerted by the surface tension is diminished. As the air space contracts to perhaps half its expanded size, the increasing concentration of the surface-active agent reduces the surface tension, balancing the Laplace equation in the other direction and again decreasing the pressure in the air space. Similarly, between

LUNGS OF FROGS contain quite large air spaces that are not threatened by pressure of surface tension created by a liquid that coats the inner lining. Therefore a surface-active agent is not necessary, and none has been found. One lung is shown in longitudinal section at right. Color indicates blood vessels. The diagrams on this page are highly schematic.

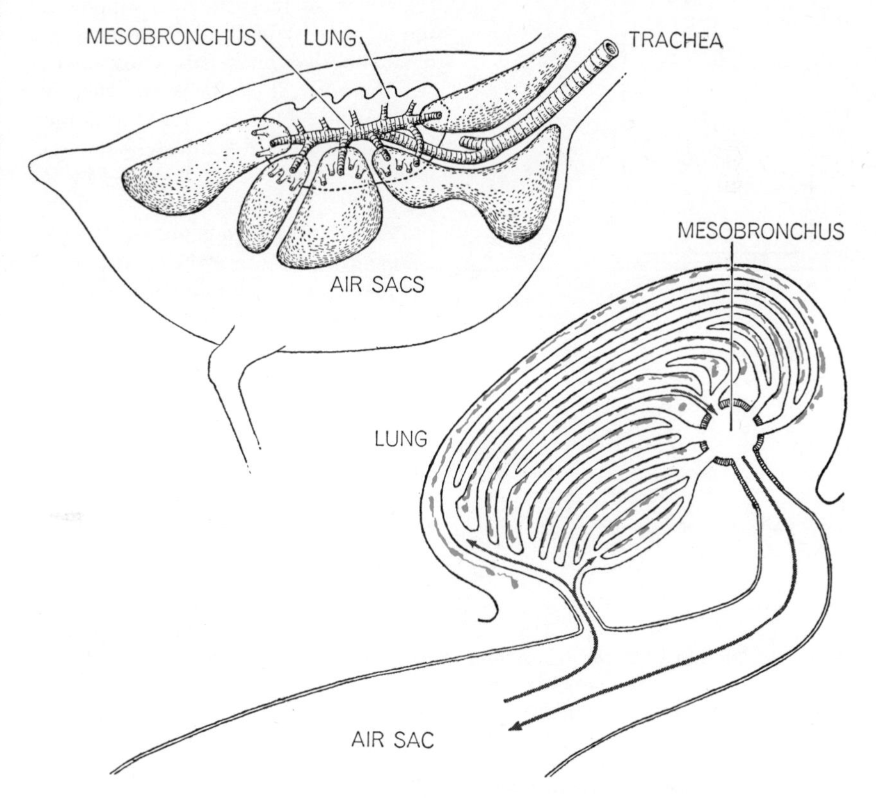

LUNGS OF BIRDS contain air capillaries, tiny tubes in which gas exchange with the blood takes place. Air passes through lungs into and out of large air sacs. Since there is little change in lung volume, surface tension does not need to be adjusted. Surface-active agent has not been found in the lungs of birds. A cross section of a bird lung is seen at right.

the smallest alveolus and the largest, which may be three or four times larger, differences in the concentration of the surface-active agent in the interface of tissue fluid and air bring about a homogeneous distribution of pressure. The performance of hundreds of millions of alveoli, of random size, is thereby smoothed and co-ordinated.

The action of the alveolar surface-active agent in balancing the forces that otherwise tend to draw fluid out of the capillaries into the alveolar air spaces is also important. The blood pressure in the capillaries, the osmotic pull of the tissue fluid in and on the alveolar membrane, and the surface tension of this fluid all work to move fluids outward. One force, the osmotic pull of the blood plasma, opposes this combination of forces. The maintenance of a favorable balance of forces is assisted by the reduction in surface tension through the action of the surface-active agent. In hyaline-membrane disease, with surface tension sustained at as much as 45 dynes per centimeter, the leakage of fluid from the capillaries blocks the exchange of oxygen and carbon dioxide between blood and air; the process is limited in the end only by engorgement of the air spaces or by their collapse.

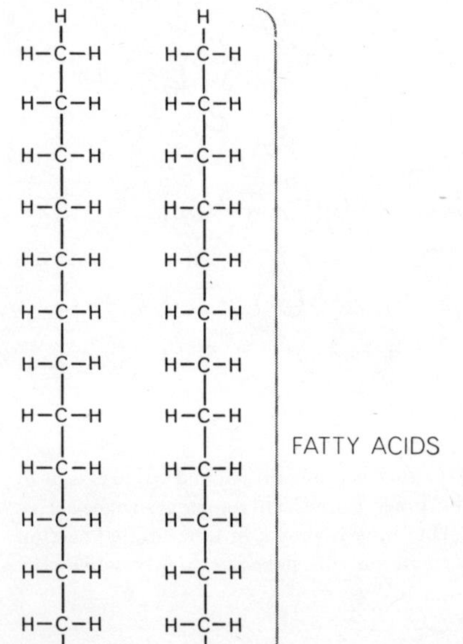

Chemical analysis has shown that the surface-active agent of the lungs is a lipoprotein, that is, a compound molecule made up of protein and fatty constituents. The latter are of an appropriately soapy kind, lecithin being the predominant component. A member of the same chemical family that has been made synthetically, a substance called dipalmitoyl lecithin, shows the same surface activity. Dipalmitoyl lecithin has even been isolated from the lung fluid. It is tempting, therefore, to attribute the surface activity of the lung material to dipalmitoyl lecithin. Against this conclusion, however, it can be shown that the active material isolated from the extract of tissue fluid is the intact lipoprotein molecule. Its activity is destroyed by attempts to segregate the lipids from the protein or to isolate any one of the lipid components from the whole. At present the most reasonable opinion is that the native material is a complex of protein and lipids, particularly dipalmitoyl lecithin, and that both are essential to its activity.

The discovery of this remarkable substance in the lung fluid of man has prompted a search for it in other animals. So far it has turned up in all the other mammals that have been tested (the mouse, rat, guinea pig, rabbit, cat, dog and cow), but not in any amphibian (frog and toad), reptile (snake and crocodile) or bird (pigeon and chicken). There appears to be some rationale for this distribution among species. Amphibians and reptiles depend on their environment to supply a major portion of the heat that sustains their metabolism; weight for weight they do not require as much exchange of respiratory gases as mammals and therefore do not need as much lung surface. Accordingly they have relatively large air spaces in their lungs, and their lung function is not seriously threatened by the action of surface tension. Birds, on the other hand, have small air spaces, more comparable in size to those in the lungs of mammals. But the bird lung is ventilated in a peculiar way. Instead of the tidal ventilation, which alternately inflates and deflates the air spaces of mammals, the exchange of respiratory gases is accomplished by drawing air through the lungs into large air sacs that are separate from the lungs. In this way the change of volume in the air spaces is minimized. The air spaces can remain at or near their maximum volume, and the lungs are stabilized by the elasticity of the lung tissue itself.

In mammals the lung tissues apparently begin to secrete the critical surface-active material late in embryonic development. This is true, at least, of the two species in which the question has been investigated. In the mouse, which has a gestation period of 20 days, surface activity in the lungs appears suddenly at 17 or 18 days. The lungs of the human fetus develop the activity somewhat more gradually, during the fifth to the seventh month of gestation. This is the interval during which prematurely born infants become increasingly viable.

Since it is now reasonably certain that the secretion of a surface-active material is an adaptation peculiar to mammals, investigators are finding new significance in an observation made in 1954 by Charles Clifford Macklin of the University of Western Ontario. He showed that the walls of the alveoli in mammalian lungs contain special cells that he called granular pneumonocytes. He even suggested that the "granules" discharged by these cells "regulated the surface tension" of the alveoli, but he did not enlarge on this idea further. Under the electron microscope it now appears that the granules of Macklin are cellular particles called mitochondria and possibly the products of mitochondria from certain cells in the alveolar membrane. Mitochondria are associated with the metabolic and synthetic activities of all cells; they appear in high concentration in those tissue cells that have specialized secretory functions [see "Energy Transformation in the Cell," by Albert Lehninger; Scientific American Offprint 69]. Some of the granules can be identified as true mitochondria, with the fine structure that characterizes them in other cells. Others appear to be mitochondria-

FATTY ACIDS

GLYCEROL

PHOSPHATE

CHOLINE

LECITHIN MOLECULE consists of long-chain fatty-acid groups that are not strongly attracted to water, as well as glycerol and electrically charged, or polar, phosphate and choline groups that are attracted to polar molecules of water. The fatty acid is thought to stand up out of the water when the molecule is part of a surface layer. Lecithin is strongly surface-active; it is a constituent of the substance that coats the alveoli.

like bodies involved in a process of transformation by which they lose their fine structure and become relatively featureless. Most remarkable of all, the electron-microscope pictures show these same forms passing through the cell membrane from the cytoplasm into the air space. This process could be the means by which the surface-active substance is secreted into the tissue fluid that coats the surface of the alveoli.

Various stages of the process have been observed in the cells of a half-dozen species of mammals but never in the amphibians or birds in which it has been looked for. In mammals, moreover, it has been found that this peculiar transformation of the mitochondria appears in the lung tissues along with surface activity at the same stage of fetal development.

With this background of evidence established by classical physiology and the most modern techniques of cell biology, the way seems to be cleared for investigation of the hormonal, neural, nutritional, environmental and genetic factors that may influence the production, function and elimination of the alveolar surface-active agent. Diabetes, for example, is associated with derangement of lipid, or fat, metabolism. In view of the importance of the lipid fraction of the surface-active agent one wonders if diabetes in the mother may not be a factor predisposing the fetus to respiratory distress at birth, particularly since the syndrome occurs more frequently among infants born to diabetic mothers. The experience with patients in heart surgery suggests another line of investigation. From the rapid decrease in surface activity and the collapse of the lung that sometimes follows the bypassing of the pulmonary circulation, it can be surmised that the production of the agent depends on blood flow and that distribution of the flow affects the distribution of air in the lungs. The question of neural control is raised by experiments with small animals, in which cutting of the vagus nerve is followed by decline in surface activity, the accumulation of fluid and finally collapse of the air spaces. Pure oxygen, atmospheric pollutants and some industrial chemicals have been shown to affect the alveolar surfaces. Animals in which hormonal activity is high—young animals, females in estrus and animals that have been treated with cortisone—are particularly subject to the toxic effects of pure oxygen on the lungs. Occasionally massive collapse of the lungs follows general anesthesia, with no indication of obstruction to the air passages. It is not too much to hope that problems of this kind can be brought within the reach of effective treatment by the next advances in the understanding of the mechanism that regulates the surface tension in the lungs.

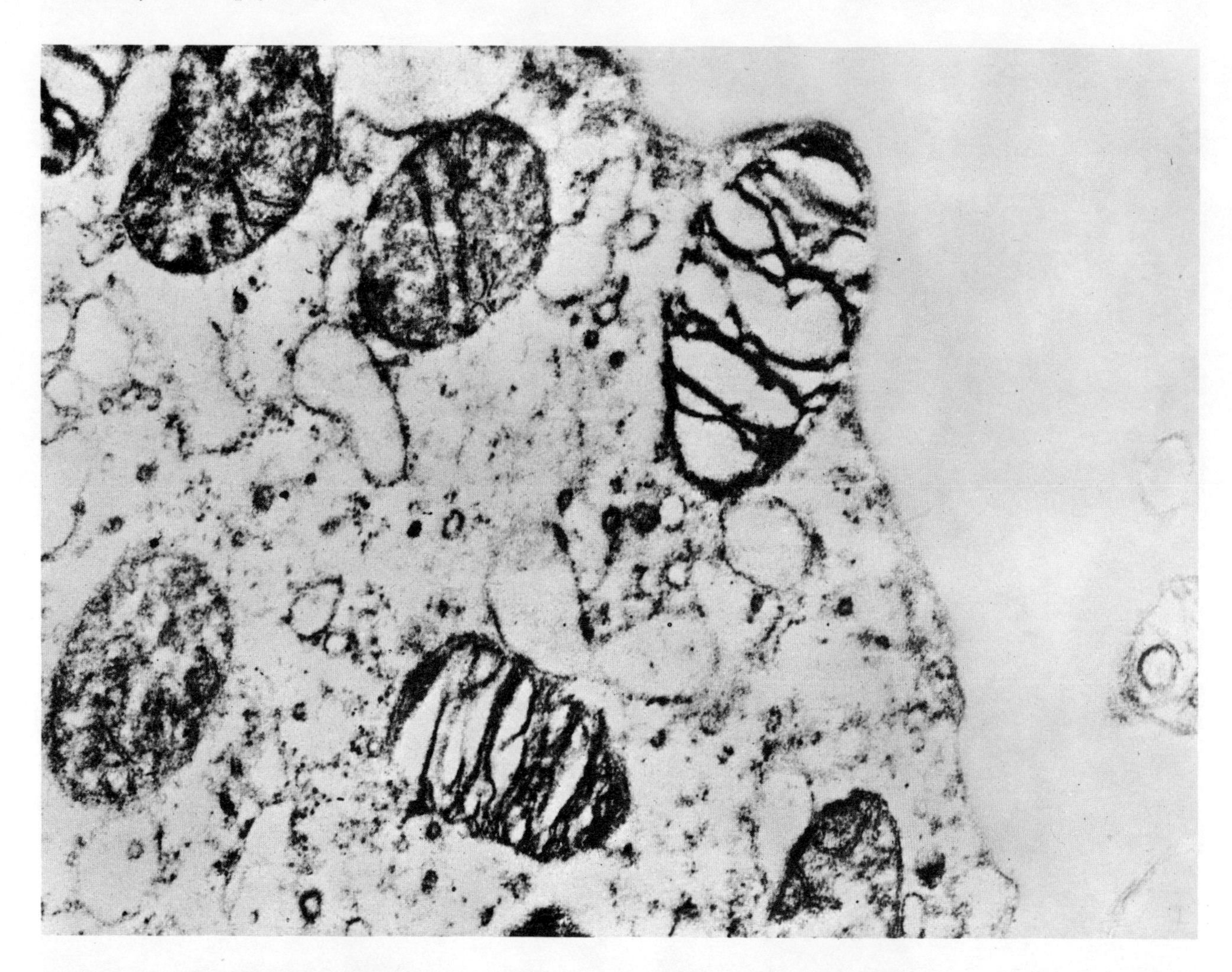

MITOCHONDRIAL TRANSFORMATION can be seen in this electron micrograph by Robert E. Brooks of the University of Oregon. It shows part of an alveolar epithelial cell enlarged approximately 45,000 diameters. Toward upper right a transformed mitochondrion seems to be emerging into the air space. To its left is a normal mitochondrion, and to the left of that, one is beginning to change, losing fine structure. Mitochondria probably produce lung agent; transformed mitochondria may carry it to the surface.

JAPANESE DIVING WOMAN was photographed by the Italian writer Fosco Maraini near the island of Hekura off the western coast of Japan. The ama's descent is assisted by a string of lead weights tied around her waist. At the time she was diving for abalone at a depth of about 30 feet. At the end of each dive a helper in a boat at the surface pulls the ama up by means of the long rope attached to her waist. The other rope belongs to another diver. The ama in this region wear only loincloths during their dives.

The Diving Women of Korea and Japan

by Suk Ki Hong and Hermann Rahn
May 1967

Some 30,000 of these breath-holding divers, called ama, are employed in daily foraging for food on the bottom of the sea. Their performance is of particular interest to the physiologist

Off the shores of Korea and southern Japan the ocean bottom is rich in shellfish and edible seaweeds. For at least 1,500 years these crops have been harvested by divers, mostly women, who support their families by daily foraging on the sea bottom. Using no special equipment other than goggles (or glass face masks), these breath-holding divers have become famous the world over for their performances. They sometimes descend to depths of 80 feet and can hold their breath for up to two minutes. Coming up only for brief rests and a few breaths of air, they dive repeatedly, and in warm weather they work four hours a day, with resting intervals of an hour or so away from the water. The Korean women dive even in winter, when the water temperature is 50 degrees Fahrenheit (but only for short periods under such conditions). For those who choose this occupation diving is a lifelong profession; they begin to work in shallow water at the age of 11 or 12 and sometimes continue to 65. Childbearing does not interrupt their work; a pregnant diving woman may work up to the day of delivery and nurse her baby afterward between diving shifts.

The divers are called ama. At present there are some 30,000 of them living and working along the seacoasts of Korea and Japan. About 11,000 ama dwell on the small, rocky island of Cheju off the southern tip of the Korean peninsula, which is believed to be the area where the diving practice originated. Archaeological remains indicate that the practice began before the fourth century. In times past the main objective of the divers may have been pearls, but today it is solely food. Up to the 17th century the ama of Korea included men as well as women; now they are all women. And in Japan, where many of the ama are male, women nevertheless predominate in the occupation. As we shall see, the female is better suited to this work than the male.

In recent years physiologists have found considerable interest in studying the capacities and physiological reactions of the ama, who are probably the most skillful natural divers in the world. What accounts for their remarkable adaptation to the aquatic environment, training or heredity or a combination of both? How do they compare with their nondiving compatriots? The ama themselves have readily cooperated with us in these studies.

We shall begin by describing the dive itself. Basically two different approaches are used. One is a simple system in which the diver operates alone; she is called *cachido* (unassisted diver). The other is a more sophisticated technique; this diver, called a *funado* (assisted diver), has a helper in a boat, usually her husband.

The *cachido* operates from a small float at the surface. She takes several deep breaths, then swims to the bottom, gathers what she can find and swims up to her float again. Because of the oxygen consumption required for her swimming effort she is restricted to comparatively shallow dives and a short time on the bottom. She may on occasion go as deep as 50 or 60 feet, but on the average she limits her foraging to a depth of 15 or 20 feet. Her average dive lasts about 30 seconds, of which 15 seconds is spent working on the bottom. When she surfaces, she hangs on to the float and rests for about 30 seconds, taking deep breaths, and then dives again. Thus the cycle takes about a minute, and the diver averages about 60 dives an hour.

The *funado* dispenses with swimming effort and uses aids that speed her descent and ascent. She carries a counterweight (of about 30 pounds) to pull her to the bottom, and at the end of her dive a helper in a boat above pulls her up with a rope. These aids minimize her oxygen need and hasten her rate of descent and ascent, thereby enabling her to go to greater depths and spend more time on the bottom. The *funado* can work at depths of 60 to 80 feet and average 30 seconds in gathering on the bottom—twice as long as the *cachido*. However, since the total duration of each dive and resting period is twice that of the *cachido*, the *funado* makes only about 30 dives per hour instead of 60. Consequently her bottom time per hour is about the same as the *cachido*'s. Her advantage is that she can harvest deeper bottoms. In economic terms this advantage is partly offset by the fact that the *funado* requires a boat and an assistant.

There are variations, of course, on the two basic diving styles, almost as many variations as there are diving locations. Some divers use assistance to ascend but not to descend; some use only light weights to help in the descent, and so on.

By and large the divers wear minimal clothing, often only a loincloth, during their work in the water. Even in winter the Korean divers wear only cotton bathing suits. In Japan some ama have recently adopted foam-rubber suits, but most of the diving women cannot afford this luxury.

The use of goggles or face masks to improve vision in the water is a comparatively recent development—hardly a century old. It must have revolutionized the diving industry and greatly increased the number of divers and the size of the harvest. The unprotected human eye suffers a basic loss of visual acuity in water because the light passing through water undergoes relatively little refraction when it enters the tissue of the cornea, so that the focal point of the image is considerably behind the retina [*see top illustration on page 111*]. Our sharp vision in air is due to the difference in the refractive index between air and the corneal tissue; this difference bends light sharply as it enters the eye and thereby helps to focus images on the retinal surface. (The lens serves for fine adjustments.) Goggles sharpen vision in the water by providing a layer of air at the interface with the eyeball.

Goggles create a hazard, however, when the diver descends below 10 feet in the water. The hydrostatic pressure on the body then increases the internal body pressures, including that of the blood, to a level substantially higher than the air pressure behind the goggles. As a result the blood vessels in the eyelid lining may burst. This conjunctival bleeding is well known to divers who have ventured too deep in the water with only simple goggles. When the Korean and Japanese divers began to use goggles, they soon learned that they must compensate for the pressure factor. Their solution was to attach air-filled, compressible bags (of rubber or thin animal hide) to the goggles. As the diver descends in the water the increasing water pressure compresses the bags, forcing more air into the goggle space and thus raising the air pressure there in proportion to the increase in hydrostatic pressure on the body. Nowadays, in place of goggles, most divers use a face mask covering the nose, so that air from the lungs instead of from external bags can serve to boost the air pressure in front of the eyes.

The ama evolved another technique that may or may not have biological value. During hyperventilation before their dives they purse their lips and emit a loud whistle with each expiration of breath. These whistles, which can be heard for long distances, have become the trademark of the ama. The basic reason for the whistling is quite mysterious. The ama say it makes them "feel better" and "protects the lungs." Various observers have suggested that it may prevent excessive hyperventilation (which can produce unconsciousness in a long dive) or may help by increasing the residual lung volume, but no evidence has been found to verify these hypotheses. Many of the Japanese divers, male and female, do not whistle before they dive.

Preparing for a dive, the ama hyperventilates for five to 10 seconds, takes a final deep breath and then makes the plunge. The hyperventilation serves to

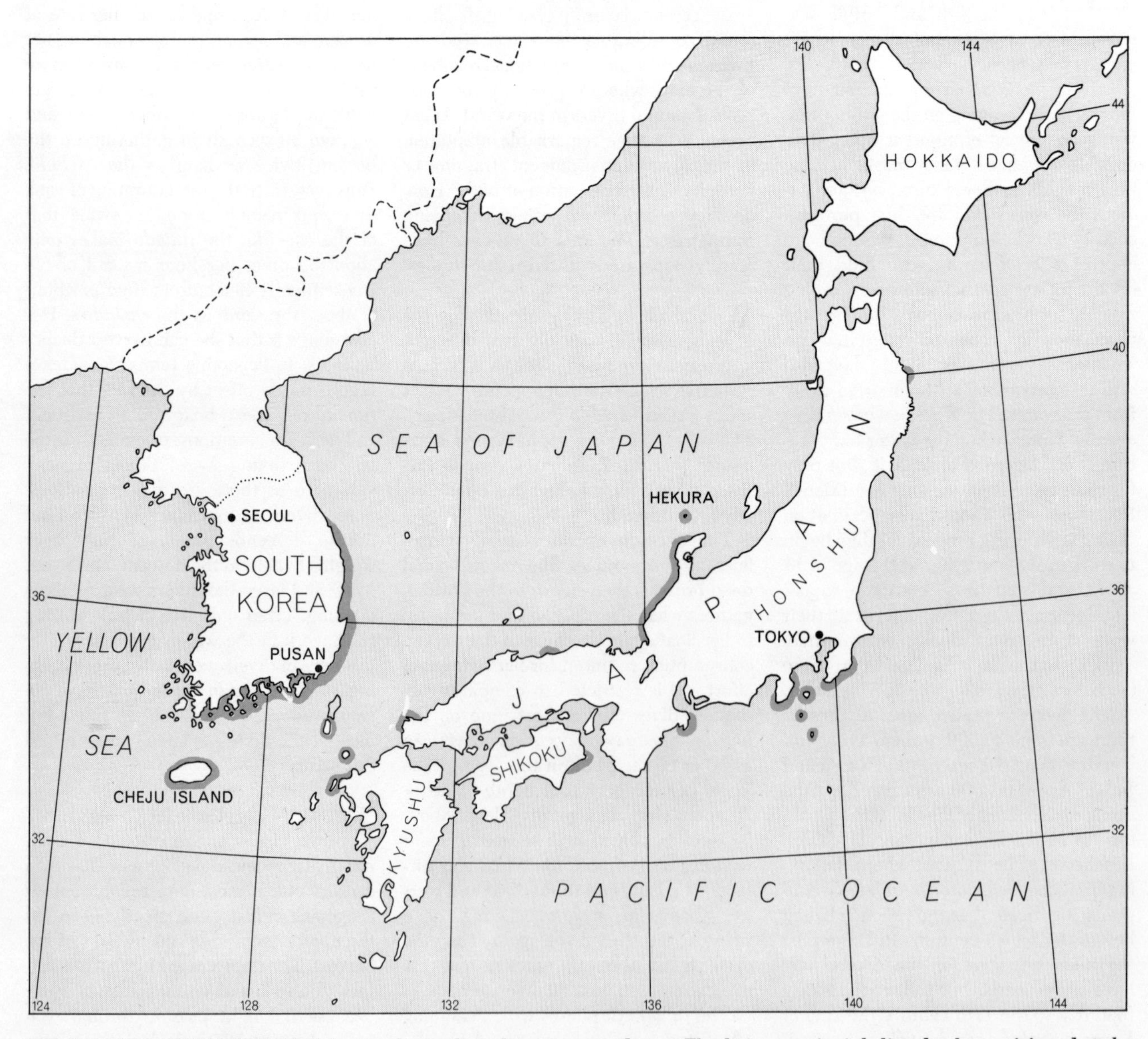

GEOGRAPHIC DISTRIBUTION of the ama divers along the seacoasts of South Korea and southern Japan is indicated by the colored areas. The diving practice is believed to have originated on the small island of Cheju off the southern tip of the Korean peninsula.

remove a considerable amount of carbon dioxide from the blood. The final breath, however, is not a full one but only about 85 percent of what the lungs can hold. Just why the ama limits this breath is not clear; perhaps she does so to avoid uncomfortable pressure in the lungs or to restrict the body's buoyancy in the water.

As the diver descends the water pressure compresses her chest and consequently her lung volume. The depth to which she can go is limited, of course, by the amount of lung compression she can tolerate. If she dives deeper than the level of maximum lung compression (her "residual lung volume"), she becomes subject to a painful lung squeeze; moreover, because the hydrostatic pressure in her blood vessels then exceeds the air pressure in her lungs, the pulmonary blood vessels may burst.

The diver, as we have noted, starts her dive with a lungful of air that is comparatively rich in oxygen and comparatively poor in carbon dioxide. What happens to the composition of this air in the lungs, and to the exchange with the blood, during the dive? In order to investigate this question we needed a means of obtaining samples of the diver's lung air under water without risk to the diver. Edward H. Lanphier and Richard A. Morin of our group (from the State University of New York at Buffalo) devised a simple apparatus into which the diver could blow her lung air and then reinhale most of it, leaving a small sample of air in the device. The divers were understandably reluctant at first to try this device, because it meant giving up their precious lung air deep under water with the possibility that they might not recover it, but they were eventually reassured by tests of the apparatus.

We took four samples of the diver's lung air: one before she entered the water, a second when she had hyperventilated her lungs at the surface and was about to dive, a third when she reached the bottom at a depth of 40 feet and a fourth after she had returned to the surface. In each sample we measured the concentrations and calculated the partial pressures of the principal gases: oxygen, carbon dioxide and nitrogen.

Normally, in a resting person out of the water, the air in the alveoli of the lungs is 14.3 percent oxygen, 5.2 percent carbon dioxide and 80.5 percent nitrogen (disregarding the rare gases and water vapor). We found that after hyperventilation the divers' alveolar air con-

KOREAN DIVING WOMAN from Cheju Island cooperated with the authors in their study of the physiological reactions to breath-hold diving. The large ball slung over her left shoulder is a float that is left at the surface during the dive; attached to the float is a net for collecting the catch. The black belt was provided by the authors to carry a pressure-sensitive bottle and electrocardiograph wires for recording the heart rate. The ama holds an alveolar, or lung, gas sampler in her right hand. The Korean ama wear only light cotton bathing suits even in the winter, when the water temperature can be as low as 50 degrees Fahrenheit.

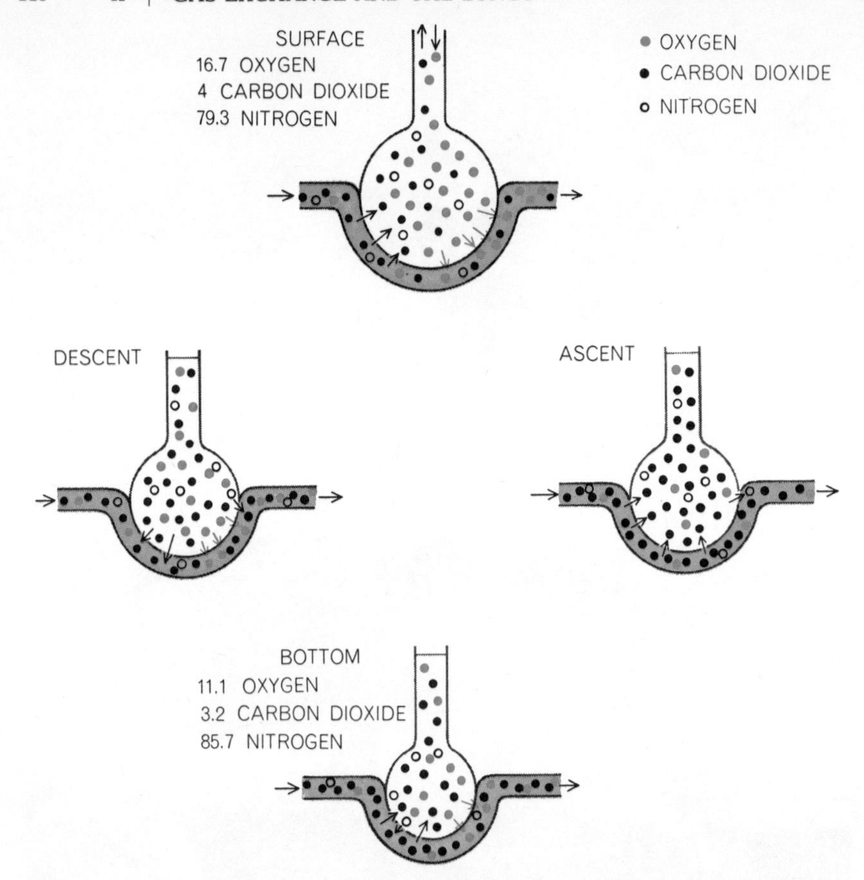

GASES EXCHANGED between a single alveolus, or lung sac, and the bloodstream are shown for four stages of a typical ama dive. The concentrations of three principal gases in the lung at the surface and at the bottom are given in percent. During descent water pressure on the lungs causes all gases to enter the blood. During ascent this situation is reversed.

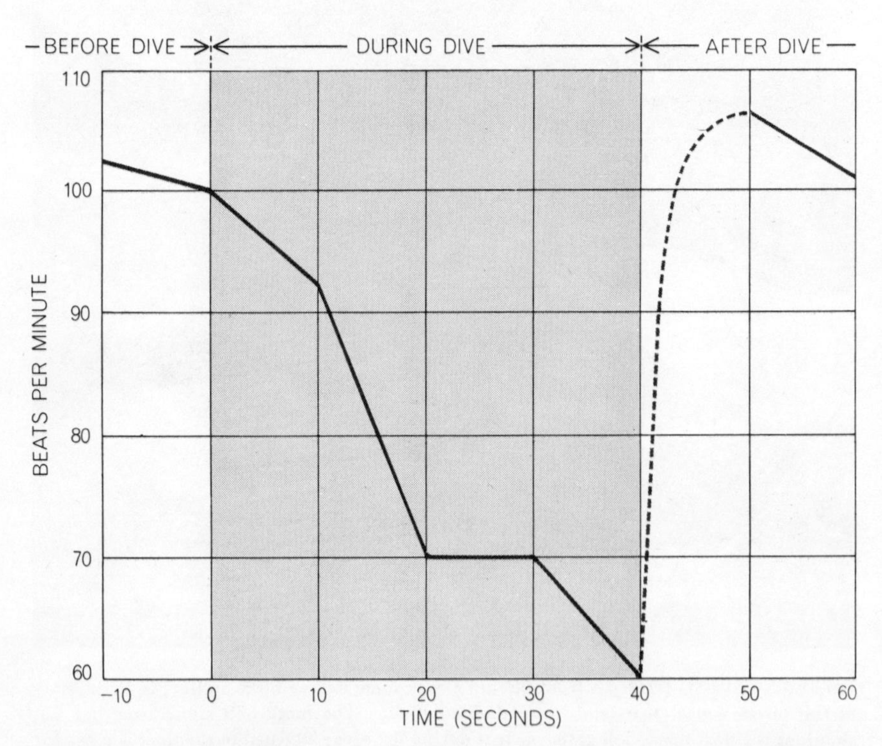

AVERAGE HEART RATE for a group of Korean ama was measured before, during and after their dives. All the dives were to a depth of about 15 feet. The average pattern shown here was substantially the same in the summer, when the water temperature was about 80 degrees Fahrenheit, as it was in winter, when water temperature was about 50 degrees F.

sists of 16.7 percent oxygen, 4 percent carbon dioxide and 79.3 percent nitrogen; translating these figures into partial pressures (in millimeters of mercury), the respective proportions are 120 millimeters for oxygen, 29 for carbon dioxide and 567 for nitrogen.

By the time the *cachido* (unassisted diver) reaches the bottom at a depth of 40 feet the oxygen concentration in her lungs is reduced to 11.1 percent, because of the uptake of oxygen by the blood. However, since at that depth the water pressure has compressed the lungs to somewhat more than half of their pre-dive volume, the oxygen pressure amounts to 149 millimeters of mercury—a greater pressure than before the dive. Consequently oxygen is still being transmitted to the blood at a substantial rate.

For the same reason the blood also takes up carbon dioxide during the dive. The carbon dioxide concentration in the lungs drops from 4 percent at the beginning of the dive to 3.2 percent at the bottom. This is somewhat paradoxical; when a person out of the water holds his breath, the carbon dioxide in his lungs increases. At a depth of 40 feet, however, the compression of the lung volume raises the carbon dioxide pressure to 42 millimeters of mercury, and this is greater than the carbon dioxide pressure in the venous blood. As a result the blood and tissues retain carbon dioxide and even absorb some from the lungs.

As the diver ascends from the bottom, the expansion of the lungs drastically reverses the situation. With the reduction of pressure in the lungs, carbon dioxide comes out of the blood rapidly. Much more important is the precipitous drop of the oxygen partial pressure in the lungs: within 30 seconds it falls from 149 to 41 millimeters of mercury. This is no greater than the partial pressure of oxygen in the venous blood; hence the blood cannot pick up oxygen, and Lanphier has shown that it may actually lose oxygen to the lungs. In all probability that fact explains many of the deaths that have occurred among sports divers returning to the surface after deep, lengthy dives. The cumulative oxygen deficiency in the tissues is sharply accentuated during the ascent.

Our research has also yielded a measure of the nitrogen danger in a long dive. We found that at a depth of 40 feet the nitrogen partial pressure in the compressed lungs is doubled (to 1,134 millimeters of mercury), and throughout the dive the nitrogen tension is sufficient to drive the gas into the blood. Lanphier has calculated that repeated dives to depths of 120 feet, such as are performed by male pearl divers in the Tuamotu Archipelago of the South Pacific, can result in enough accumulation of nitrogen in the blood to cause the bends on ascent. When these divers come to the surface they are sometimes stricken by fatal attacks, which they call *taravana.*

The ama of the Korean area are not so reckless. Long experience has taught them the limits of safety, and, although they undoubtedly have some slight anoxia at the end of each dive, they quickly recover from it. The diving women content themselves with comparatively short dives that they can perform again and again for extended periods without serious danger. They avoid excessive depletion of oxygen and excessive accumulation of nitrogen in their blood.

As far as we have been able to determine, the diving women possess no particular constitutional aptitudes of a hereditary kind. The daughters of Korean farmers can be trained to become just as capable divers as the daughters of divers. The training, however, is important. The most significant adaptation the trained diving women show is an unusually large "vital capacity," measured as the volume of air that can be drawn into the lungs in a single inspiration after a complete expiration. In this attribute the ama are substantially superior to nondiving Korean women. It appears that the divers acquire this capacity through development of the muscles involved in inspiration, which also serve to resist compression of the chest and lung volume in the water.

A large lung capacity, or oxygen intake, is one way to fortify the body for diving; another is conservation of the oxygen stored in the blood. It is now well known, thanks to the researches of P. F. Scholander of the Scripps Institution of Oceanography and other investigators, that certain diving mammals and birds have a built-in mechanism that minimizes their need for oxygen while they are under water [see "The Master Switch of Life," by P. F. Scholander; SCIENTIFIC AMERICAN, May, 1967]. This mechanism constricts the blood vessels supplying the kidneys and most of the body muscles so that the blood flow to these organs is drastically reduced; meanwhile a normal flow is maintained to the heart, brain and other organs that require an undiminished supply of oxygen. Thus the heart can slow down, the rate of removal of oxygen from the blood by tissues is reduced,

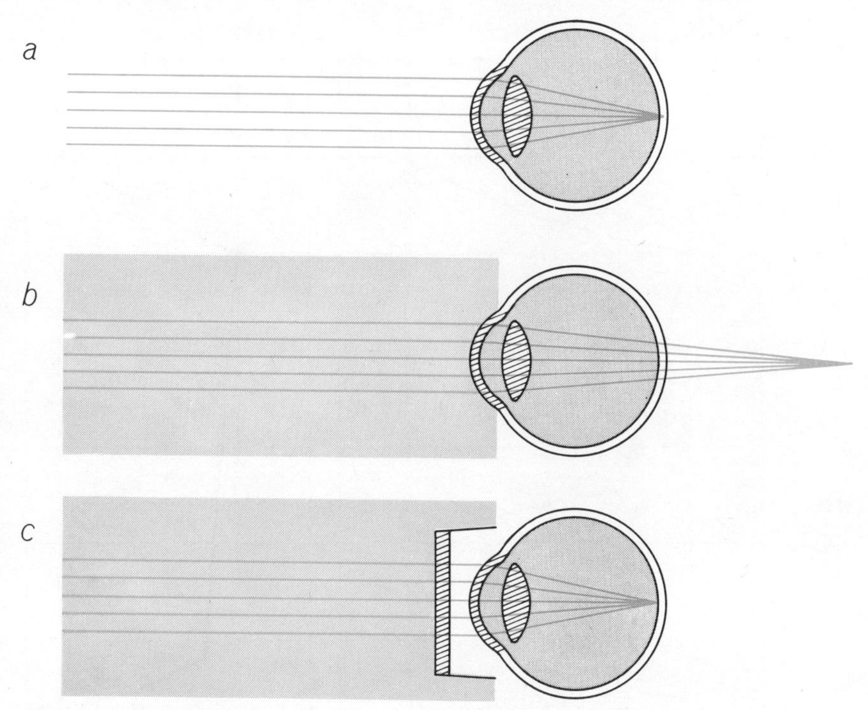

GOGGLES SHARPEN VISION under water by providing a layer of air at the interface with the eyeball (*c*). Vision is normally sharp in air because the difference in refractive index between air and the tissue of the cornea helps to focus images on the retinal surface (*a*). The small difference in the refractive index between water and corneal tissue causes the focal point to move considerably beyond the retina (*b*), reducing visual acuity under water.

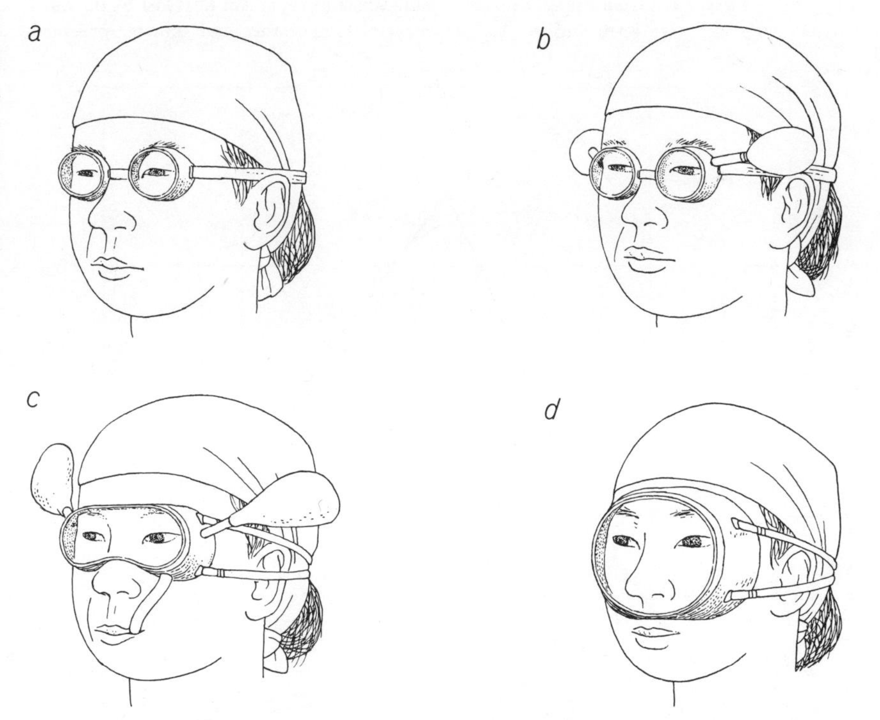

EVOLUTION OF GOGGLES has resulted in several solutions to the problem presented by the increase in hydrostatic pressure on the body during a dive. The earliest goggles (*a*) were uncompensated, and the difference in pressure between the blood vessels in the eyelid and the air behind the goggles could result in conjunctival bleeding. The problem was first solved by attaching air-filled, compressible bags to the goggles (*b*). During a dive the increasing water pressure compresses the bags, raising the air pressure behind the goggles in proportion to the increase in hydrostatic pressure on the body. In some cases (*c*) the lungs were used as an additional compensating gas chamber. With a modern face mask that covers the nose (*d*) the lungs provide the only source of compensating air pressure during a dive.

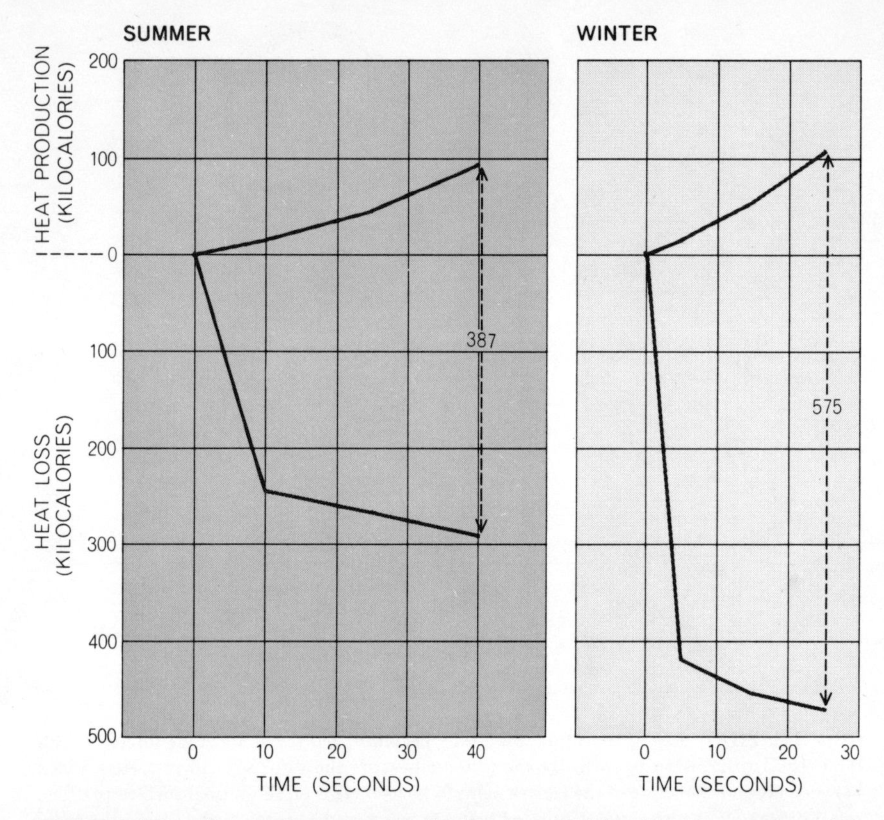

BODY HEAT lost by ama divers was found to be about 400 kilocalories in a summer shift (*left*) and about 600 kilocalories in a winter shift (*right*). The curves above the abscissa at zero kilocalories represent heat generated by swimming and shivering and were estimated by the rate of oxygen consumption. The curves below abscissa represent heat lost by the body to the water and were estimated by changes in rectal temperature and skin temperature.

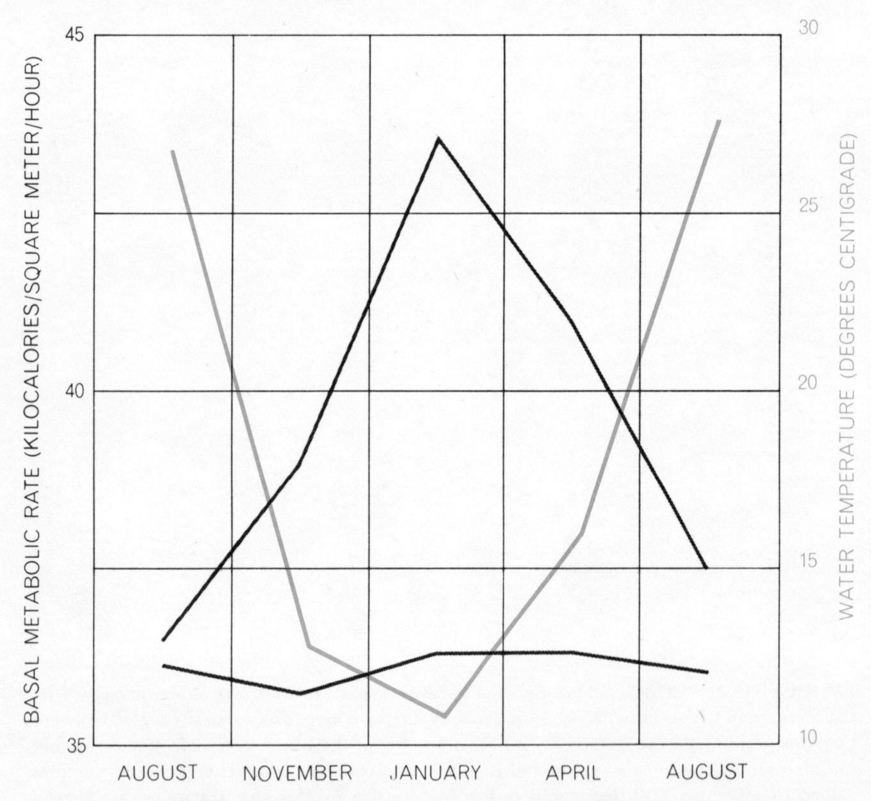

BASAL METABOLIC RATE of ama women (*top gray curve*) increases in winter and decreases in summer. In nondiving Korean women (*bottom gray curve*) basal metabolic rate is constant throughout the year. The colored curve shows the mean seawater temperatures in the diving area of Pusan harbor for the same period covered by the other measurements.

and the animal can prolong its dive.

Several investigators have found recently that human subjects lying under water also slow their heart rate, although not as much as the diving animals do. We made a study of this matter in the ama during their dives. We attached electrodes (sealed from contact with the seawater) to the chests of the divers, and while they dived to the bottom, at the end of a 100-foot cable, an electrocardiograph in our boat recorded their heart rhythms. During their hyperventilation preparatory to diving the divers' heart rate averaged about 100 beats a minute. During the dive the rate fell until, at 20 seconds after submersion, it had dropped to 70 beats; after 30 seconds it dropped further to some 60 beats a minute [*see bottom illustration on page 110*]. When the divers returned to the surface, the heart rate jumped to slightly above normal and then rapidly recovered its usual beat.

Curiously, human subjects who hold their breath out of the water, even in an air pressure chamber, do not show the same degree of slowing of the heart. It was also noteworthy that in about 50 percent of the dives the ama showed some irregularity of heartbeat. These and other findings raise a number of puzzling questions. Nevertheless, one thing is quite clear: the automatic slowing of the heart is an important factor in the ability of human divers to extend their time under water.

In the last analysis the amount of time one can spend in the water, even without holding one's breath, is limited by the loss of body heat. For the working ama this is a critical factor, affecting the length of their working day both in summer and in winter. (They warm themselves at open fires after each long diving shift.) We investigated the effects of their cold exposure from several points of view, including measurements of the heat losses at various water temperatures and analysis of the defensive mechanisms brought into play.

For measuring the amount of the body's heat loss in the water there are two convenient indexes: (1) the increase of heat production by the body (through the exercise of swimming and shivering) and (2) the drop in the body's internal temperature. The body's heat production can be measured by examining its consumption of oxygen; this can be gauged from the oxygen content of the lungs at the end of a dive and during recovery. Our measurements were made on Korean diving women in Pusan harbor at two seasons of the year: in August,

UNASSISTED DIVER, called a *cachido*, employs one of the two basic techniques of ama diving. The *cachido* operates from a small float at the surface. On an average dive she swims to a depth of about 15 to 20 feet; the dive lasts about 25 to 30 seconds, of which 15 seconds is spent working on the bottom. The entire diving cycle takes about a minute, and the diver averages 60 dives per hour.

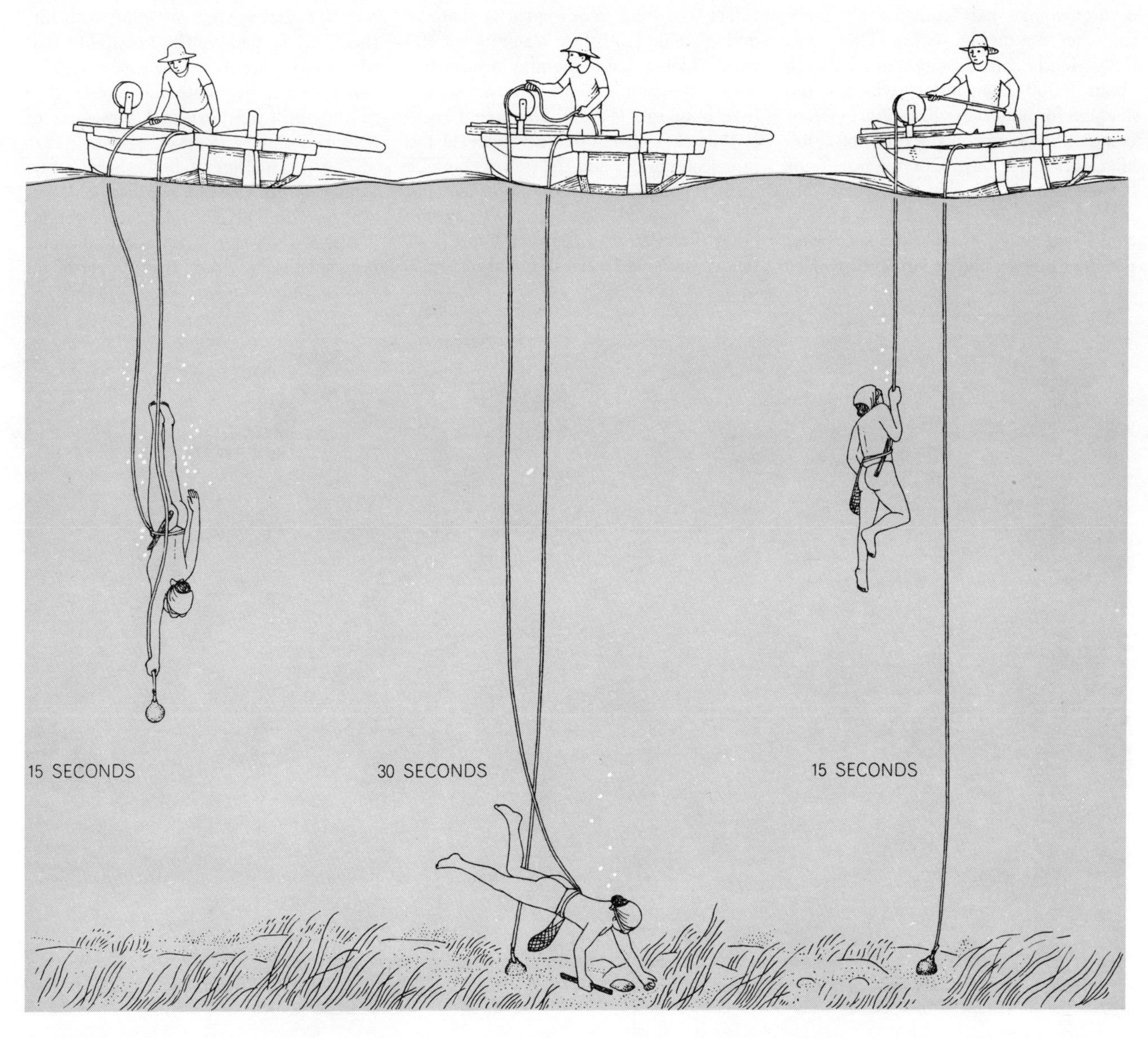

ASSISTED DIVER, called a *funado*, uses a counterweight to descend passively to a depth of 60 to 80 feet. She averages 30 seconds in gathering on the bottom but makes only about 30 dives per hour. At the end of each dive a helper in the boat pulls her up.

when the water temperature was 80.6 degrees F., and in January with the water temperature at 50 degrees.

In both seasons at the end of a single diving shift (40 minutes in the summer, 25 minutes in winter) the deep-body temperature was found to be reduced from the normal 98.6 degrees F. to 95 degrees or less. Combining this information with the measurements of oxygen consumption, we estimated that the ama's body-heat loss was about 400 kilocalories in a summer shift and about 600 kilocalories in a winter shift. On a daily basis, taking into consideration that the ama works in the water for three long shifts each day in summer and only one or two short shifts in winter, the day's total heat loss is estimated to be about the same in all seasons: approximately 1,000 kilocalories per day.

To compensate for this loss the Korean diving woman eats considerably more than her nondiving sisters. The ama's daily food consumption amounts to about 3,000 kilocalories, whereas the average for nondiving Korean women of comparable age is on the order of 2,000 kilocalories per day. Our various items of evidence suggest that the Korean diving woman subjects herself to a daily cold stress greater than that of any other group of human beings yet studied. Her extra food consumption goes entirely into coping with this stress. The Korean diving women are not heavy; on the contrary, they are unusually lean.

It is interesting now to examine whether or not the diving women have developed any special bodily defenses against cold. One such defense would be an elevated rate of basal metabolism, that is, an above-average basic rate of heat production. There was little reason, however, to expect to find the Korean women particularly well endowed in this respect. In the first place, populations of mankind the world over, in cold climates or warm, have been found to differ little in basal metabolism. In the second place, any elevation of the basal rate that might exist in the diving women would be too small to have much effect in offsetting the large heat losses in water.

Yet we found to our surprise that the diving women did show a significant elevation of the basal metabolic rate—but only in the winter months! In that season their basal rate is about 25 percent higher than that of nondiving women of the same community and the same economic background (who show no seasonal change in basal metabolism). Only one other population in the world has been found to have a basal metabolic rate as high as that of the Korean diving women in winter: the Alaskan Eskimos. The available evidence indicates that the warmly clothed Eskimos do not, however, experience consistently severe cold stresses; their elevated basal rate is believed to arise from an exceptionally large amount of protein in their diet. We found that the protein intake of Korean diving women is not particularly high. It therefore seems probable that their elevated basal metabolic rate in winter is a direct reflection of their severe exposure to cold in that season, and that this in turn indicates a latent human mechanism of adaptation to cold that is evoked only under extreme cold stresses such as the Korean divers experience. The response is too feeble to give the divers any significant amount of protection in the winter water. It does, however, raise an interesting physiological question that we are pursuing with further studies, namely the possibility that severe exposure to winter cold may, as a general rule, stimulate the human thyroid gland to a seasonal elevation of activity.

The production of body heat is one aspect of the defense against cold; another is the body's insulation for retaining heat. Here the most important factor (but not the only one) is the layer of fat

BETWEEN DIVES the ama were persuaded to expire air into a large plastic gas bag in order to measure the rate at which oxygen is consumed in swimming and diving to produce heat. The water temperature in Pusan harbor at the time (January) was 50 degrees F. One of the authors (Hong) assists. Data obtained in this way were used to construct the graph at the top of page 112.

under the skin. The heat conductivity of fatty tissue is only about half that of muscle tissue; in other words, it is twice as good an insulator. Whales and seals owe their ability to live in arctic and antarctic waters to their very thick layers of subcutaneous fat. Similarly, subcutaneous fat explains why women dominate the diving profession of Korea and Japan; they are more generously endowed with this protection than men are.

Donald W. Rennie of the State University of New York at Buffalo collaborated with one of the authors of this article (Hong) in detailed measurements of the body insulation of Korean women, comparing divers with nondivers. The thickness of the subcutaneous fat can easily be determined by measuring the thickness of folds of skin in various parts of the body. This does not, however, tell the whole story of the body's thermal insulation. To measure this insulation in functional terms, we had our subjects lie in a tank of water for three hours with only the face out of the water. From measurements of the reduction in deep-body temperature and the body's heat production we were then able to calculate the degree of the subject's overall thermal insulation. These studies revealed three particularly interesting facts. They showed, for one thing, that with the same thickness of subcutaneous fat, divers had less heat loss than nondivers. This was taken to indicate that the divers' fatty insulation is supplemented by some kind of vascular adaptation that restricts the loss of heat from the blood vessels to the skin, particularly in the arms and legs. Secondly, the observations disclosed that in winter the diving women lose about half of their subcutaneous fat (although nondivers do not). Presumably this means that during the winter the divers' heat loss is so great that their food intake does not compensate for it sufficiently; in any case, their vascular adaptation helps them to maintain insulation. Thirdly, we found that diving women could tolerate lower water temperatures than nondiving women without shivering. The divers did not shiver when they lay for three hours in water at 82.8 degrees F.; nondivers began to shiver at a temperature of 86 degrees. (Male nondivers shivered at 88 degrees.) It appears that the diving women's resistance to shivering arises from some hardening aspect of their training that inhibits shiver-triggering impulses from the skin. The inhibition of shivering is an advantage because shivering speeds up the emission of body heat. L. G. Pugh, a British physiologist who has studied long-distance swimmers, discovered the interesting additional fact that swimmers, whether fat or thin, lose heat more rapidly while swimming than while lying motionless in the water. The whole subject of the body's thermal insulation is obviously a rather complicated one that will not be easy to unravel. As a general conclusion, however, it is very clearly established that women are far better insulated than men against cold.

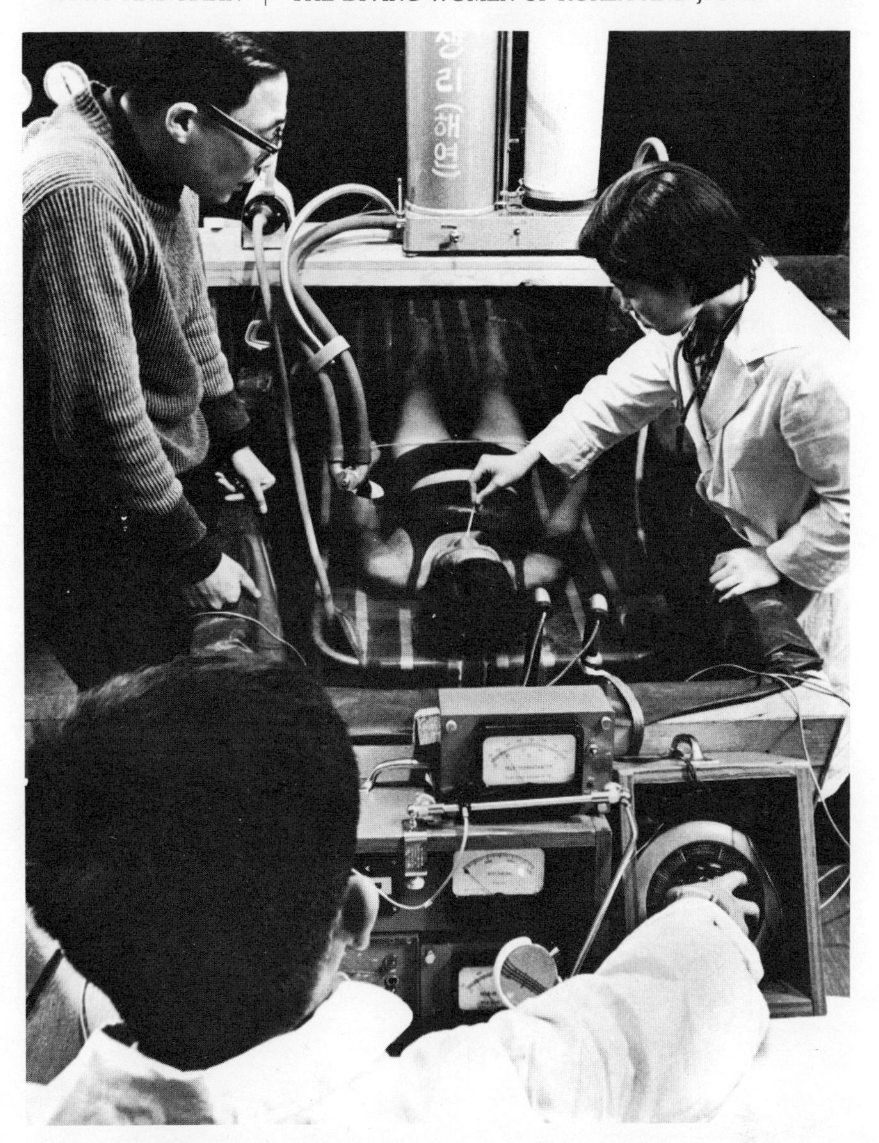

AMA'S THERMAL INSULATION (mainly fat) was measured by having the subjects lie in a tank of water for three hours. From measurements of the reduction of deep-body temperature and the body's heat production the authors were able to calculate the degree of the subject's overall thermal insulation. Once again Hong (*left*) keeps a close eye on the operation.

As a concluding observation we should note that the 1,500-year-old diving occupation in Korea and Japan is now declining. The number of divers has dwindled during the past few decades, and by the end of this century the profession may disappear altogether, chiefly because more remunerative and less arduous ways of making a living are arising. Nonetheless, for the 30,000 practitioners still active in the diving profession (at least in summer) diving remains a proud calling and necessary livelihood. By adopting scuba gear and other modern underwater equipment the divers could greatly increase their production; the present harvest could be obtained by not much more than a tenth of the present number of divers. This would raise havoc, however, with employment and the economy in the hundreds of small villages whose women daily go forth to seek their families' existence on the sea bottom. For that reason innovations are fiercely resisted. Indeed, many villages in Japan have outlawed the foam-rubber suit for divers to prevent too easy and too rapid harvesting of the local waters.

III

WATER BALANCE AND ITS CONTROL

In Africa the greater part of the wild animals do not drink at all in summer, owing to lack of rains for which reason Libyan mice in captivity die if given drink. The perpetually dry parts of Africa produce the antelope, which owing to the nature of the region goes without drink in quite a remarkable fashion, for the assistance of thirsty people, as the Gaetulian brigands rely on their help to keep going, bladders containing extremely healthy liquid being found in their body.

Pliny
NATURAL HISTORY, V, XCIV

III

WATER BALANCE AND ITS CONTROL

INTRODUCTION

The body of higher vertebrates is about 60 percent water. Somewhat less than half of the water is extracellular, and in it are dissolved a variety of organic and inorganic molecules. Because animals are open systems in that they take in and give off water and many other types of molecules each day, a regulating system that maintains the constancy of the body water and its constituents has evolved. As summarized by A. V. Wolf ("Body Water," *Scientific American,* November 1958), the relation among the quantity of water an animal drinks, the volume of urine it produces, and evaporative loss is so controlled that the total amount of body water is kept within a few percent of its average value. Control of salt content and the balance between different salt ions is equally strict. The core of the system is, of course, the kidney and its physiology.

The functional units of the kidney—the nephrons—appeared early in vertebrate evolution. Each nephron was originally a duct that drained the body cavity (the coelom) to the exterior. In later forms, a cluster of blood capillaries that adjoined a portion of the duct evolved, and the connection between the duct and coelom was closed off. Thereafter, the ducts drained fluid and molecules that filtered from the blood capillaries into the cavity of the ducts. The most primitive kidney we know today is in lamprey larvae, in which a glomus (a tangle of blood vessels) is located in the wall of the coelom near the opening of each nephron. Similar glomi and nephrons are present in adult hagfish (like the lamprey, a cyclostome), whose body fluids have the same osmotic pressure as sea water. Some regulation within the hagfish occurs, however, because the ratio between various ions in the body is different from the ratio between those ions in sea water. In adult lampreys and all higher vertebrates, the glomus is inserted into the end, or funnel, of the nephron, and in these organisms it is called the glomerulus. The adult lampreys and all fresh-water fishes require an active regulatory system because their blood is hypertonic to the fresh water in which they live (that is, it has osmotic pressure higher than the fresh water). This system must eliminate the water that tends to enter the body because of the osmotic gradient, and yet retain the salts, which tend to flow out from the body (because they tend to flow from a region of high concentration—the body fluids—to one of low concentration—the watery habitat).

The glomerular portion of the nephron is the site at which water, salts, and some organic molecules are forced through the wall of the glomerular capillary and nephron capsule. This filtration is directly dependent upon blood pressure in the glomerulus. As the filtrate flows from the capsule and through the tubular portion of the nephron, water, glucose, small proteins, chloride ions, and some other molecules are reabsorbed to varying degrees, whereas other substances—such as urea, magnesium ions, and potassium ions—are secreted into the filtrate by tubule cells. The end product, of course, is the urine. The system functions at a remarkable rate in humans; about 175 quarts of fluid per day pass from the blood into the capsules of the kidney nephrons and all but one or two quarts are resorbed by the tubules. Homer W. Smith points out in "The Kidney" (*Scientific American* Offprint 37) that 2.5 pounds of salt enter the capsules with the water, but only a third of an ounce is lost in urine.

The ability of the mammalian kidney to produce urine hypertonic to blood is dependent upon a countercurrent flow system that transfers sodium ions (Na^+) from one portion of the kidney tubule to the other; consequently, high sodium concentration can be maintained in the medullary (central) part of the kidney. Refer to Figure 1 for help in visualizing the following relationships. First, for clarity, we must trace the route of urine flow: nephron capsule, proximal tubule, descending loop, ascending loop, distal tubule, and, finally, collecting duct. Most of the active transport of sodium ions takes place across the thick-walled ascending loops. The sodium ions removed from these loops are added to the intercellular fluid and then transferred back to the descending loops; thus, the sodium ions tend to move in a circle—descending loop to ascending loop to intercellular fluid to descending loop. The result is to create a very high concentration of sodium ions within and between these loops, that is, in this whole "medullary" portion of the kidney. Consider the effects on water flowing down the kidney tubule: because of the high sodium ion concentration, water tends to move through the walls of both the descending and ascending loops; the result is to reduce the *volume* of urine that ultimately leaves the top of the ascending loop and enters the distal tubule and collecting duct. Furthermore, since sodium ions have also left the urine as it passes upward through the ascending loop, the concentration of the urine is lowered; that is, it becomes dilute. Next, the urine flows down through the thin-walled collecting ducts that actually pass among the medullary descending and ascending loops; recall that the extracellular sodium ion concentration is very high there (four times that of blood, in humans). The result is to remove more water from the urine in the collecting ducts. This, of course, raises the concentration of the urine (until it is hypertonic to the blood). In the end, a small volume of highly concentrated urine is produced and sent on to the bladder.

The control system for kidney function is marvelously complex and depends upon activity of both nerve and endocrine cells. Primary control resides in the brain, where osmoreceptor cells of the hypothalamus control the release of antidiuretic hormone (vasopressin). These nerve cells respond to alterations in

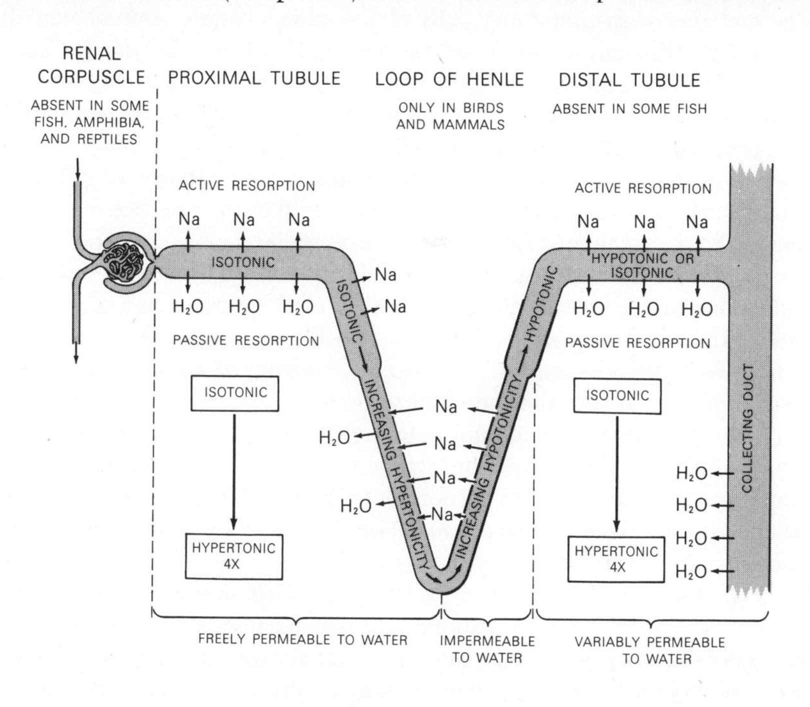

Figure 1. **A diagrammatic representation of a nephron unit, showing the manner in which hypertonic urine is produced. [After Robert F. Pitts, *The Physiological Basis of Diuretic Therapy*, Charles C. Thomas, 1959.]**

the osmotic pressure of blood plasma flowing through the brain and make compensatory adjustments in quantities of vasopressin released in the posterior pituitary.

The hormone affects mammalian kidney nephrons (particularly distal tubules and collecting ducts) and the bladder wall to increase resorption of water. In other organisms, it also decreases glomerular filtration rate. Interestingly, this peptide hormone can work physiologically only if it is applied to the connective tissue or basal side of the tissue. The mechanism of action appears to include stimulated production of cyclic-AMP (adenosine-3′,5′-phosphate), aggregation of intramembranous particles at sites where water will be resorbed, and altered cell permeability. In amphibians vasopressin activates a sodium pump and so causes increased retention of salts (and so, indirectly, of osmotic water).

An important element of chemical regulation of physiological processes is the short half-life of chemical mediators in the body; if the half-life were not short, the control machinery would be relatively insensitive to altering conditions in the body. Vasopressin of various vertebrates has a half-life that varies between 1 and 24 minutes; under most circumstances the shorter times (1 to 4 minutes) apply. The half-life is governed by destruction of the peptide; two-thirds is removed from the blood and inactivated by the kidney and the remaining one-third is destroyed in the liver.

Several other coordinating agents control urine content. Aldosterone, produced by the cortex of the adrenal gland, controls resorption of sodium ions by the distal tubules of the nephrons. Aldosterone secretion is regulated by angiotensin II, a substance produced in the kidney when an enzyme called renin is released from cells of the juxtaglomerular apparatus. Renin acts on α-2-globulin, one of the proteins in serum, to change it into angiotensin I. Then another blood protein, the "converting enzyme," splits two amino acids from this decapeptide to produce the active octapeptide, angiotensin II. Besides affecting aldosterone release, this agent causes contraction of smooth muscle cells in the walls of blood capillaries, both in the kidney and elsewhere in the body. It seems significant that the posterior pituitary hormones vasopressin and oxytocin, both of which produce contraction of smooth muscles in various other organs of the body, also are composed of eight amino acids.

Angiotensin II has a vital function in a feedback-control loop of the mammalian nephron. The apparatus of this loop is composed of the first portion of the distal tubule (see Figure 1), the cells of which are called the macula densa cells, and the renin-containing cells of the blood vessels approaching the glomerulus. Although in the figure the two kinds of cells appear far apart, in an intact kidney the various parts of the nephron are in fact twisted about one another and, specifically, the macula densa cells of the tubule and the renin-containing cells of the blood vessels really interdigitate with one another; thus, they can interact easily. The sodium concentration of either the tubular fluid (the urine) near the macula densa cells or the resorbed fluid in the same regions is somehow measured. If, for instance, sodium concentrations rise in urine within the distal tubule, then renin is released, thereby leading to production of angiotensin II and accumulation of aldosterone. As a result, more sodium is resorbed through the distal tubule walls (because of aldosterone) and blood flow to the glomerulus is decreased (because of angiotensin II); this reduces glomerular filtration rate and so, sodium loss. Therefore, the loop is complete: the quantity of sodium entering the distal tubular portion of the nephron ultimately determines the filtration rate back in the glomerulus. Thus, a sodium-sensitive feedback loop coordinates glomerular and tubular function, and operates through proteins produced by the kidney (renin) and by the liver (globulin and converting enzyme).

It is often said that the length of the loop of Henle (comprising the ascending and descending loops described above) is the unique feature of mammalian and avian nephrons that permits production of urine that is hypertonic to blood. Recent observations have disproved that generalization. A number of terrestrial birds (such as the Savannah sparrow and the house finch) that lack functional salt glands (see the article "Salt Glands" by Knut Schmidt-Nielsen) have been investigated, and they possess extraordinarily large numbers of relatively short loops of Henle. Yet hypertonic urine is produced by these kidneys. Thus the same physiological ability can be derived from either anatomical arrangement—the long loops or the large numbers of short loops.

Those birds and reptiles that cannot produce a hypertonic urine possess a salt-secreting gland near the eyes. Work by Schmidt-Nielsen and his collaborators shows that this gland is necessary because of an indirect effect of the final water resorption that occurs in the cloaca (the final chamber through which kidney and intestinal products pass). Water is taken back into the body through the walls of the cloaca, thus drying out the feces and uric acid. Cations tend to move with the water and they must be excreted elsewhere; this excretion is the function of the salt glands of the head. Physiologically the salt-secreting glands act like kidney tubules in response to vasopressin and hypothalamic control.

The biochemical basis for salt gland function is beginning to be understood. A Na^+,K^+,-dependent ATPase appears to be functional at the luminal (outer) end of salt-gland cells and as a result Na^+ ions are pumped out of the cell and the body. What is surprising about salt-gland function is that the *continuous* presence of the neurotransmitter acetylcholine is required for active pumping. The system works like this: receptors in the heart detect an elevation in blood tonicity caused by the drinking of sea water; that information is sent to the brain whence nerve discharge at the salt gland liberates acetylcholine, and Na^+ pumping starts. It is not yet known whether the acetylcholine acts directly on the ATPase enzyme or via cyclic AMP, as is often the case in other situations.

With these comments on salt glands as background, it is worth turning to the closely related case of fish adapted for life in the sea. Salts tend to enter their bodies and must also be excreted, just as in marine birds. Interestingly, the same sort of Na^+,K^+,-dependent ATPase appears in the gill "chloride" cells and functions to pump Na^+ ions from the blood, against the concentration gradient, into the sea.

Fish in fresh water have the opposite problem, one that is solved by a shift in the type of enzyme activities present. Thus, the Na^+,K^+,-dependent ATPase is lost; a new Na^+,-dependent ATPase activity appears, and the direction of pumping is altered. Na^+ ions are pumped from fresh water, once again opposite to the concentration gradient, into the blood. In a steelhead trout which migrates back and forth between fresh and salt water, or in brackish-water fishes, both enzyme activities apparently are present.

We see in these cases important supplementary adaptations that complement the kidneys. Another interesting adaptation of various vertebrates is the ability to retain large quantities of urea in the body fluids. Sharks, their relatives, and various amphibia that live in saline environments build up high concentrations of urea in order to counteract the osmotic gradient resulting from higher salt concentrations in the sea outside the body. The urea molecules act as osmotic particles that raise internal osmotic pressure and reduce the tendency for water to leave the body.

In the crab-eating frog *(Rana cancrivora),* immersion in brackish sea water leads to an actual change in metabolism, so that the urea-cycle enzymes are activated, and the urea produced is not excreted. The high levels of urea in these amphibians and in the elasmobranchs (200 to 500 mM) are sufficient to cause marked alterations in the secondary and tertiary structure of proteins characteristic of organisms with "normal," low urea levels. But high urea levels have no effect on enzymes, on heart or skeletal muscle, and on other organs in creatures adapted to using high levels of urea as an osmotic aid. The means by which proteins and cells become tolerant of high urea concentrations is unknown.

Although these introductory remarks have concentrated on properties of the kidney, there is much more than the kidney involved in permitting survival of

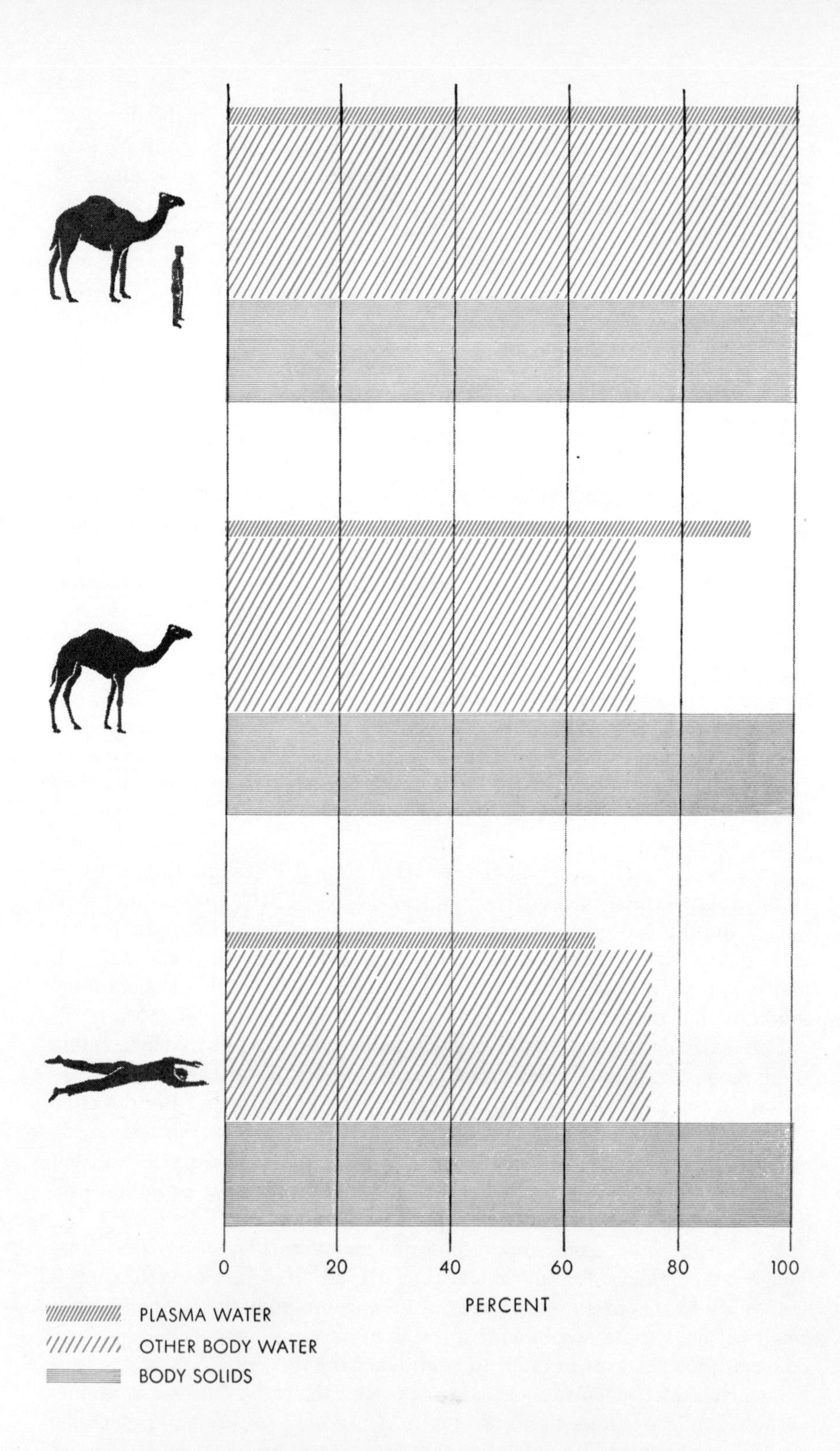

Figure 2. **Camel survives dehydration partly by maintaining the volume of its blood. Under normal conditions (*top*) plasma water in both the camel and the human accounts for about a twelfth of total body water. In a camel that has lost about a fourth of its body water (*center*) the blood volume will drop by less than a tenth. Under the same conditions the human's blood volume will drop by a third (*bottom*). The viscous blood circulates too slowly to carry the human's body heat outward to the skin, so that the human's temperature soon rises to a fatal level.**

[From "The Physiology of the Camel" by Knut Schmidt-Nielsen. Copyright © 1959 by Scientific American, Inc. All rights reserved.]

vertebrates under extreme environmental conditions. Studies of desert animals (see, for instance, Knut Schmidt-Nielsen's "The Physiology of the Camel," *Scientific American* Offprint 1096) have revealed a variety of adaptations to extreme heat and low humidity. An intuitively confusing adaptation relates to insulation. Animals such as camels have thick coats of hair, particularly on their backs. Even during the hottest months of the year in the desert, when the thinner "summer" coat is found over most of the body, the back retains its thick insulating fur. The reason for this is that the insulation is helping to keep heat *out* of the body; consequently the camel can sweat less to maintain a given body temperature. For this same reason, people inhabiting the deserts of the world commonly clothe themselves in several layers of loose-fitting, thick garments. Experience has shown that baring the skin is not a cool move under the desert sun!

Schmidt-Nielsen and his collaborators also made an important discovery concerning blood volume under dehydrating conditions. Their findings are summarized in Figure 2. A human being exposed to the desert sun sweats copiously and becomes severely dehydrated. When about 12 percent of the body water is lost, the individual most likely undergoes "explosive heat death." This may result in large part from the fact that the blood volume—total plasma water—decreases even more precipitously, thus causing a drastic rise in blood viscosity. Even when a camel has lost nearly one-quarter of total body water, its plasma water content falls less than 10 percent; the blood viscosity is not changed appreciably, and so heat death is avoided. How this maintenance of water is achieved in one of the body's compartments is another mystery of vertebrate biology.

C. R. Taylor, in "The Eland and the Oryx," discusses other adaptations for desert survival. In the oryx, we see a good example of a rete bed being used for cooling (see the article by Mary Ann Baker in Section I). This kind of cooling process no doubt provides an increased opportunity for life under conditions of temperature stress.

ELAND is the largest of all African antelopes. An average adult bull weighs more than half a ton and may measure six feet at the shoulder. A gregarious and docile member of the family Bovidae, the eland can thrive in drought-ridden rangeland unfit for cattle.

ORYX, another large African antelope, is four feet high at the shoulder. It is even better adapted to arid lands than the eland and is found in barren desert. The oryx, however, is far from docile. It wields its long horns readily and has been known to kill lions.

The Eland and the Oryx

by C. R. Taylor
January 1969

These large African antelopes can survive indefinitely without drinking. Their feat is made possible by strategems of physiology that minimize the amount of water they lose through evaporation

When travelers' accounts of snow on the Equator first reached 19th-century London, learned members of the Royal Geographical Society ridiculed the reports. To many zoologists today the existence of antelopes in the deserts of Africa may seem as much of a surprise as equatorial snow was to 19th-century geographers. Such an environment for any member of the family Bovidae is simply unreasonable. If the reports were accompanied by the statement that the animals survive without drinking, some zoologists would replace "unreasonable" with an emphatic "impossible." Yet it is now well known that snow covers the peaks of many equatorial mountains, and it is consistently reported by naturalists, hunters and local people that certain desert antelopes can survive indefinitely without drinking.

Why should desert survival without drinking seem impossible? It is certainly impossible for humans; in the course of a hot day in the desert a man can lose as much as three gallons of water as a result of sweating and evaporation. Without drinking he could not survive one such day. On the other hand, the kangaroo rat and other desert rodents thrive without drinking at all, even when eating dry food. Their water requirements are met by the very small amounts of free water in their food and by the additional water the food yields when it is oxidized in the process of metabolism. Rodents, however, are small; they can escape the high temperatures of the desert day by burrowing underground. No such shelter is available to the larger mammals. In order to regulate their body temperature during the heat of the day they must evaporate substantial quantities of water. Even the camel, probably the best-known desert animal, has the same problem. As the physiologist Knut Schmidt-Nielsen and his collaborators discovered, the camel has an unusual ability to limit its loss of water by evaporation, but it still must drink in order to survive [see "The Physiology of the Camel," by Knut Schmidt-Nielsen; SCIENTIFIC AMERICAN Offprint 1096]. If the antelopes of the African desert did not have similar abilities, survival would be impossible.

Numerous eyewitness accounts identify two antelopes—the eland and the oryx—as animals that do not need drinking water. The eland (*Taurotragus*) is a large, tractable animal that occupies a variety of East African habitats, including the edge of the Sahara. It is easily domesticated, and it has often been proposed as a means of utilizing rangeland that during droughts is too arid for cattle.

Although the eland does not require water, it does not penetrate the most barren deserts. The oryx, on the other hand, is truly a desert species. Unlike the eland, it does not seek shelter during the midday heat but remains exposed to the hot sun throughout the day. The oryx is as aggressive as the eland is tractable. It wields its rapier-like horns with great facility; those who study its physiology get physical as well as mental exercise.

Some years ago, with the help of Charles P. Lyman of Harvard University, I set out to see if these two antelopes really did live in the African deserts without drinking and, if so, how. I had three simple questions. First, do the eland and the oryx possess any unusual mechanisms for conserving water? Second, if they do, how much water do they require when the mechanisms are operating? Third, can they get this amount of water in some way other than drinking? I had the opportunity to investigate these problems at the East African Veterinary Research Organization at Muguga in Kenya, where the directors (initially Howard R. Binns and later Marcel Burdin) generously provided the needed laboratory space and equipment.

The first step was to establish a laboratory environment that would simulate a hot desert and make it possible to find out how much water the antelopes lost.

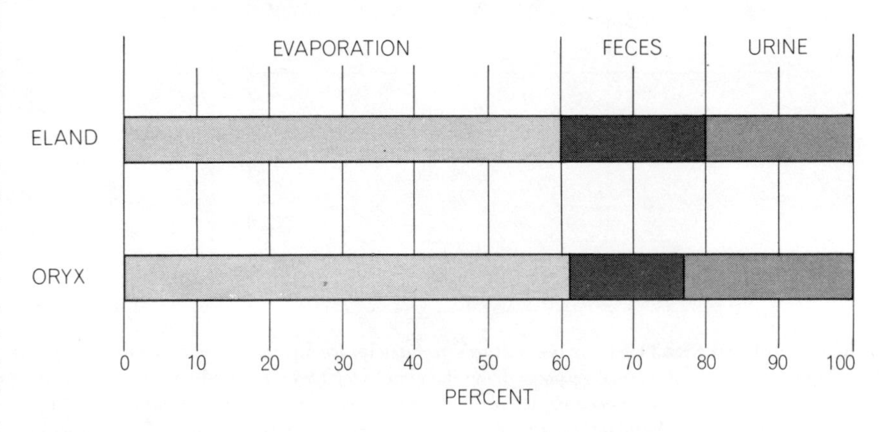

EVAPORATION from the skin and respiratory tract proved to be the major avenue of water loss for antelopes allowed to drink freely in a simulated desert environment. The only means of saving significant amounts of water in the desert heat is to reduce evaporation.

I did not attempt to duplicate the desert's day and night temperature extremes, but I maintained the average daytime temperature for 12 hours and the average nighttime temperature for the same period. With heating and air-conditioning equipment I was able to rapidly raise the temperature of the animals' room to 40 degrees Celsius (104 degrees Fahrenheit) or lower it to 22 degrees C. (72 degrees F.), thus simulating the desert's average day and night temperatures.

In this environment I measured the various ways the animals lost water. I found that loss by evaporation was the most important [*see illustration on preceding page*]. The function served by evaporation is to prevent the animal from overheating. Most mammals maintain their body temperature at a nearly constant level, usually about 37 degrees C. (98.6 degrees F., the "normal" point on a clinical thermometer). If their body temperature rises too high, they die; for most mammals a body temperature of 43 degrees C. (109 degrees F.) for a few hours would be fatal.

When the temperature of the environment is lower than the animal's temperature, which is usually the case, heat flows from the animal to the surroundings (by conduction and radiation). The maintenance of a constant body temperature in these circumstances requires that the animal's metabolic machinery generate an amount of heat equal to the net outward heat flow. When the environmental temperature is higher than the animal's, however, the direction of heat flow reverses and heat is transferred from the environment to the animal. If under these circumstances the body temperature is to remain constant, the heat gained from the environment, as well as the heat generated by metabolism, must be dissipated by evaporation. Each gram of water that is evaporated carries away .58 kilocalorie of heat, and the combined heat load on a man may be so great that he is obliged to evaporate more than a liter (about a quart) of water an hour.

One way the body can reduce evaporation under heat stress is to abandon the maintenance of a constant body temperature. Schmidt-Nielsen and his collaborators found that when the camel is confronted with a shortage of water, its body temperature rises during the course of the day by as much as seven degrees C. To see if the eland or the oryx had the same ability, I recorded their rectal temperature during the laboratory's hot 12-hour day. The animals had all the water they could drink, so that nothing prevented the maintenance of a constant body temperature by evaporation. Nonetheless, during 12 hours at 40 degrees C. the eland's temperature on occasion rose by more than seven degrees (from 33.9 to 41.2 degrees C.) and the oryx's by more than six degrees (from 35.7 to 42.1 degrees) before increased evaporation prevented any further rise (although usually the temperature rise was less extreme). Thus instead of spending water to maintain a constant body temperature, the animals "stored" heat in their bodies.

In an eland weighing 500 kilograms a 7.3-degree rise in temperature means that the animal has managed to store some 3,000 kilocalories. To dissipate the same amount of heat by evaporation would cost it more than five liters of water. In the wild, as the lower night temperature allows a reversal of heat flow from the animal back to the environment, this stored heat is dissipated by conduction and radiation rather than by evaporation.

I found that when the eland and the oryx were exposed to the high experimental temperature and had water available to drink, their body temperature usually increased by three or four degrees during a 12-hour day. After three or four hours' exposure the animals' evaporative processes accelerated sufficiently to prevent a further rise in their temperature, which remained below the temperature of the laboratory even at the end of the full 12 hours. In the wild such a pattern of gradual warming before an increase in evaporation means that the animals might get past the hottest hours

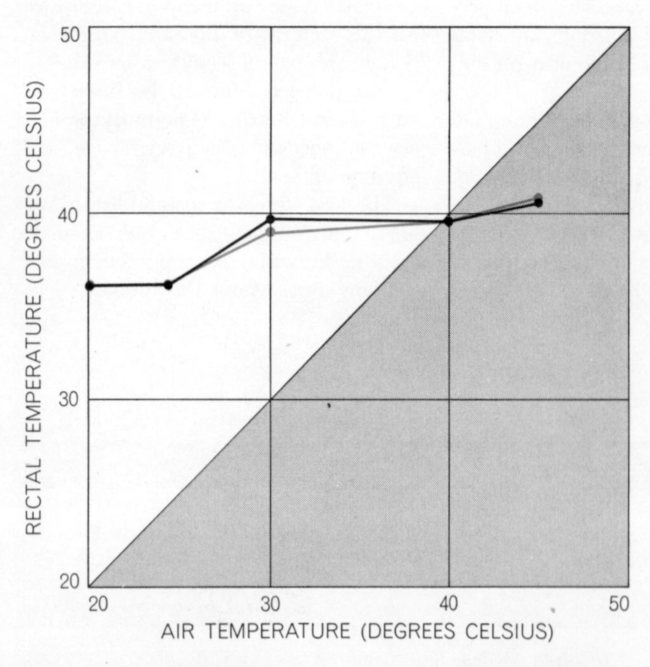

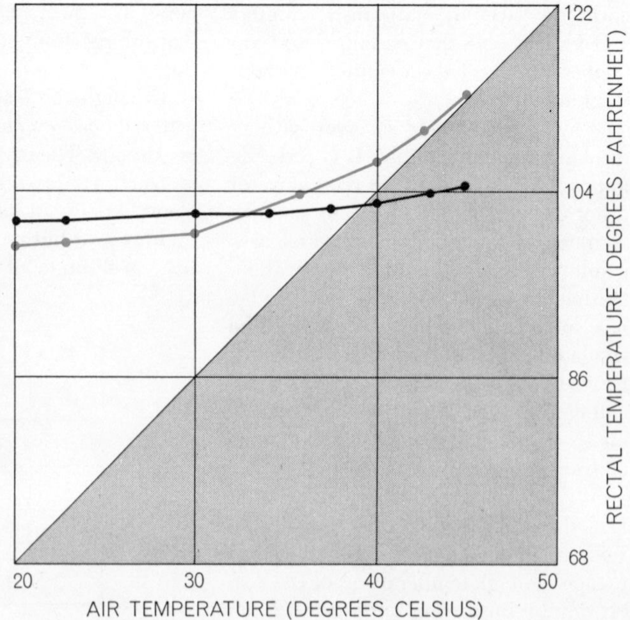

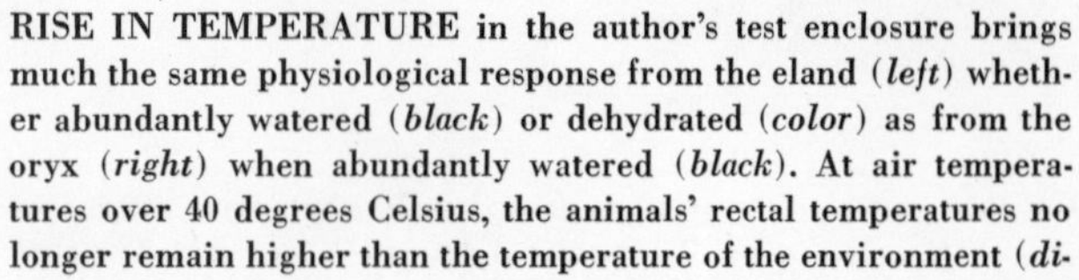

RISE IN TEMPERATURE in the author's test enclosure brings much the same physiological response from the eland (*left*) whether abundantly watered (*black*) or dehydrated (*color*) as from the oryx (*right*) when abundantly watered (*black*). At air temperatures over 40 degrees Celsius, the animals' rectal temperatures no longer remain higher than the temperature of the environment (*diagonal connects points where air and rectal temperature are equal*). Instead evaporative cooling keeps rectal temperature below air temperature, so that heat flows from the environment to the animal. The temperature of the dehydrated oryx (*color*), however, continues to rise, and heat flows from animal to environment. The dehydrated oryx therefore gains no heat from its surroundings.

of the day before expending precious water for cooling.

These first observations showed that in desert-adapted antelopes, as in the camel, water loss was reduced by the increase in body temperature. So far, however, I had measured the responses of the antelopes only under circumstances where they had been free to drink as much water as they wanted; this was a long way from testing their reported ability to get along without drinking at all. Before making my next series of measurements, therefore, I restricted the animals' water intake until they became dehydrated and lost weight. They were given just enough water to keep them at the point where they maintained their body weight at 85 percent of their original weight. I then exposed them to the same experimental 12-hour hot days as before. The body temperature of the dehydrated eland still remained below the temperature of the environment, even after 12 hours at 40 degrees C., and of course this can only be achieved by maintaining a high rate of evaporation. The temperature of the dehydrated oryx, in contrast, routinely exceeded that of the environment by a wide enough margin for metabolic heat to be lost by conduction and radiation; evaporation did not increase during the entire 12-hour exposure.

Although I had selected 40 degrees C. as the temperature of the hot periods, I knew that the desert air temperature is higher for a few hours every day, and that the solar radiation at midday near the Equator is literally searing. During the worst heat of the day the eland moves into the shade but the oryx appears to be oblivious of the intense heat. Oryx species with the lightest-colored coat are the ones that penetrate farthest into the desert; the coat's reflectivity must help to reduce the radiant heat gain. Unless the coat is perfectly reflective, however, an oryx standing in the hot desert sun should absorb far more heat than one in the laboratory at 40 degrees C.

I decided to test the animals' reactions to an even more severe heat load by raising the laboratory temperature to 45 degrees C. (113 degrees F.). Dehydrated or not, the eland managed to maintain its temperature some five degrees below the new high. So did the oryx, when it was supplied with water. The dehydrated oryx's physiological response was quite different. Its temperature rose until it exceeded the temperature of the laboratory and remained above 45 degrees C. for as long as eight hours without evident

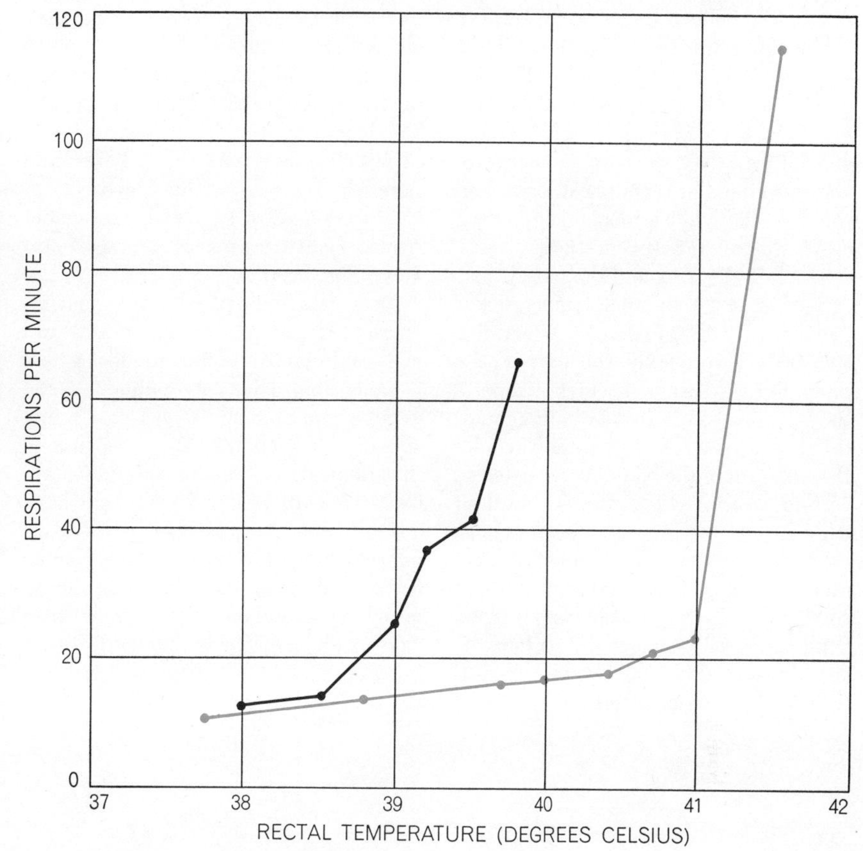

EVAPORATIVE COOLING through panting rather than through sweating is characteristic of the dehydrated oryx. When abundantly watered (*black*), an oryx not only sweats but also starts panting as its temperature reaches 39 degrees C. Although dehydrated oryx (*color*) does not sweat, it pants vigorously once its temperature exceeds 41 degrees C.

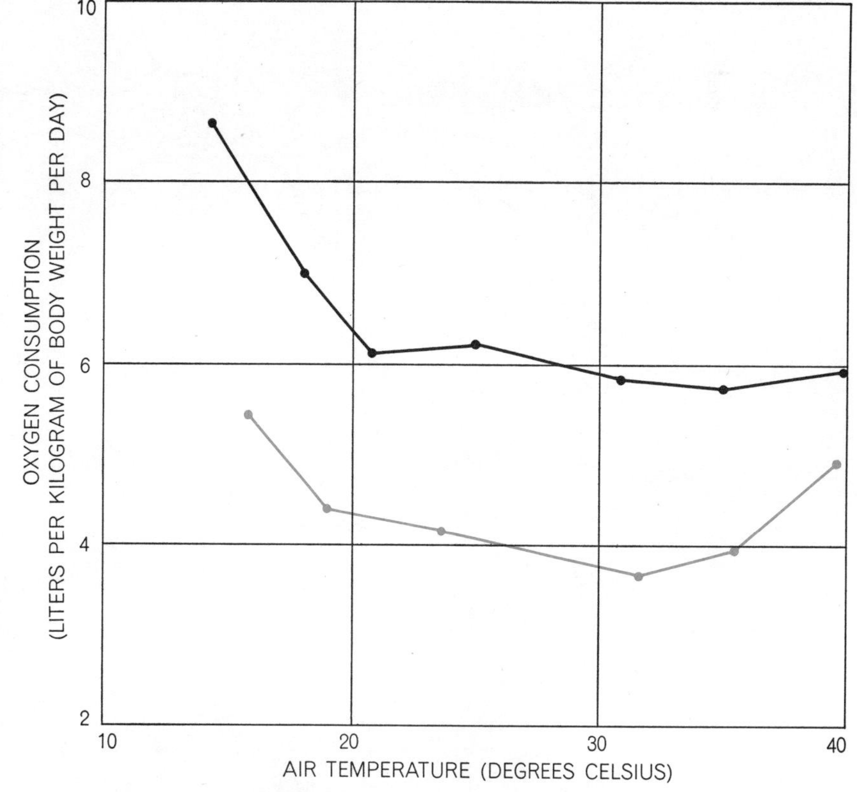

ORYX'S OXYGEN CONSUMPTION varies with the temperature of the environment and is drastically reduced by dehydration (*color*). As a result the dehydrated oryx generates much less metabolic heat and the quantity of heat to be lost by evaporation is reduced.

ill effect. I believe such a high continuous temperature has been observed in only one other mammal: the small desert gazelle *Gazella granti*. This hyperthermia in both the oryx and the gazelle enables them to save large amounts of water even under severe heat loads, and this is probably the critical factor in their survival under desert conditions.

How do the oryx and the gazelle survive these high internal temperatures? The brain, with its complex integrative functions, is probably the part of the body most sensitive to high temperatures. It is possible that in both animals the brain remains substantially cooler than the rest of the body. As the external carotid artery, which supplies most of the blood to the brain in these animals, passes through the region called the cavernous sinus, it divides into hundreds of small parallel arteries. Cool venous blood from the nasal passages drains into the sinus, presumably reducing the temperature of the arterial blood on its way to the brain. Evidence for this view is that, when temperature readings are taken during exercise, the brain of a gazelle proves to be cooler than the arterial blood leaving the heart by as much as 2.9 degrees C. Mary A. Baker and James N. Hayward of the University of California at Los Angeles have demonstrated that such a mechanism also operates in the sheep.

The manner in which an animal increases evaporation to keep cool can make a difference in the amount of heat it gains from a hot environment. Some animals pant, some sweat and some spread saliva on the body. An animal that depends on sweating or salivation for loss of body heat will necessarily have a skin temperature that is lower than its internal temperature. The blood must flow rapidly to the skin, carrying the internal heat to the evaporative surface. Conversely, a high skin temperature and a low flow of blood to the skin reduces the rate of heat flow from the hot environment to the animal. If the animal can pant rather than sweat, it can dispose of body heat by respiratory evaporation and at the same time have a higher skin temperature that minimizes its accumulation of environmental heat. Accordingly I wanted to find out whether the oryx under heat load increased its evaporation by sweating, by panting or by both. I also wondered what effect dehydration might have on the relative importance of the two evaporative routes. When I measured sweating and panting in an oryx freely supplied with water, I found that the evaporation rate increased in both routes but that evaporation from the skin accounted for more than 75 percent of the total. When the animal was deprived of water, it did not sweat at all in response to heat, but it began to pant when its body temperature exceeded 41 degrees C.

David Robertshaw of the Hannah Dairy Research Institute in Scotland and I had previously found that the sweat

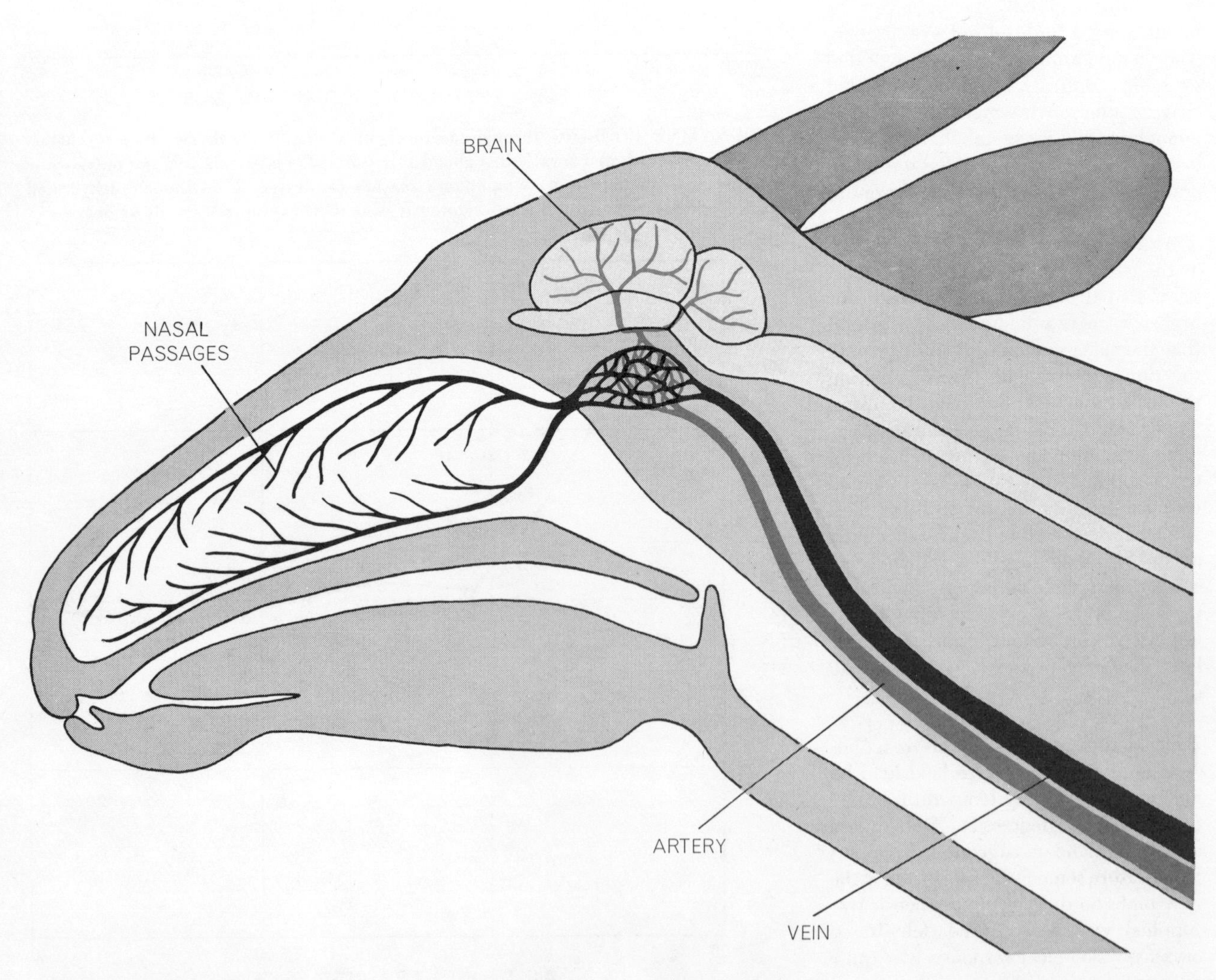

COUNTERCURRENT COOLING of arterial blood on its way from heart to brain occurs in the cavernous sinus, where the carotid artery ramifies into hundreds of smaller vessels (*color*). There venous blood (*black*) from the oryx's nasal passages, cooled by respiratory evaporation, lowers the arterial blood temperature. A brain cooler than the body temperature may be vital to desert survival.

glands of the oryx are controlled by nerve cells that release adrenalin. For example, an oryx can be made to sweat by means of a small intravenous injection of adrenalin. When we gave the same doses of adrenalin to a dehydrated oryx, the animal also sweated. The response makes it clear that the sweat glands of a dehydrated oryx can still function, but that the animal's nervous system has simply stopped stimulating them.

Increases in the temperature of the body and of the skin are not the only stratagems that minimize evaporation. A lowering of the animal's usual metabolic rate would reduce the amount contributed to the total heat load by the animal itself. To see whether or not this potential saving was being accomplished, I measured the metabolic rate of both the eland and the oryx over a broad range of temperatures. The metabolic rate of the dehydrated eland was somewhat reduced, but in the oryx the reduction was much greater. At 40 degrees C. the observed reduction in the metabolic rate of the oryx was sufficiently low to reduce its evaporation by 17 percent as compared to evaporation from animals freely supplied with water.

Taken together, these findings indicate that when water is scarce, the eland and the oryx reduce their rate of evaporation during the hot desert day in various ways. Both animals store heat; both decrease the heat flow from the environment, the eland by seeking shade and the oryx by accommodating to an extreme body temperature; both decrease the amount of metabolic heat they produce.

What about water loss during the night? When the sun goes down, of course, the antelopes are no longer under a heat load from the environment. The heat generated by their own metabolic processes is easily lost to the cooler surroundings by means other than evaporative cooling. Nonetheless, some evaporation, both from the respiratory tract and from the skin, continues at night.

The water lost through the skin at night is not lost through sweating. It is probably lost through simple diffusion. Skin is slightly permeable to water; even apparently dry-skinned reptiles lose appreciable amounts of water through the skin. So do the eland and the oryx. During the 12-hour cool night in the laboratory I found that both animals, when they had free access to water, lost about half a liter of water per 100 kilograms of body weight through the skin. The water loss was reduced when the animals were dehydrated; their skin seemed drier and

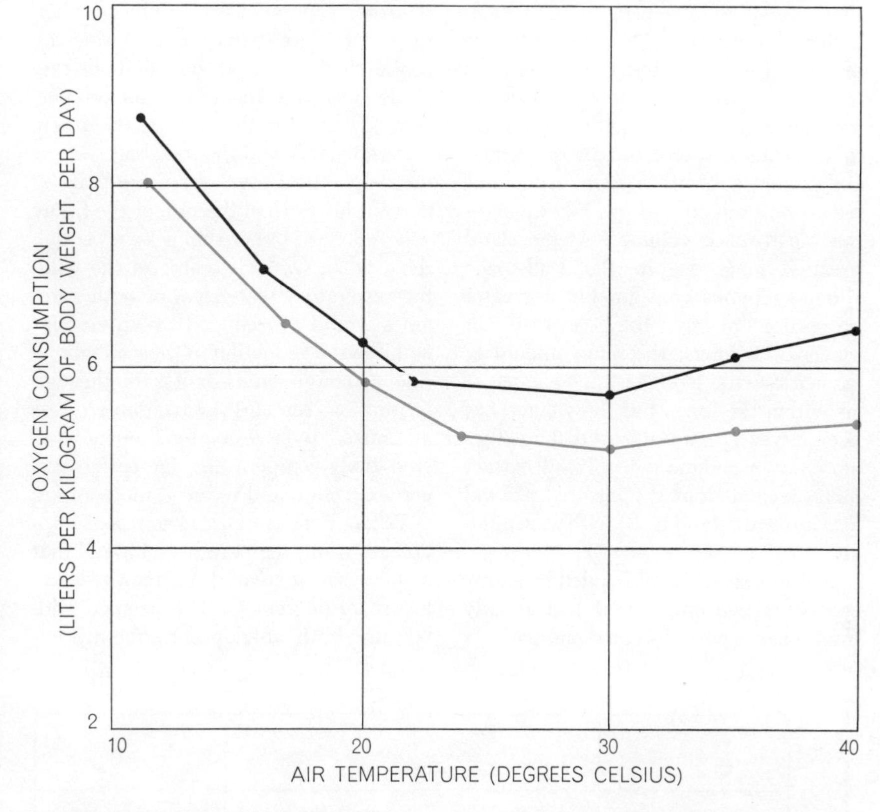

ELAND'S OXYGEN CONSUMPTION also varies with the temperature of the environment but is only slightly reduced by dehydration (*color*). Like the oryx, the eland increases its oxygen consumption at the low temperatures of the desert night. The increase threatens both animals with a net loss of water if night feeding produces less than 10 percent water.

less permeable. The water loss from the dehydrated eland was 30 percent less than when it could drink freely. In the dehydrated oryx the water loss from the skin was reduced by nearly 60 percent.

I wondered if the loss of water from the respiratory tract could also be reduced. As a mammal breathes, the inhaled air is warmed to body temperature and is saturated with water vapor in the respiratory tract before it reaches the lungs. Normally most mammals then exhale saturated air that is still at body temperature. Donald C. Jackson of the University of Pennsylvania School of Medicine and Schmidt-Nielsen have observed, however, that two species of small rodents manage to exhale air that is much cooler than their body temperature. The rodents apparently recondense some water vapor within the respiratory tract. This is one way water loss through respiratory evaporation could be minimized by antelopes. Two other possible means of water economy are related to oxygen requirements. First, if more oxygen can be extracted from each breath, an animal does not need to move as great a volume of air through its respiratory tract, thus reducing the loss of water to the respiratory air. Second, if the animal's oxygen consumption is lowered, the volume of inhaled air (and therefore the loss of water) is also reduced.

It is known that domesticated cattle exhale saturated air at body temperature. It is probable that the eland and the oryx, with their large nasal passages and relatively slow rate of respiration, do the same, so that water economy by recondensation is not available to them. Both animals have a lower body temperature at night than during the day, however, and the difference is enough to significantly reduce the amount of water needed to saturate the respiratory air. Air saturated at 39 degrees C., a typical daytime temperature for the eland, contains some 48 milligrams of water per liter. Air at the typical night temperature of 33.8 degrees contains some 25 percent less water.

When I investigated the animals at nighttime temperatures, I found that both the eland and the oryx extracted more oxygen from the air and breathed more slowly when they had a low body temperature. When the eland's temperature is at a nighttime low of 33.8 degrees C., about twice as much oxygen is extracted from each liter of air that it breathes. As oxygen extraction increases, the amount of air inspired with each breath also increases. Only part of the air an animal breathes actually reaches the lung, where oxygen and carbon dioxide are exchanged. The rest fills the respiratory passages, where the air is warmed and saturated with water vapor but where no exchange of gas takes place—the "dead-space volume." As the eland breathes more deeply, the dead-space volume remains constant but a greater proportion of the total inspired air reaches the lungs: the same amount of oxygen is extracted from each volume of air within the lungs but the volume has been increased. Any animal that breathes more slowly and more deeply will extract more oxygen from the inspired air and lose less water (and heat) with its expired air.

Is water economy also aided by a lower oxygen consumption? I had already found that in both the eland and the oryx the rate of metabolism is reduced when the animals are dehydrated. During the cool nighttime period the metabolic rate of the dehydrated eland was about 5 percent lower than the rate of the freely watered eland, and the metabolic rate of the dehydrated oryx was more than 30 percent lower than the rate of the freely watered oryx. Other things being equal, then, when water is scarce in the wild, the respiratory water loss of both antelopes would be reduced by an amount equal to the reduction in their metabolic rate. Although some loss of water through respiration remains unavoidable, it is minimized by the combination of lowered body temperature, increased oxygen extraction and reduced metabolism.

When I measured oxygen consumption at various temperatures, I found that metabolism increased at temperatures below 20 degrees C. This seemed odd, because both antelopes frequently encounter temperatures below 20 degrees C. at night. The increase, which only serves to keep the animals warm, would not be necessary if the animals had a slightly thicker fur, which would also reduce the heat gain during the hot day. It seems possible that such an adaptation has not appeared because the increased metabolism at night means the difference between a net loss or gain of water. If an animal increases its metabolism at night, it also eats more, takes in more free water with its food and generates more oxidation water. At the same time additional water is lost by breathing more air to get the necessary additional oxygen. When one calculates the amount of water needed in the food to offset a net loss of water through increased metabolism at night, it works out at about 10 percent water content in the food. Thus if the eland and the oryx feed at night on plants with a water content higher than 10 percent, they achieve a net gain of water. As we shall see, they favor plants that have a water content considerably above this level.

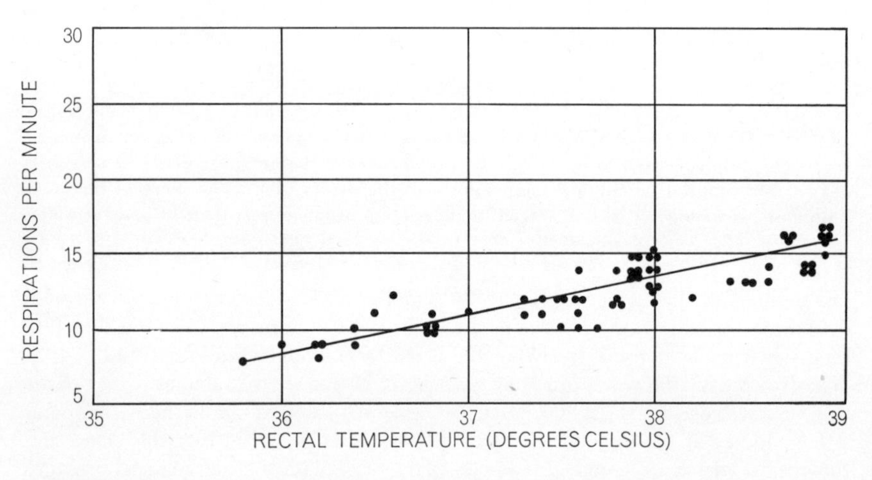

RESPIRATION RATE of an eland is controlled by the animal's temperature, decreasing as the temperature falls. When it breathes more slowly, the eland also breathes more deeply, so that a greater part of each breath reaches the lung areas where gas exchange occurs.

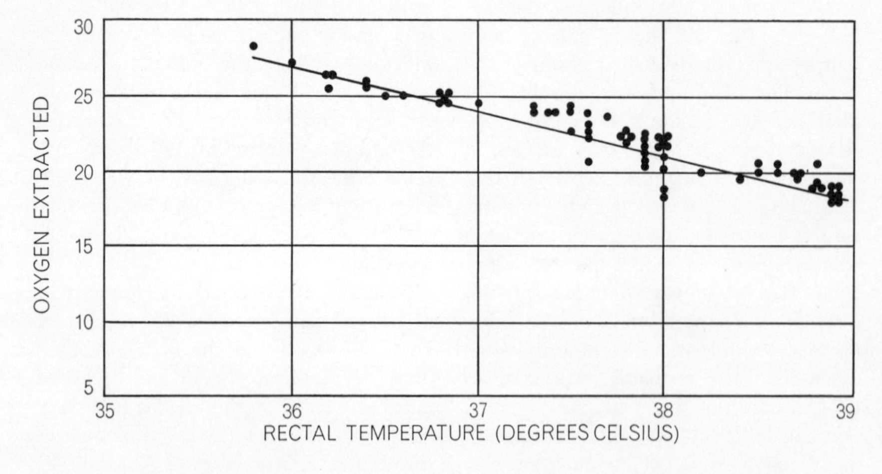

OXYGEN EXTRACTION from each breath of air increases as the eland's temperature falls and the animal breathes more slowly and deeply (*see upper illustration*). Because the eland's oxygen needs are met by a lesser volume of air, respiratory water loss is reduced.

Having found that the two animals do indeed possess unusual mechanisms for conserving water, I next undertook to find out how much water—or rather how little—each required. To do this I kept dehydrated animals in two contrasting laboratory environments. One was the usual alternation of 12-hour hot days and cool nights; the other was constantly cool. In the cool environment the dehydrated eland managed to stay at an even 85 percent of its original weight when its total water intake (free water in food, oxidation water and drinking water) was slightly in excess of 3.5 liters per 100 kilograms of body weight per day. The dehydrated oryx got by on scarcely half that amount: a little less than two liters per 100 kilograms. On a regime of cool nights and hot days the eland's water requirement increased to nearly 5.5 liters and the oryx's to three liters. These findings brought me to a final question: Could the two animals obtain this minimum amount of water without drinking?

The eland not only finds shade beneath acacia trees; acacia leaves are one of its favored foods. I collected acacia leaves and measured their moisture content. Even during a severe drought they contained an average of 58 percent water. Calculating the weight of acacia leaves that an eland would ingest in a day to meet its normal metabolic requirements, I found that the leaves would provide about 5.3 liters of water per 100 kilograms of body weight, or almost ex-

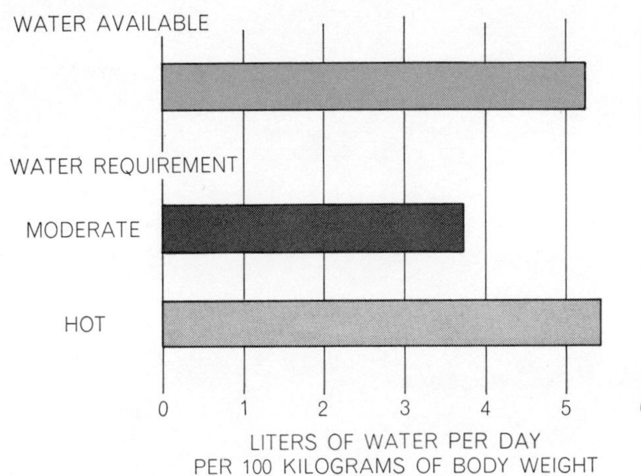

SURVIVAL WITHOUT DRINKING is possible for the oryx and the eland because their food contains almost all the water they need. Even in droughts the leaves of the acacia (*left*), the eland's preferred fodder, are 58 percent water. The leaves of a shrub, *Disperma*, and other fodder preferred by the oryx (*right*) contain little water by day but may average 30 percent water at night. Thus the amount of water each animal can obtain by feeding (*color*) is more than the animal needs for survival in a moderate environment (*black*) when dehydrated and closely approaches the quantities necessary for both antelopes' survival under desert conditions (*gray*).

actly the amount needed by the dehydrated eland for survival. An eland that obtains this much water by browsing only can probably live indefinitely without drinking.

The oryx favors grasses and shrubs, particularly a shrub of the genus *Disperma*. In the daytime these plants are so dry their leaves fall apart when they are touched; my measurements showed that they contain as little as 1 percent water. At first this seemed an impossible contradiction. Nonetheless, there is a way for the oryx to get all the water it needs by grazing. As long ago as 1930 the British naturalist Patrick A. Buxton observed that at night dry grass collects moisture from the desert air, even when there is no dew. The reason is that the drop in nighttime temperatures raises the relative humidity of the desert air and that the dry plant material can absorb moisture. To determine whether or not this mechanism was of importance to the oryx, I exposed some of the plants to laboratory air of the same average temperature and humidity as desert night air. Within 10 hours the formerly parched plants had acquired a water content of 42 percent.

In the wild, of course, the plants would not always contain this much water. At sunset their water content would be less, but later at night the plants could be substantially cooler than the surrounding air because of radiation to the night sky and thus might collect more water than the plants in the laboratory experiment. It seems entirely possible that by eating mainly at night the oryx could take in food containing an average of 30 percent water. If this is the case, the oryx, which needs only half as much water as the eland, would also be independent of drinking water as it roams the desert.

Hence we see that both eland and oryx have unusual physiological and behavioral adaptations for life in an arid environment. It is therefore tempting to conclude that ranching eland and oryx in arid regions would be an excellent way to expand Africa's meat supply. The conservation of beautiful and interesting species would be an additional benefit. Serious problems, however, exist in getting antelope protein off the hoof and to the market at a price competitive with beef (equivalent to about 28 cents per pound in most of East Africa). The success of the antelopes in arid regions with sparse vegetation depends on their low density per square mile; this makes it difficult to locate them in the vast areas where they live. To harvest them economically would require the development of inexpensive ways to find, kill, butcher and transport them from the isolated deserts to the cities and towns of Africa. The alternative to wild ranching —domestication, fencing and concentrated feeding—dissipates the physiological and behavioral advantages of antelopes over cattle. In fact, there is every reason to believe man's intensive breeding of cattle for meat production has produced an animal superior to antelopes under these conditions. Tapping the potential of antelope meat awaits economists and agriculturists who can solve these seemingly insurmountable problems.

Salt Glands

12

by Knut Schmidt-Nielsen
January 1959

A special organ which eliminates salt with great efficiency enables marine birds to meet their fluid needs by drinking sea water. Similar organs have been found in marine reptiles

As the writers of stories about castaways are apt to point out, a man who drinks sea water will only intensify his thirst. He must excrete the salt contained in the water through his kidneys, and this process requires additional water which is taken from the fluids of his body. The dehydration is aggravated by the fact that sea water, in addition to common salt or sodium chloride, also contains magnesium sulfate, which causes diarrhea. Most air-breathing vertebrates are similarly unable to tolerate the drinking of sea water, but some are not so restricted. Many birds, mammals and reptiles whose ancestors dwelt on land now live on or in the sea, often hundreds of miles from any source of fresh water. Some, like the sea turtles, seals and albatrosses, return to the land only to reproduce. Whales, sea cows and some sea snakes, which bear living young in the water, have given up the land entirely.

Yet all these animals, like man, must limit the concentration of salt in their blood and body fluids to about 1 per cent—less than a third of the salt concentration in sea water. If they drink sea water, they must somehow get rid of the excess salt. Our castaway can do so only at the price of dehydrating his tissues. Since his kidneys can at best se-

PETREL EJECTS DROPLETS of solution produced by its salt gland through a pair of tubes atop its beak, as shown in this high-speed photograph. The salt-gland secretions of most birds drip from the tip of the beak. The petrel, however, remains in the air almost continuously and has apparently evolved this "water pistol" mechanism as a means of eliminating the fluid while in flight.

crete a 2-per-cent salt solution, he must eliminate up to a quart and a half of urine for every quart of sea water he drinks, with his body fluids making up the difference. If other animals drink sea water, how do they escape dehydration? If they do not drink sea water, where do they obtain the water which their bodies require?

The elimination of salt by sea birds and marine reptiles poses these questions in particularly troublesome form. Their kidneys are far less efficient than our own: a gull would have to produce more than two quarts of urine to dispose of the salt in a quart of sea water. Yet many observers have seen marine birds drinking from the ocean. Physiologists have held that the appearance of drinking is no proof that the birds actually swallow water, and that the low efficiency of their kidneys proves that they do not. Our experiments during the past two years have shown that while the physiologists are right about the kidneys, the observations of drinking are also correct. Marine birds do drink sea water. Their main salt-eliminating organ is not the kidney, however, but a special gland in the head which disposes of salt more rapidly than any kidney does. Our studies indicate that all marine birds and probably all marine reptiles possess this gland.

The obvious way to find out whether birds can tolerate sea water is to make them drink it. If gulls in captivity are given only sea water, they will drink it without ill effects. To measure the exact amount of sea water ingested we administered it through a stomach tube, and found that the birds could tolerate large quantities. Their output of urine increased sharply but accounted for only a small part of the salt they had ingested. Most of the salt showed up in a clear, colorless fluid which dripped from the tip of the beak. In seeking the source of this fluid our attention was drawn to the so-called nasal glands, paired structures of hitherto unknown function found in the heads of all birds. Anatomists described these organs more than a century ago, and noted that they are much larger in sea birds than in land birds. The difference in size suggested that the glands must perform some special function in marine species. Some investigators proposed that the organs produce a secretion akin to tears which serves to rinse sea water from the birds' sensitive nasal membranes.

We were able to collect samples of the secretion from the gland by inserting a thin tube into its duct. The fluid turned out to be an almost pure 5-per-cent solution of sodium chloride—many times saltier than tears and nearly twice as salty as sea water. The gland, it was plain, had nothing to do with rinsing the nasal membranes but a great deal to do with eliminating salt. By sampling the output of other glands in the bird's head, we established that the nasal gland was the only one that produces this concentrated solution.

The nasal glands can handle relatively enormous quantities of salt. In one experiment we gave a gull 134 cubic centimeters of sea water—equal to about a tenth of the gull's body weight. In man this would correspond to about two gallons. No man could tolerate this much sea water; he would sicken after drinking a small fraction of it. The gull, however, seemed unaffected; within three hours it had excreted nearly all the salt. Its salt glands had produced only about two thirds as much fluid as its kidneys, but had excreted more than 90 per cent of the salt.

The fluid produced by the salt gland is about five times as salty as the bird's blood and other body fluids. How does the organ manage to produce so concentrated a solution? Microscopic examination of the gland reveals that it consists of many parallel cylindrical lobes, each composed of several thousand branching tubules radiating from a central duct like bristles from a bottle brush. These tubules, about a thousandth of an inch in diameter, secrete the salty fluid.

A network of capillaries carries the blood parallel to the flow of salt solution in the tubules, but in the opposite direction [*see illustration on opposite page*]. This arrangement brings into play the principle of counter-current flow, which seems to amplify the transfer of salt from the blood in the capillaries to the fluid in the tubules. A similar arrangement in the kidneys of mammals appears to account for their efficiency in the concentration of urine [see "'The Wonderful Net," by P. F. Scholander; SCIENTIFIC AMERICAN, April, 1957]. No such provision for counter-current flow is found in the kidneys of reptiles, and it is only slightly developed in birds.

Counter-current flow, however, does not of itself account for the gland's capacity to concentrate salt. The secret of this process lies in the structure of the tubules and the cells that compose them.

The microscopic structure of a salt-gland tubule resembles a stack of pies with a small hole in the middle. Each "pie" consists of five to seven individual

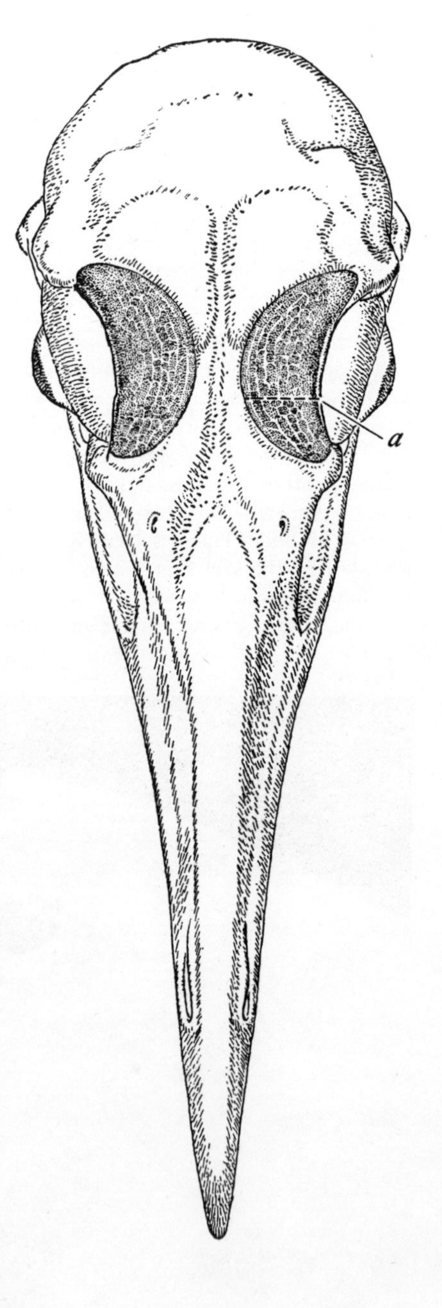

STRUCTURE of salt gland is essentially the same in all sea birds. In the gull the glands lie above the bird's eyes, as shown at left. Cross section of a gland (*a*) shows that it consists of many lobes (*b*). Each of these

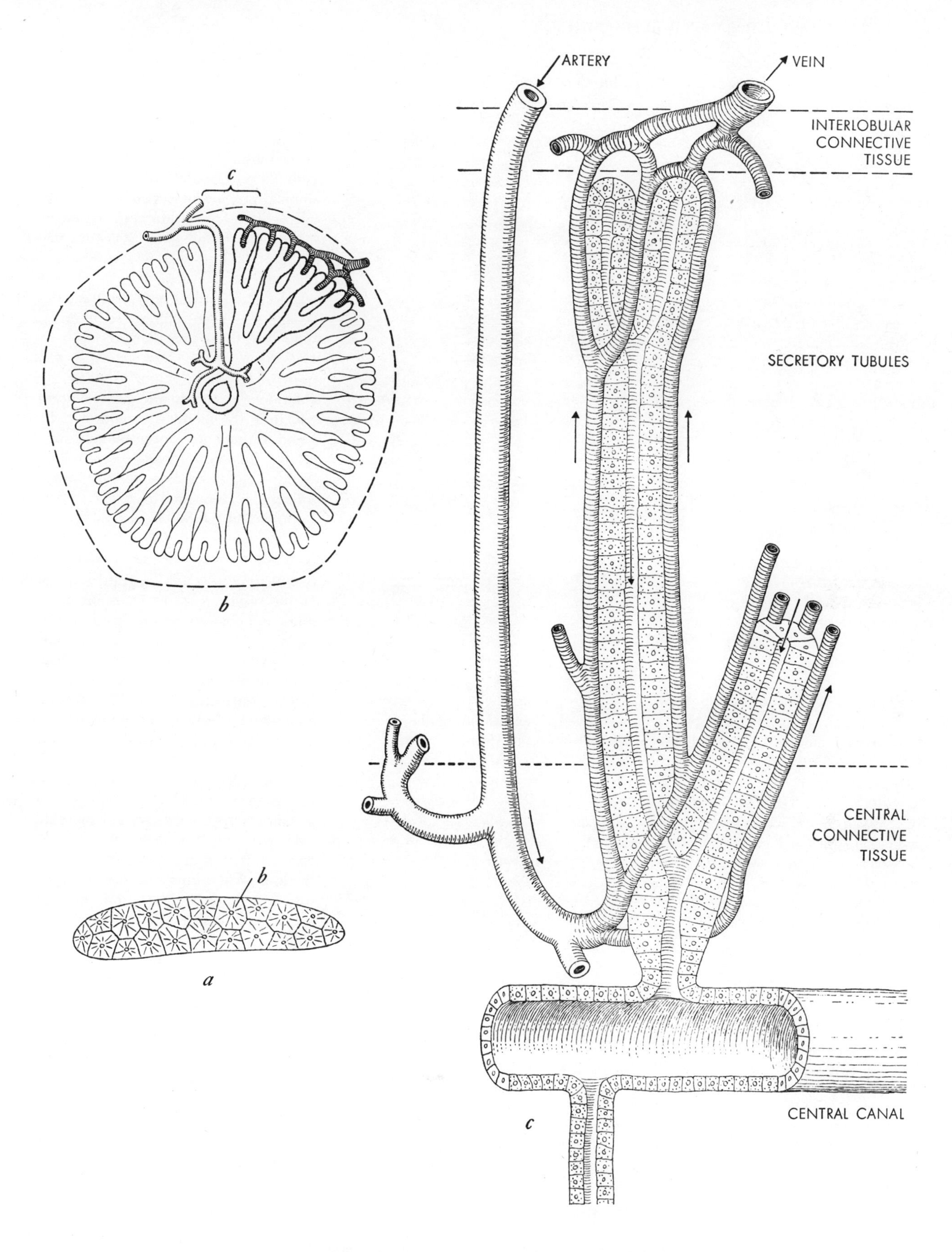

lobes contains several thousand branching tubules which radiate from a central duct like the hairs of a bottle brush. Enlargement of a single tubule (*c*) reveals that it is surrounded by capillaries in which blood flows counter to the flow of salt secretion in the tubule. This counter-current flow, which also occurs in the kidneys of mammals, facilitates the transfer of salt from the blood to the tubule. The tubule wall, only one cell thick, consists of rings of five to seven wedge-shaped cells. These rings, stacked one on top of another, encircle a small hole, or lumen, through which the salty secretion flows from the tubule into the central canal of the lobe.

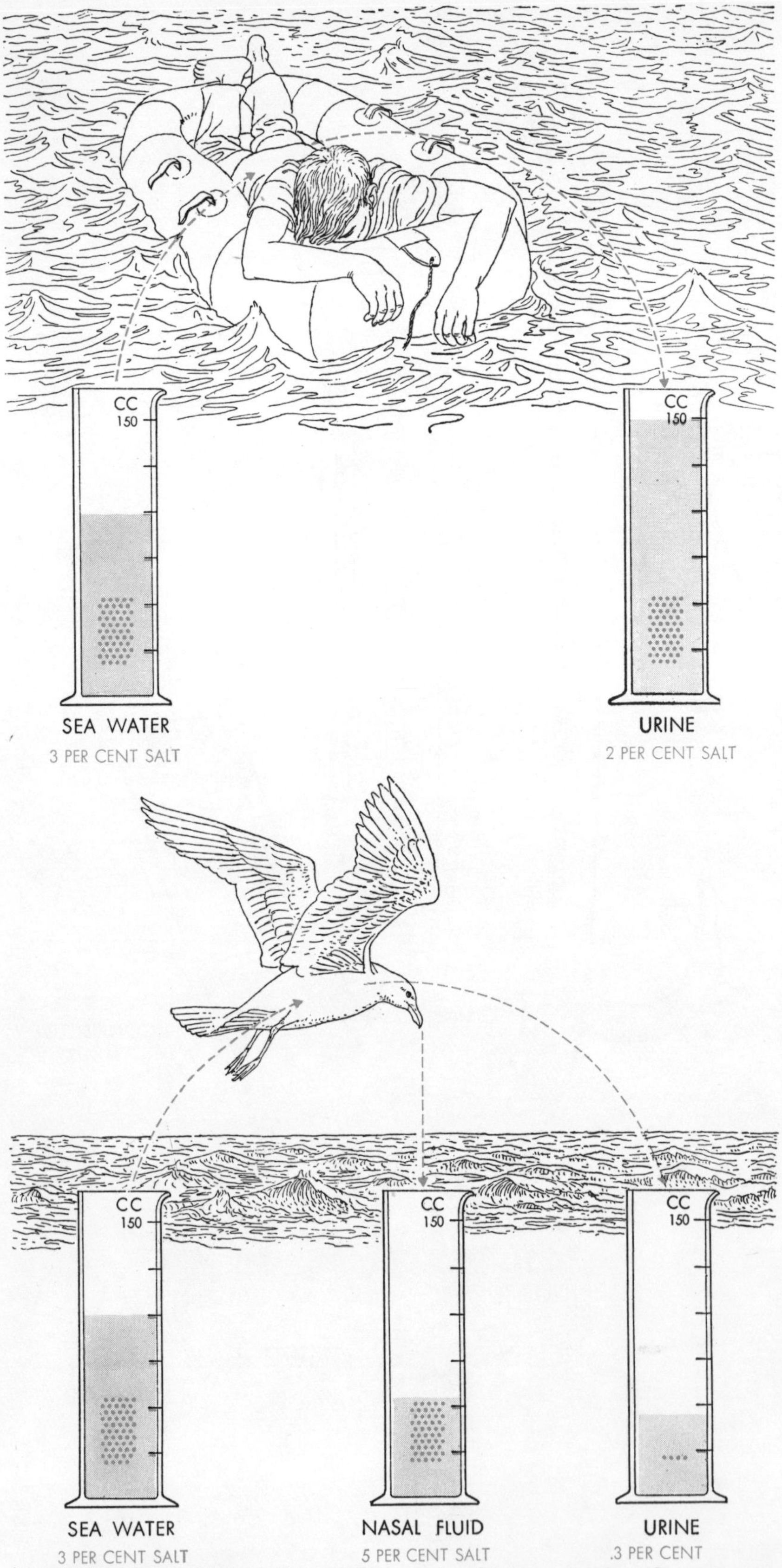

SEA WATER
3 PER CENT SALT

URINE
2 PER CENT SALT

SEA WATER
3 PER CENT SALT

NASAL FLUID
5 PER CENT SALT

URINE
.3 PER CENT

SALT EXCRETION IN MEN AND BIRDS is compared in these drawings. Castaway at top cannot drink sea water because in eliminating the salt it contains (*colored dots*) he will lose more water than he has drunk. His kidney secretions have a salt content lower than that of sea water. Gull (*below*) can drink sea water even though its kidneys are far less efficient than a man's. It eliminates salt mainly through its salt, or "nasal," glands. These organs, more efficient than any kidney, secrete a fluid which is nearly twice as salty as sea water.

cells arranged like wedges. The hole, or lumen, funnels the secretion into the central duct. When we inject dye into the lumen, colored fluid seeps out into a system of irregular crevices in the walls of the tubule. More detailed examination with the electron microscope reveals a similar, interlocking system of deep folds which extend inward from the outer surface of the tubule. This structure may be important in that it greatly multiplies the surface area of the cell. It is worth noting that cells with similar, though shallower, folds are found in the tubules of the mammalian kidney.

Evidently some physiological mechanism in the cell "pumps" sodium and chloride ions against the osmotic gradient, from the dilute salt solution of the blood to the more concentrated solution in the lumen. Nerve cells similarly "pump" out the sodium which they absorb when stimulated [see "The Nerve Impulse and the Squid," by Richard D. Keynes; SCIENTIFIC AMERICAN Offprint 58]. Of course the mechanisms in the two processes may be quite different. In the tubule cells the transport of sodium and chloride ions seems to involve the mitochondria, the intracellular particles in which carbohydrates are oxidized to produce energy.

The similarities between the salt gland and the mammalian kidney should not obscure their important differences. For one thing, the salt gland is essentially a much simpler organ. The composition of its secretions, which apart from a trace of potassium contain only sodium chloride and water, indicates that its sole function is to eliminate salt. In contrast, the kidney performs a variety of regulatory and eliminative tasks and produces a fluid of complex and variable composition, depending on the animal's physiological needs at a particular time.

The salt gland's distinctive structure, elegantly specialized to a single end, enables it to perform an almost unbelievable amount of osmotic work in a short time. In one minute it can produce up to half its own weight of concentrated salt solution. The human kidney can produce at most about a twentieth of its weight in urine per minute, and its normal output is much less.

Another major difference between the two glands is that the salt gland functions only intermittently, in response to the need to eliminate salt. The kidney, on the other hand, secretes continuously, though at a varying rate. The salt gland's activity depends on the concentration of salt in the blood. The injection of salt solutions into a bird's bloodstream causes the gland to secrete, indicating that some center, probably in the brain, responds to the salt concentration. The gland responds to impulses in a branch of the facial nerve, for electric stimulation of this nerve causes the gland to secrete.

While the structure and function of the salt gland is essentially the same in all sea birds, its location varies. In the gull and many other birds the glands are located on top of the head above the eye sockets [*see illustrations on this page*]; in the cormorant and the gannet they lie between the eye and the nasal cavity. The duct of the gland in either case opens into the nasal cavity. The salty fluid flows out through the nostrils of most species and drips from the tip of the beak, but there are some interesting variations on this general scheme. The pelican, for example, has a pair of grooves in its long upper beak which lead the fluid down to the tip; the solution would otherwise trickle into the pouch of the lower beak and be reingested. In the cormorant and the gannet the nostrils are nonfunctional and covered with skin; the fluid makes its exit through the internal nostrils in the roof of the mouth and flows to the tip of the beak.

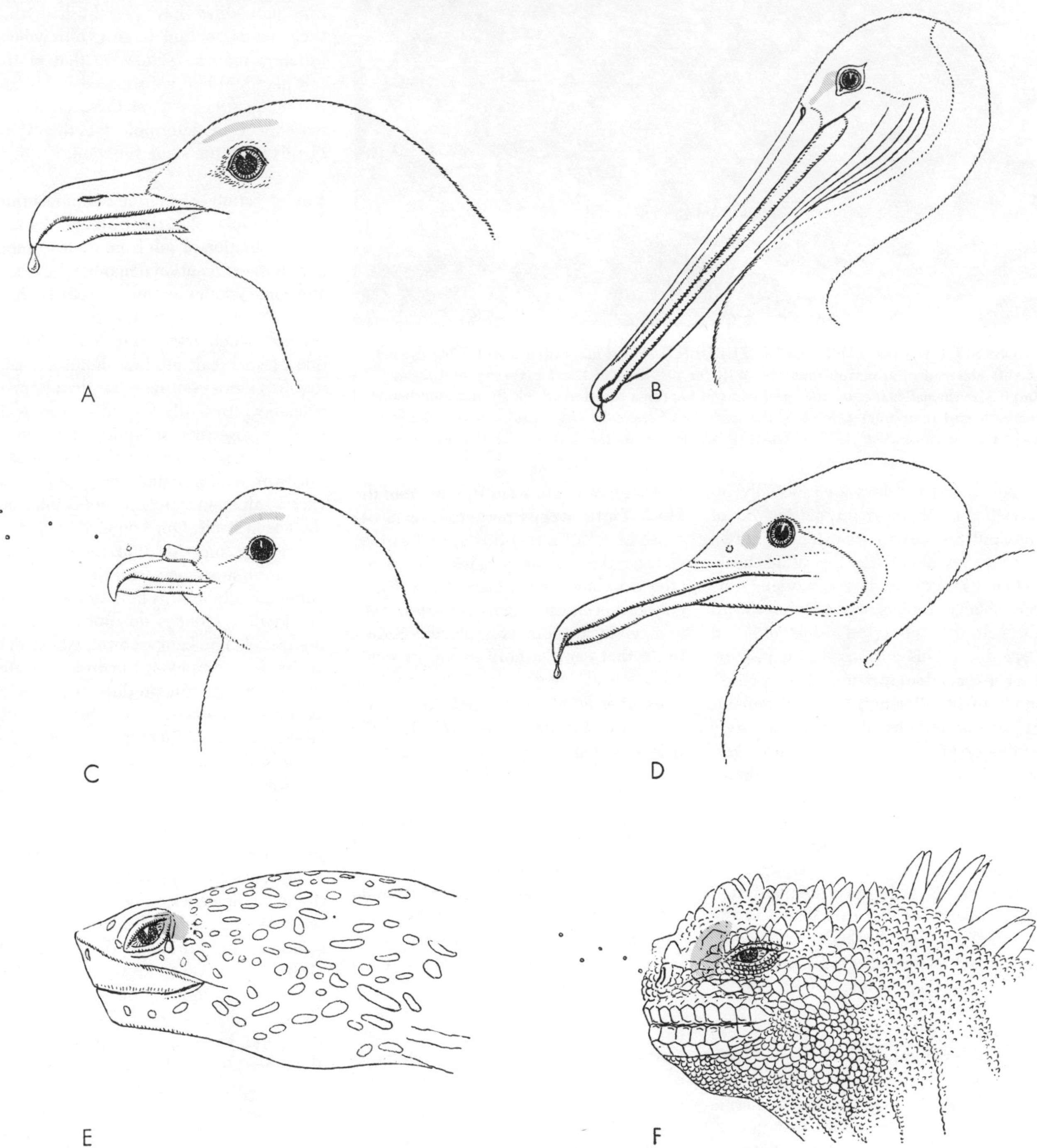

LOCATION OF SALT GLAND (*color*) varies in different species of marine birds and reptiles. In the gull (*A*) the gland's secretions emerge from the nostril and drip from the beak; in the cormorant (*D*) the fluid flows along the roof of the mouth. The pelican (*B*) has grooves along its upper beak which keep the fluid from dripping into its pouch; the petrel (*C*) ejects the fluid through tubular nostrils. In the turtle (*E*) the gland opens at the back corner of the eye; in the marine iguana (*F*) it opens into the nasal cavity.

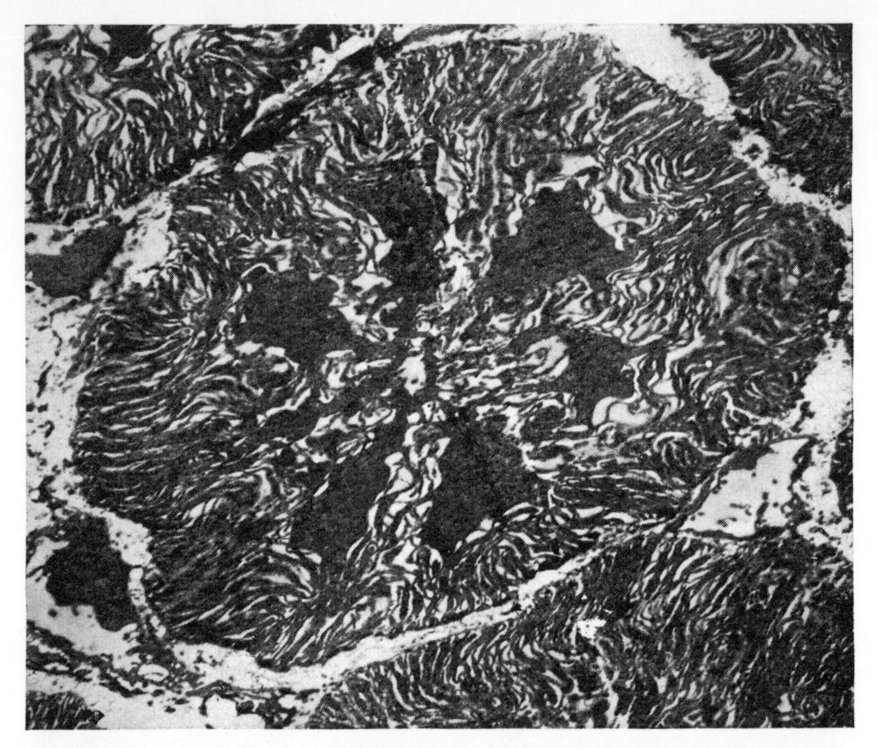

CROSS SECTION OF SALT-GLAND TUBULE is shown magnified about 5,700 diameters in this electron micrograph made by William L. Doyle of the University of Chicago. To emphasize the cell-structure the specimen was kept in a solution which shrank and distorted the cells and their nuclei. Most of the material of the cells lies in folded, leaflike layers; cells with a somewhat similar structure are found in the kidney tubules of mammals.

The petrel displays an especially interesting mechanism for getting rid of the fluid. Its nostrils are extended in two short tubes along the top of its beak. When its salt glands are working, the bird shoots droplets of the fluid outward through the tubes [*see illustration on page 133*]. This curious design may reflect a special adaptation to the petrel's mode of life. Though the bird remains at sea for months at a time, it rarely settles down on the water to rest. Presumably the airstream from its almost continuous flight would hamper the elimination of fluid from the bird's nostrils, were it not for the water-pistol function of the tubes.

Our studies so far have demonstrated the existence of the salt gland in the herring gull, black-backed gull, common tern, black skimmer, guillemot, Louisiana heron, little blue heron, double-crested cormorant, brown pelican, gannet, petrel, albatross, eider duck and Humboldt penguin. These species, from a wide variety of geographical locations, represent all the major orders of marine birds. There is little doubt that this remarkable organ makes it possible for all sea birds to eliminate salt and live without fresh water.

The discovery of the salt gland in sea birds prompted us to look for a similar organ in other air-breathing sea animals. In *Alice's Adventures in Wonderland* the Mock Turtle weeps perpetual tears because he is not a real turtle; real turtles, at least the marine species, also weep after a fashion. A. F. Carr, Jr., a distinguished specialist in marine turtles, gives us a vivid account of a Pacific Ridley turtle that came ashore to lay its eggs. The animal "began secreting copious tears shortly after she left the water, and these continued to flow after the nest was dug. By the time she had begun to lay, her eyes were closed and plastered over with tear-soaked sand and the effect was doleful in the extreme." Thus Carr makes it clear that the turtle's tears do not serve to wash its eyes free of sand, an explanation that otherwise might seem reasonable. The suggestion that the turtle weeps from the pangs of egg-laying is even wider of the mark.

With the loggerhead turtle as our subject, we have found that the sea turtle's tears come from a large gland behind its eyeball. The tears have much the same composition as that of the salt-gland secretions of the sea bird. Thus it would seem more than likely that the turtle's "weeping" serves to eliminate salt. The salt gland of the turtle has a structure similar to that of the gland in sea birds, with tubules radiating from a central duct, and it seems that this structure is essential for the elaboration of a fluid with a high salt concentration. The similarity is the more striking because the location of the gland in the turtle indicates that it has a different evolutionary origin. Still a third independent line of evolution may be represented by the salt gland in the Galápagos marine iguana, the only true marine lizard.

Anatomical studies of the other marine reptiles—the sea snakes and the marine crocodiles—have established that their heads contain large glands whose function may be similar to that of the salt gland. When we succeed in obtaining living specimens of these creatures, we expect to determine whether their glands have the same function.

Investigations of marine mammals thus far indicate that these animals handle the elimination of salt from their systems in a more conventional manner. The seal and some whales apparently satisfy their need for water with the fluids of the fish on which they feed. The elimination of such salt as these fluids contain requires kidneys of no more than human efficiency. But other whales, and walruses, whose diet of squid, plankton or shellfish is no less salty than sea water, must surely eliminate large quantities of excess salt even if they do not drink from the ocean itself. Our knowledge of their physiology suggests that their kidneys, which are more powerful than ours, can eliminate all the salts in their food. Some mammalian kidneys do function at this high level. The kangaroo rat, whose desert habitat compels it to conserve water to the utmost, can produce urine twice as salty as the ocean, and thrives in the laboratory on a diet of sea water and dried soybeans [see "The Desert Rat," by Knut and Bodil Schmidt-Nielsen; SCIENTIFIC AMERICAN Offprint 1050].

We should like to study salt excretion in whales, but these animals are obviously not easy to work with. We have undertaken, however, some pilot studies on seals. When we injected them with salt solutions that stimulate the salt glands of birds and reptiles, they merely increased their output of urine. Methacholine, a drug which also stimulates the salt gland, gave equally negative results. Whatever the seal's need to eliminate salt, its kidneys are evidently adequate to the task. We must therefore assume that the salt gland has evolved only in the birds and reptiles, animals whose kidneys cannot produce concentrated salt solutions.

IV

TEMPERATURE ADAPTATIONS

The males [bears] lie in hiding for periods of forty days and the females four months. If they have not got caves, they build rain-proof dens by heaping up branches and brushwood, with a carpet of soft foliage on the floor. For the first fortnight they sleep so soundly that they cannot be aroused even by wounds; at this period they get fat with sloth to a remarkable degree. . . . As a result of these days of sleep they shrink in bulk and they live by sucking their forepaws. . . . No evidence of food and only the smallest amount of water is found in the belly at this stage, and . . . there are only a few drops of blood in the neighborhood of the heart and none in the rest of the body.

Pliny
NATURAL HISTORY, VIII, liv

IV TEMPERATURE ADAPTATIONS

INTRODUCTION

Fishes and amphibians are poikilotherms—animals whose temperature varies with that of the environment. They are adapted in a variety of ways to their normal environmental temperature range, and, as we shall see, they show ability to become acclimated to different temperatures by a number of biochemical modifications. However, as is so often the case, the evolutionary versatility of the vertebrate stock provides us with exceptions to the generality that fishes are "cold-blooded." As explained by Francis Carey in "Fishes with Warm Bodies," heat conservation mechanisms have evolved quite independently in certain large bony fish (the tuna, for example) and sharks (the mako). The result is internal body temperatures that can be maintained for long periods at levels considerably above those of the surrounding sea water. An interesting problem with respect to the "bends" arises in the heat exchangers of these warm-bodied fishes. One might anticipate that, as oxygen-loaded hemoglobin goes from the cool gills to the warm heat exchanger, the typical vertebrate response would occur: affinity for oxygen would drop markedly, and oxygen would be released, even enough to form bends-producing bubbles! But, again "nature's forethought," as Pliny called it, has operated: blue-fin tuna hemoglobin shows virtually no change in affinity as a function of temperature; hence it does not give up its oxygen precipitously, and the bends are avoided. This is still another example of a situation in which evolutionary modification at the anatomical level in one system (blood vessels) is accompanied by changes at the biochemical level in another system (the red cells).

Reptiles, too, are poikilotherms, but as has been pointed out by a number of biologists (see Charles M. Bogert, "How Reptiles Regulate Their Body Temperature," *Scientific American*, April 1959), they may behave in special ways in order to raise and regulate their body temperatures at least part of the time. The hormonal and neuronal mechanisms contributing to this capacity in reptiles are still present in birds and mammals and are part of the more complex control machinery used for true homeothermy. (Homeotherms are animals whose temperature is high and is held constant despite fluctuations in environmental temperature.) Most important of the reptilian properties is a group of cells, apparently located in the hypothalamus, that responds to temperature by altering blood pressure and the rate of heart beat. In fact, the vascular system occupies a critical position among those adaptations that permit temperature regulation. Reptiles are sometimes called "ectotherms" because they derive most of their heat from outside the body; the blood fluid transports heat from the surface to the internal regions of the body. Because of its ubiquitous flow, moving blood fluid is also the best means of providing the organism with a continuous, rapidly responding source of information about overall body temperature. In birds and mammals, control of vasoconstriction of peripheral blood vessels has become the primary means of regulating the degree of heat conservation (the insulation derived from feathers or fur is not considered a regulator).

Although contemporary reptiles are poikilotherms, some paleontologists believe that large dinosaurs were true "endotherms." This conclusion is based on bone structure, stance, and some highly controversial data interpreted to reflect prey–predator ratios, all of which suggest that some large-bodied dinosaurs may have had high body temperatures (see Robert T. Bakker's "Dinosaur Renaissance," *Scientific American* Offprint 916). These creatures lived in what apparently were warm latitudes of the earth; it is maintained that once their body temperatures were elevated to about 37 degrees centigrade either by daytime warming or by internal generation of heat by their muscles, their large mass in relation to relatively small surface area would have made it easier for them not to lose much heat at night. Thus they have been called "inertial homeotherms" by B. K. McNab and W. Auffenberg (see the Bibliography), to reflect their capacity for carrying over into the night the heat produced during the day. Inertial homeothermy is a much simpler type of endothermy than that of today's birds and mammals, which can generate heat by changing rates of ATP hydrolysis in a process not associated with musclework, called nonshivering thermiogenesis (which will be described later in this Introduction). Inasmuch as dinosaurs lacked insulation in the form of feathers or fur, they would not have had that means of retaining heat at night or when the weather cooled. This theory is unsatisfactory in some regards. For example, the presence of "growth rings" in dinosaur teeth would indicate, in fact, the sort of metabolism and growth rate characteristic of ectotherms like fishes, not endotherms like mammals. Such growth rings are described in the article by P. A. Johnston that is listed in the Bibliography at the end of this book. Furthermore, it is not clear how newly hatched or small dinosaurs—with greater ratios of surface area to mass—could have functioned as endotherms without insulation. Nor is it clear whether the changing climate of the late Cretaceous was responsible for the demise of terrestrial, aquatic, and aerial dinosaurs because the earth cooled, because it overheated owing to the greenhouse effect resulting from massive carbon dioxide accumulation in the atmosphere (for a discussion of the earth's climate during the Mesozoic era, see the article by D. M. McClean listed in the Bibliography), or because the weather simply became highly variable (see the chapter by F. H. Pough in *Vertebrate Life,* also listed in the Bibliography).

To understand the origins of mammalian and avian homeothermy it may be more profitable to consider the rudimentary temperature-regulating arsenal of living reptiles than to speculate about fossils. Development of the capacity to govern temperature seems to have involved utilizing a number of existing organ systems for the secondary purpose of participating in thermal control. A lizard, for example, when it is frightened, shows a brief shivering tremor as a means of increasing heat production. It never shivers when exposed to chilling conditions, but the same sort of shivering is used by birds and mammals to generate heat in cold environments. The biologist V. H. Hutchison has shown that the Indian python goes one step further. The female coils about a cluster of eggs and incubates them at elevated temperature by continually contracting her body musculature. The heat generated is sufficient to maintain a body and egg temperature of as much as 7 degrees centigrade above ambient. To perform this feat the female functions like a homeotherm as environmental temperature falls: it increases its consumption of oxygen and so is able to produce the extra heat needed for constancy of internal temperature. Nonincubating pythons and all other poikilotherms show rather strict proportionality of oxygen utilization to temperature, so that the quantities of oxygen used fall rapidly as the environment cools.

Still another example of reptiles employing a temperature-regulatory activity has been described by George A. Bartholomew (see the Bibliography). Several types of lizards breathe faster and faster as their bodies warm (thus meeting the need for more oxygen). In addition, some large lizards take special deep, rapid breaths, utilizing in particular the floor of the mouth and the neck,

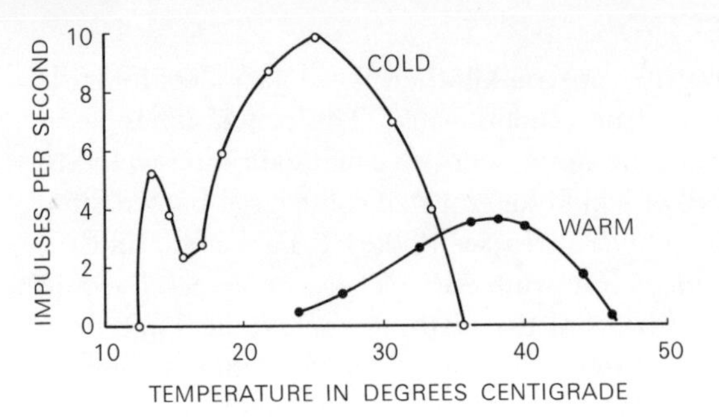

Figure 1. **The rates of discharge of mammalian hot and cold temperature-sensing cells. Note that the cold receptor is virtually quiet when 37 degrees centigrade is approached. [After Y. Zotterman, *Ann. Rev. Physiol.* 15, 1953.]**

as temperature rises above 38 degrees centigrade. This increases the volume of air pumped and the amount of evaporation from the respiratory system. Obviously, here is an equivalent to panting, one of the emergency measures used by birds and mammals for cooling the body. There are many such neuronal or endocrine-controlled processes, which originated in reptiles but are now included in the temperature-regulating arsenal of the homeotherm.

From a study of the monotremes (the echidna or spiny anteater, or the duckbill platypus *Ornithorhynchus*), the most primitive living mammals, we attain further insight into the development of homeothermy. These "Prototherian" animals diverged from the early mammals long before the marsupial and placental insectivores appeared. An echidna can maintain its body temperature near 30 degrees centigrade as the ambient environmental temperature falls. It is able to do this by increasing its rate of heat production by shivering: its muscles generate heat. It has no brown fat (see "The Production of Heat by Fat" by Michael J. R. Dawkins and David Hull) and apparently cannot increase its rate of so-called nonshivering thermiogenesis. Hence, even though it can generate more heat and conserve it because of its fur insulation, the echidna lacks crucial components of the "Eutherian" (placental mammal's) temperature-regulating machinery.

What happens to an echidna if ambient temperature is raised? The animal does not salivate, can't pant, and sweats little, if at all (but the other monotreme, the duckbilled platypus, can sweat profusely). Monotremes do reduce heat production under such circumstances; but that is not of much help and the animals may die of heat apoplexy at 35 degrees centigrade. In nature echidnas use behavioral means to avoid exposure to such high environmental temperatures.

These observations have led M. L. Augee to regard the monotreme's temperature-regulating system as representing an early stage in the evolution of the Eutherian system (see Augee's article, listed in the Bibliography for this book). In particular, the lack of "mammalian-type energetics"—the absence of nonshivering thermiogenesis—sets these animals off as an interesting remnant of an early stage in Eutherian evolution.

What then of marsupials—the "Metatheria"? Kangaroos, opposums, and their kin are much like the Eutherians in that they have a higher average body temperature (about 33 degrees centigrade) and are able to control it over a wider range of ambient temperature. Sweat glands, panting, and peripheral vasoconstriction are all present, so that control in both upper and lower danger zones is feasible. In all probability, nonshivering thermiogenesis in Metatherians is much like that in Eutherians.

Properties of the Eutherian temperature-regulatory system are described by Laurence Irving in "Adaptations to Cold" and by H. Craig Heller, Larry I. Crawshaw, and Harold T. Hammel in "The Thermostat of Vertebrate Animals." A crucial aspect of the vertebrate thermoregulatory system is the ability of certain neurons to respond to altered temperature. The output from the medulla of the brain that elicits a coordinated response to compensate for such a temperature change is dependent upon sensory information from the body periphery, from deep body temperature sensors, and from the hypothalamic sensor cells themselves.

The activity of temperature-sensing nerves (see Figure 1) is typified by the following data on the responses of cold receptors in the skin of monkeys. It has been found that nerve impulses are discharged from the cold receptors at a rate that is different for different temperatures: each temperature generates a characteristic discharge rate. Under conditions of constant ambient temperature this rate is nearly maximal when surface temperature of the skin is between 20 and 35 degrees centigrade. Discharge falls to zero as temperature rises from 35 to 40 degrees centigrade. When the temperature drops from about 40 degrees centigrade to 20 degrees, a peak discharge of 150 impulses per second is attained. A change as small as 0.5 degrees centigrade alters the discharge rates of these nerves, and even smaller increments affect the rates of some other temperature sensors. It is not clear how changing temperature evokes nerve discharge in such cells as these, or in the temperature sensors of the central nervous system.

A variety of biochemical adaptations to temperature are seen in various vertebrates. The amazing heat-generating brown fat of immature or hibernating mammals is described in "The Production of Heat by Fat" by Dawkins and Hull. This type of heat production is dependent upon the activity of different kinds of differentiated cells (brown or white fat cells) of an organism. In work performed since the appearance of the article on brown fat, it has become clear that the release of epinephrine from nerve endings in the brown-fat tissue results in activation of adenyl-cyclase, production of cyclic AMP, and action of that "second messenger" on the lipase enzyme.

A great deal is being learned about the biochemical control of heat production. Perhaps one of the most significant achievements is the demonstration that the hormone thyroxine (mentioned near the beginning of "An Essay on Vertebrates") acts on muscle, liver, and kidney cells to activate a Na^+,K^+,-dependent ATPase enzyme associated with the cell membranes. The enzyme splits ATP, and heat is generated as a byproduct. This process may well be the basis for nonshivering thermiogenesis—that is, the basic heat production under nonemergency conditions. As is emphasized by Peter Hochachka and George Somero in their *Strategies of Biochemical Adaptation*, this is a superb example of a case where a truly ancient enzyme has been used for a completely new purpose by relative newcomers to the biological world, the birds and mammals.

Some biochemical adaptations to temperature may vary between closely related species. An example of this type is seen in lizards that "prefer" different normal body temperatures ranging from 30 to 38.8 degrees centigrade. Differences are seen in myosin ATPase, an important enzyme in ATP metabolism during muscle contraction. Paul Licht of the University of California has shown that the temperature optimum for the enzyme differs between the species and correlates nicely with preferred body temperatures. Interestingly, lizards preferring high normal temperatures tend to have narrow preferred temperature ranges and the optima for their enzyme is similarly narrow; in contrast, animals preferring cooler conditions have more widely ranging preferred body temperatures and myosin ATPases with broad optima. It seems reasonable to assume that these functional differences in the enzymes are probably due to minor variations in the structure of the protein chains (that is, they reflect the sorts of mutations discussed already for vertebrate hemoglobins).

Another type of temperature adaptation is the alteration of proteins within single cells. It can be seen in animals that are placed at new temperatures in the laboratory so that they become acclimated. In rainbow trout, for instance, electrophoretically distinct forms of brain acetyl cholinesterase are present at 2 degrees and 17 degrees centigrade. Peter Hochachka of the University of British Columbia has shown that at 12 degrees centigrade, on the other hand, both forms of the enzyme are present. These differences are seen in nature during seasonal temperature fluctuations. Besides this sort of variation in

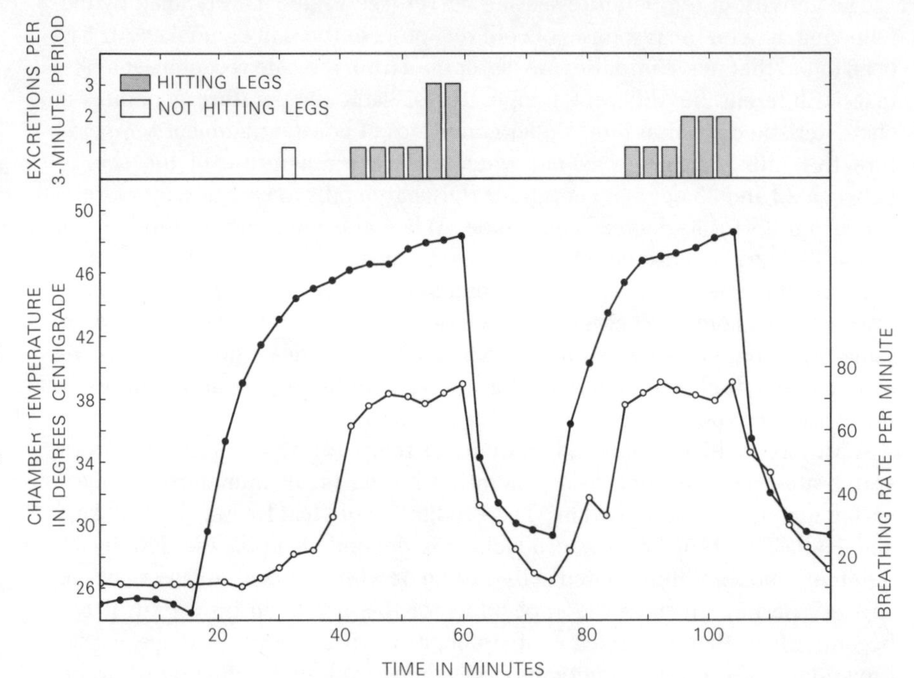

Figure 2. **Responses of a stork to altering environmental temperature. As room temperature (indicated by solid circles) rises, the respiratory rate (indicated by open circles) rises to increase evaporative cooling. In addition, the stork begins to urinate on its legs (see the bars at the top) so that evaporation from these uninsulated regions can aid the cooling process.**
[After M. P. Kahl, Jr., *Physiol. Zool.* 36, 1963. The University of Chicago Press.]

different enzyme forms, other strategies have appeared during evolution: for instance, in certain goldfish the activity of succinic dehydrogenase is markedly altered by the type of lipid present. Thus, changes in membrane lipids, which might occur as part of acclimatizing to different environmental temperatures, may be capable of affecting the basic enzymatic machinery of cells.

In addition to these specific sorts of biochemical adaptations, there is a general difference between the basic catalytic efficiency of enzymes obtained from ectotherms (poikilotherms) and endotherms: enzymes from the former animals are the most efficient in terms of turnover number of substrate or other physicochemical indexes. Why this should be so—why the enzymes of mammals should have become *less* efficient as mammals evolved from ectothermal reptiles—is a mystery for the future to solve.

In his article "Adaptations to Cold," Irving discusses a variety of biochemical and structural adaptations of vertebrates that permit survival under conditions of cold stress. A new line of investigation concerns vertebrate fishes that live in water at subfreezing temperatures. One strategy is used by marine fishes that live in shallow waters in Arctic regions. During the coldest months, such fish live in a "supercooled" condition by staying near the bottom of the sea and avoiding conditions that will act to "seed" ice-crystal formation in their bodies. A quite different tactic is used by other fishes that synthetize special "antifreeze" glycoprotein molecules. Among different species of one genus, the quantity of these antifreeze molecules is directly related to the depth at which the fish live; most molecules are present in pelagic fishes where micro-sized ice crystals are encountered frequently; progressively fewer molecules are found with increasing depth and decreased likelihood of encountering ice seeding. Although the structure of the glycoprotein is known, it is still a mystery how the glycoprotein is rendered an antifreeze by its regularly spaced hydroxyl groups.

Irving mentions several specializations of the extremities that permit survival in cold environments. In warm environments, the extremities, as sites of controlled heat loss, are equally important. The hairless tail of a muskrat, the horns of a goat, the tail of a beaver, the flippers of seals, the legs of birds, and the huge ears of a jackrabbit all function in this way. In these organs, insulation tends to be sparse, vasoconstriction and rete mirabile nets are well controlled, and biochemical specializations are present. A bird's body lacks sweat glands, and little water leaks through the skin. Hyperventilation and air sac evaporation, and radiation from the featherless legs are the ways in which birds lower their body heat. Interestingly, an overheated wood stork will urinate on its long legs as often as once a minute, using evaporation from the warm legs as a means of lowering body heat (see Figure 2). A final example of interest is the dog, which of course lacks sweat glands. The tongue hanging from the mouth of an overheated dog is one obvious way that evaporative cooling can occur But in addition special nasal glands that open just within the nostrils copiously secrete fluid when a dog pants and, in fact, may account for up to 36 percent of the water that is evaporated for cooling purposes. These glands are in a sense the sweat glands of a dog.

In "The Thermostat of Vertebrate Animals," Heller, Crawshaw, and Hammel discuss the process of hibernation as a strategy used by some mammals to avoid attempting to maintain normal high body temperatures during the coldest winter months. A parallel to hibernation is seen in many small birds and mammals that suspend high temperature maintenance each night and become torpid. This torpor is necessitated by the large relative surface areas of these animals and the resultant heat loss during periods of inactivity. Bats and hummingbirds become poikilotherms, and they show the same basic regulatory phenomena and awakening sequence, including repeated muscle flexion, as is seen in hibernators (see Figure 3). The amount of energy saved by torpor is significant: for a hummingbird, 10.3 kilocalories per 24 hours would be lost if sleeping took place at normal body temperatures, but only 7.6 kilocalories are lost in 24 hours if the torpid state is entered at night. If hummingbirds did not lower their nighttime temperature they would be unable to survive unless they awoke and fed during the night. This, of course, would require an additional set of special adaptations for navigation and food gathering in the dark. Such additional adaptations and continual activity at high metabolic rates would seem to be of little value for survival to reproduce; instead, small birds and mammals utilize torpidity and limited times of activity.

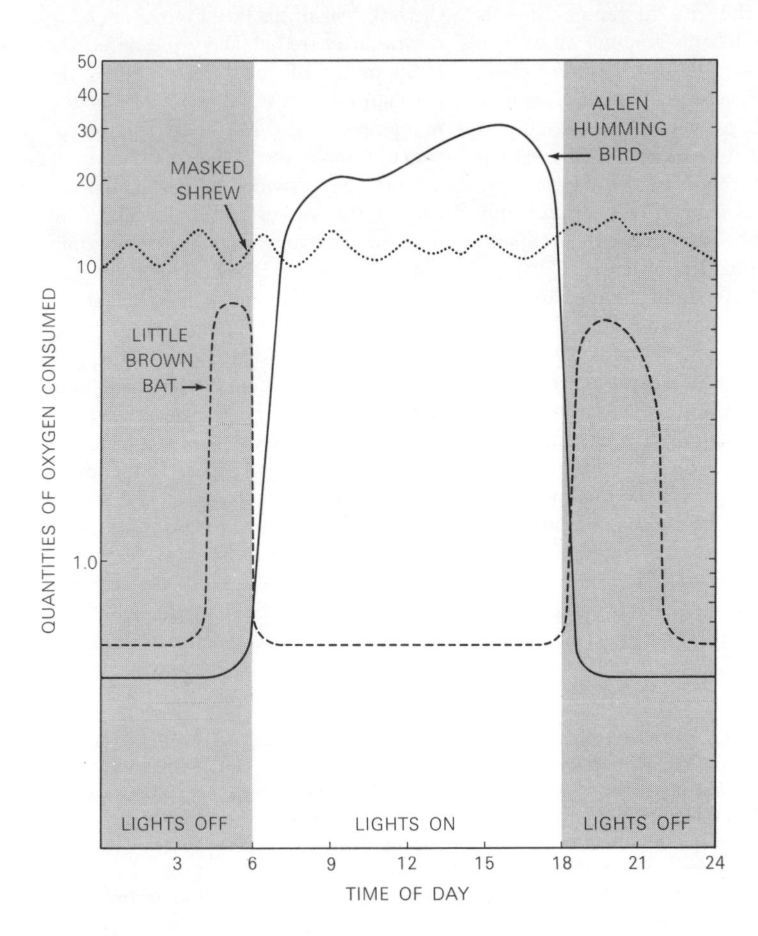

Figure 3. **The pattern of metabolism (as reflected in quantities of oxygen consumed) of three small homeotherms (each weighs from 3 to 6 grams and so might be expected to have similar heat-balance problems). The animals were kept on a regular schedule of light and dark periods. The shrew maintains its high metabolic rate at all times. The hummingbird enters the state of torpor soon after the lights go off. The bat shows two activity periods, one in the "evening," soon after the lights are extinguished, and the other in the early "morning," just before they come on again. Note that the oxygen consumed is expressed on a logarithmic scale, and that the hummingbird's consumption falls more than thirtyfold during torpor.**
[After G. Bartholomew; in M. S. Gordon, ed., *Animal Function: Principles and Adaptations*, Macmillan, 1968.]

13 Fishes with Warm Bodies

by Francis G. Carey
February 1973

Not all fishes are cold-blooded. In some fast-swimming species of tuna and mackerel shark a "wonderful net" of arteries and veins conserves the heat of metabolism to increase the power of the swimming muscles.

Fishes are regarded as cold-blooded animals, that is, animals whose body temperature is the same as the temperature of their surroundings. The fact is that not all fishes are cold-blooded. The first man to write about a warm fish was a British physician named John Davy. He was voyaging in the Tropics in 1835, and the ship's company was supplementing its rations by fishing. Davy was intrigued by the copious blood and mammal-like red flesh of one of the fishes brought aboard, a species of tuna known as the skipjack. He took the temperature of several skipjack and discovered that they were warmer than the water they swam in by 10 degrees Celsius (18 degrees Fahrenheit). In the years since Davy's observation most other species of tuna have proved to be warm-blooded and so has one family of sharks: the mackerel sharks. At the Woods Hole Oceanographic Institution my colleagues and I have long been fascinated by these warm-blooded fishes.

What is it that prevents most fishes from maintaining a body temperature that is more than negligibly higher than the temperature of the surrounding water? It is not, as might be supposed, the loss of surface heat that comes from being immersed in a cool medium. Rather it is the heat loss implicit in the fishes' mode of respiration. All land animals and some aquatic ones obtain the oxygen required to support their metabolism by breathing air, which is rich in oxygen and has a low capacity for absorbing heat. Water, from which fish must obtain their oxygen, contains scarcely 2.5 percent of the oxygen found in an equal volume of air but has 3,000 times the capacity of air for absorbing heat. As a result the heat lost in extracting oxygen from water can be 100,000 times greater than the loss in extracting the same amount of oxygen from air.

The oxygen in the water flowing through the gills of a fish diffuses into venous blood pumped to the gills by the heart. The blood may reach the gills at a somewhat elevated temperature as a result of having accumulated metabolic heat. Heat, however, diffuses 10 times faster than oxygen molecules do, so that by the time the blood in the gills is saturated with oxygen its temperature has dropped to the temperature of the water.

The oxygenated blood now flows back through the fish's circulatory system, and the oxygen is utilized in the body tissues. It is this process that generates the metabolic heat; the rise in temperature is proportional to the amount of oxygen extracted from the blood. In any given volume of oxygenated blood the volume of extractable oxygen is less than 20 percent. Taking into account the usual caloric yield of a fish's foodstuffs, that limited supply of oxygen will support only enough metabolism to raise the temperature of the blood one degree C. Moreover, even that slight temperature increase is lost to the water when the blood next passes through the gills, so that no heat accumulates to keep the fish warmer than the surrounding water.

Changes in the fish's rate of metabolism do not affect the cycle, because any exercise or other activity that produces more metabolic heat also demands more oxygen. An increased demand of oxygen means a greater loss of heat through the gills, so that whether the fish is at rest or exercising violently it will always be cold. Indeed, for a fish to raise its body temperature by metabolic processes alone would seem to be an impossibility.

Tunas and mackerel sharks manage to stay warm in apparent defiance of nature because they possess a special structure in their circulatory system. It is the *rete mirabile*, or "wonderful net," a tissue composed of closely intermingled veins and arteries that was first noted in vertebrates by students of anatomy in the 19th century. The French naturalist Georges Cuvier observed the presence of the *rete* in the tuna in 1831, the last year of his life. Four years later, the same year that Davy measured the temperature of the skipjack tuna, the German anatomist Johannes Müller formally described the structures in the viscera of the bluefin.

The *rete* provides a thermal barrier against the loss of metabolic heat. The mass of fine veins and arteries permits the free flow of blood for transport of oxygen and other molecules, but at the same time it short-circuits the flow of heat from the body tissues to the gills and shunts the accumulated heat back to the tissues. In the *rete* an artery, carrying cool, oxygen-rich blood from the gills to the body tissues, branches to form a mass of small vessels. The arterial network runs parallel to and is intermingled with a similar mass of fine veins carrying warm, oxygen-depleted blood from the tissues back to the gills. The system of closely associated arteries and

RETE MIRABILE, or "wonderful net," of the bluefin tuna is seen in cross section. The _rete_ is a system of parallel small arteries and veins that supplies and drains the band of dark-colored muscle that is used for sustained swimming effort. The system constitutes a countercurrent heat exchanger: venous blood warmed by metabolism gives up its heat to cold, newly oxygenated arterial blood fresh from the fish's gills. The effect is to increase the temperature and thus the power of the muscle. In this bluefin _rete_ sample the vessels have been visualized by the injection of latex: red latex in the arteries and blue latex in the veins.

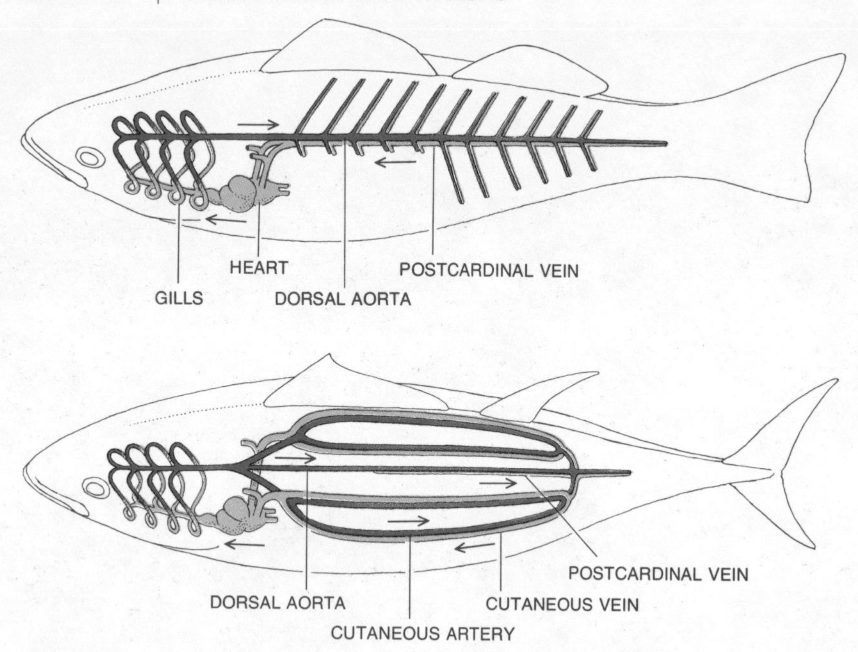

VASCULAR SYSTEMS of cold-blooded (*top*) and warm (*bottom*) fishes are different. The main vessels in most fishes, the cold-blooded ones, run along the backbone and radiate outward to the small vessels (*not shown*) that supply the muscle. In warm fishes the central vessels are smaller; most of the blood flows through cutaneous vessels under the skin and thence through countercurrent nets (*not shown*) that supply the muscle. This arrangement puts the cold end of the exchanger near the skin and the warm end in warm tissues.

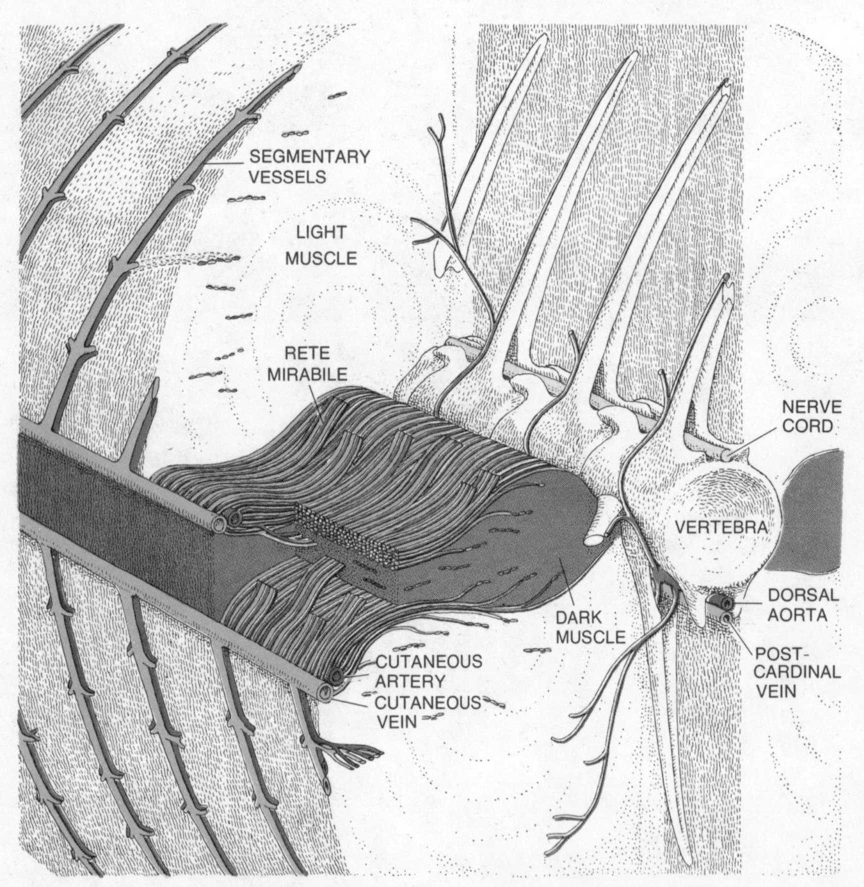

BLOOD SUPPLY TO MUSCLE of the bluefin tuna is primarily through four pairs of cutaneous vessels. A multitude of smaller vessels branch from them, forming the thick vascular slab above and below the dark muscle, the *rete*, from which the dark muscle is supplied. The light muscle is supplied by bands of alternating arteries and veins that slant through the muscle from the segmental vessels, thus serving as two-dimensional heat exchangers. Other, non-heat-exchanging arteries and veins run out from the central vessels.

veins acts as a countercurrent heat exchanger. There are large areas of contact between the two kinds of fine vessel, and heat is readily transferred through the vessels' thin walls. Therefore the warm venous blood is cooled as it flows through the *rete*, so that little or no heat is lost when the venous blood at last reaches the gills. At the same time the cold arterial blood is warmed, so that when it reaches the interior of the fish, it has nearly reached the temperature of the warm body tissues.

A *rete* of this type serves exclusively as a countercurrent heat exchanger. Its vessels, although small, are too large and their walls are too thick to allow any significant diffusion of oxygen molecules from the arterial blood to the venous blood. Other *retia* that do serve as gas exchangers are found in many species of fish [see "'The Wonderful Net,'" by P. F. Scholander; SCIENTIFIC AMERICAN, April, 1957]. The tiny vessels that comprise them are capillaries, and so oxygen molecules readily pass from one to another. Such *retia* are found in the swim bladder and the eye.

A major adaptive advantage of an elevated body temperature is greatly enhanced muscle power. If the difference in temperature between two otherwise equivalent muscles is 10 degrees C., the warmer muscle is able to contract and relax three times more rapidly, with no reduction in the force applied at each contraction; thus it can generate three times as much power. This kind of increase in power is evidently essential in flying animals such as birds and bats; their body temperatures are characteristically high. Even some large insects with a heavy wing loading elevate their body temperature when they fly. For example, cicadas and locusts are often unable to take off until they have warmed up by shivering or by basking in the sun.

Like flying, high-speed swimming is a demanding form of locomotion. Water is a dense and viscous medium, and to move through it rapidly requires not only fine streamlining but also substantial muscle power. The high body temperature of tunas and mackerel sharks apparently helps to provide the extra power needed for high-speed swimming: the most important of the vascular heat exchangers are those that serve the dark-colored swimming muscles.

Vertebrate muscle is generally a mixture of dark-colored fibers and light-colored ones. The dark fibers contain high concentrations of the oxygen-transporting pigments and enzymes associated with oxidative metabolism; they are also richly supplied with blood. The light-colored fibers usually have a less well-developed blood supply. They can function without oxygen during activity by metabolizing foodstuff through fermentation, recovering later when they are at rest. Muscles that are continuously active, for example the heart, are made up largely or entirely of dark fibers. Muscles that alternate between periods of rest and short intervals of activity contain few, if any, dark fibers.

As anyone who has seen a fish steak knows, the two kinds of muscle fiber are sharply segregated in the axial musculature of fishes. For example, in many tunas a large mass of very dark muscle, well supplied with blood, lies in a broad band between the backbone and the skin along the midplane of the body. This dark muscle can operate continuously; its contractions propel the fish as it swims at ordinary cruising speeds. The blood that supplies the muscle reaches the muscle fibers after passing through a large heat-exchanging *rete*. As a result the muscle is warmer than any other part of the fish's body.

In order to provide for this large heat exchanger the arrangement of the tuna's circulatory system has been drastically altered. Most fishes have a central distribution system for the blood. The main blood supply to the muscle flows outward through a large artery, the dorsal aorta, and returns through the postcardinal vein; both are deep inside the fish just below the backbone. From these central blood vessels segmental arteries and veins radiate out to the periphery.

In the tuna the normal pattern of circulation to the muscle has been almost entirely reversed. The major blood vessels are no longer the central artery and vein. Instead four artery-vein pairs just under the skin, two on each side of the fish, provide the main blood supply [*see top illustration on opposite page*]. From these cutaneous vessels large numbers of tiny blood vessels, each only a tenth of a millimeter in diameter, branch off. The tiny vessels intermingle to form slabs of vascular tissue, the *retia*, that lie close to the upper and lower surfaces of the dark muscle [*see bottom illustration on opposite page*]. These *retia* are the heat exchangers that ensure the warmth of the dark muscle. The blood to the light muscle also comes from the large cutaneous vessels by way of pairs of segmental arteries and veins that run up and down over the surface of the muscle and send numerous branches into it. These branches are in the form of vascular bands: ribbons of alternating arteries and veins that act as heat exchangers.

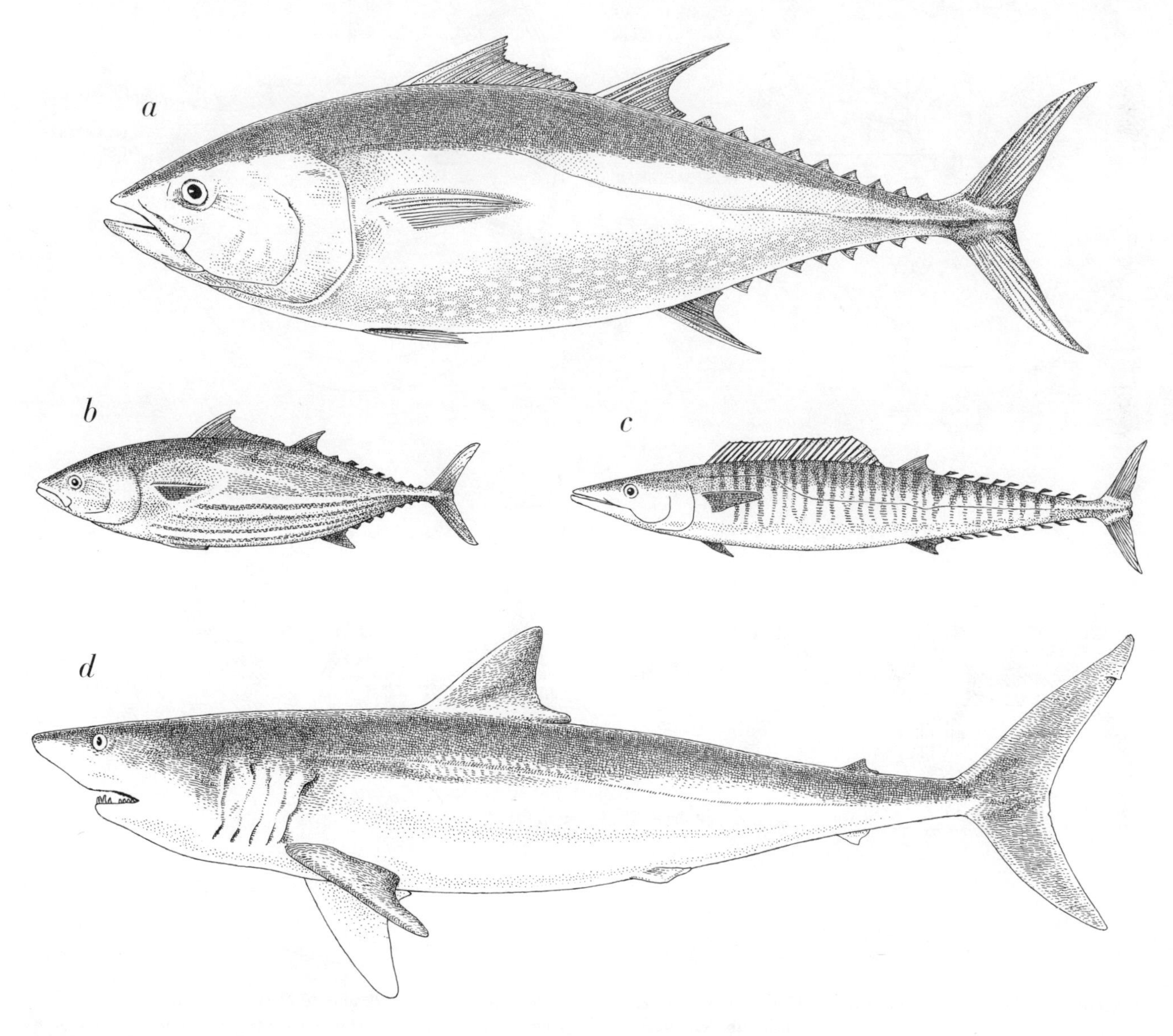

FAST-SWIMMING FISHES with warm bodies have streamlined bodies and heavily muscled tails with crescent-shaped caudal fins. The ones illustrated here are three tunas: the bluefin (*a*), the skipjack (*b*) and the wahoo (*c*), and a mackerel shark, the mako (*d*).

The area of thermal contact between arteries and veins in these two-dimensional structures is smaller than it is in the massive heat exchangers serving the dark muscle, but apparently it is adequate for the typically low rate of blood flow to the light muscle and ensures that much of this tissue too is warm.

One result of the inside-out arrangement of the main blood vessels in a warm fish is that the large arteries and veins that give rise to the *retia* are just under the skin. This puts the cool end of the heat exchanger near the surface of the fish and the warm end deep in the interior, an arrangement that minimizes surface heat loss.

By measuring the temperature of various parts of freshly caught bluefin tuna we have learned that the warmest regions are within the dark muscle [*see illustration on page 150*]. The hot spots are not, as one might expect, in the deepest part of the muscle. The reason is that the deep interior receives some of its blood supply from the dorsal aorta through a system that has no specialized heat exchangers. The highest muscle temperature is found near the center of the dark-muscle band on each side of the fish, but substantial areas of muscle tissue, both dark and light, are considerably warmer than the temperature of the water from which the fish are taken.

The bigeye tuna and the albacore have circulatory systems that closely resemble the bluefin's. Other tuna species, particularly the skipjack and to a lesser extent the yellowfin, elevate the temperature of the muscle in a different way. Although the pairs of cutaneous arteries and veins are still present, the heat exchangers that arise from them may contain no more than a single layer of fine blood vessels. In these fishes the principal heat exchanger is connected to the central artery-vein pair below the backbone. This central *rete*, a rodlike vascular mass that may be larger in diameter than the backbone itself, extends along the vertebrae in the region above the

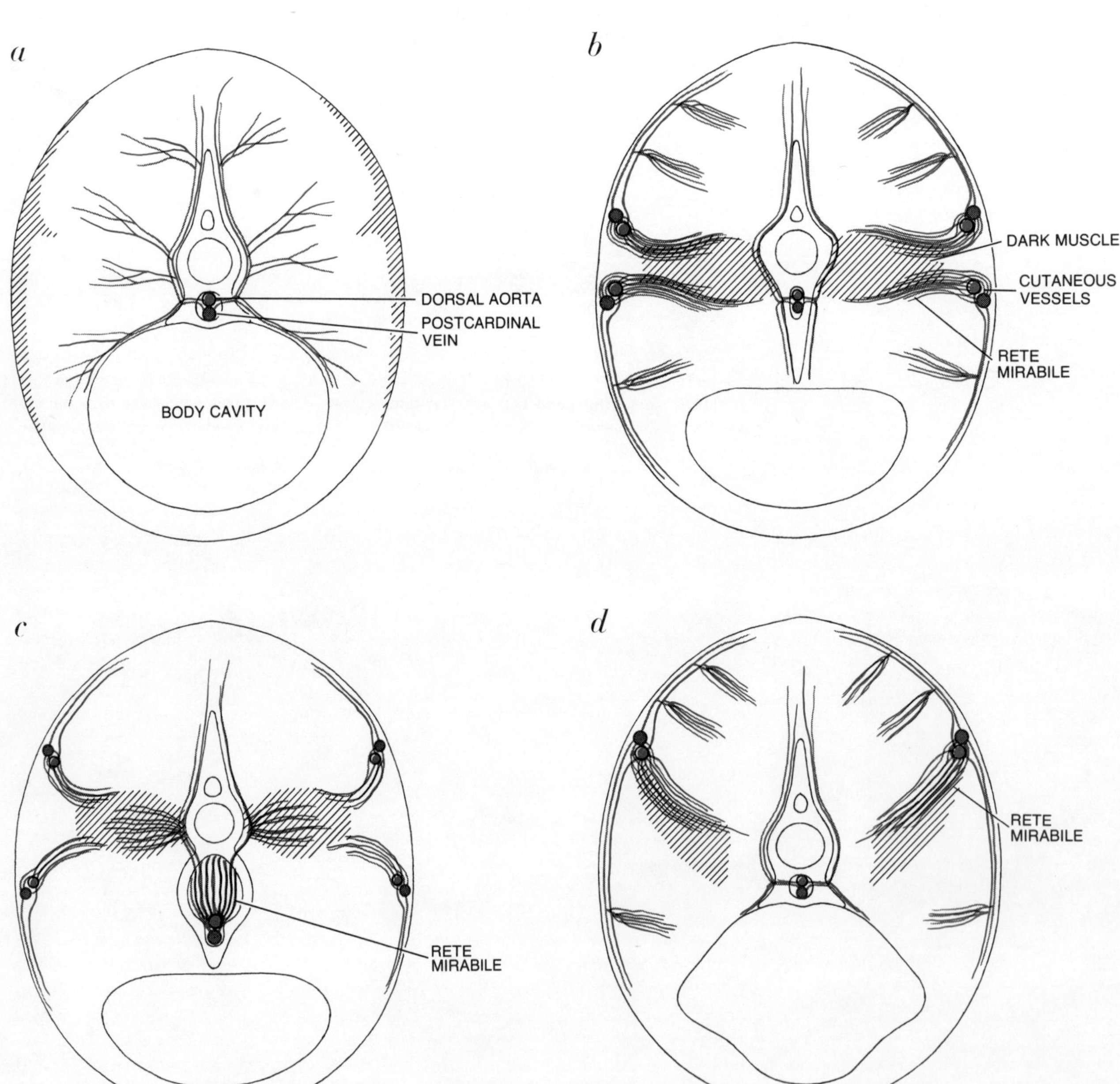

MAJOR BLOOD VESSELS of a typical cold-blooded fish (*a*) radiate from the large central vessels. In warm fishes the vascular system is rearranged so that warmed blood is supplied to the dark muscle. In the bluefin (*b*) the *retia* are slabs of vascular tissue above and below the dark muscle. In the skipjack (*c*), on the other hand, the main *rete* is a net of small vessels running vertically in a cavity below the backbone. The mako shark (*d*) has a single cutaneous blood-vessel pair, supplying a single massive *rete*, on each side.

body cavity [*see illustration on opposite page*]. Body-temperature measurements of skipjack attest to the efficiency of the rod-shaped heat exchanger: these little fish are often 10 degrees C. warmer than the water and their warmest muscle is adjacent to the backbone.

It is an interesting coincidence that the mackerel sharks, a group that is evolutionarily far removed from higher bony fishes such as the tunas, should have developed a heat-exchange system that is so much like the bluefin's. The deep interior blood vessels along a mackerel shark's backbone are small. The major blood supply circulates through a single artery-vein pair running just below the skin along each side of the fish. The paired cutaneous blood vessels give rise to a single massive *rete* that warms the shark's dark muscle. In some mackerel sharks, such as the mako shark, the heat exchanger is a solid slab of blood-vessel tissue resembling the bluefin's [*see illustration on opposite page*]. In others, such as the porbeagle shark and the white shark, the *rete* is diffuse, with many bundles of fine blood vessels extending into the dark muscle. In mackerel sharks the body-temperature distribution closely resembles the distribution in the bluefin; most of the sharks' muscle is warmer than the water they swim in, and the warmest parts of all are twin regions within the dark muscle.

Tunas and mackerel sharks possess heat exchangers that are not associated with muscle tissue but serve the organs of the body cavity. In the albacore, bigeye and bluefin tunas dense bundles of intermingled fine arteries and veins are found on the surface of the liver; in a large bluefin the bundles may be five centimeters in diameter. The arrangement in the mackerel sharks is different because the artery that in other fishes carries the main blood supply to the viscera is either very small or nonexistent. Instead the visceral blood supply is provided by arteries that are insignificant in most other fishes. These arteries penetrate into a much enlarged venous space known as the hepatic sinus. Venous blood flows through this space on its way from the viscera back to the heart. The arteries branch and branch again within the hepatic sinus until the space is virtually filled with a spongelike mass of tiny vessels; the cool blood within them is warmed by the bath of venous blood.

Visceral *retia* do not keep the organs of the body cavity warm continuously; the temperature can vary from a level equal to the warmest muscle to one only slightly above the water temperature.

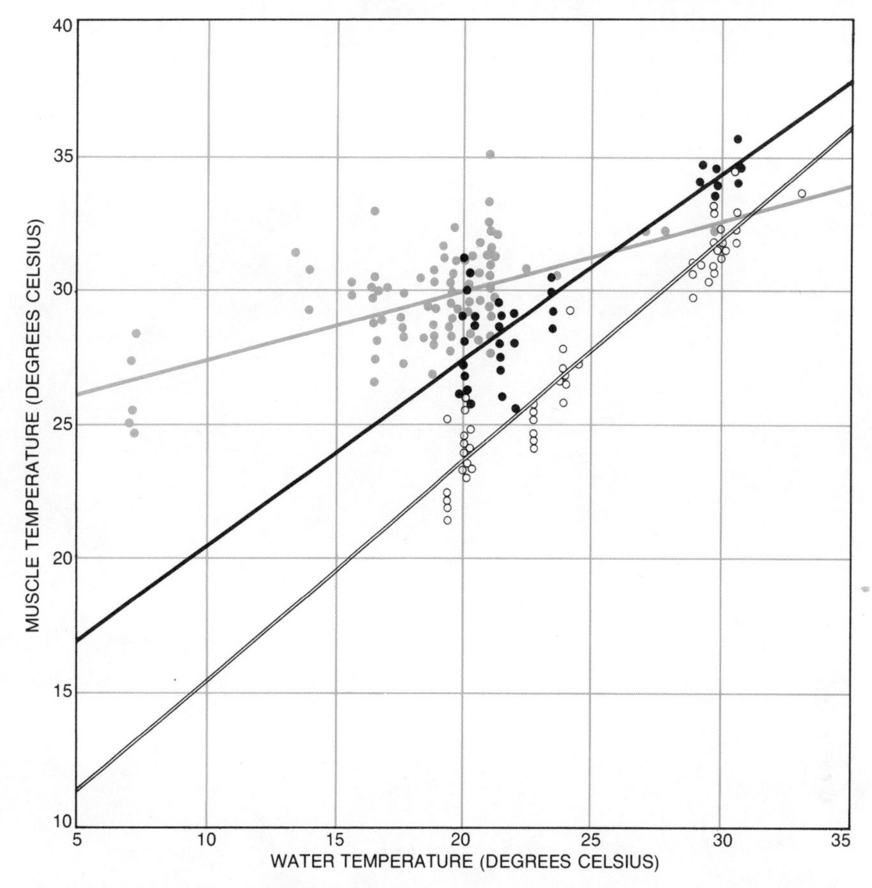

MAXIMUM MUSCLE TEMPERATURES are plotted against water temperature for bluefin (*color*), skipjack (*solid black*) and yellowfin (*open black*). The bluefin regulates its temperature almost independently of water temperature. The skipjack is warmer than the yellowfin but both tend to maintain a fixed temperature difference above water temperature.

The variations may be correlated with the mackerel sharks' and tunas' digestive activity. The tunas in particular have remarkably small body cavities, and the mass of their visceral organs is limited. Raising the temperature of the viscera may speed digestion. Conversely, during times of digestive inactivity the viscera may be allowed to cool.

Since tunas are warm, we wondered whether they could control their temperature at a constant level and thus be independent of water temperature or whether they warmed and cooled as the water temperature changed. Other investigators had noted that yellowfin tuna and skipjack tend to maintain a temperature a fixed number of degrees above that of the water—something that might be accomplished by an unregulated heat exchanger. These two species are found only within a rather narrow range of water temperature, however, and so we turned our attention to the bluefin tuna, which is equally at home in tropical and in subpolar waters. We measured the maximum temperature in the muscle of bluefin taken along the coast from the Bahamas to Newfoundland. Comparison of their muscle temperature with the temperature of the water in which they were captured suggested that they control their temperature quite well. Over a 24-degree C. range of water temperature the average muscle temperature changed only six degrees; bluefin from the 30-degree waters of the Bahamas were only a few degrees warmer than the water, whereas those from seven-degree northern waters were 20 degrees above the water temperature.

There are two ways an animal might achieve such control. One is by means of the swift thermoregulatory adjustment to environmental temperature changes that is characteristic of birds and mammals, which can vary their rate of heat loss in several ways and also generate extra heat by increasing their metabolism. The other is the much slower process of acclimatization, by which cold-blooded animals adjust their activities to changing temperature and which involves cellular and enzyme changes over an extended period of time. In order to discover which of these mechanisms the bluefin were using we decided to try a field experiment in which we would monitor the body tem-

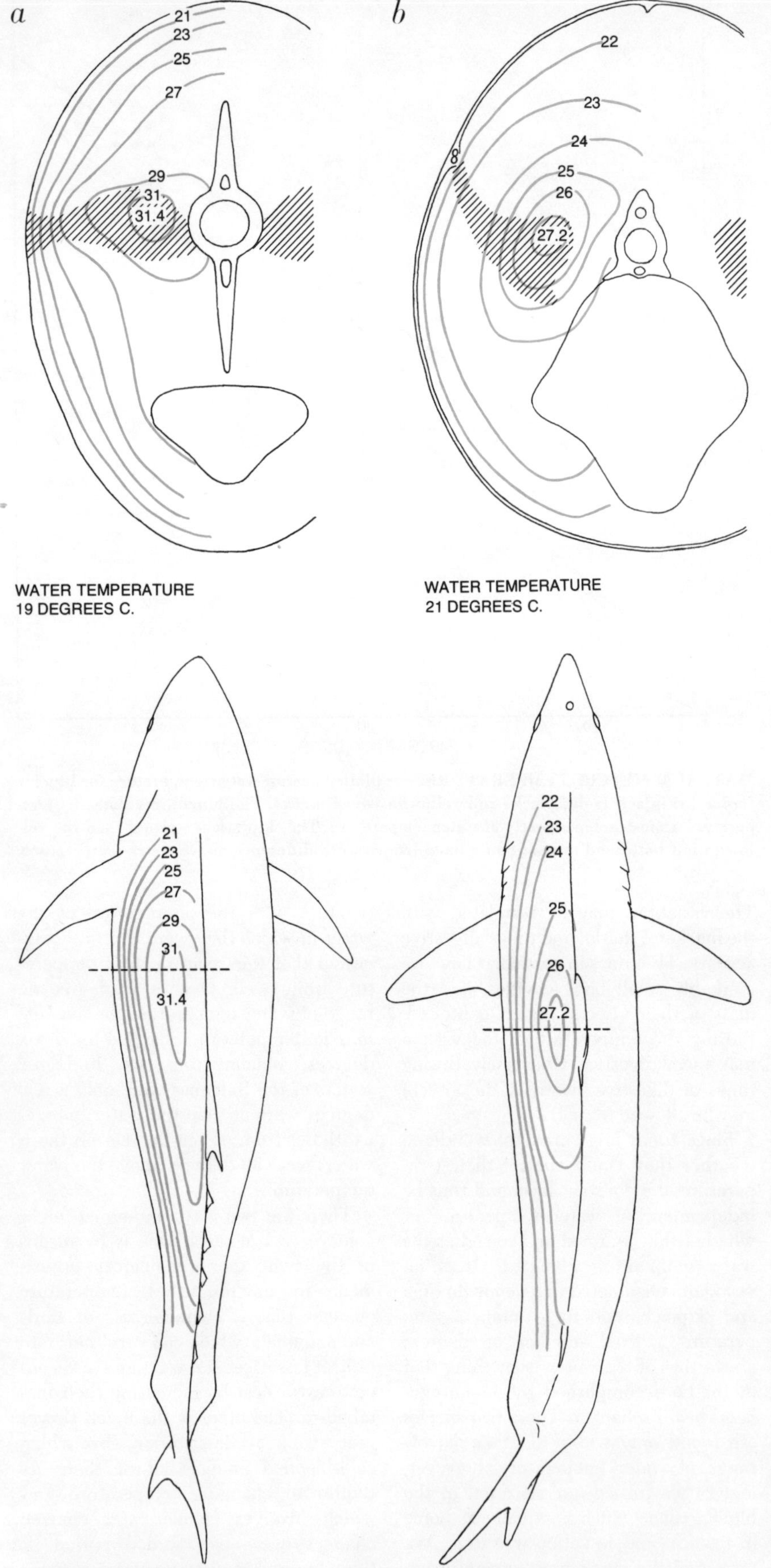

TEMPERATURES were recorded at various points in bluefin tuna (*a*) and mako sharks (*b*). Isotherms show that the warmest parts of the fishes are within the dark muscle on each side of the body. Much of the rest of the muscle is also considerably warmer than the water.

perature of fish as they swam from warm into cold water. If the bluefin were capable of rapid thermoregulation of the kind shown by mammals, their temperature should remain constant during a sudden change in water temperature.

We found that the coast of Nova Scotia was an ideal setting for our project. The Coolen brothers of Fox Point operate pound nets in St. Margaret's Bay, and they frequently catch bluefin tuna in their nets. Moreover, the coastal waters of Nova Scotia are characterized by a marked thermocline (a sharp drop in temperature as the depth increases), so that a free-swimming fish might encounter a wide range of water temperatures.

We used a small harpoon equipped with two thermistors in our first attempts to record bluefin temperatures. One thermistor sensed water temperature and the other muscle temperature; they were connected to an indicator by a 1,000-yard length of field-telephone wire that could unreel as the harpooned fish swam away. We failed on almost every attempt: the fish died or the harpoon was pulled out when the wire was snagged by kelp or by lobster-pot lines. It was at this point that my colleague John Kanwisher suggested that we telemeter the information from the fish.

The telemetering devices we selected were battery-powered acoustical transmitters. Their service life was from one day to three days and their range was as much as five miles. The instrument package was mounted on a small harpoon that could be driven through the fish's thick skin with a minimum of injury to the muscle. The thermistor that sensed the muscle temperature was in the tip of the harpoon and the one that sensed the water temperature was attached to the transmitter, outside the fish's body on the harpoon shaft. For other experiments, in which we measured the stomach temperature, an instrument package was pushed down the bluefin's throat and into its stomach; the thermistor that measured the water temperature was at the end of a wire we led out through the tuna's gill slit. The transmitters broadcast the temperature of the fish and the temperature of the water during alternate one-minute periods.

Our tracking vessel was equipped with a directional hydrophone that enabled us to follow the tuna. We rotated the hydrophone until the telemetered signal was at a maximum and then steered the vessel on that heading. At hourly intervals we lowered a bathythermograph to measure the temperature of the water at various depths. By comparing these readings with the telemetered temperature of the water surrounding the fish we could estimate the bluefin's swimming depth, and we could approximate the fish's course by plotting our own course as we followed the signals.

We placed transmitters on or in 14 bluefin tuna. The longest we tracked a fish was 54 hours; by then the bluefin had reached a position 130 miles offshore. We soon found that, as commercial fishermen had told us, tuna avoid changes in water temperature if they can. Most of our specimens remained near the surface or at least on the warm side of the steep thermocline that separated the upper waters from the cold depths. Some of the bluefin would dive through the thermocline, but they spent only a few minutes in the colder water before returning to the warm side.

Near the end of our efforts we were lucky enough to get a most satisfactory result. The specimen was a 600-pound bluefin with a transmitter in its stomach. The fish had been handled quite roughly while the instrument was being installed. Perhaps for this reason as soon as it was released from the pound at about 9:00 A.M. it swam down through the thermocline into water 14 degrees C. colder than the surface water in the pound. When the fish was released, its stomach temperature registered 21 degrees C. During its four-hour stay in the five-degree water the temperature of its stomach gradually fell to about 19 degrees. Early in the afternoon the fish returned to the warm side of the thermocline and remained in water that registered between 13 and 14 degrees for the rest of the day. In spite of a change in water temperature of nearly 10 degrees C. the temperature of the fish's stomach remained around 18 degrees. The fact that the deep-body temperature of the fish remained nearly constant over extended periods in both cold water and warm indicates that the bluefin was indeed thermoregulating. Just how the fish do this we do not know. Presumably it is by somehow varying the efficiency of the heat-exchanging *retia*.

We were not fortunate enough in our muscle-temperature experiments to have a bluefin stay on the cold side of the thermocline for any substantial length of time. One of the muscle-tagged fish, however, did swim for hours in water that was gradually decreasing in temperature. The water temperature dropped four degrees C. in one 90-minute period. During this interval the muscle temperature of the bluefin slowly rose from a

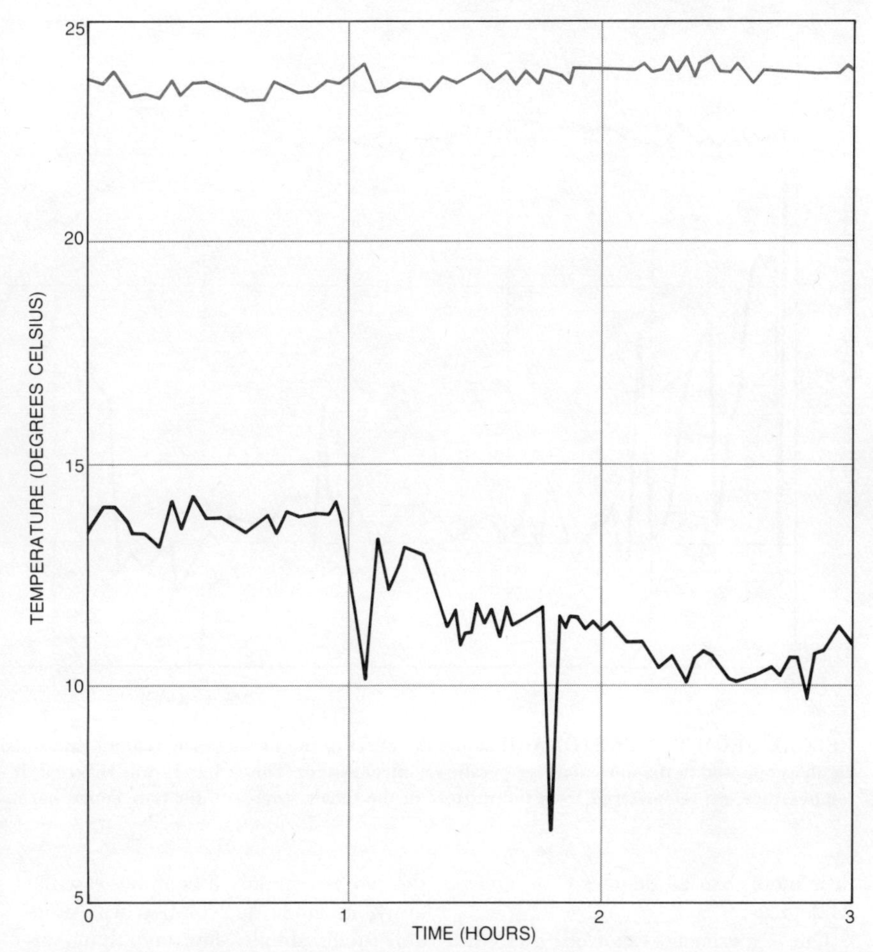

TELEMETRY RECORD compares the temperature in the muscle of a bluefin (*color*) with that of the water (*black*) through which it swam for three hours. The muscle temperature was held constant as the water temperature declined gradually from about 14 to 10 degrees.

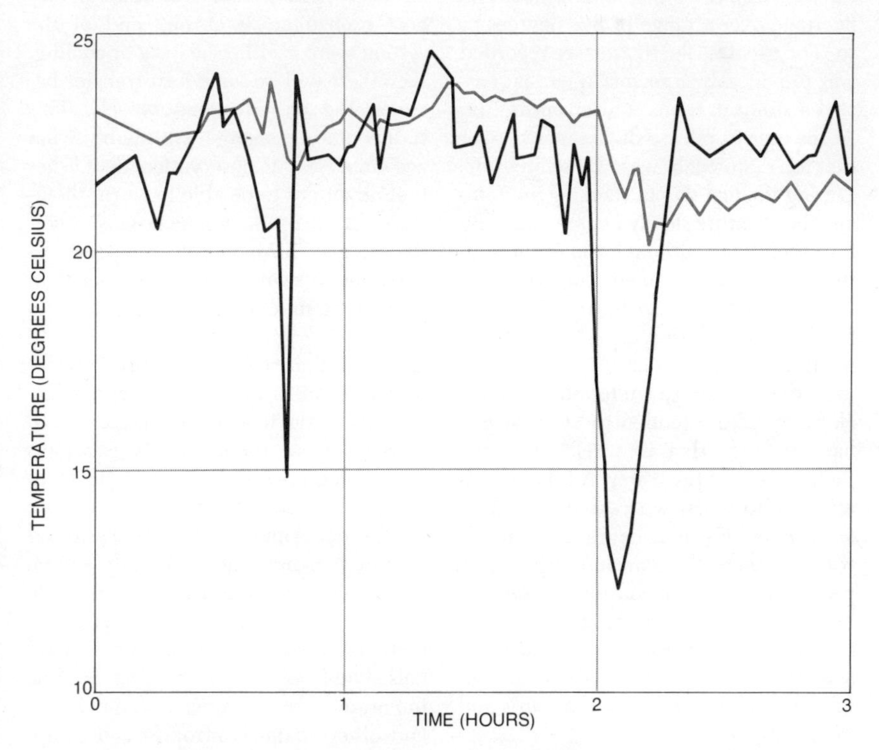

RECORD OF TEMPERATURES from the muscle of a dusky shark (*color*) and the water it swam in (*black*) shows a different relation. The dusky shark is a cold-blooded fish and its muscle temperature stayed close to that of the water, dipping during even short dives.

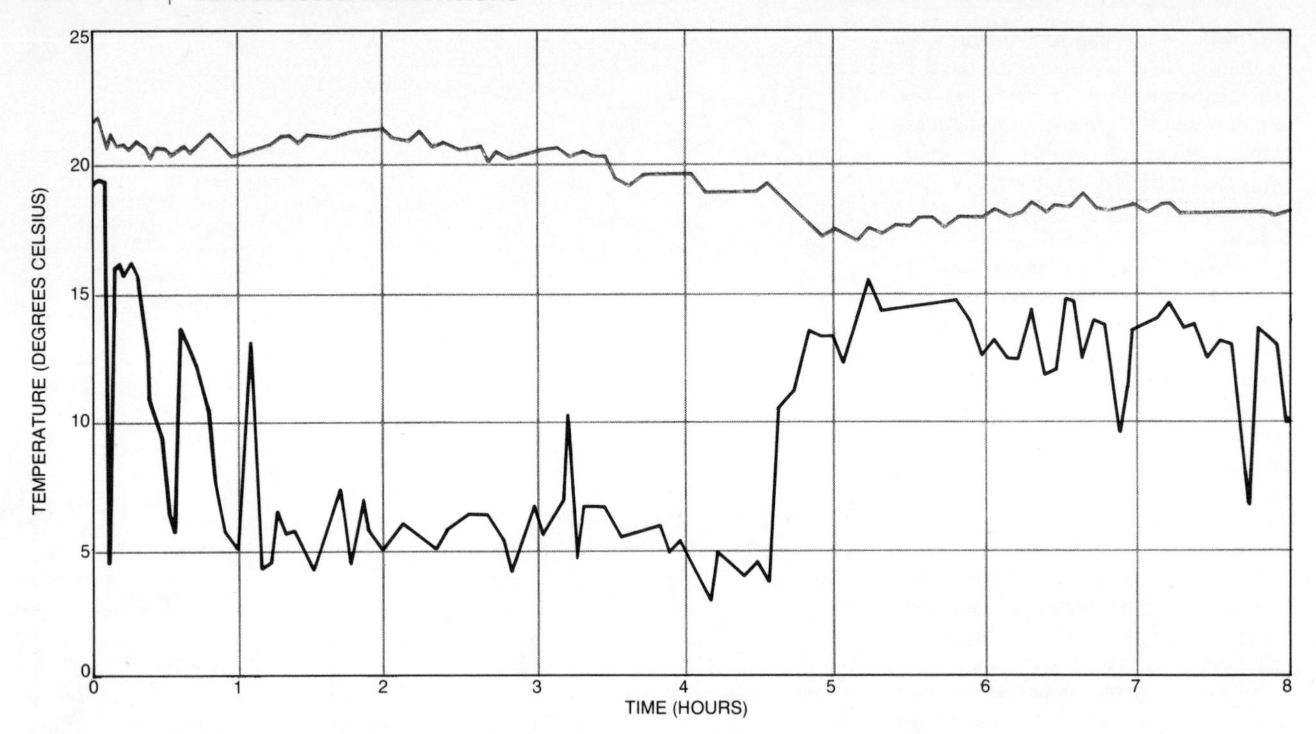

RECORD FROM TUNA'S STOMACH shows the effect of the visceral, as opposed to the muscular, temperature-control system. The temperature was telemetered from thermistors in the tuna's stomach (*color*) and outside the fish in the water (*black*). When the tuna was released, it swam down into cold water and stayed there for four hours, but its stomach temperature decreased only a little.

little more than 23 degrees C. to above 24 degrees.

Our experiments showed that the bluefin does not maintain its body temperature at a constant level as, for example, human beings do. The temperatures of fish from the same school can fluctuate over a range of five degrees or so. The greatest fluctuation we recorded was provided by a second tuna that carried a stomach temperature transmitter. At the time of release the temperature of the fish's stomach was 19 degrees C. During the first day of tracking its stomach temperature slowly began to rise. By the second day, although the water temperature had remained constant, the fish's stomach was registering 26 degrees C. Perhaps the fish was trying to digest the transmitter!

In other readings we found that muscle temperature tended to cycle over a narrow range that is unrelated to the temperature of the water. All these variations tell us that whereas the bluefin's various organs are being kept much warmer than the surrounding water, precise temperature control is either not necessary or not possible. In effect, the fish has a sloppy thermostat. Nonetheless, these fish can be said to thermoregulate in the same sense that birds and mammals do.

How did the tunas and mackerel sharks come to develop these remarkably efficient heat exchangers? Although the *rete* is certainly a complex vascular array, its mode of evolution was probably quite simple. For one thing, arteries and veins generally follow the same path as they travel between organs and within tissues. Every such pairing of an artery and a vein is a rudimentary heat exchanger; if at one end of the system there is either heating or cooling, then there will be some heat transfer between the paired blood vessels. This rudimentary exchange is probably what accounts for our observation that fishes that ought not to be able to warm themselves at all by metabolic means nevertheless may have muscle temperatures from one to two degrees C. higher than the water temperature.

Suppose that such a slight rise in tissue temperature offers a genuine selective advantage to an evolving species. In such an event there already exists (as part of the normal embryonic development of the circulatory system) a basis for the development of more advanced heat exchangers. The circulatory system of the developing embryo first takes the form not of discrete blood vessels but of beds of interconnecting spaces and channels. Later, as certain channels become important, these routes become larger and other channels atrophy and disappear. Moreover, there is a tendency toward the development of multiple channels. Several such channels may form within a single bed, with the result that the vascular system of the adult will have some duplicate components. Therefore in order to form the mass of parallel blood vessels that is an efficient heat exchanger, only a modification of the normal embryonic pattern is required, not a radical genetic change.

A fast-swimming predatory fish finds an abundance of food awaiting it in the form of fast-swimming squid, herring and mackerel that slower predators cannot capture. In terms both of streamlining and of muscle power the tunas are probably the swiftest predators of the open ocean. Yellowfin tuna and wahoo tuna are noted for their speed: they have been observed to reach speeds in excess of 40 miles per hour during 10- to 20-second sprints. Mackerel sharks (at least the mako and porbeagle) probably also enjoy the benefits of fast swimming. Such a shark, with its bulky, muscular body, its streamlining and its narrow, crescent-shaped tail, looks remarkably like a tuna. That the two unrelated groups of fishes independently evolved the same means of raising their body temperature must surely be connected to the adaptive advantage of increased swimming speed; the extra power available from warm muscle must have been decisive in achieving that speed. It is a classic example of parallel adaptations that evidently gave both groups access to an underexploited source of food.

Adaptations to Cold

14

by Laurence Irving
January 1966

One mechanism is increased generation of heat by a rise in the rate of metabolism, but this process has its limits. The alternatives are insulation and changes in the circulation of heat by the blood

All living organisms abhor cold. For many susceptible forms of life a temperature difference of a few degrees means the difference between life and death. Everyone knows how critical temperature is for the growth of plants. Insects and fishes are similarly sensitive; a drop of two degrees in temperature when the sun goes behind a cloud, for instance, can convert a fly from a swift flier to a slow walker. In view of the general hostility of cold to life and activity, the ability of mammals and birds to survive and flourish in all climates is altogether remarkable.

It is not that these animals are basically more tolerant of cold. We know from our own reactions how sensitive the human body is to chilling. A naked, inactive human being soon becomes miserable in air colder than 28 degrees centigrade (about 82 degrees Fahrenheit), only 10 degrees C. below his body temperature. Even in the Tropics the coolness of night can make a person uncomfortable. The discomfort of cold is one of the most vivid of experiences; it stands out as a persistent memory in a soldier's recollections of the unpleasantness of his episodes in the field. The coming of winter in temperate climates has a profound effect on human well-being and activity. Cold weather, or cold living quarters, compounds the misery of illness or poverty. Over the entire planet a large proportion of man's efforts, culture and economy is devoted to the simple necessity of protection against cold.

Yet strangely enough neither man nor other mammals have consistently avoided cold climates. Indeed, the venturesome human species often goes out of its way to seek a cold environment, for sport or for the adventure of living in a challenging situation. One of the marvels of man's history is the endurance and stability of the human settlements that have been established in arctic latitudes.

The Norse colonists who settled in Greenland 1,000 years ago found Eskimos already living there. Archaeologists today are finding many sites and relics of earlier ancestors of the Eskimos who occupied arctic North America as long as 6,000 years ago. In the middens left by these ancient inhabitants are bones and hunting implements that indicate man was accompanied in the cold north by many other warm-blooded animals: caribou, moose, bison, bears, hares, seals, walruses and whales. All the species, including man, seem to have been well adapted to arctic life for thousands of years.

It is therefore a matter of more than idle interest to look closely into how mammals adapt to cold. In all climates and everywhere on the earth mammals maintain a body temperature of about 38 degrees C. It looks as if evolution has settled on this temperature as an optimum for the mammalian class. (In birds the standard body temperature is a few degrees higher.) To keep their internal temperature at a viable level the mammals must be capable of adjusting to a wide range of environmental temperatures. In tropical air at 30 degrees C. (86 degrees F.), for example, the environment is only eight degrees cooler than the body temperature; in arctic air at −50 degrees C. it is 88 degrees colder. A man or other mammal in the Arctic must adjust to both extremes as seasons change.

The mechanisms available for making the adjustments are (1) the generation of body heat by the metabolic burning of food as fuel and (2) the use of insulation and other devices to retain body heat. The requirements can be expressed quantitatively in a Newtonian formula concerning the cooling of warm bodies. A calculation based on the formula shows that to maintain the necessary warmth of its body a mammal must generate 10 times more heat in the Arctic than in the Tropics or clothe itself in 10 times more effective insulation or employ some intermediate combination of the two mechanisms.

We need not dwell on the metabolic requirement; it is rarely a major factor. An animal can increase its food intake and generation of heat to only a very modest degree. Moreover, even if metabolic capacity and the food supply were unlimited, no animal could spend all its time eating. Like man, nearly all other mammals spend a great deal of time in curious exploration of their surroundings, in play and in family and social activities. In the arctic winter a herd of caribou often rests and ruminates while the young engage in aimless play. I have seen caribou resting calmly with wolves lying asleep in the snow in plain view only a few hundred yards away. There is a common impression that life in the cold climates is more active than in the Tropics, but the fact is that for the natural populations of mammals, including man, life goes on at the same leisurely pace in the Arctic as it does in warmer regions; in all climates there is the same requirement of rest and social activities.

The decisive difference in resisting cold, then, lies in the mechanisms for conserving body heat. In the Institute of Arctic Biology at the University of Alaska we are continuing studies that have been in progress there and elsewhere for 18 years to compare the

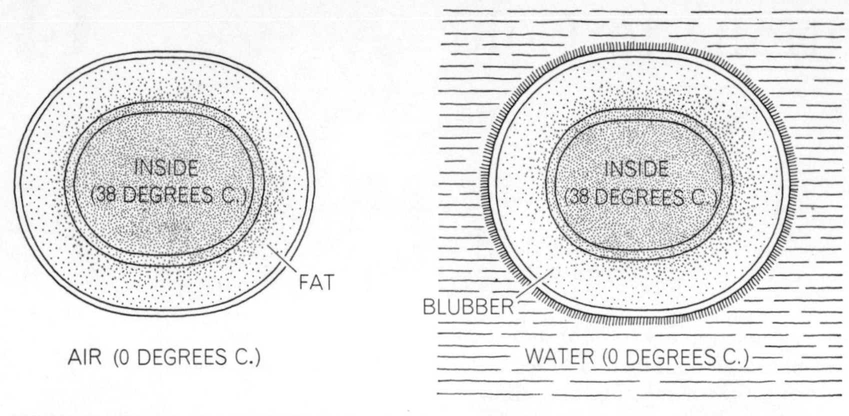

TEMPERATURE GRADIENTS in the outer parts of the body of a pig (*left*) and of a seal (*right*) result from two effects: the insulation provided by fat and the exchange of heat between arterial and venous blood, which produces lower temperatures near the surface.

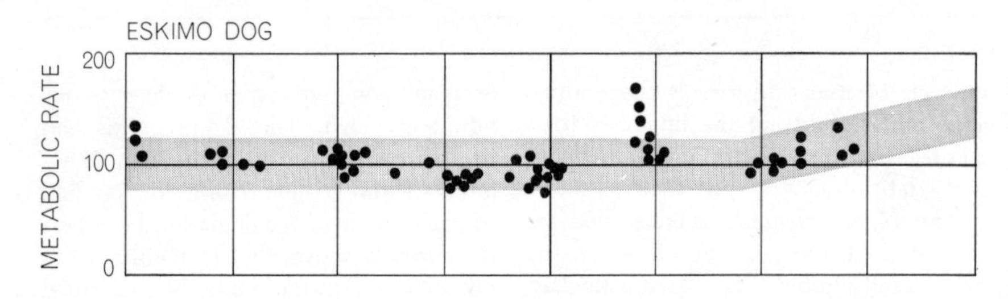

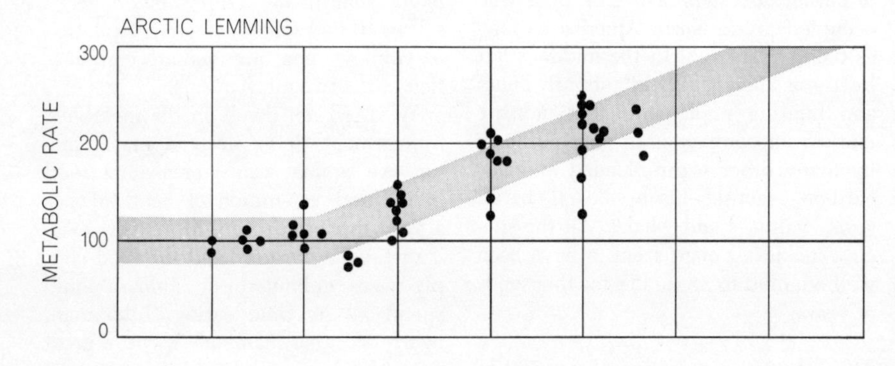

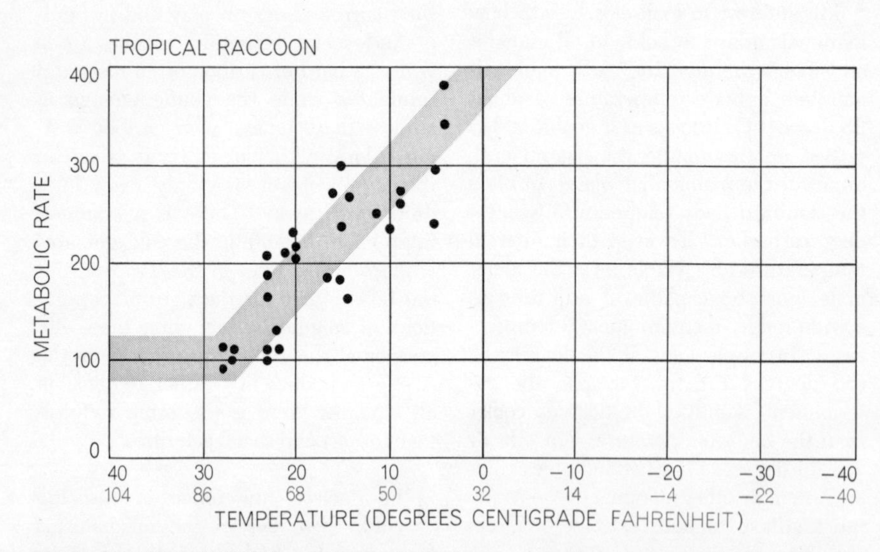

RATE OF METABOLISM provides a limited means of adaptation to cold. The effect of declining temperatures on the metabolic rate is shown for an Eskimo dog (*top*), an arctic lemming (*middle*) and a tropical raccoon (*bottom*). Animals in warmer climates tend to increase metabolism more rapidly than arctic animals do when the temperature declines.

mechanisms for conservation of heat in arctic and tropical animals. The investigations have covered a wide variety of mammals and birds and have yielded conclusions of general physiological interest.

The studies began with an examination of body insulation. The fur of arctic animals is considerably thicker, of course, than that of tropical animals. Actual measurements showed that its insulating power is many times greater. An arctic fox clothed in its winter fur can rest comfortably at a temperature of −50 degrees C. without increasing its resting rate of metabolism. On the other hand, a tropical animal of the same size (a coati, related to the raccoon) must increase its metabolic effort when the temperature drops to 20 degrees C. That is to say, the fox's insulation is so far superior that the animal can withstand air 88 degrees C. colder than its body at resting metabolism, whereas the coati can withstand a difference of only 18 degrees C. Naked man is less well protected by natural insulation than the coati; if unclothed, he begins shivering and raising his metabolic rate when the air temperature falls to 28 degrees C.

Obviously as animals decrease in size they become less able to carry a thick fur. The arctic hare is about the smallest mammal with enough fur to enable it to endure continual exposure to winter cold. The smaller animals take shelter under the snow in winter. Weasels, for example, venture out of their burrows only for short periods; mice spend the winter in nests and sheltered runways under the snow and rarely come to the surface.

No animal, large or small, can cover all of its body with insulating fur. Organs such as the feet, legs and nose must be left unencumbered if they are to be functional. Yet if these extremities allowed the escape of body heat, neither mammals nor birds could survive in cold climates. A gull or duck swimming in icy water would lose heat through its webbed feet faster than the bird could generate it. Warm feet standing on snow or ice would melt it and soon be frozen solidly to the place where they stood. For the unprotected extremities, therefore, nature has evolved a simple but effective mechanism to reduce the loss of heat: the warm outgoing blood in the arteries heats the cool blood returning in the veins from the extremities. This exchange occurs in the *rete mirabile* (wonderful net), a network of small arteries and veins near the junction between the trunk of the animal and the extremity [see " 'The Wonderful Net,' " by P. F. Scholander; SCIENTIFIC AMERICAN, April, 1957]. Hence the extremities can become much colder than the body without either draining off body heat or losing their ability to function.

This mechanism serves a dual purpose. When necessary, the thickly furred animals can use their bare extremities to release excess heat from the body. A heavily insulated animal would soon be overheated by running or other active exercise were it not for these outlets. The generation of heat by exercise turns on the flow of blood to the extremities so that they radiate heat. The large, bare flippers of a resting fur seal are normally cold, but we have found that when these animals on the Pribilof Islands are driven overland at their laborious gait, the flippers become warm. In contrast to the warm flippers, the rest of the fur seal's body surface feels cold, because very little heat escapes through the animal's dense fur. Heat can also be dissipated by evaporation from the mouth and tongue. Thus a dog or a caribou begins to pant, as a means of evaporative cooling, as soon as it starts to run.

In the pig the adaptation to cold by means of a variable circulation of heat in the blood achieves a high degree of refinement. The pig, with its skin only thinly covered with bristles, is as naked as a man. Yet it does well in the Alaskan winter without clothing. We can read the animal's response to cold by its expressions of comfort or discomfort, and we have measured its physiological reactions. In cold air the circulation of heat in the blood of swine is shunted away from the entire body surface, so that the surface becomes an effective insulator against loss of body heat. The pig can withstand considerable cooling of its body surface. Although a man is highly uncomfortable when his skin is cooled to 7 degrees C. below the internal temperature, a pig can be comfortable with its skin 30 degrees C. colder than the interior, that is, at a temperature of 8 degrees C. (about 46 degrees F.). Not until the air temperature drops below the freezing point (0 degrees C.) does the pig increase its rate of metabolism; in contrast a man, as I have mentioned, must do so at an air temperature of 28 degrees C.

With thermocouples in the form of needles we have probed the tissues of pigs below the skin surface. (Some pigs, like some people, will accept a little

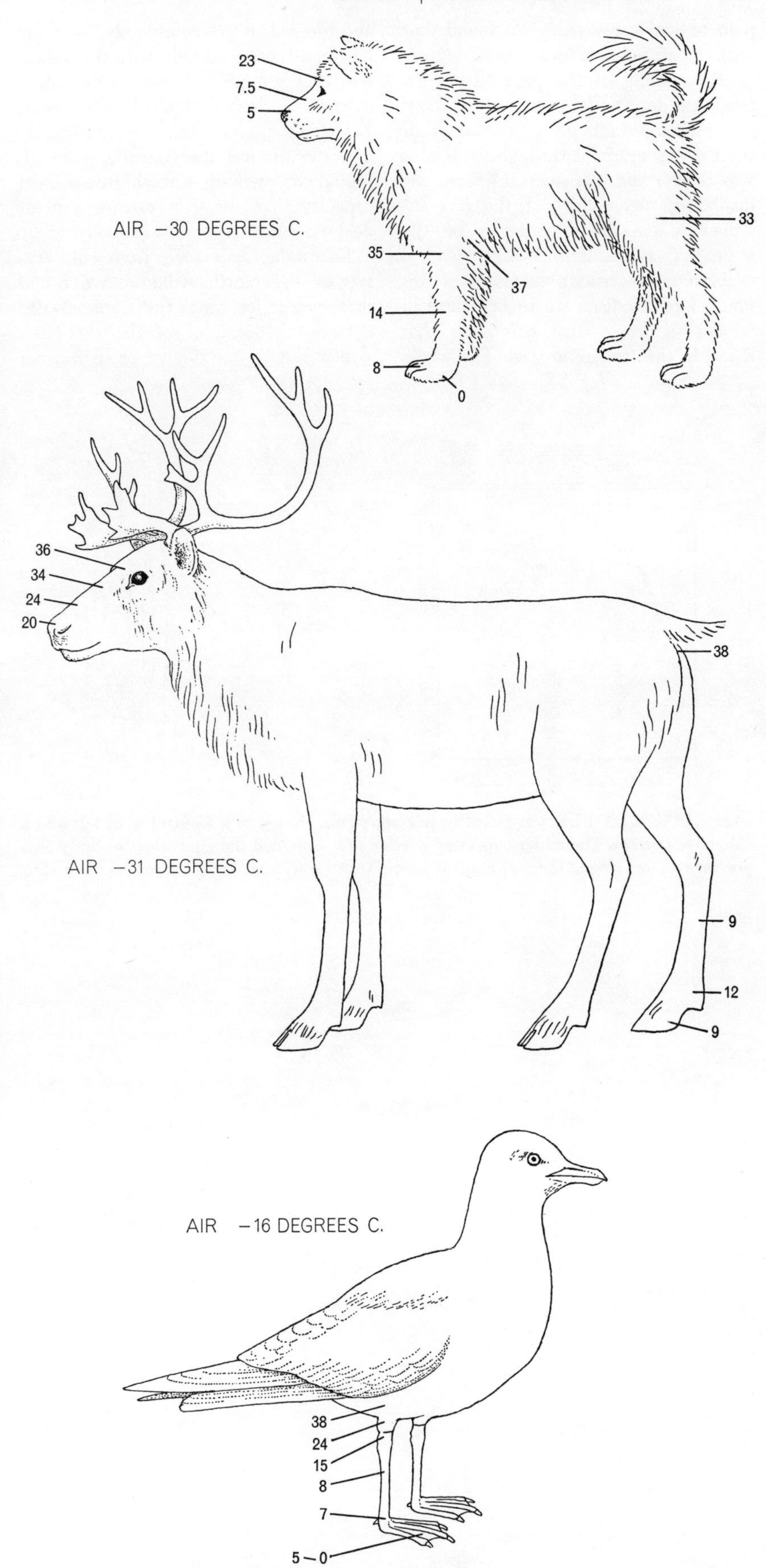

TEMPERATURES AT EXTREMITIES of arctic animals are far lower than the internal body temperature of about 38 degrees centigrade, as shown by measurements made on Eskimo dogs, caribou and sea gulls. Some extremities approach the outside temperature.

pain to win a reward.) We found that with the air temperature at −12 degrees C. the cooling of the pig's tissues extended as deep as 100 millimeters (about four inches) into its body. In warmer air the thermal gradient through the tissues was shorter and less steep. In short, the insulating mechanism of the hog involves a considerable depth of the animal's fatty mantle.

Even more striking examples of this kind of mechanism are to be found in whales, walruses and hair seals that dwell in the icy arctic seas. The whale and the walrus are completely bare; the hair seal is covered only with thin, short hair that provides almost no insulation when it is sleeked down in the water. Yet these animals remain comfortable in water around the freezing point although water, with a much greater heat capacity than air, can extract a great deal more heat from a warm body.

Examining hair seals from cold waters of the North Atlantic, we found that even in ice water these animals did not raise their rate of metabolism. Their skin was only one degree or so warmer than the water, and the cooling effect extended deep into the tissues—as much as a quarter of the distance through the thick part of the body. Hour after hour the animal's flippers all the way through would remain only a few degrees above freezing without the seals' showing any sign of discomfort. When the seals were moved into warmer water, their outer tissues rapidly warmed up. They would accept a transfer from warm water to ice water with equanimity and with no diminution of their characteristic liveliness.

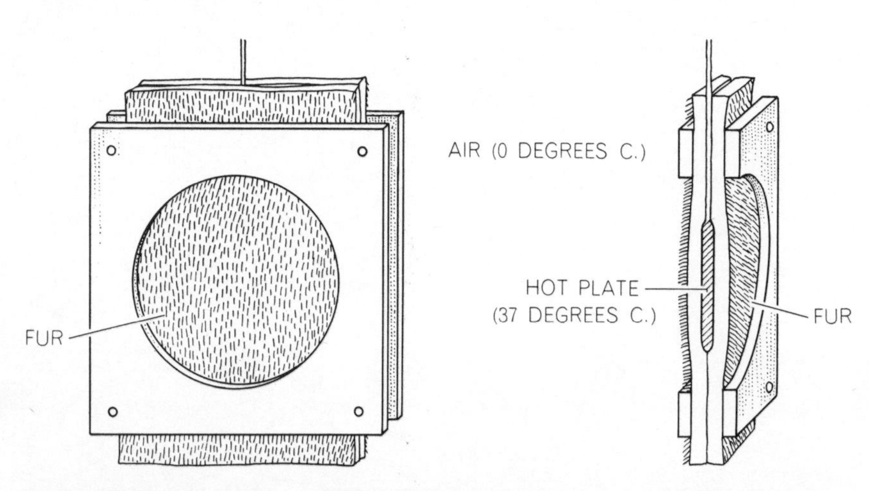

INSULATION BY FUR was tested in this apparatus, shown in a front view at left and a side view at right. The battery-operated heating unit provided the equivalent of body temperature on one side of the fur; outdoor temperatures were approximated on the other side.

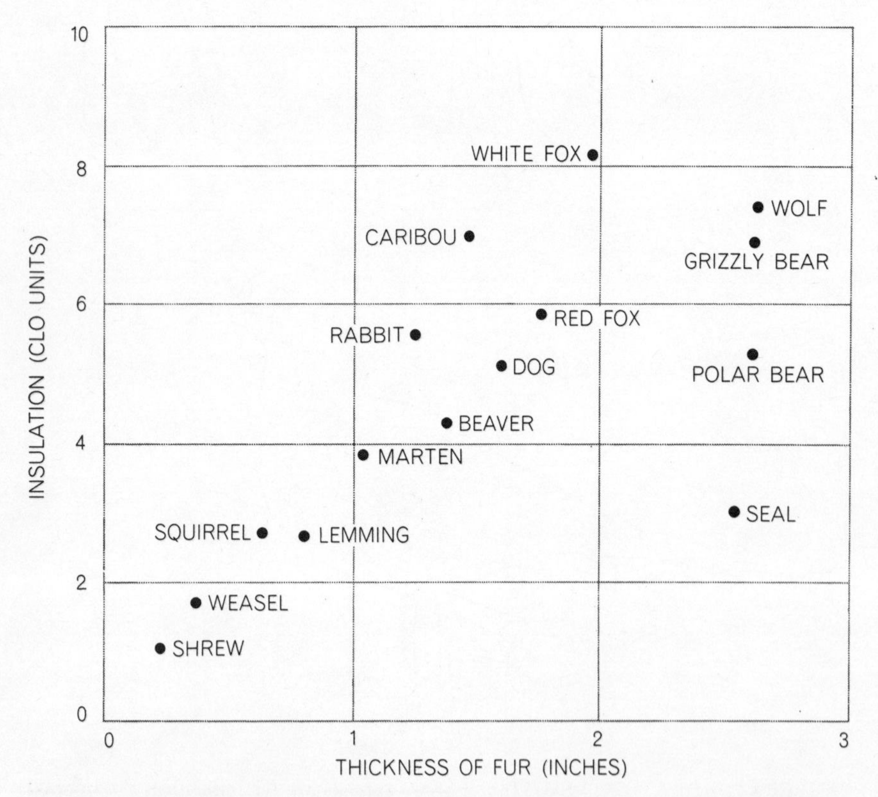

INSULATING CAPACITY of fur is compared for various animals. A "clo unit" equals the amount of insulation provided by the clothing a man usually wears at room temperature.

How are the chilled tissues of all these animals able to function normally at temperatures close to freezing? There is first of all the puzzle of the response of fatty tissue. Animal fat usually becomes hard and brittle when it is cooled to low temperatures. This is true even of the land mammals of the Arctic, as far as their internal fats are concerned. If it were also true of extremities such as their feet, however, in cold weather their feet would become too inflexible to be useful. Actually it turns out that the fats in these organs behave differently from those in the warm internal tissues. Farmers have known for a long time that neat's-foot oil, extracted from the feet of cattle, can be used to keep leather boots and harness flexible in cold weather. By laboratory examination we have found that the fats in the bones of the lower leg and foot of the caribou remain soft even at 0 degrees C. The melting point of the fats in the leg steadily goes up in the higher portions of the leg. Eskimos have long been aware that fat from a caribou's foot will serve as a fluid lubricant in the cold, whereas the marrow fat from the upper leg is a solid food even at room temperature.

About the nonfatty substances in tissues we have little information; I have seen no reports by biochemists on the effects of temperature on their properties. It is known, however, that many of the organic substances of animal tissues are highly sensitive to temperature. We must therefore wonder how the tissues can maintain their serviceability over the very wide range of temperatures that the body surface experiences in the arctic climate.

We have approached this question by studies of the behavior of tissues at various temperatures. Nature offers many illustrations of the slowing of tissue functions by cold. Fishes, frogs and water insects are noticeably slowed down by cool water. Cooling by 10 degrees

RANGE OF TEMPERATURES to which warm-blooded animals must adapt is indicated. All the animals shown have a body temperature close to 100 degrees Fahrenheit, yet they survive at outside temperatures that, for the arctic animals, can be more than 100 degrees cooler. Insulation by fur is a major means of adaptation to cold. Man is insulated by clothing; some other relatively hairless animals, by fat. Some animals have a mechanism for conserving heat internally so that it is not dissipated at the extremities.

C. will immobilize most insects. A grasshopper in the warm noonday sun can be caught only by a swift bird, but in the chill of early morning it is so sluggish that anyone can seize it. I had a vivid demonstration of the temperature effect one summer day when I went hunting on the arctic tundra near Point Barrow for flies to use in experiments. When the sun was behind clouds, I had no trouble picking up the flies as they crawled about in the sparse vegetation, but as soon as the sun came out the flies took off and were uncatchable. Measuring the temperature of flies on the ground, I ascertained that the difference between the flying and the slow-crawling state was a matter of only 2 degrees C.

Sea gulls walking barefoot on the ice in the Arctic are just as nimble as gulls on the warm beaches of California. We know from our own sensations that our fingers and hands are numbed by cold. I have used a simple test to measure the amount of this desensitization. After cooling the skin on my fingertips to about 20 degrees C. (68 degrees F.) by keeping them on ice-filled bags, I tested their sensitivity by dropping a light ball (weighing about one milligram) on them from a measured height. The weight multiplied by the distance of fall gave me a measure of the impact on the skin. I found that the skin at a temperature of 20 degrees C. was only a sixth as sensitive as at 35 degrees C. (95 degrees F.); that is, the impact had to be six times greater to be felt.

We know that even the human body surface has some adaptability to cold. Men who make their living by fishing can handle their nets and fish with wet hands in cold that other people cannot endure. The hands of fishermen, Eskimos and Indians have been found to be capable of maintaining an exceptionally vigorous blood circulation in the cold. This is possible, however, only at the cost of a higher metabolic production of body heat, and the production in any case has a limit. What must arouse our wonder is the extraordinary adaptability of an animal such as the hair seal. It swims in icy waters with its flippers and the skin over its body at close to the freezing temperature, and yet under the ice in the dark arctic sea it remains sensitive enough to capture moving prey and find its way to breathing holes.

Here lies an inviting challenge for all biologists. By what devices is an animal able to preserve nervous sensitivity in tissues cooled to low temperatures? Beyond this is a more universal and more interesting question: How do the warm-blooded animals preserve their overall stability in the varying environments to which they are exposed? Adjustment to changes in temperature requires them to make a variety of adaptations in the various tissues of the body. Yet these changes must be harmonized to maintain the integration of the organism as a whole. I predict that further studies of the mechanisms involved in adaptation to cold will yield exciting new insights into the processes that sustain the integrity of warm-blooded animals.

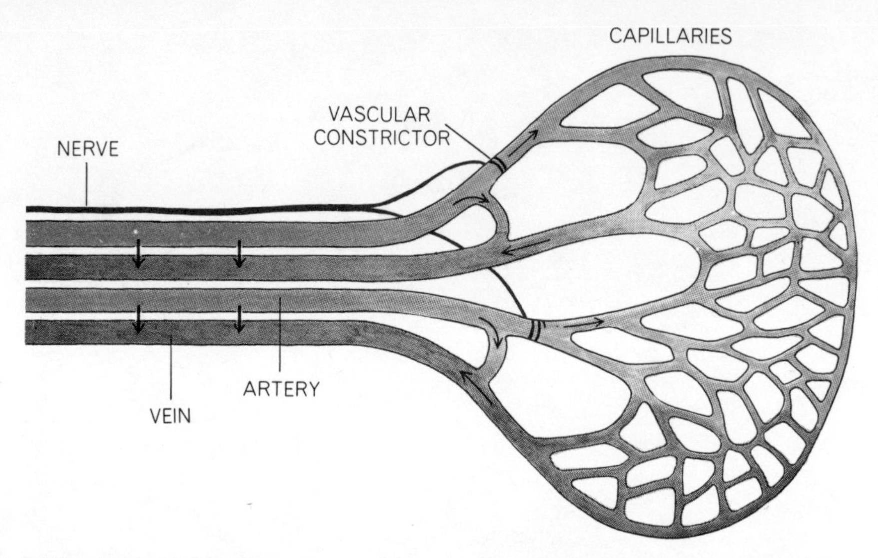

ROLE OF BLOOD in adaptation to cold is depicted schematically. One mechanism, indicated by the vertical arrows, is an exchange of heat between arterial and venous blood. The cold venous blood returning from an extremity acquires heat from an arterial network. The outgoing arterial blood is thus cooled. Hence the exchange helps to keep heat in the body and away from the extremities when the extremities are exposed to low temperatures. The effect is enhanced by the fact that blood vessels near the surface constrict in cold.

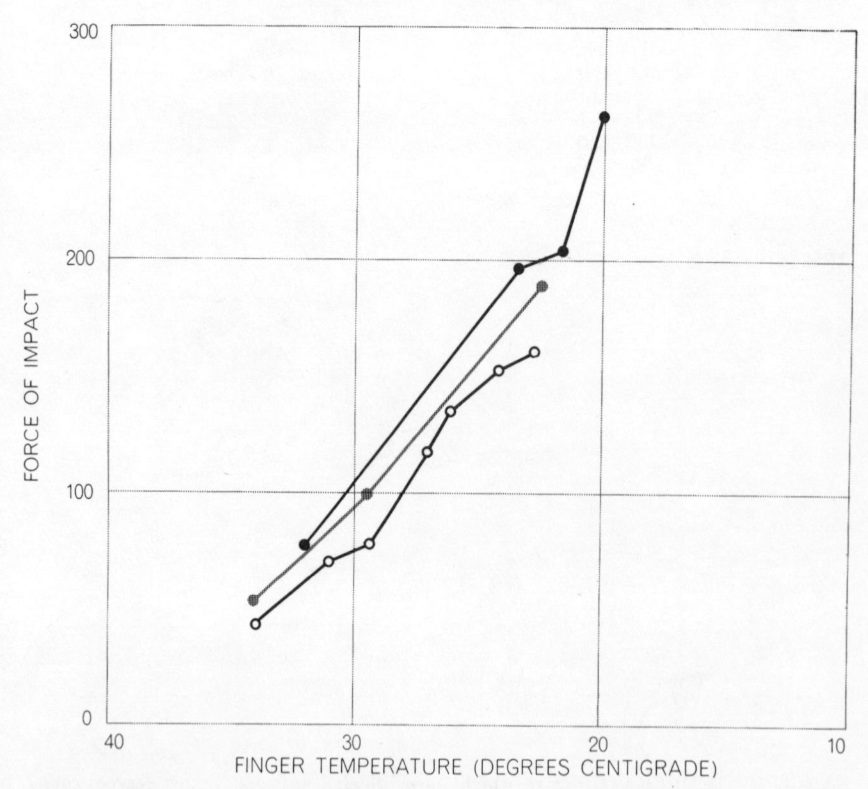

FINGER EXPERIMENT performed by the author showed that the more a finger was chilled, the farther a one-milligram ball had to be dropped for its impact to be felt on the finger. The vertical scale is arbitrary but reflects the relative increase in the force of impact.

The Thermostat of Vertebrate Animals

by H. Craig Heller, Larry I. Crawshaw
and Harold T. Hammel
August 1978

The hypothalamus, a structure at the base of the brain, monitors body temperature in a wide variety of animals and maintains it at an optimal level by controlling thermoregulatory mechanisms

Animals with backbones function most efficiently within a narrow range of body temperatures. The upper limit for survival is about 45 degrees Celsius (113 degrees Fahrenheit), when proteins begin to denature and become inactivated; the lower limit is slightly below zero degrees C. (32 degrees F.), when intracellular water begins to form ice crystals that rupture and kill cells. Even within these extremes slight shifts in body temperature away from an optimal level can have adverse effects because each physiological process involves many integrated biochemical reactions, the rates of which are dependent in a highly specific way on temperature. When the temperature changes, the total integration of the reactions may be disrupted. In spite of these physiological realities vertebrate animals thrive in virtually every habitat on the earth, from the icy waters of polar seas to the fierce heat of equatorial deserts. How vertebrates maintain a favorable internal temperature in such hostile thermal environments is the subject of this article.

Commonly but, as we shall see, inaccurately the vertebrates are separated into two major groups: the "cold-blooded" animals and the "warm-blooded" animals. Biologists refer to these same two groups of vertebrates as ectotherms or endotherms in recognition of their major sources of heat. The ectotherms (fishes, amphibians and reptiles) have poor body insulation and low rates of metabolic heat production. Their body temperature is largely dependent on heat from their environment, and their most important thermoregulatory mechanism is the selection of a suitable environment or "microclimate." In the absence of a strong source of radiant energy such as the sun ectotherms prefer to remain at an ambient temperature almost identical with their optimal body temperature.

In contrast, the endotherms (birds and mammals) are usually well insulated with fur, feathers or fat and have a resting metabolic rate at least five times higher than that of ectotherms of comparable size at a similar body temperature. When endotherms are exposed to a cold environment, they elevate their rate of metabolic heat production and thereby maintain their body temperature at an optimal level. Because of their high metabolic rate endotherms must lose a significant amount of heat to the environment in order to prevent their body temperature from rising. Therefore endotherms always prefer an ambient temperature somewhat below their optimal body temperature so that heat can be lost passively down a temperature gradient. For example, a clothed human being maintains an internal body temperature of about 37 degrees C. (98.6 degrees F.) but prefers an ambient temperature of about 22 degrees C. (71.6 degrees F.).

Endotherms have evolved physiological mechanisms in addition to variable metabolic rates in order to regulate their body temperature. These additional mechanisms control the rate at which the animal exchanges heat with its environment. The insulative value of fur or feathers can be varied by muscles controlling their attitude. In addition, the insulative value of the skin is variable. Heat is transported rapidly by the blood from the core of the body to the skin, where it is lost to the environment; the rate of heat loss can be controlled by the dilation or constriction of the arterioles of the skin. In many species the loss of body heat from a poorly insulated appendage is controlled by heat exchange between the vessels carrying blood into the appendage and those carrying blood out of it.

Another way of losing heat to the environment is through the evaporation of water from the skin or the respiratory surfaces. In many species this kind of heat loss is enhanced by the mechanisms of panting and sweating. For evaporation to be effective the animal must direct the appropriate amount of blood to the area being cooled.

Biologists did not fully appreciate the thermoregulatory capabilities of ectotherms until field measurements of body temperature were made on a variety of Temperate Zone reptiles. Surprisingly, regardless of the air temperature the body temperature of active lizards was found to be generally within the range of 34 to 40 degrees C. (93.2 to 104 degrees F.). Since then it has been shown that a wide variety of ectotherms can regulate their body temperature with a great deal of precision through their behavior. For example, John R. Brett, working at the Fisheries Research Board of Canada Biological Station in Nanaimo, B.C., discovered that if young sockeye salmon are given a choice, they will select water at 15 degrees C. (59 degrees F.), the temperature at which their growth, heart output and sustainable swimming speed are all maximal. When a variety of ectotherms (including sharks, bony fishes, amphibians and lizards) are given a choice between two nonoptimal temperatures, they regulate their body temperature at the preferred level with remarkable accuracy by moving back and forth between the two compartments.

Additional thermoregulatory effector mechanisms have recently been documented in ectotherms. Eugene C. Crawford and Billy J. Barber of the University of Kentucky found that the chuckwallah lizard exploits panting as a mechanism to lose heat if its skin temperature or deep-body temperature exceeds a certain level, and Marvin L. Riedesel of the University of New Mexico showed that box turtles can hold their body temperature at 10 degrees C. (50 degrees F.) below the ambient tempera-

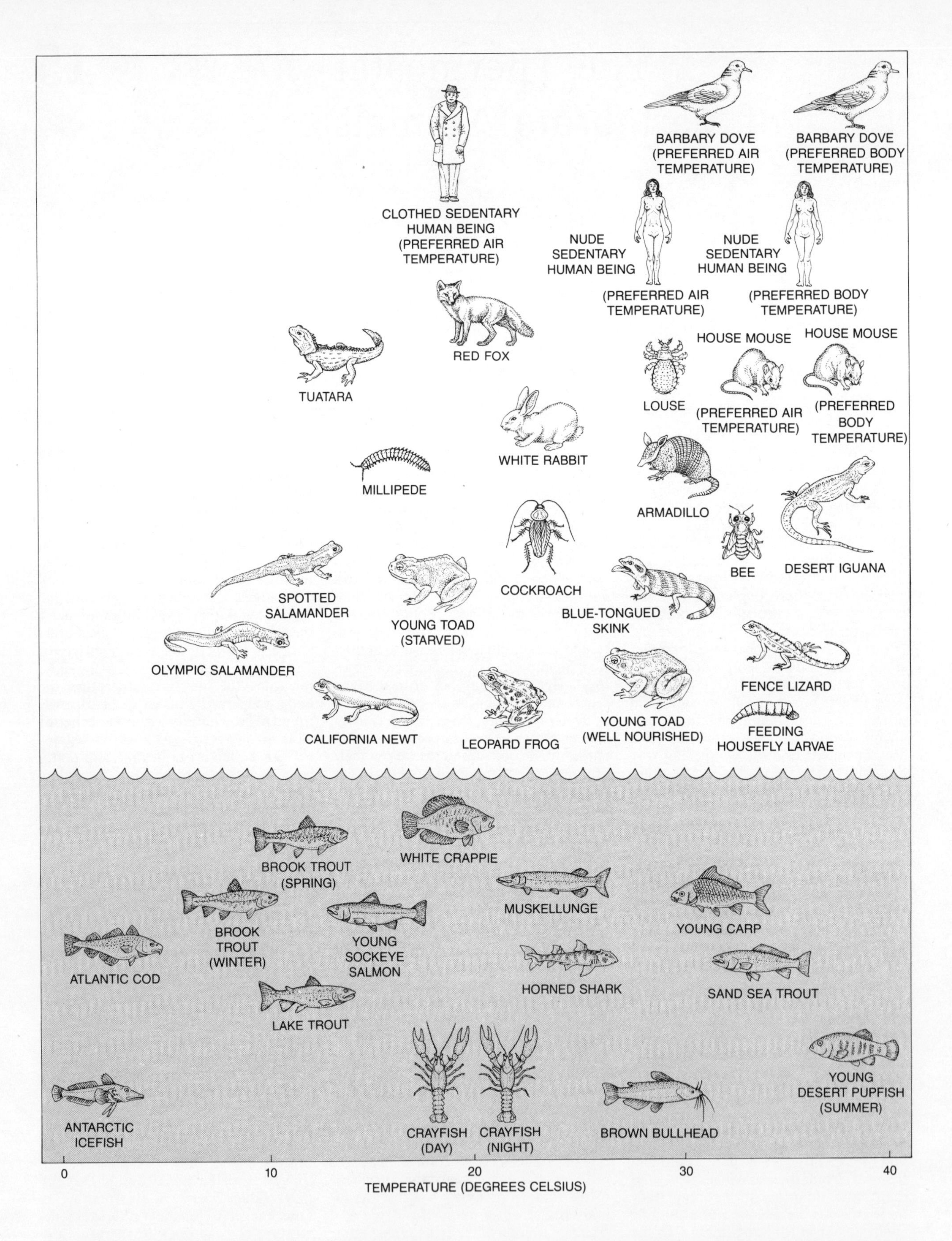

PREFERRED AMBIENT AND BODY TEMPERATURES for vertebrate animals vary greatly and depend on the thermal environment to which the animal is adapted. For ectotherms (animals that obtain most of their body heat from their surroundings) the preferred internal and ambient temperatures are the same. For endotherms (animals that obtain most of their body heat from internal metabolic processes) the preferred ambient temperature is considerably below the preferred body temperature because internally generated heat must be continually lost down a temperature gradient. The crayfish demonstrates that preferred temperature may change on a daily basis, the trout demonstrates seasonal change and the toad provides an example of preferred temperature changing with nutritional state.

ture by panting. When ectotherms can no longer find environmental heat sufficient to maintain a high body temperature, they generally seek a cool place safe from predators and become inactive, allowing their body temperature to follow the ambient temperature. This response may also result when the food supply is inadequate, since a lowering of the body temperature means a reduction of the metabolic rate and hence the conservation of vital energy stores. As we shall see, a similar strategy is exploited by some endotherms faced with notably harsh environmental conditions.

In the 1880's Charles Richet in France and Isaac Ott in the U.S. observed that the localized destruction of tissue in the hypothalamus at the base of the brain in dogs gave rise to elevated body temperature. In classic experiments conducted by Henry G. Barbour in 1912 silver thermodes were implanted in the hypothalamus. When the thermodes were cooled, the body temperature rose; when the thermodes were heated, the body temperature fell. Subsequent experiments on a variety of mammals by numerous investigators indicated that the region of the hypothalamus located just over the optic chiasm (the place where the optic nerves from the two eyes converge) is essential for the regulation of body temperature. It is now known that thermoregulatory functions are centered in this part of the brain in all classes of vertebrates.

What is the nature of this internal "thermostat," and how does it regulate body temperature? On the basis of control theory one can postulate that the thermostat must include several features. First, it must have information about the actual temperature of the body by means of at least one feedback circuit. Second, the thermostat must "know" the optimal body temperature in that it must be programmed with some kind of reference point. Third, the thermostat must be able to compare the actual body temperature at any given moment with the optimal setting and, if a difference exists, trigger the appropriate behavioral and physiological thermoregulatory mechanisms.

As we have seen, the hypothalamus is sensitive to temperature. When the hypothalamus of an endotherm is warmed, the animal reacts with heat-dissipating responses such as sweating, panting and dilation of the peripheral blood vessels; when the hypothalamus is cooled, the animal reacts with heat-generating responses such as shivering, erection of the fur and constriction of the peripheral blood vessels. The most basic thermoregulatory mechanism—the seeking of a suitable thermal environment—is shared by all vertebrates. When the hypothalamus of ectotherms and endotherms is warmed, they behave as if they are too hot and select a cool environment that serves to lower their body temperature. Cooling the hypothalamus elicits the opposite response.

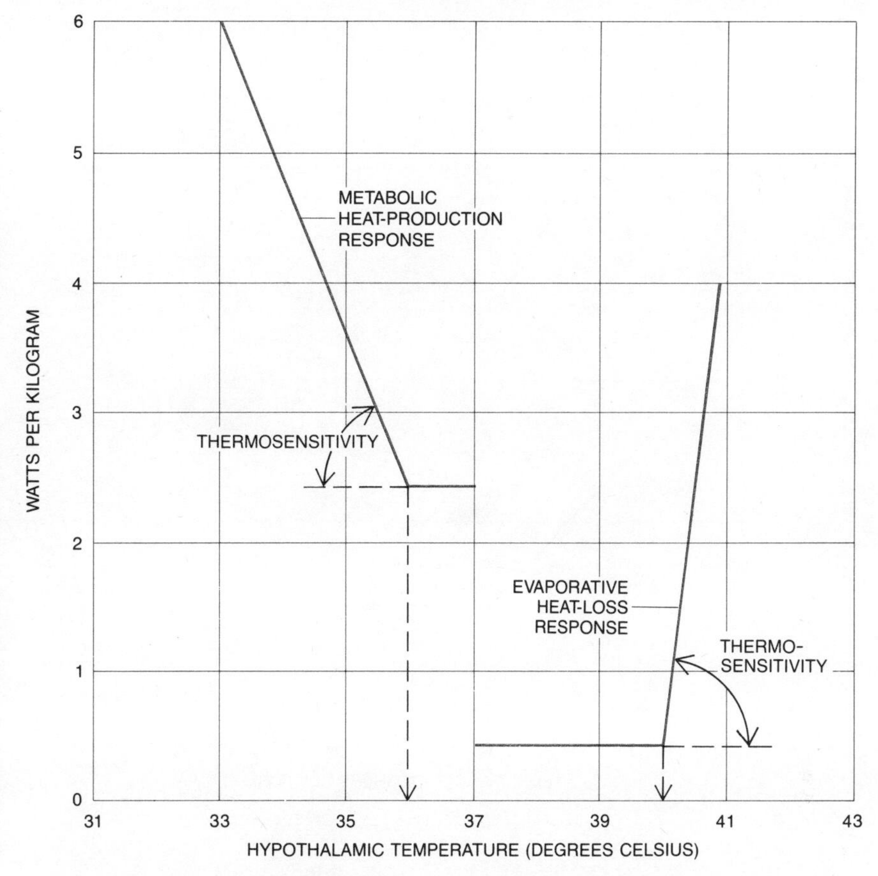

THERMOREGULATORY RESPONSES are triggered when the temperature of the hypothalamus either falls below or rises above specific threshold temperatures. The thresholds may be influenced by a variety of factors including ambient temperature, level of exercise, sleep and microorganisms that cause fever. In the graph shown here (for a dog with an optimal body temperature of 37 degrees Celsius and an ambient temperature of 25 degrees) the hypothalamic threshold for the metabolic heat response is 36 degrees and that for the evaporative heat-loss response is 40 degrees. As is indicated by the slopes of the curves, the intensity of each thermoregulatory response is proportional to the difference between the actual temperature of the hypothalamus and the threshold temperature at which that particular response is initiated.

One of us (Hammel), who was then working at the John B. Pierce Foundation laboratory at Yale University, was able to quantitatively measure the thermoregulatory responses of dogs as a function of hypothalamic temperature. These experiments were made possible by the development of small stainless-steel thermodes that could be permanently implanted around the hypothalamus of intact animals. The thermodes were perfused with water of a specific temperature in order to heat or cool the hypothalamus. When the hypothalamic temperature of dogs was systematically changed, temperature thresholds for the initiation of each thermoregulatory response were observed: shivering was induced when the hypothalamus was cooled below a particular temperature and panting was induced when the hypothalamus was heated above another temperature. Moreover, at temperatures above the threshold level the intensity of the thermoregulatory response was proportional to the difference between the threshold temperature and the actual hypothalamic temperature. These characteristics of the temperature sensitivity of the hypothalamus are qualitatively the same in all the mammalian species tested.

Does the temperature of the hypothalamus provide all the feedback information the system requires to monitor the thermal condition of the body? Two simple observations suggest that the story is more complex. If one enters a cold environment, one begins to shiver almost immediately, before there is any fall in the hypothalamic temperature. Similarly, if one enters a hot environment such as a sauna bath, one begins to sweat before there is any rise in hypothalamic temperature. It is as if the thermostat predicts a change in the internal body temperature and takes corrective action immediately. One explanation for the fact might be that peripheral temperature sensors at the surface of the body are providing the central thermostat with information about rapid changes in the thermal environment.

This hypothesis was tested by measur-

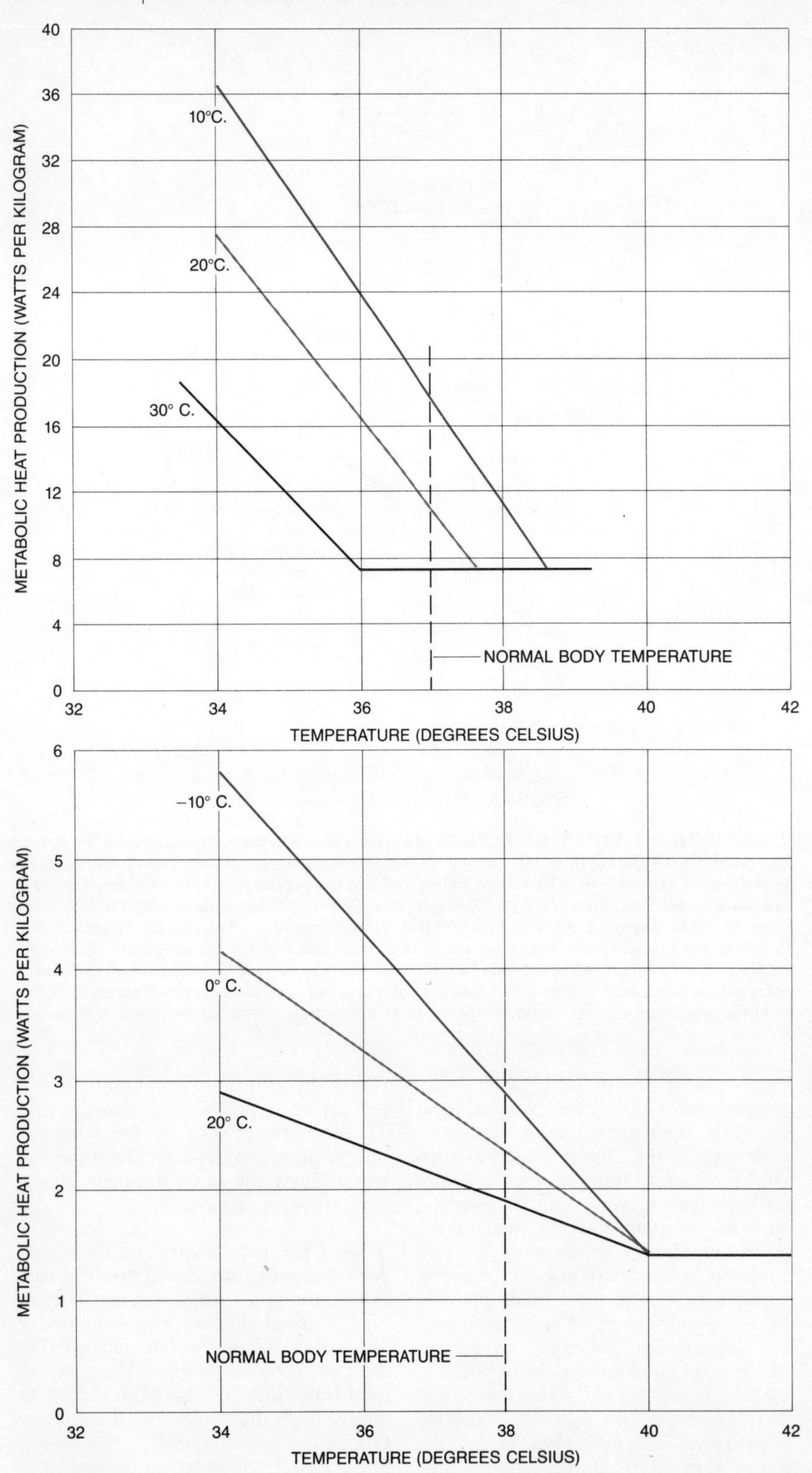

EFFECT OF AMBIENT TEMPERATURE on the rate of metabolic heat production reveals two ways that inputs from peripheral temperature sensors can alter the characteristics of the central thermostat. In the kangaroo rat (*top graph*) the hypothalamic threshold temperature for the metabolic heat response rises as ambient temperature declines. (Response curves at three different ambient temperatures are shown.) This increase of threshold means that a fall in actual core-body temperature is not required at low ambient temperatures to stimulate increased metabolic heat production. Thus the thermostat anticipates future changes in body temperature and makes compensatory responses before those changes actually occur. In the harbor seal (*bottom graph*) a decrease in ambient temperature does not change the hypothalamic threshold temperature for the metabolic heat-production response but instead alters the sensitivity of the thermostat, so that more heat is generated for each unit of temperature below the hypothalamic threshold. This shift in sensitivity is reflected in the changing slope of the response curves.

ing thermoregulatory responses to the heating and cooling of the hypothalamus at different ambient temperatures. Experiments with some species of mammals have shown that as the ambient temperature decreases, the threshold temperature for the metabolic heat-production response increases. In effect at low ambient temperatures the characteristics of the thermostat are modified so that the metabolic heat-production response is maintained at an elevated level without any change in the hypothalamic temperature. In this way the thermoregulatory response occurs before the internal body temperature has declined to a dangerously low level. Changes in the ambient temperature may also affect the thermostat by altering its thermosensitivity, that is, the magnitude of the thermoregulatory response evoked by a unit change in the hypothalamic temperature. This change in sensitivity is reflected by a change in slope of the curve relating the response to hypothalamic temperature.

The effect of ambient temperature on the thermostat differs from species to species. In the harbor seal a decrease in the ambient temperature results in a large increase in the thermosensitivity of the hypothalamus but no change in the threshold temperature for the metabolic heat-production response. The antelope ground squirrel, on the other hand, shows a normal inverse relation between the ambient temperature and the hypothalamic threshold temperature for the metabolic heat-production response but a maximal hypothalamic thermosensitivity at high ambient temperatures. The antelope ground squirrel is a desert animal that is active during the day, so that it is highly adaptive for it to be most sensitive to changes in its deep body temperature at high ambient temperatures.

To sum up, the central thermostat of vertebrate animals is able to obtain information about ambient temperature from peripheral sensors at the body surface; these inputs then modify the characteristics of the thermostat so that a different thermoregulatory output is achieved without a change in the hypothalamic temperature. It is the ability to respond to rapid changes in the ambient temperature that provides the vertebrate thermostat with its predictive capability: the animal is able to anticipate a change in body heat content and take corrective action before that change actually occurs.

There are, however, a few situations where actual changes in the hypothalamic temperature activate thermoregulatory responses in mammals. For example, at the onset of a fever there is an increase in the regulated body temperature and at the onset of sleep there is a decrease in the body temperature.

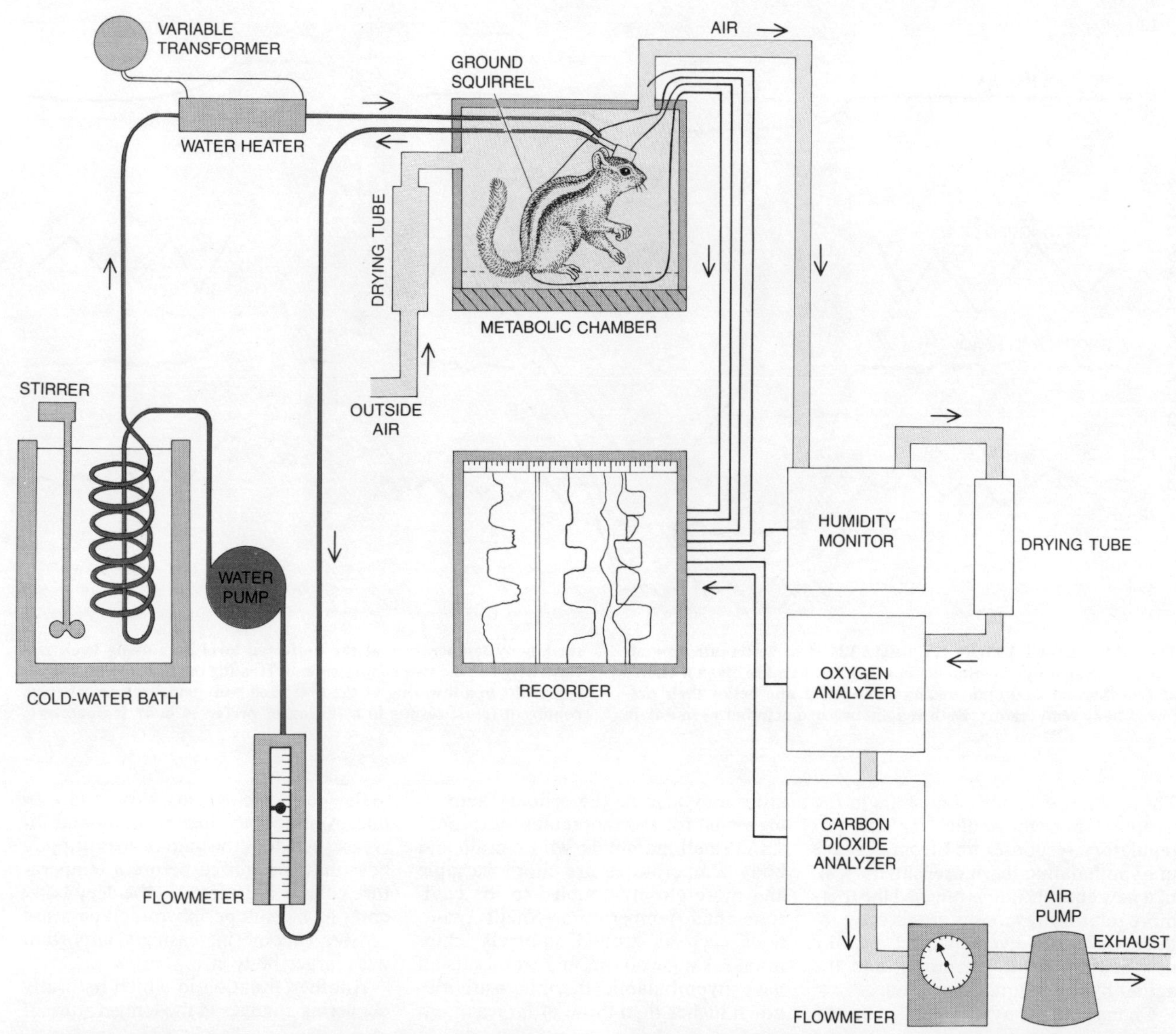

EXPERIMENTAL APPARATUS diagrammed here is used to measure the temperature sensitivity of the hypothalamus in small mammals. The animal is placed in the sealed metabolic chamber and the temperature of its hypothalamus is manipulated by means of thermodes: implanted stainless-steel tubes through which water is circulated at controlled temperatures. At the same time the animal's rate of metabolism and evaporative water loss are measured by drawing dry outside air through the chamber at a known rate and analyzing the effluent air for its content of water, oxygen and carbon dioxide. Temperatures, measured by thermocouples, are recorded on chart.

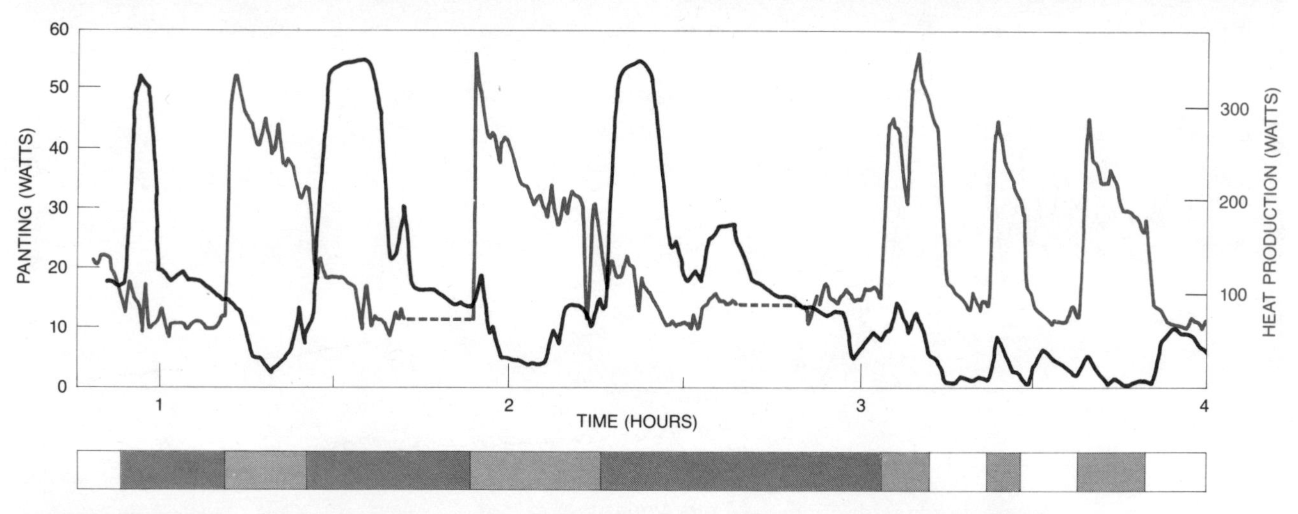

CHART RECORDING shows the results of an experiment in which the hypothalamus of a Labrador retriever was alternately warmed (*color bars*) and cooled (*gray bars*) by means of implanted thermodes as the rates of two thermoregulatory responses were continuously measured. Cooling of the hypothalamus resulted in a rapid loss of body heat by evaporative water loss, primarily from panting (*black curve*), whereas warming of the hypothalamus resulted in shivering and consequently an increase in the metabolic rate (*color curve*).

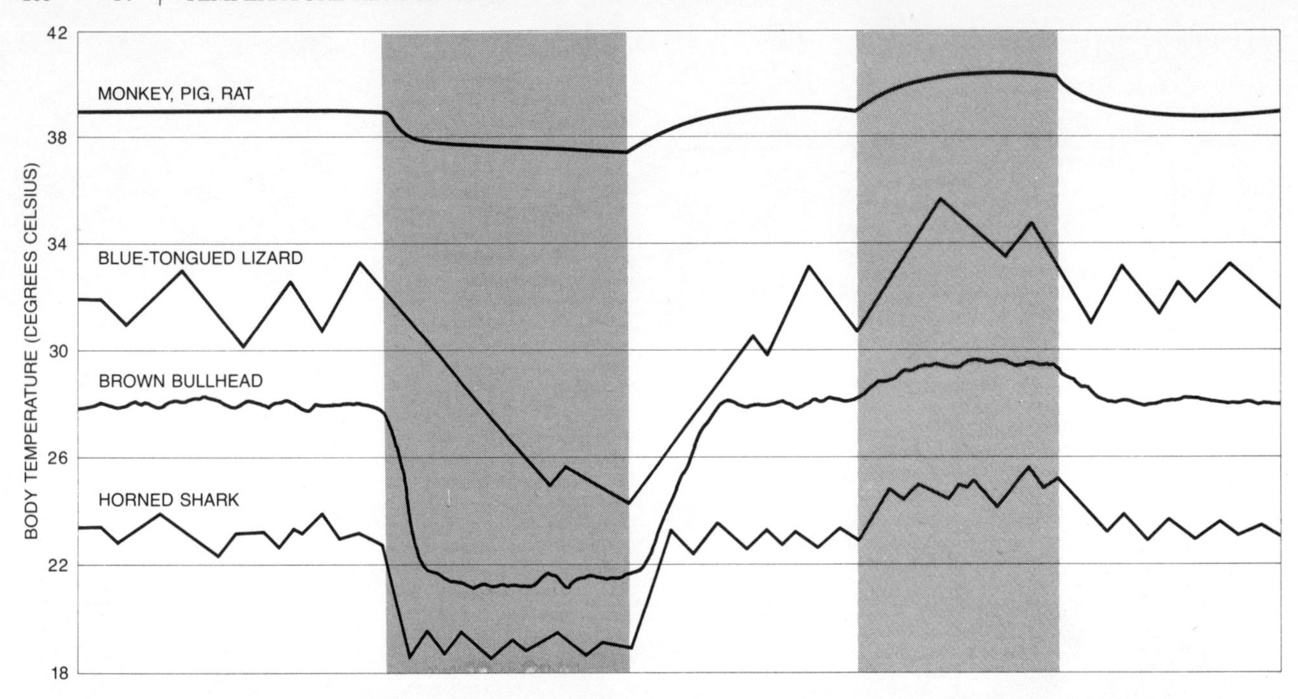

THERMOREGULATION BY BEHAVIOR in vertebrate animals is demonstrated by experiments in which animals are given a choice of two thermal environments, one above and one below their preferred body temperature. Both endotherms and ectotherms maintain their body temperature at the preferred level by moving back and forth between the two compartments. Heating the hypothalamus (*color*) results in a lowering of the preferred body temperature, whereas cooling it (*gray*) results in a raising of preferred body temperature.

These changes result from shifts in the hypothalamic thresholds for thermoregulatory responses and from changes in hypothalamic thermosensitivity. Until a new equilibrium is reached the thermoregulatory responses are directly influenced by the divergence between the new setting of the thermostat and the actual hypothalamic temperature.

Changes in the hypothalamic temperature may also be the primary activating signal for thermoregulatory responses in small mammals, whose brain and body temperatures are more variable and more closely coupled to the environmental temperature. Small mammals such as ground squirrels, chipmunks, kangaroo rats and wood rats all have hypothalamic thermosensitivities much higher than those of larger mammals such as seals, cats, dogs and rabbits. As the body size of an animal increases, so does the gain or loss of body heat that is required before a temperature change is detected in the deep body core. As a result peripheral temperature sensors become increasingly important with larger body size.

Another situation in which naturally occurring changes in the temperature of

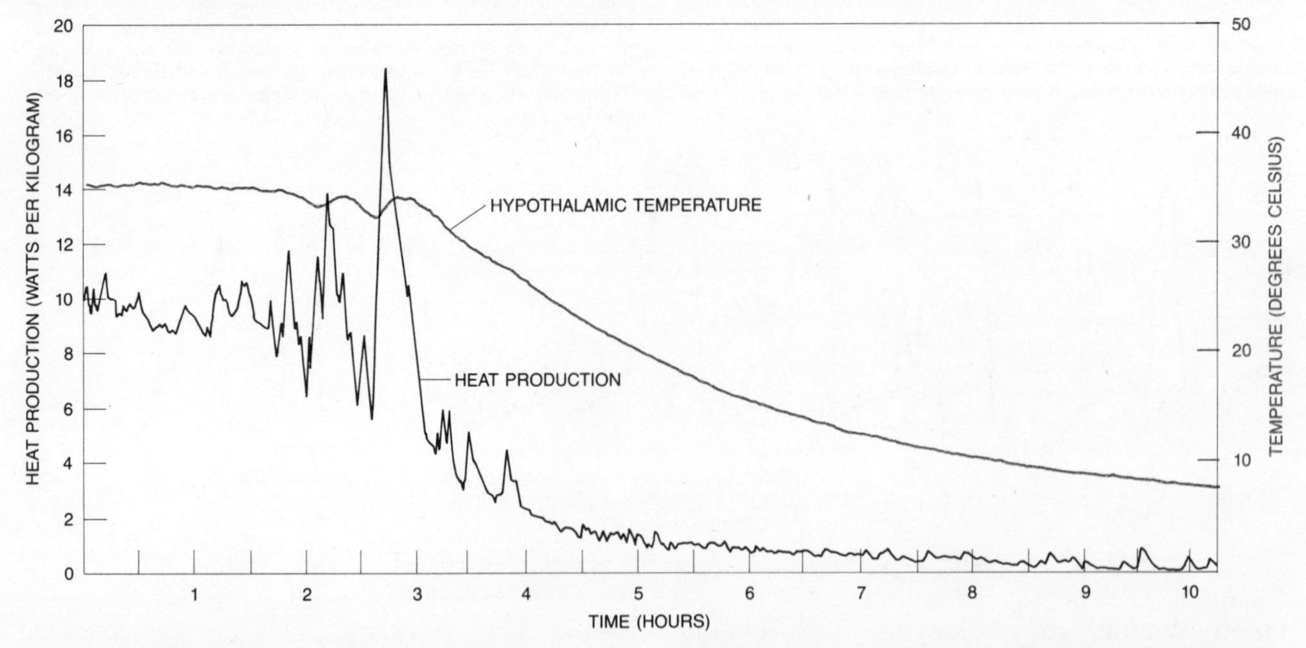

ENTRANCE INTO HIBERNATION of the golden-mantled ground squirrel is monitored in this recording, which shows the rate of metabolic heat production and the hypothalamic temperature of the animal over a period of time. The otherwise passive cooling curve is interrupted by bursts of metabolic heat production whenever the body temperature of the animal temporarily falls below the declining set of the thermostat. This finding and others suggest that the thermostat is not "switched off" during hibernation but rather is "turned down."

the body core are primarily responsible for inducing thermoregulatory responses is found in physical exertion. In man, as everyone knows, sustained exertion is accompanied by an increase in the rate of sweating. Ethan R. Nadel and his colleagues at the John B. Pierce Foundation laboratory demonstrated, however, that at a given ambient temperature the human sweat rate is poorly correlated with the level of physical exertion but is closely correlated with the temperature of the body core. (It is not known whether the rise in body temperature is having its primary effect on the hypothalamus or on other parts of the nervous system, such as the spinal cord.) In the dog, on the other hand, the hypothalamus apparently receives information directly from sensory receptors in the joints and the muscles, so that panting is initiated soon after the exercise begins and before the temperature of the body core rises.

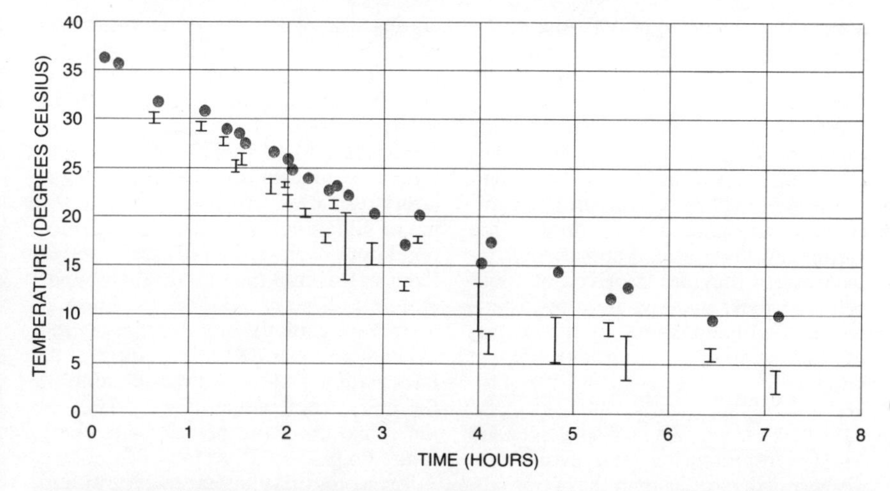

RESETTING OF THE THERMOSTAT during the entrance into hibernation of the golden-mantled ground squirrel was demonstrated by a series of experiments in which the hypothalamic temperature of the animal was manipulated. The dots represent actual hypothalamic temperatures at specific times during entrance. Manipulations of hypothalamic temperature showed that the threshold for the metabolic heat-production response was somewhere in the range indicated by the vertical line below each hypothalamic temperature point. It can be seen that entrance into hibernation involves a progressive and continuous resetting of thermostat.

Hibernation in mammals is an intriguing phenomenon from the point of view of temperature regulation. When an animal is ready to hibernate, it retires to a secluded nest or burrow and becomes inactive. Its temperature then falls to the point where it may be very close to the temperature of the environment. For a long time many biologists assumed that hibernating mammals temporarily abandoned thermoregulation and returned to the more primitive ectothermic state of their reptilian ancestors. Recent studies have revealed that hibernation is not an abandonment of thermoregulation but rather a precisely regulated lowering of the central thermostat for the purpose of conserving energy.

That hibernation is a regulated condition was first suggested by the observation that although ectotherms need an outside source of heat to recover from a cold-induced torpor, hibernating mammals are capable of rousing and returning to normal body temperature even when the ambient temperature is low. For example, the golden-mantled ground squirrel rouses spontaneously many times during the hibernation season; its periods of torpor may last from less than a day to two weeks. Moreover, if the ambient temperature falls to dangerously low levels, the animal will rouse from torpor, a "fail-safe" mechanism that prevents it from freezing to death. The alarm temperature appears to be sensed by the hypothalamus, because if the ambient temperature is held at a level above the alarm temperature and the hypothalamus of the hibernating animal is cooled by means of thermodes, the animal will rouse. Conversely, if the hypothalamus is slightly warmed, the temperature of the rest of the body can fall below the alarm level without inducing arousal. When the warming is stopped, however, arousal occurs.

Charles P. Lyman of the Harvard Medical School has studied sensitivity to low temperatures in three other species of hibernating mammal: the thirteen-lined ground squirrel, the Turkish hamster and the garden dormouse. He found that the ground squirrel and the hamster rouse from hibernation when their head is cooled by an external thermode but that the animals continue to hibernate if the head is warmed and the body is cooled by lowering the ambient temperature. In contrast, the garden dormouse can be aroused from torpor by applying a cold stimulus to its feet; this species appears to depend on peripheral temperature sensors to warn it of dangerously low ambient temperatures. In view of the differences between the alarm systems of the two species it is interesting to note that the garden dormouse hibernates on its back with its feet sticking up, whereas ground squirrels and hamsters hibernate with their feet tucked under them.

The arousal from deep hibernation involves a large increase in metabolic heat production. When the golden-mantled ground squirrel is aroused from hibernation by a cooling of the hypothalamus, the ensuing increase in metabolic rate can be suppressed by a heating of the hypothalamus. The longer the interval following the initiation of arousal is, the higher the hypothalamus must be heated to suppress the thermoregulatory response. These findings suggest that the hibernating animal's thermostat is not simply switched on or off but is capable of a wide range of settings.

In addition to the alarm response to excessive cooling of the hypothalamus, the hibernating ground squirrel may exhibit proportional thermoregulatory heat-production responses without rousing from torpor when its hypothalamus is cooled within a range of temperatures just above the alarm threshold. Whereas the alarm response is characterized by an abrupt increase in the threshold temperature of the thermostat, there is no such change in threshold following the proportional responses. The thermosensitivity of the thermostat, as is indicated by the degree to which a change in body temperature elicits a thermoregulatory response, is much lower when the ground squirrel is hibernating than when it is active, but the temperature-sensitive characteristics of the thermostat are qualitatively the same in the active animal and in the hibernating one.

Could the same thermostat be operative over the entire range of body temperatures experienced by the hibernator? We tested this hypothesis by heating and cooling the hypothalamus for short periods while the animal was entering hibernation. At discrete time intervals during entrance we ascertained the highest hypothalamic temperature that induced a metabolic heat-production response and the lowest hypothalamic temperature that did not induce the response. These experiments demonstrated the continuity of the operation of the thermostat over the wide range of body temperatures experienced by the hibernator and showed that entrance into hibernation is associated with a progressive lowering of the thermostat.

When the animal is going into hibernation, the decline in body temperature and metabolic rate is not always smooth. There are sometimes bursts of metabolic heat production and shivering, which causes the decline in body temperature to stop or reverse slightly. Such bursts appear to occur when the temperature of the hypothalamus of the hibernating animal has fallen temporarily below the declining setting of the thermostat.

A variety of ectotherms also become torpid under certain conditions: snakes and lizards seek a cool burrow at the end of the day, the brook trout prefers cooler water in winter than it does in the spring and toads seek a cooler environment when they are deprived of food. Whether these animals are also experiencing regulated shifts in the setting of their central thermostat is not yet known.

The birds, which are also descended from reptilian ancestors, evolved endothermy in parallel with the mammals. What little is known about the avian thermostat suggests that it is somewhat different from the mammalian one. For one thing, although the hypothalamus of birds plays an important role in integrating the thermoregulatory responses, it is virtually insensitive to temperature. Instead temperature-sensitive cells in the spinal cord provide the major source of information about the temperature of the body core. Werner Rautenberg and his colleagues at the University of the Ruhr have shown that cooling the spinal cord of pigeons by pumping water through a very thin tube inserted into the spinal canal results in the heat-generating and heat-conserving responses of shivering and constriction of peripheral blood vessels. Warming the spinal cord induces dilation of the peripheral blood vessels and eventually panting. Manipulations of spinal-cord temperature in mammals also elicit thermoregulatory responses, which suggests that the major difference between the thermoregulatory system of birds and that of mammals is that the mammalian hypothalamus directly monitors the internal temperature of the body, whereas the avian hypothalamus receives most of its information about the thermal state of the body from temperature sensors in the spinal cord and perhaps elsewhere in the body.

Fishes too may integrate central and peripheral thermal information in the hypothalamus to control certain physiological responses. Because of the high thermal conductivity of water and the thermal heterogeneity of many bodies of water fishes often experience rapid changes in body temperature. Such changes result in fluctuations of many vital parameters such as metabolic rate. Although the effector mechanisms available to a fish cannot provide it with a constant internal temperature, they may be controlled to anticipate the physiological changes that inevitably accompany thermal changes. For example, a rise in a fish's body temperature means an increase in its metabolic rate and hence in its oxygen requirement. The fish could either wait until an oxygen deficit occurs and then respond by increasing its gill ventilation, or it could use its temperature-sensing capabilities to relay projected changes in oxygen demand to the respiratory centers. In the latter case alterations in arterial oxygen concentration would be avoided, ensuring a constant supply of oxygen to the tissues.

One of us (Crawshaw) studied the respiratory response in the scorpion fish by manipulating the temperature of its hypothalamus with the aid of implanted thermodes. Heating the hypothalamus caused the fish to ventilate its gills faster, whereas cooling the hypothalamus caused it to ventilate them more slowly. Gill ventilation also appears to be influenced by thermal input from the periphery. On encountering a rapid increase in water temperature carp quickly increase the ventilation of their gills for about 30 seconds; the response then diminishes. Thus fishes appear to utilize both central and peripheral thermal information to project changes in oxygen demand. This regulatory system may well be the evolutionary antecedent of the physiological control system in endotherms that maintains internal body temperature at close to optimal levels.

On the basis of what is now understood about the characteristics of the vertebrate thermostat, it is possible to speculate on its neural design. Neurophysiological investigations have revealed the existence of neurons, or nerve cells, in the hypothalamus that could serve as components of a thermostat. Two populations of thermosensitive neurons have been identified: one that responds to a local warming of the brain tissue and another that responds to a local cooling. The heat-sensitive neurons increase the frequency of their discharge of impulses in direct proportion to the degree that hypothalamic temperature is raised above the normal value for the animal's body. Similarly, the cold-sensitive neurons respond with an increase in the frequency of impulses when the hypothalamic temperature falls below the normal value. It is not yet known whether such hypothalamic neurons are really part of the central thermostat, but this assumption is supported by studies showing that these neurons sometimes also respond to changes in the temperature of the skin and of the spinal cord.

Wolf Wünnenberg and his colleagues at the University of Kiel have shown that the hypothalamic temperature-sensitive neurons in a hibernating mammal and those in a nonhibernating mammal differ in their range of temperature sensitivity, as would be expected if the neurons are actually components of the thermostat. The hypothalamic neurons

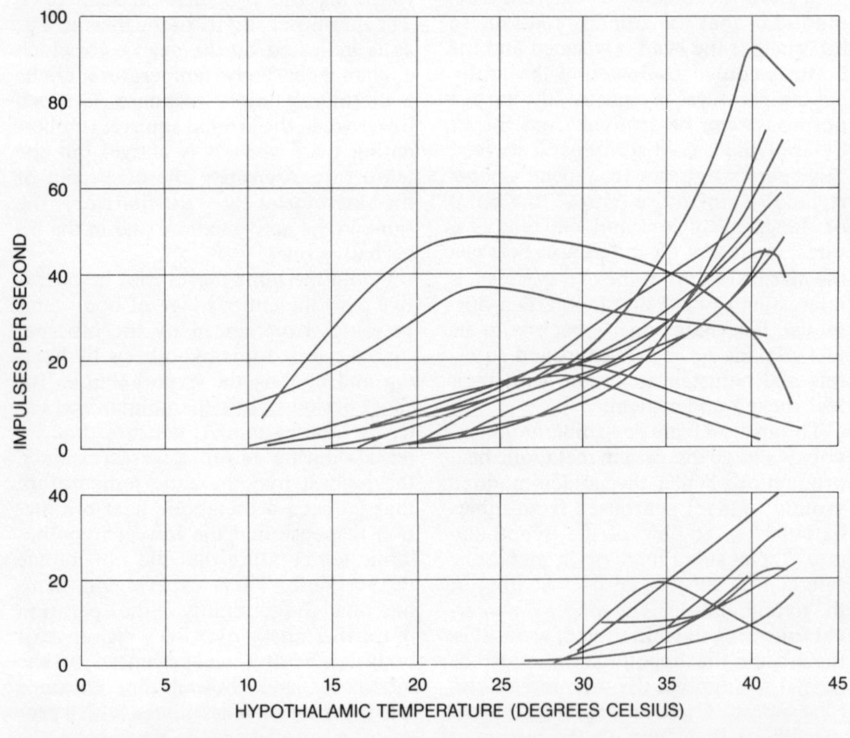

TEMPERATURE-SENSITIVE NEURONS, or nerve cells, in the mammalian hypothalamus have been detected with the aid of implanted recording electrodes. These two graphs show the firing rate of individual neurons as a function of hypothalamic temperature in a mammal that hibernates, the golden hamster (*top*), and in a mammal that cannot hibernate, the guinea pig (*bottom*). The temperature-sensitive neurons in the hamster respond to a broad range of body temperatures that would be experienced during the entrance into hibernation, whereas the hypothalamic neurons in the guinea pig are silent below a temperature of 30 degrees C. (86 degrees F.). Experiments shown here were performed by Wolf Wünnenberg of the University of Kiel.

of the nonhibernator had a narrow range of temperature sensitivity and were mostly silent below a temperature of 30 degrees C. (86 degrees F.), whereas many hypothalamic neurons of the hibernator had continuous temperature-response curves over a much broader range.

A simple model proposed by one of us (Hammel) suggests how neurons in the hypothalamus could be interconnected to achieve the regulation of body temperature. According to the model, when the activity of the heat-sensitive neurons in the hypothalamus prevails over the activity of the cold-sensitive ones, heat-loss mechanisms are activated. Conversely, when the activity of the cold-sensitive neurons predominates, heat-generating and heat-conservation measures are activated. The thermoregulatory thresholds of the hypothalamus are determined by the relative activity of the two neuronal populations.

As we have seen, under most conditions the body temperature of an endothermic animal remains constant; thermoregulatory responses are evoked by rapid changes in peripheral temperature. How is the thermal information from the temperature sensors in the skin integrated by the central thermostat? One hypothesis is that heat-sensitive nerve endings in the skin increase their firing rate as the ambient temperature increases and thereby excite the heat-sensitive neurons in the hypothalamus. The cold-sensitive nerve endings could innervate the central cold-sensitive neurons in a similar fashion. Hence if you walk out of a warm room into a cold environment, the peripheral cold sensors could increase the firing rate of the cold-sensitive neurons in the hypothalamus, moving the point of intersection with the population of heat-sensitive neurons to a higher temperature. Because this new intersection is above the actual hypothalamic temperature of 37 degrees C. the peripheral cold stimulus triggers heat-generating responses without requiring a change in hypothalamic temperature. This simple model can therefore account for the predictive capability of the thermostat.

The hypothalamus receives many inputs besides those from peripheral temperature sensors in the skin, and any detailed model of the thermostat will have to account for the integration of thermal signals from the spinal cord and the abdominal viscera, for inputs from tension receptors in the muscles and joints (which signal the level of physical exercise) and for inputs from the reticular formation of the brain stem (which is an important part of the neural system controlling the animal's level of arousal). All these diverse signals are integrated by the hypothalamus and provide information about the temperature of the body core and about the more rapid thermal changes occurring at the body surface.

The value of comparative studies of a wide variety of vertebrate species facing very different thermal problems and having diverse strategies for dealing with them is that they illuminate the general features of the vertebrate thermoregulatory system as well as its specialized features. Continued investigations should enable us to propose, test and refine hypothetical models so that they will better reflect how the thermostat is really designed.

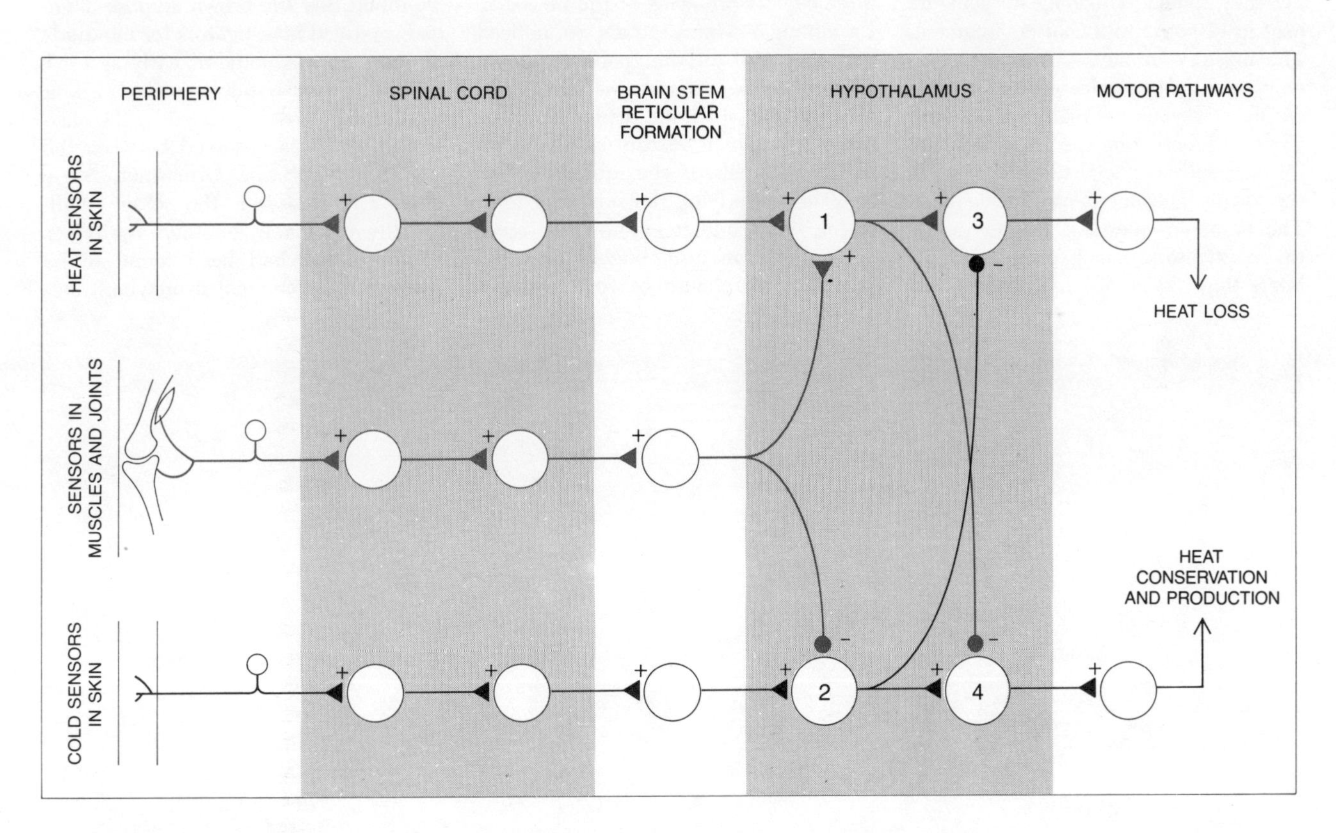

THEORETICAL MODEL outlined here shows how neurons in the central nervous system might be interconnected to result in a functional thermostat. Four basic neuronal populations thought to exist in the hypothalamus are represented diagrammatically by single neurons. Populations *1* and *2* are respectively sensitive to warming or cooling. These temperature-sensitive populations facilitate and inhibit populations *3* and *4*, which serve to trigger thermoregulatory responses. The relative firing rates of populations *1* and *2* determine the hypothalamic threshold temperatures for particular thermoregulatory responses. These thresholds can be modified by diverse neural inputs: information from peripheral temperature sensors in the skin, from movement and position sensors in the muscles and joints, from temperature-sensitive neurons in the spinal cord and from neurons in the reticular formation of brain stem that control sleep and wakefulness.

16

The Production of Heat by Fat

by Michael J. R. Dawkins and David Hull
August 1965

In addition to normal "white" fat, many newborn mammals and adults of hibernating species have "brown" fat deposits. It is metabolism in brown fat cells that increases heat output as a response to cold

When a warm-blooded animal is exposed to cold, it increases its production of body heat by shivering. Mammals face a cool environment for the first time at birth, however, and many newborn mammals (including human infants) do not shiver. Yet they somehow manage to generate heat in response to a cool environment. The mystery of how this is done has only recently been cleared up. It turns out that the young of many species (and adults of hibernating species as well) are fortified with a special tissue that is exceptionally efficient in producing heat. The tissue in question, long a puzzle to investigators, has become a highly interesting object of physiological and chemical study within the past few years.

Our own interest in it was aroused in 1963 by a chance observation at the Nuffield Institute for Medical Research of the University of Oxford, where we were working in a group studying physiological problems of the newborn. Examining newborn rabbits, we noticed that they had striking pads of brown adipose tissue around the neck and between the shoulder blades. Adipose tissue is a salient feature of all warm-blooded animals. It constitutes the layer of fat underlying the skin over most of the body, and it is known to serve not only as an insulating blanket but also as a storehouse of food and energy. In the adult animal the adipose tissue is almost entirely of the white variety. The large deposits of adipose tissue we saw in the newborn rabbits were in the brown form, and this was a phenomenon that called for explanation.

Looking back through the literature, we found that the brown adipose tissue had mystified investigators for hundreds of years. It was noted as early as 1551 by the Swiss naturalist Konrad von Gesner, who was impressed by the mass of this tissue he observed between the shoulder blades of a marmot. Some observers confused the tissue with the thymus gland, another mysterious structure that had been found to be particularly prominent in newborn ani-

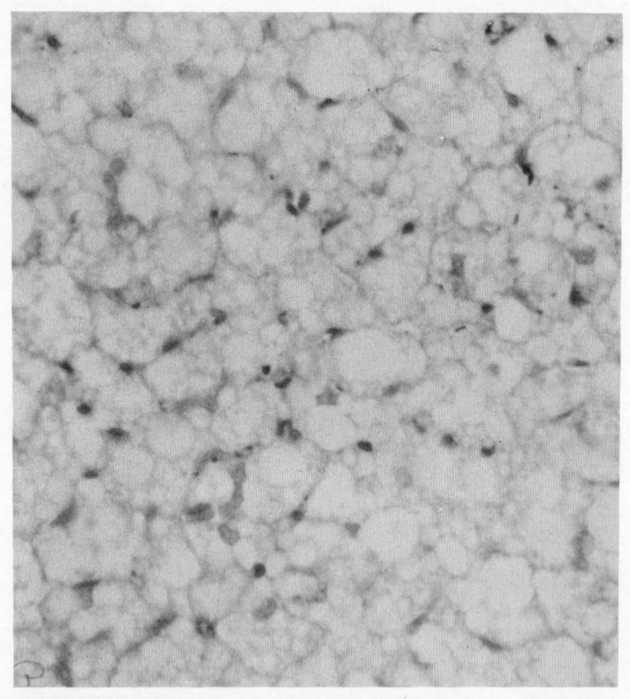

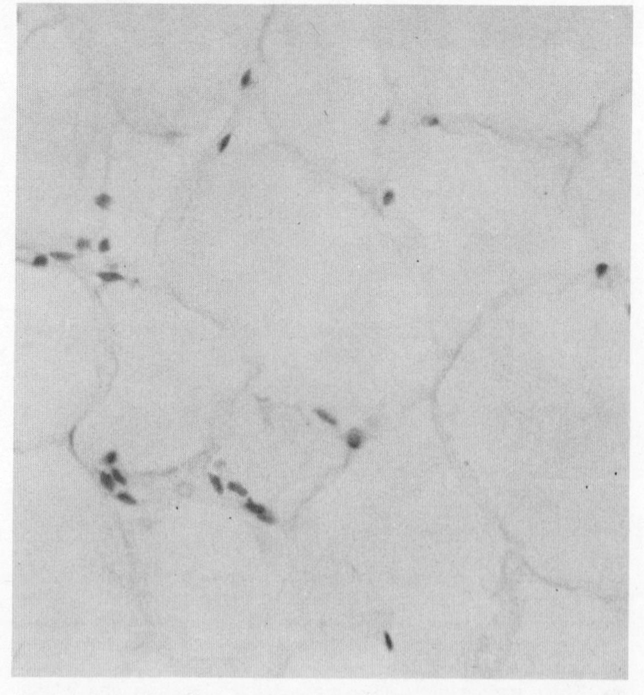

ADIPOSE TISSUE is enlarged 425 diameters in these photomicrographs made by the authors. Brown fat cells (*left*) from between the shoulder blades of a newborn rabbit have numerous small droplets of fat suspended in the stained cytoplasm. White fat cells (*right*) from an adult rabbit have large droplets of fat surrounded by narrow rims of cytoplasm. The stain was hematoxylin and eosin.

mals. Other zoologists, noting that a brown adipose mass was typical of hibernating animals, called it the hibernation gland. In this century more modern theories were advanced. Some physiologists suggested that brown adipose tissue had something to do with the formation of blood cells; others, that it was an endocrine gland. It does, in fact, contain hormones similar to those secreted by the adrenal cortex, but experiments in administering extracts from the tissue failed to show any consistent evidence of hormonal effects.

In 1961 two physiologists independently suggested a more plausible hypothesis. George F. Cahill, Jr., of the Harvard Medical School, noting that adipose tissue has an active metabolism that must generate heat as a by-product, proposed that the layer of white fat clothing the body should be regarded "not merely as a simple insulating blanket but perhaps as an electric blanket." And Robert E. Smith of the University of California School of Medicine at Los Angeles specifically called attention to the high heat-producing potentiality of brown adipose tissue, whose oxidative metabolism he had found to be much more active than that of white adipose tissue.

The cells of adipose tissue are characterized by droplets of fat in the cytoplasm (the part of the cell that lies outside the nucleus). In the white adipose cell there is a single large droplet, surrounded by a small amount of cytoplasm. The brown adipose cell, on the other hand, has many small droplets of fat, suspended in a considerably larger amount of cytoplasm. With the electron microscope one can see that the brown fat cells contain many mitochondria, whereas the white fat cells have comparatively few. Mitochondria, the small bodies sometimes called the powerhouses of cells, carry the enzymes needed for oxidative metabolism. What gives the brown fat cells their color is a high concentration of iron-containing cytochrome pigments—an essential part of the oxidizing enzyme apparatus—in the mitochondria.

It is easy to show by experiment that brown fat cells, loaded as they are with mitochondria, have a large capacity for generating energy through oxidation of substrates. Tested, for example, on succinic acid, an intermediate product in the Krebs energy-producing cycle, the brown fat cells of rabbits prove to have a capacity for oxidizing this substance that is 20 times greater than the oxidative capacity of white fat cells

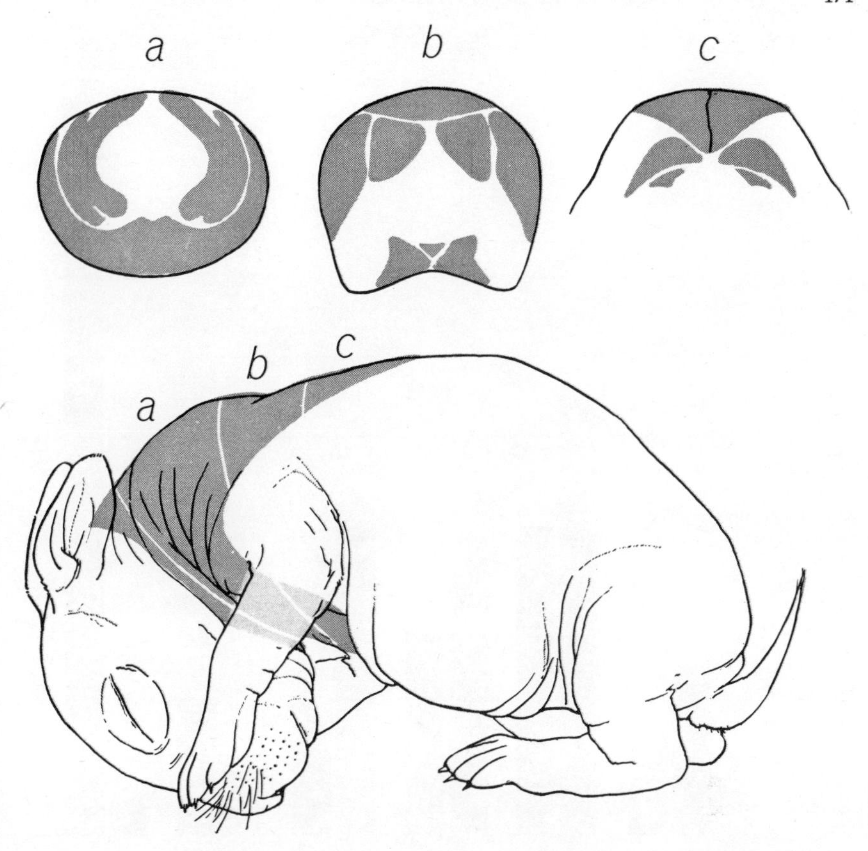

BROWN FAT accounts for 5 or 6 percent of the body weight of the newborn rabbit. It is concentrated, as shown in sections, around the neck and between the shoulder blades.

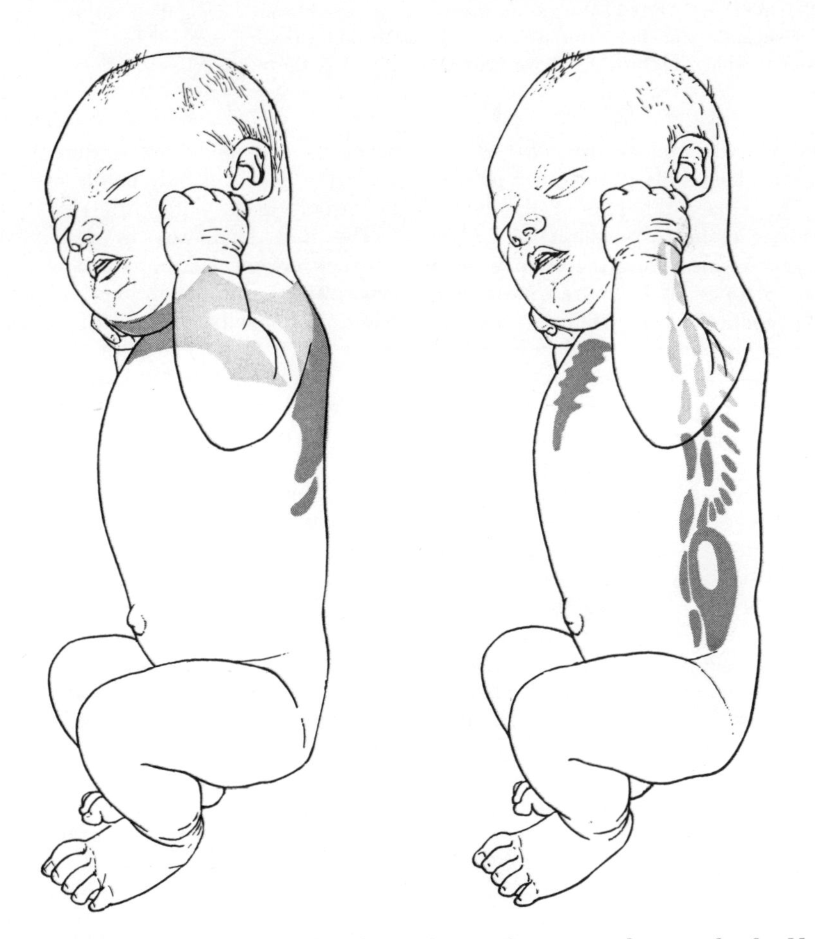

HUMAN INFANT at birth has a thin sheet of brown adipose tissue between the shoulder blades and around the neck, and small deposits behind the breastbone and along the spine.

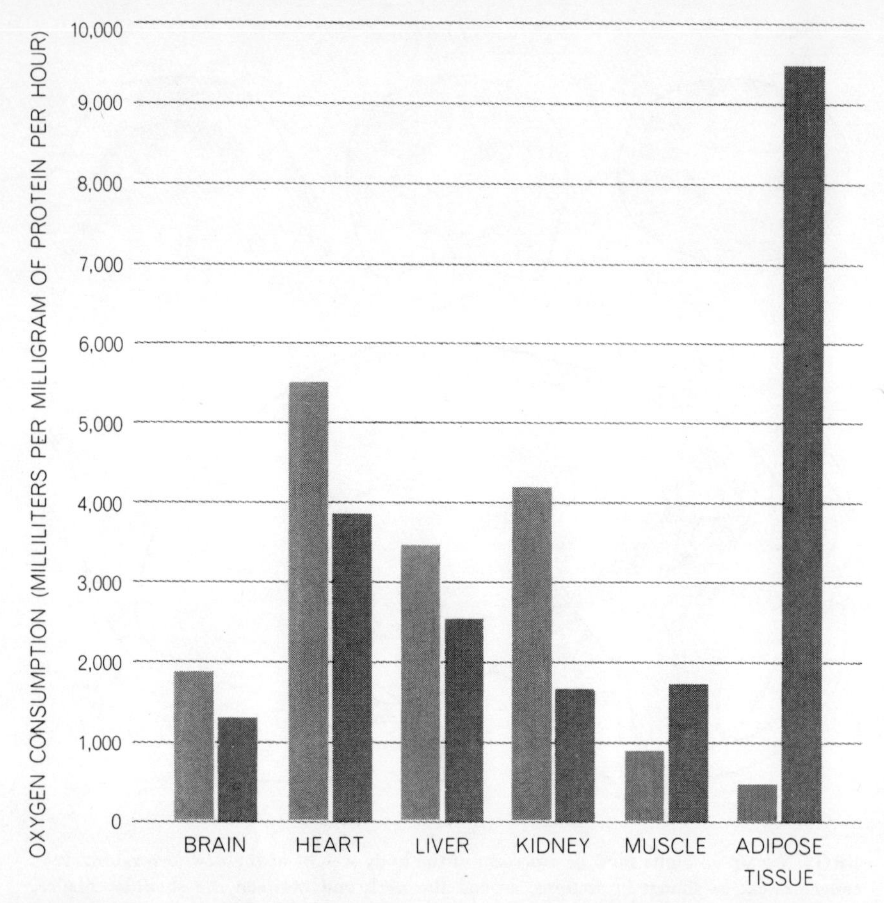

OXIDATIVE CAPACITY of various tissues is compared by measuring their ability to oxidize succinic acid. The colored bars are for adult rabbit tissues, the grey bars for tissues from newborn rabbits. Newborn brown fat (*right*) is the most active oxidizer by far.

and is even greater than that of the hardworking cells of the heart muscle.

To explore the role of the brown adipose tissue in newborn rabbits we began by measuring the animals' total heat production. An indirect measure of this production is the animal's consumption of oxygen: each milliliter of oxygen consumed is equivalent to about five calories of heat in the body. We found that at an environmental temperature of 35 degrees centigrade (95 degrees Fahrenheit) newborn rabbits produced heat at a minimal rate. When their environment was cooled to 25 degrees C. (77 degrees F.), they trebled their heat output. At 20 degrees C. (68 degrees F.) their heat production reached a peak: 400 calories per kilogram of body weight per minute. If newborn rabbits were as fully protected against heat loss and as large as adults are, this rate of production would be sufficient to maintain their internal body temperature at a normal level even in a cold environment as low as 30 degrees below zero C. A newborn rabbit, however, has little or no fur and a large surface area in relation to its body mass; hence its deep-body temperature falls when the animal is only mildly chilled by the outside air.

The next step was to determine whether or not heat production was concentrated in the brown adipose tissue. In a newborn rabbit this tissue makes up 5 to 6 percent of the total body weight and is localized, as we have noted, around the neck and between the shoulder blades. We inserted a fine thermocouple under the skin next to the brown adipose tissue to measure any change in the temperature of that tissue, and for comparison we inserted a second thermocouple in back-muscle tissue at the same distance from the skin and a third in the colon to record the deep-body temperature. At the neutral environmental temperature of 35 degrees C. the temperatures at all three sites in the body were the same. When the environmental temperature was lowered to 25 degrees C., differences developed: the temperature at the brown adipose tissue then was 2.5 degrees higher than that in muscle tissue in the back and 1.3 degrees higher than the deep-body temperature [*see illustration on the next page*]. The temperature difference persisted for many hours, until the fat stored in the brown tissue was almost completely exhausted.

This clear indication that the brown adipose tissue produced heat was strengthened by an experiment in which the newborn animals were deprived of oxygen, the oxygen content of the air in the experimental chamber being reduced from the normal 21 percent to 5 percent. Deprived of the oxygen required for oxidative metabolism, the brown adipose tissue promptly cooled to the same low temperature as the muscle tissue. When the oxygen concentration in the air was restored to the normal 21 percent, the brown adipose tissue immediately warmed up again, with the muscle tissue and deep-body temperature trailing after it in recovery.

Is the brown adipose tissue solely responsible for the newborn animal's increase in heat production in response to cold? We examined this question in a series of experiments with Malcolm M. Segall collaborating. The experiments consisted simply in observing the effect of excising most of the brown adipose tissue (amounting to a few grams) from newborn rabbits. When 80 percent of this tissue was removed (by surgery under anesthesia), the animals no longer increased their heat production in response to exposure to cold. In short, removal of the few grams of this specific tissue practically abolished the newborn rabbit's ability to multiply its oxygen consumption threefold and step up its heat production correspondingly. Evidently, then, the brown adipose tissue was entirely, or almost entirely, responsible for this ability.

Our results did not necessarily mean that all the metabolic heat in response to cold was produced within the brown fat cells themselves. Those cells might release fat in some form into the bloodstream for transport to other tissues, where it might be oxidized. Fortunately this question too could be investigated experimentally.

The fat in the droplets in adipose cells is in the form of triglyceride molecules. A triglyceride consists of a glycerol molecule with three long-chain fatty acids attached [*see middle illustration on page 174*]. Before the triglyceride can be oxidized it must be split into smaller, more soluble units—that is, into glycerol and free fatty acids. Glycerol cannot be used for metabolism in a fat cell, because that type of cell does not contain the necessary enzymes. Consequently all the glycerol molecules freed by the splitting of triglycerides in fat cells are discharged into the bloodstream. The glycerol level in the blood therefore provides an index of the rate of breakdown of triglycerides in fat cells. Now, if the level of free fatty acids in the blood corresponds to the glycerol level, we can assume that fatty acids also are released from these cells in substantial amounts for distribution to other tissues.

We examined the blood of newborn rabbits from this point of view. To begin with, at the neutral incubation temperature of 35 degrees C. the level of free fatty acids in the blood was slightly higher than that of glycerol. When the environmental temperature was lowered to 20 degrees, the glycerol level in the blood increased threefold. The concentration of free fatty acids in the blood, however, rose only a little. This showed that most of the fatty acid molecules freed by the splitting of triglycerides in fat cells must have remained in the cells and been metabolized there. Studies of adipose tissue in the test tube indeed demonstrated that less than 10 percent of the freed fatty acid is released from the cell. We can conclude that the brown fat cells are the main site of cold-stimulated heat production.

How is the heat produced? Brown fat cells are admirably suited, by virtue of their abundance of mitochondria, for generating heat by means of the oxidation of fatty acids. In this process a key role is played by adenosine triphosphate (ATP), the packaged chemical energy that powers all forms of biological work, from the contraction of muscle to the light of the firefly. The probable cycle of reactions that turns chemical energy into heat in adipose tissue cells is shown in the illustration on page 175.

Triggered by the stimulus of cold, the brown fat cell splits triglyceride molecules into glycerol and fatty acids. The glycerol and a small proportion of the free fatty acids are released into the bloodstream for metabolism by other tissues (probably liver and muscle). More than 90 percent of the fatty acid molecules remain, however, in the fat cell. They combine with coenzyme *A*, the energy for this combination being donated by ATP. Since the donation involves the splitting of high-energy bonds in ATP, with its consequent hydrolysis to adenosine monophosphate (AMP), the cell has to regenerate ATP. This is accomplished by oxidative phosphorylation: the addition of inorganic phosphate to AMP with the simultaneous oxidation of a substrate.

Now, some molecules of the fatty acid–coenzyme *A* compound formed with the help of ATP are oxidized to provide energy for the regeneration of ATP. But most of this complex is reconverted, by combination with alpha-glycerol phosphate, to the original triglyceride. In short, there is an apparently purposeless cycle that breaks triglyceride down to fatty acids only

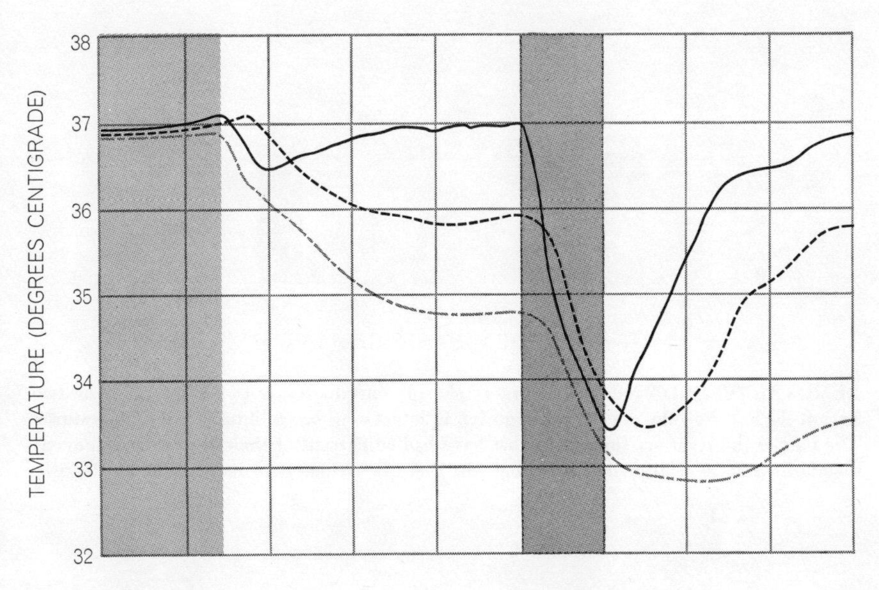

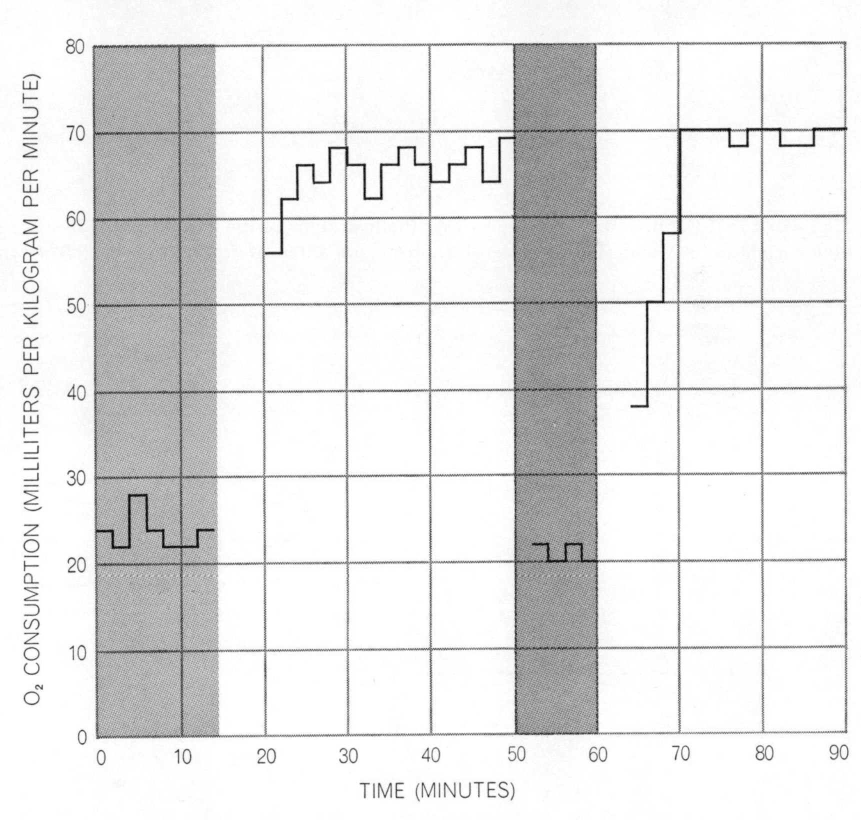

SITE AND OXYGEN DEPENDENCE of heat production are established by data from newborn rabbits subjected to cold and temporarily deprived of oxygen. The top curves are for body temperature measured near brown fat (*solid black line*), in muscle (*broken gray*) and in the colon (*broken black*). The bottom curves trace metabolic activity. In the period covered by the colored band the environmental temperature was 35 degrees centigrade; thereafter it was 25 degrees. The gray band marks a period during which the oxygen concentration was cut from 21 percent to 5 percent. Apparently brown-fat metabolism in the presence of adequate oxygen accounts for a rabbit's ability to respond to a drop in temperature.

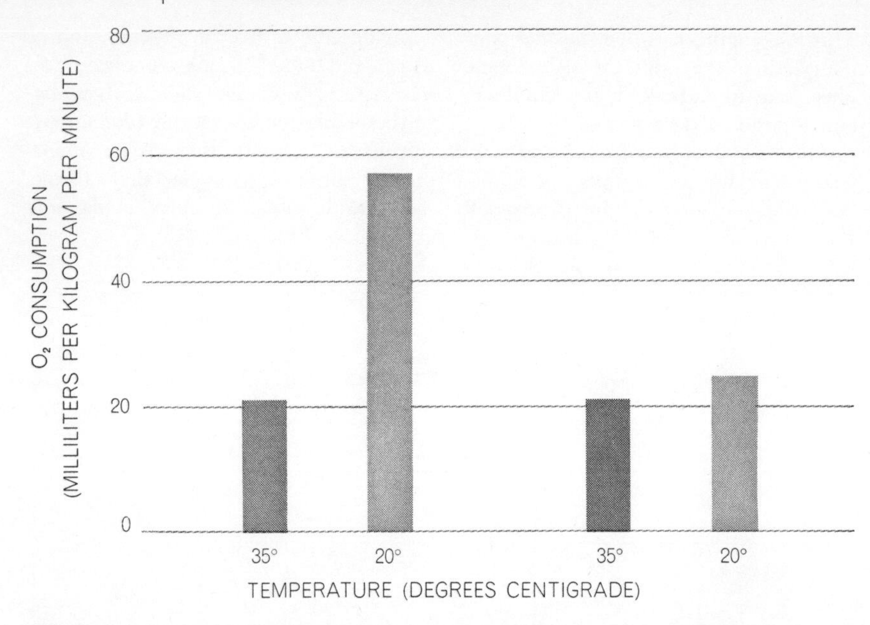

REMOVAL OF BROWN FAT by surgery sharply curtails the response to cold. The two bars at the left show the oxygen consumption in intact newborn rabbits at two temperatures. The bars at the right are for rabbits that have had 80 percent of their brown fat removed: metabolism at 35 degrees is unchanged but there is virtually no increase at 20 degrees.

$3H_2O$

$CH_2O - CO - (CH_2)_n - CH_3$ $\qquad CH_2OH \qquad HOOC - (CH_2)_n - CH_3$

$CHO - CO - (CH_2)_n - CH_3 \longrightarrow CHOH \; + \; HOOC - (CH_2)_n - CH_3$

$CH_2O - CO - (CH_2)_n - CH_3$ $\qquad CH_2OH \qquad HOOC - (CH_2)_n - CH_3$

METABOLISM OF BROWN FAT begins with the hydrolysis of the triglyceride molecule (*left*), yielding one molecule of glycerol (*center*) and three of free fatty acid (*right*).

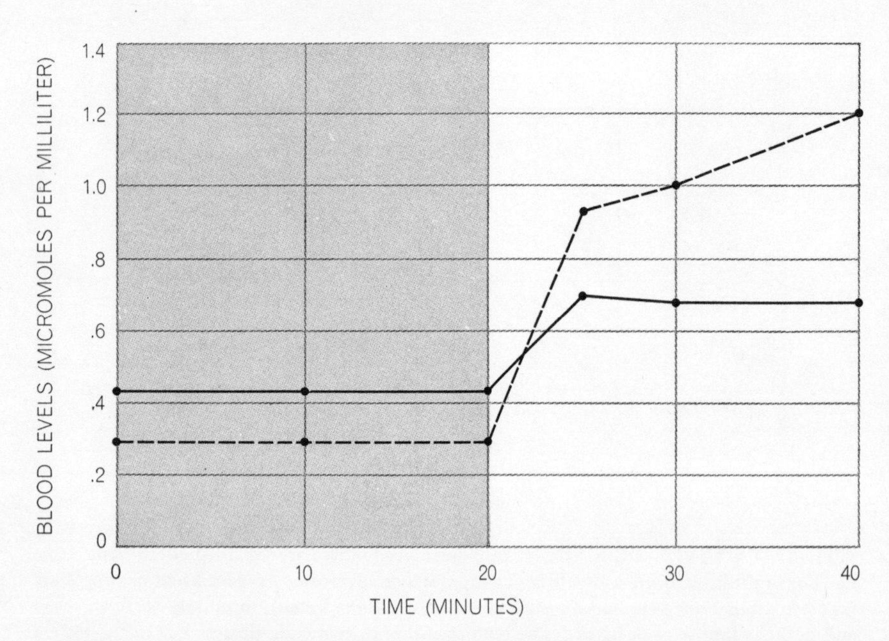

GLYCEROL CONCENTRATION (*broken line*) in the blood of a newborn rabbit rises sharply when the environmental temperature drops from 35 degrees centigrade (*colored area*) to 20 degrees (*white area*). The concentration of fatty acids in the blood rises only slightly, however (*solid line*). Apparently fatty acids are largely metabolized in the cell.

to resynthesize the latter back to triglyceride. Although the cycle seems pointless in chemical terms, it is clearly significant in terms of work. The cycle is, in fact, a device for turning the chemical-bond energy of fatty acids into heat. The energy driving the cycle comes fundamentally from the oxidation of the fatty acids, and the fact that the cycle is exceptionally active in brown adipose tissue is demonstrated by that tissue's high consumption of oxygen. Judging from the proportion of free fatty acids retained by the cells of brown adipose tissue, and from the effects of surgical removal, this tissue accounts for more than 80 percent of the increased body heat produced by a newborn rabbit in the cold.

The heat must of course be distributed to the rest of the body by the bloodstream. The newborn animal's brown adipose tissue has an extremely rich blood supply. During exposure to cold the blood flow through this tissue may increase to several times its normal rate, and indirect calculations suggest that as much as a third of the total cardiac output is directed through the tissue.

How does cold stimulate the brown adipose tissue to generate heat? There are two possible means by which the body's sensation of cold may be communicated to the tissue: by nerve impulses and by hormones, the chemical "messengers" of the body. We found that the hormone noradrenaline has a specific stimulating effect on the brown adipose tissue. An intravenous infusion of noradrenaline in a newborn rabbit will bring about a large increase in the animal's oxygen consumption and heat production in its brown fat. If the brown fat is removed, the hormone no longer produces any increase in the body's oxygen consumption. The question remains: Is the hormone delivered to the intact tissue by way of the bloodstream or by release at sympathetic nerve endings, which are known to secrete noradrenaline close to cells? Several clues suggest that the nerve endings, rather than the bloodstream, are the agent of delivery. For one thing, the adipose tissue's rapid response to cold indicates that the message travels via the nerves. Second, experiments show that drugs that block the action of noradrenaline circulating in the blood do not block the tissue's response to cold. Third and conclusively, direct electrical stimulation of the sympathetic nerves going to the tissue will cause the brown adipose tissue to produce heat, whereas when the sympathetic nerves are cut, the tissue can no longer burn its fat when the animal is exposed to cold.

Various findings indicate that the overall system controlling the production of heat by brown adipose tissue is probably as follows: The temperature receptors in the skin, on sensing cold, send nerve impulses to the brain. The brain's temperature-regulating center then relays impulses along the sympathetic nerves to the brown adipose tissue, where the nerve endings release noradrenaline. The hormone activates an enzyme that splits triglyceride molecules into glycerol and free fatty acids and thereby triggers the heat-producing cycle. Thus the rate of heat production is controlled by the sympathetic nervous system.

Among the animals that have brown adipose tissue at birth, the amount varies considerably from species to species. As in the rabbit, there are large deposits between the shoulder blades in the newborn guinea pig and the coypu (a water rodent). In the cat, dog and sheep at birth there are sheets of brown adipose tissue between the muscles of the trunk and around the kidneys. The human infant has well-marked deposits of such tissue [*see bottom illustration on page 171*], and recent studies indicate that this tissue is a source of heat for a baby as it is for other newborn animals. When a baby is exposed to cold, its blood shows a small but definite rise in the glycerol level with no significant change in the level of fatty acids; on prolonged exposure to cold the fat in its brown adipose tissue is used up.

In most species of animals born with brown adipose tissue the tissue appears to be largely converted to the white form by the time the animal has reached adulthood. Certain animals retain at least some tissue in the brown form, however. The adult rat, for example, has small amounts of brown adipose tissue in the shoulder blade region and elsewhere. The rat's venous system indicates that there is a rich flow of blood from this region to the plexus of veins around the spinal cord; this suggests that the brown adipose tissue in the adult rat may serve particularly to warm vital structures in the animal's body core during exposure to cold. When a laboratory rat is kept in a cold environment, it develops additional brown adipose tissue and an increased ability to produce heat without shivering.

For hibernating animals brown adipose tissue is an all-important necessity throughout life. These animals possess large amounts of the tissue, and direct studies have now shown that the brown adipose tissue is responsible for the animals' rapid warming and awakening from the torpid hibernating state.

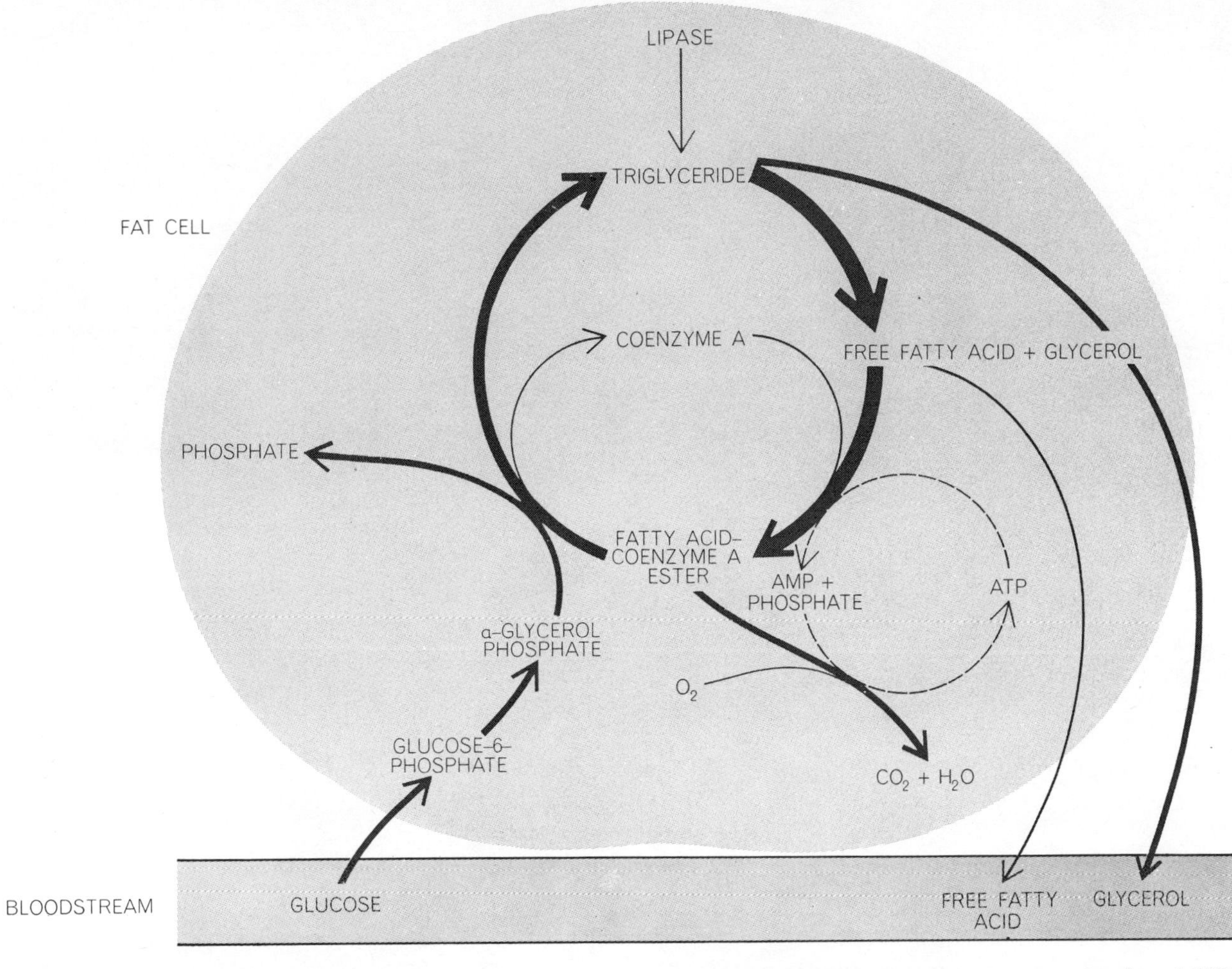

HEAT IS PRODUCED in brown fat cells by the oxidation of fatty acids. In response to cold an enzyme, lipase, splits triglyceride. The glycerol leaves the cell. The fatty acid forms an ester with coenzyme *A*, the reaction acquiring energy from the splitting of adenosine triphosphate (*ATP*) into adenosine monophosphate (*AMP*) and inorganic phosphate. Some of the ester is oxidized to regenerate ATP; the rest goes to resynthesize triglyceride. The effect of the cycle, then, is to turn chemical-bond energy into heat energy.

V

HORMONES AND INTERNAL REGULATION

There is another remarkable fact about songbirds; they usually change their colour and note with the season, and suddenly become different—which among the larger class of birds only cranes do, for these grow black in old age. The blackbird changes from black to red; and it sings in the summer, and chirps in winter, but at midsummer is silent; also the beak of yearling blackbirds, at all events the cocks, is turned to ivory colour. Thrushes are of a speckled colour round the neck in summer but self-coloured in winter.

Pliny
NATURAL HISTORY, x, xli

V HORMONES AND INTERNAL REGULATION

INTRODUCTION

Research during the past decade has truly revolutionized our understanding of the hormonal system of vertebrates. Two major areas of information involve, first, the central role of the hypothalamus in controlling endocrine activities, and, second, the mode of hormone action at the cell level.

There are two groups of neurosecretory cells in the hypothalamus. Those of the first group have cell bodies at various points in that portion of the brain and neurosecretory endings in the median eminence or stalk region near the anterior pituitary gland (the adenohypophysis). The other neurosecretory cells have axonal terminations in the posteror pituitary gland (the neurohypophysis). "The Hormones of the Hypothalamus" by Roger Guillemin and Roger Burgus summarizes the ways that the first group of cells produces, stores, and releases specific molecules (all of which may be polypetides) that act as releasing agents for anterior pituitary hormones: growth hormone, luteinizing hormone, corticotropin (ACTH), follicle-stimulating hormone, and thyrotropin are all released by these agents. Nerve discharges in the hypothalamus apparently stimulate appropriate neurosecretory discharge of the releasing agents which travel by a special blood capillary bed to the adenohypophysis where they cause secretion of anterior pituitary hormones into the blood.

The second group of neurosecretory cells, extending into the posterior pituitary, produces, stores, and releases vasopressin (ADH or antidiuretic hormone) and oxytocin. It is worth noting that there is little strong evidence favoring the two-step secretory mechanism illustrated as part *C* of the diagram on page 190. Instead, the direct release from nerve terminals, as in part *D*, may be the more common mode. Within neurosecretory cells, both vasopressin and oxytocin are apparently bound to a protein called neurophysin, so that discharge of the hormones requires dissociation from this complex. When nerve impulses reach the secretory endings, calcium ions (Ca^{++}) enter the cells and somehow participate in the release of the hormones.

An important revision in thinking concerning pituitary-brain relations stems, in part, from observations with the scanning electron microscope of the various blood vessels and capillaries of the median eminence and parts of the pituitary. Figure 1 fails to make clear that connections between the capillary bed of the adenohypophysis and the veins that drain blood from the pituitary are *not* extensive. Instead, it seems likely that a significant proportion of blood containing newly secreted anterior pituitary hormones may pass to the neurohypophysis (the posterior pituitary) and from there either to the brain or to the body. The rapid appearance of labelled adenohypophyseal hormones in the brain may result from such shunting. The important point of these findings is that we should think of the brain as being an immediate and important target of certain pituitary hormones; thus, the endocrine system and nervous system may be linked and coordinated in function.

It is not certain whether the anterior or posterior pituitary system is more primitive. Although the posterior, neurohypophyseal system appears to be simpler anatomically and functionally, it has proven difficult to assign specific and consistent functions to ADH or to oxytocin in the fishes. ADH, for instance, seems to affect ion balance (particularly sodium) in fishes, but only influences water balance in terrestrial vertebrates. Recall too that the preoral glands of close vertebrate relatives (amphioxi, sea squirts, hemichordates; see "An Essay on Vertebrates") may be the precursors of the adenohypophysis. The endocrinologists Aubrey Gorbman and Howard A. Bern have suggested that, in the earliest vertebrate endocrine system, hypothalamic neurosecretion may have acted on the adenohypophysis to cause it to secrete hormone. Later, hypothalamic cell secretions began acting directly elsewhere in the body. The cells producing these neurosecretions (ADH, oxytocin) could next have acquired a more efficient means of secreting directly into the systemic circulation; thus, the discrete anatomical area, the neurohypophysis and its blood supply, appeared.

Study of the chemical structure of pituitary hormones suggests that there are four "families" (see the reference by M. Wallis listed in the Bibliography of this book):

1. The neurohypophyseal hormones oxytocin and vasopressin.
2. Corticotrophin (ACTH); melanotrophin (MSH); and lipotrophin (LPH).
3. Growth hormone and prolactin.
4. Thyroid stimulating hormone (TSH); the gonadotrophins (LH, FSH).

There is also a degree of similarity between releasing factors of the median eminence and posterior pituitary hormones; for instance, vasopressin (ADH) and the ACTH-releasing polypeptide have sequence homologies.

Let us consider the first family of hormones in greater detail. S. Archer, of the University of Paris, has summarized the data in Table 1. It seems very likely that a gene coding for the ancestral molecule was duplicated and that each copy subsequently evolved separately by modifications in the amino acids located at positions 3, 4, and 8 of the peptide. Thus, an oxytocin "line" and a vasopressin "line" can be traced through the vertebrates (neither hormone has been detected yet in cyclostomes). The table shows the amino acid variations in the posterior pituitary hormones of several classes of vertebrate animals (the numbers in superscript refer to the positions in the peptide chains). In order to use the table, assume that, for the "oxytocin" line, position number 3 is isoleucine in all forms, with the indicated variations occurring in positions 4 and 8; for the "vasopressin" line, assume glutamine occupies position 4 throughout, and variations occur at positions 3 and 8. The invariability of the other amino acids suggests their critical nature in the shape of the molecules, whereas the substitutions at positions 3, 4 and 8 are obviously sources of difference in physiological action.

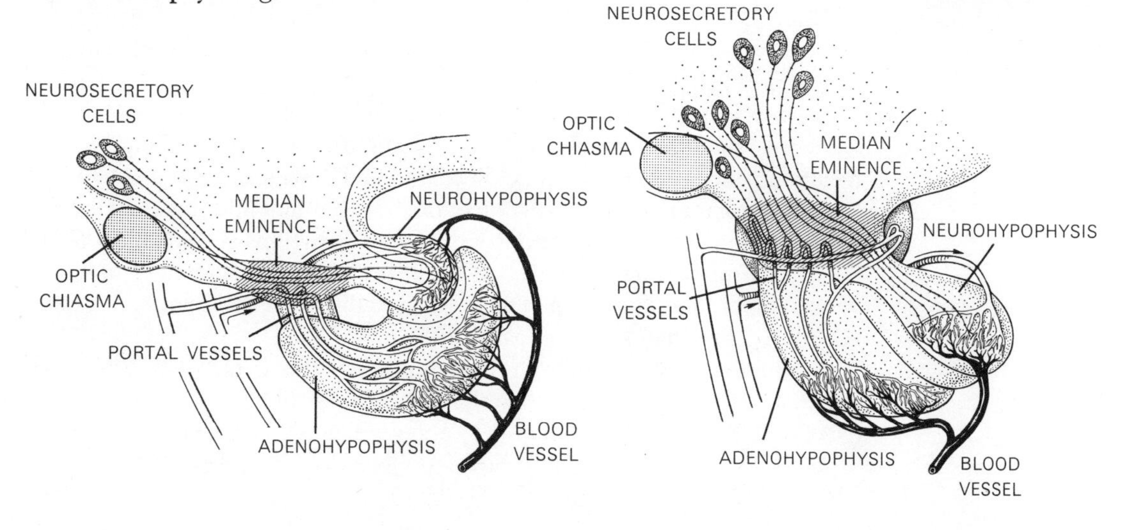

Figure 1. **Representations of a pituitary gland of a bird (left) and of that of a mammal (right). The median eminence is the source of the various "releasing factors." In a bird, none of the neurosecretory cells originating in the hypothalamus end in the median eminence; instead, blood vessels course near the axons of the neurosecretory cells and then carry the blood and releasing factors downward to the adenohypophysis. In contrast, in a mammal, some neurosecretory cells actually terminate near the capillary loops that pick up the releasing factor peptides. Although for clarity's sake it is not shown here, flow reversal may occur, so blood leaving the adenohypophyseal capillaries may go to the neurohypophysis and then (vessels not shown) to the brain (see the article by R. M. Bergland and R. B. Page listed in the Bibliography).**

Table 1.

position	1	2	3	4	5	6	7	8	9
amino acid	cystine—	tyrosine—	☐ —	☐	—asparagine—	cystine—	proline—	☐	—glycine

(cystine 1 and cystine 6 are joined by a bridge)

	OXYTOCIN LINE		VASOPRESSIN LINE	
bony fish	*isotocin*	serine^{4}, isoleucine^{8}	*vasotocin*	isoleucine^{3}, arginine^{8}
amphibians	*mesotocin*	glutamine^{4}, isoleucine^{8}	*vasotocin*	isoleucine^{3}, arginine^{8}
mammals	*oxytocin*	glutamine^{4}, leucine^{8}	*vasopressin*	phenylalanine^{3}, arginine^{8}
pig	*oxytocin*	glutamine^{4}, leucine^{8}	*vasopressin*	phenylalanine^{3}, lysine^{8}

The presence of identical amino acid sequence homologies in each of the four pituitary hormone families suggests that processes of gene duplication and rearrangement have gone on. An important consequence of the similarities in structure is overlap in physiological function. For example, injected ADH can cause release of ACTH; or injection of ACTH will cause pigment cells to darken (recall that MSH and a portion of ACTH are structurally identical). If in fact molecules with very similar structures are used in many different ways, then it seems likely that a critical feature of the endocrine system is the sensitivity and specificity of target cells and organs.

The second major area of new information about hormones and the endocrine system centers on the mode of hormone action on cells. On the one hand, peptide and protein hormones appear to elicit formation of cyclic AMP in responsive cells; this "second messenger" then evokes an appropriate physiological response to the original hormone. A typical instance of this process is discussed in the article "Calcitonin" by Howard Rasmussen and Maurice M. Pechet. In contrast to data concerning peptide hormones, increasingly conclusive data implies that steroid hormones may actually be transported into cells or even into nuclei where direct effects on the genetic apparatus or other cellular organelles are possible. These observations on the interactions of peptide and steroid hormones with cells emphasize the crucial role of receptor molecules. The presence or absence of receptor molecules may well be the key to "specificity" in the endocrine system, the reason why only certain cells respond even though hormones travel everywhere in the blood system and body fluids.

Two points must be made to modify "The Pineal Gland" by Richard J. Wurtman and Julius Axelrod. First, it is now known that the control of MSH release is complicated by the role of melatonin. When the pineal discharges melatonin, stores of MSH in the adenohypophysis are depleted. Therefore, both the hypothalamic releasing agent and melatonin can control MSH levels in the blood. Second, there are apparently significant differences among vertebrates with respect to the relationship between the "biological clock" and the pineal. In birds, the pineal is clearly associated with clock-driven phenomena; in mammals, however, an area of the brain above the optic chiasma may perform equivalent functions.

The biochemical basis for pineal involvement in clock-related phenomena has been studied profitably in birds. In these animals, cyclical release of melatonin elicits clock-dependent behavior, and in fact, the German biologist E. Gwinner has found that daily injections of melatonin into sparrows lacking pineals will continue to synchronize the activity rhythms of those birds. For a description of Gwinner's work, see the Introduction to Section V in *Vertebrates: Adaptation*, the companion volume to this book (Norman K. Wessells, ed., W. H. Freeman and Company, 1980).

Melatonin is synthesized from the neurotransmitter serotonin in a two-step reaction, the first of which is catalyzed by the enzyme *N*-acetyltransferase (NAT). S. Binkley, (in "A Time Keeping Enzyme in the Pineal Gland,"

Scientific American Offprint 1426) has summarized data showing a 27- to 70-fold difference in levels of NAT between day and night in birds, the high levels being present in the dark when most serotonin is being used in melatonin synthesis. Levels of NAT continue to fluctuate on about a 24-hour cycle in birds kept in constant darkness and even in blind chickens (of course, the pineal has light receptor characteristics itself). In fact, NAT levels even fluctuate in a pineal gland cultured in a dish, and exposed to light–dark variations. Only future research will tell us whether synthesis and secretion of melatonin by the pineal or the suprachiasmatic nucleus of the hypothalamus of mammals is indeed the general means for controlling clock-based behavior in vertebrates.

Our knowledge of the intricate workings of calcium and phosphate regulation has been modified drastically by work on the hormone calcitonin, as is discussed by Rasmussen and Pechet in their article "Calcitonin." For years it has been clear that abnormally low calcium levels in the blood lead to release of parathyroid hormone from the parathyroid gland (see "The Parathyroid Hormone" by H. Rasmussen, *Scientific American*, April 1961). Parathyroid hormone acts on bone cells called osteoclasts to stimulate their bone-destroying activities, one result being an increase in calcium ions in the blood. Calcitonin acts in just the reverse sense: when calcium levels are too high, calcitonin is released so that it can inhibit osteoclast function. As a result calcium levels fall.

The system to regulate calcium ions is more complex than this brief treatment implies. For instance, some evidence suggests that a thyrocalcitonin releasing factor comes from the parathyroid gland to act upon the thyroid and cause discharge of active calcitonin; if so, this relationship is like that between the hypothalamus and the anterior pituitary cells. In summary, calcium regulation results from a balance of positive (parathormone) and negative (thyrocalcitonin) hormones acting concordantly with vitamin D.

One of the main obstacles to our understanding of this regulatory complex is its high variability among different vertebrates. A dog or a pigeon is in immediate distress after the parathyroids are removed, but a rat or a rooster is hardly affected. Different amphibia are equally variable, and no parathyroids are known in fishes. In fishes, however, another derivative of the anterior embryonic gut is present and may be a source of a calcium-regulating hormone. This tissue is called the ultimobranchial body, and it arises as a pouch, as do the thyroid and parathyroid glands. In terrestrial vertebrates the ultimobranchial body may regress (in birds), or it may persist (in mammals) and become a part of the region in which thyrocalcitonin-secreting cells are found. Thyrocalcitonin has been isolated from several birds, a reptile, and a dogfish shark. The presence of a discrete ultimobranchial body in lower vertebrates, and the probable positive correlation between its presence and the production of thyrocalcitonin, suggest that the negative regulation by thyrocalcitonin may be the primitive means of controlling calcium levels in vertebrates, and that the parathyroid gland and parathormone came later in terrestrial forms as a refinement in control. Verification of this hypothesis must wait until thyrocalcitonin has been identified in various types of bony fishes.

An interesting sidelight on vitamin D action in humans is suggested by W. Farnsworth Loomis in his article, "Rickets." Vitamin D can, in fact, be manufactured in vertebrate skin when ultraviolet light of the correct wavelength and intensity strikes the body; thus, the compound might be called a "hormone," since it is produced by one cell type and acts on another. Loomis provides several correlations suggesting that the varying degrees of pigmentation found among different human populations may be related to supranormal, normal, or subnormal manufacture of vitamin D, and thus, indirectly, to diseases such as rickets. It is sobering to contemplate that the immense suffering of some human beings as a result of prejudice based on skin color may have been an indirect consequence of the selection for more lightly pigmented skins among ancient human populations that happened to migrate toward the higher latitudes. Interestingly, some Neanderthal fossils have been unearthed in Europe which give evidence of rickets; perhaps the skin and hair of those ancient relatives of *Homo sapiens* was so heavily pigmented that insufficient vitamin D was made and rickets resulted.

Although Loomis's arguments are controversial, they offer a logical and cohesive explanation of pigmentation differences among humans. They also suggest an answer to the long-standing enigma of pigmentation and heat absorption. It has never been clear why darkly pigmented skin, which absorbs heat so well, should be found in the tropics, whereas white skin, which tends to reflect heat rather than absorbing it, should be found in cooler climates. Just the opposite might be expected. However, if ultraviolet absorption and regulation of vitamin D synthesis is more critical than problems of heat balance, then an explanation is available.

Although these arguments are persuasive, other plausible explanations for heavy melanization of skin in tropical regions have been offered. For instance, a number of vitamins and metabolites are sensitive to ultraviolet light, and may be protected by pigmented skin. Other data suggest a lesser susceptibility to skin cancers if dark melanin pigments are present. Probably a combination of these factors have operated during human evolution to help generate the differing degrees of pigmentation of human skin.

Three articles in this Section deal with hormones and reproduction. In "How an Eggshell Is Made," T. G. Taylor emphasizes the complexity of interactions between bones, oviduct, pituitary, and gonads. The inescapable conclusion is that the evolutionary appearance of the cleidoic egg (see "An Essay on Vertebrates") must have involved numerous mutational events taking place over eons—not a precipitous acquisition of the new mode of reproduction.

In "Milk," by Stuart Patton, we examine the means by which the food of the mammalian newborn is manufactured and secreted. Milk is believed to be an "early" invention of mammals, since it is made by the most primitive of living mammals, the monotremes. The mammary gland of the Monotremata is "simpler" than that of marsupials or placentals, however, in that the collecting ducts do not drain to the surface at one site, the teat. Instead, when a tiny echidna or platypus suckles, milk is expressed onto the surface of the skin in association with some hairs. It has been proposed, in fact, that mammary glands are derivatives of special skin glands normally linked to hair follicles (so-called apocrine sweat glands or sebaceous glands). Since hair, as insulation, and sweat glands, as means of lowering body temperature, both have to do with temperature regulation, we might speculate that the unique use of mammary glands and milk in mammalian reproduction is an indirect consequence of the earlier evolution of homeothermy.

The final article in this Section considers the roles hormones play in complex reproductive behavior patterns. It is well known that the seasonal reproductive behavior of vertebrates is under endocrine control. This is obvious with respect to the gonads and sperm and egg production. It is also the case with respect to various secondary sexual characteristics and behaviors.

For instance, each spring, changing gonadal hormone levels in adult male birds cause a change in the plumage from the drab colors of winter to the brighter mating colors and patterns of spring. There are alterations in the voice box (syrinx), too, and the singing sequence is initiated in the form of stages called subsong, plastic song, and full song. If the testes are removed midway through this sequence, the male achieves only the plastic song; this is the case even if the male is an old bird that had developed and sung the full song thousand of times in previous springtimes. Even though its brain had generated the correct motor program for full song, and even though the bird had heard itself sing the full song again and again in previous years, it cannot sing that full song again unless testosterone levels rise sufficiently high. Other complex behavior patterns, such as nest building and migration, are hormone-dependent in birds as they are in salmon and many other vertebrates.

As is the case with the chameleons described by David Crews in "The Hormonal Control of Behavior in a Lizard," radioactively labelled gonadal hormones localize in specific sites in the avian brain. One such place, the nucleus intercollicularis of the midbrain, is a center which if stimulated electrically elicits vocalizations by the bird. In lizards as well as in various mammals that have been tested, the hormone linked to a particular behavior appears to localize preferentially in the portions of the brain responsible for the behavior. Perhaps the most important message from Crew's article and other studies is that the nervous and endocrine systems are intimately coupled and mutually dependent for certain behavioral repertoires.

It is interesting to contemplate in this regard the point made in "An Essay on Vertebrates," namely, that humans are unique among vertebrates in the dissociation of ovulation from copulation in females. Perhaps the phenomenal expansion of the human cerebral cortex has permitted our nervous system to have this extra degree of freedom, so that mating is not coupled to the changing hormone levels that trigger ovulation. It is certainly the case, however, that human sexual behavior is dependent upon gonadotropins and gonadal hormones, and in that respect we are like our kin, other vertebrates.

17 The Hormones of the Hypothalamus

by Roger Guillemin and Roger Burgus
November 1972

The anterior pituitary gland, which controls the peripheral endocrine glands, is itself regulated by "releasing factors" originating in the brain. Two of these hormones have now been isolated and synthesized

The pituitary gland is attached by a stalk to the region in the base of the brain known as the hypothalamus. Within the past year or so, after nearly 20 years of effort in many laboratories throughout the world, two substances have been isolated from animal brain tissue that represent the first of the long sought hypothalamic hormones. Because the molecular structure of the new hormones is fairly simple the substances can readily be synthesized in large quantities. Their availability and their high activity in humans has led physiologists and clinicians to consider that the hypothalamic hormones will open a new chapter in medicine.

It has long been known that the pituitary secretes several complex hormones that travel through the bloodstream to target organs, notably the thyroid gland, the gonads and the cortex of the adrenal glands. There the pituitary hormones stimulate the secretion into the bloodstream of the thyroid hormones, of the sex hormones by the gonads and of several steroid hormones such as hydrocortisone by the adrenal cortex. The secretion of the thyroid, sex and adrenocortical hormones thus has two stages beginning with the release of pituitary hormones. Studies going back some 50 years culminated in the demonstration that the process actually has three stages: the release of the pituitary hormones requires the prior release of another class of hormones manufactured in the hypothalamus. It is two of these hypothalamic hormones that have now been isolated, chemically identified and synthesized.

One of the hypothalamic hormones acts as the factor that triggers the release of the pituitary hormone thyrotropin, sometimes called the thyroid-stimulating hormone, or TSH. Thus the hypothalamic hormone associated with TSH is called the TSH-releasing factor, or TRF. The other hormone is LRF. Here again "RF" stands for "releasing factor"; the "L" signifies that the substance releases the gonadotropic pituitary hormone LH, the luteinizing hormone. A third gonadotropic hormone, FSH (follicle-stimulating hormone), may have its own hypothalamic releasing factor, FRF, but that has not been demonstrated. It is known, however, that the hypothalamic hormone LRF stimulates the release of FSH as well as LH.

Studies are continuing aimed at characterizing several other hypothalamic hormones that are known to exist on the basis of physiological evidence but that have not yet been isolated. One of them regulates the secretion of adrenocorticotropin (ACTH), the pituitary hormone whose target is the adrenal cortex. Another hormone (possibly two hormones with opposing actions) regulates the release of prolactin, the pituitary hormone involved in pregnancy and lactation. Still another hormone (again possibly two hormones with opposing actions) regulates the release of the pituitary hormone involved in growth and structural development (growth hormone).

That the hypothalamus and the pituitary act in concert can be suspected not only from their physical proximity at the base of the brain but also from their development in the embryo. During the early embryological development of all mammals a small pouch forms in the upper part of the developing pharynx and migrates upward toward the developing brain. There it meets a similar formation, resembling the finger of a glove, that springs from the base of the primordial brain. Several months later the first pouch, now detached from the upper oral cavity, has filled into a solid mass of cells differentiated into glandular types. At this point the second pouch, still connected to the base of the brain, is rich with hundreds of thousands of nerve fibers associated with a modified type of glial cell, not too unlike the glial cells found throughout the brain. The two organs are now enclosed in a single receptacle that has formed as an open spherical cavity within the sphenoid bone, on which the brain rests.

This double organ, now ensconced in the sphenoidal bone, is the pituitary gland, or hypophysis. The part that migrated from the brain is the posterior lobe, or neurohypophysis; the part that migrated from the pharynx is the anterior lobe, or adenohypophysis. Both parts of the gland remain connected to the brain by a common stalk that goes through the covering flap of the sphenoidal cavity. For many years after the double embryological origin of the pituitary gland was recognized the role of the gland was no more clearly understood than it had been in the old days. Indeed, the name "pituitary" had been given to it in the 16th century by Vesalius, who thought that the little organ had to do with secretion of *pituita:* the nasal fluid.

We know now that the anterior lobe of the pituitary gland controls the secretion and function of all the "periph-

HYPOTHALAMIC FRAGMENTS of sheep brains were the source from which the authors' laboratory extracted one milligram of TRF, the first hypothalamic hormone to be characterized and synthesized. The photograph is of about 30 frozen hypothalamic fragments; some five million such fragments, dissected from 500 tons of sheep brain tissue, were processed over a period of four years.

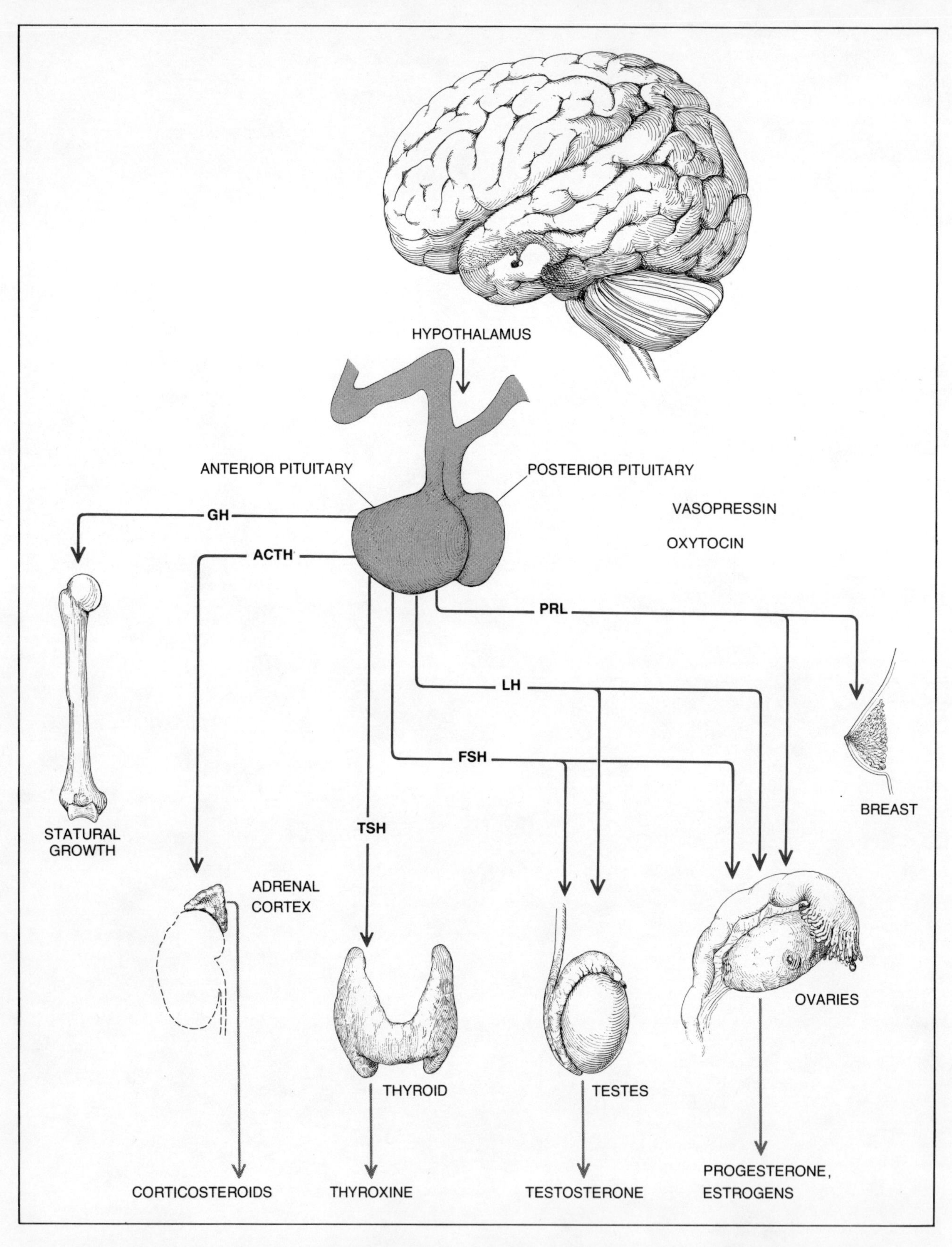

PITUITARY GLAND, connected to the hypothalamus at the base of the brain, has two lobes and two functions. The posterior lobe of the pituitary stores and passes on to the general circulation two hormones manufactured in the hypothalamus: vasopressin and oxytocin. The anterior lobe secretes a number of other hormones: growth hormone (GH), which promotes statural growth; adrenocorticotropic hormone (ACTH), which stimulates the cortex of the adrenal gland to secrete corticosteroids; thyroid-stimulating hormone (TSH), which stimulates secretions by the thyroid gland, and follicle-stimulating hormone (FSH), luteinizing hormone (LH) and prolactin (PRL), which in various combinations regulate lactation and the functioning of the gonads. Several of these anterior pituitary hormones are known to be controlled by releasing factors from the hypothalamus, two of which have now been synthesized.

eral" endocrine glands (the thyroid, the gonads and the adrenal cortex). It also controls the mammary glands and regulates the harmonious growth of the individual. It accomplishes all this by the secretion of a series of complex protein and glycoprotein hormones. All the pituitary hormones are manufactured and secreted by the anterior lobe. Why should this master endocrine gland have migrated so far in the course of evolution (a journey recapitulated in the embryo) to make contact with the brain? As we shall see, recent observations have answered the question.

The posterior lobe of the pituitary has been known for the past 50 years to secrete substances that affect the reabsorption of water from the kidney into the bloodstream. These secretions also stimulate the contraction of the uterus during childbirth and the release of milk during lactation. In the early 1950's Vincent Du Vigneaud and his co-workers at the Cornell University Medical College resolved a controversy of many years' standing by showing that the biological activities of the posterior lobe are attributable to two different molecules: vasopressin (or antidiuretic hormone) and oxytocin. The two molecules are octapeptides: structures made up of eight amino acids. Du Vigneaud's group showed that six of the eight amino acids in the two molecules are identical, which explains their closely related physicochemical properties and similar biological activity. Both hormones exhibit (in different ratios) all the major biological effects mentioned above: the reabsorption of water, the stimulation of uterine contractions and the release of milk.

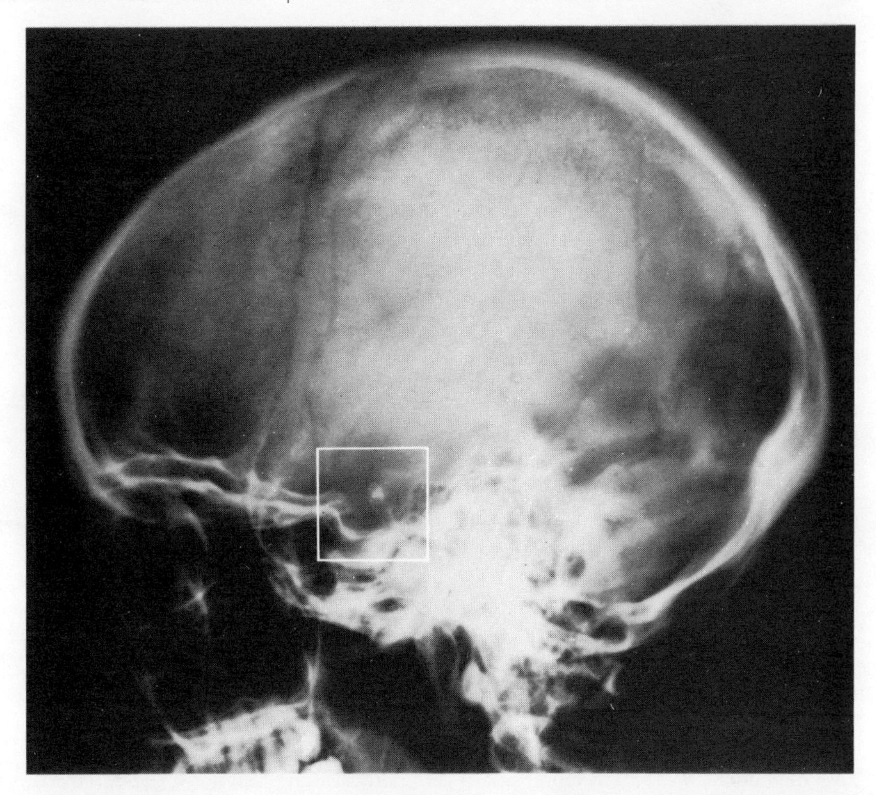

BONY RECEPTACLE in which the pituitary gland is enclosed is a cavity in the sphenoid bone, on which the base of the brain rests. White rectangle shows area diagrammed below.

As early as 1924 it was realized that the hormones secreted by the posterior lobe of the pituitary are also found in the hypothalamus: that part of the brain with which the lobe is connected by nerve fibers through the pituitary stalk. Later it was shown that the two hormones of the posterior pituitary are actually manufactured in some specialized nerve cells in the hypothalamus. They flow slowly down the pituitary stalk to the posterior pituitary through the axons, or long fibers, of the hypothalamic nerve cells [*see top illustration on page 189*]. They are stored in the posterior pituitary, which is now reduced to a storage organ rather than a manufacturing one. From it they are secreted into the bloodstream on the proper physiological stimulus.

These observations had led several

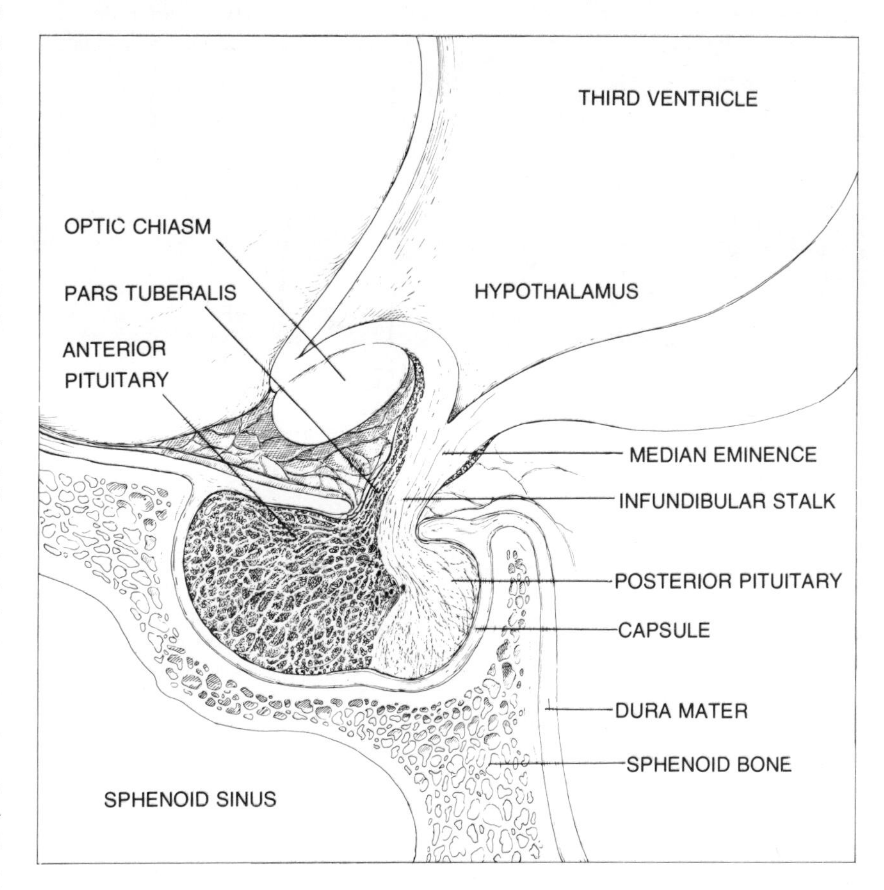

HYPOTHALAMUS AND PITUITARY are connected by a stalk that passes through the membranous lid of the receptacle in the sphenoid bone in which the pituitary rests. The double embryological origin of the two lobes of the pituitary is reflected in their differing tissues and functions and in the different ways that each is connected to the hypothalamus.

biologists, notably Ernst and Berta Scharrer, to the striking new concept of neurosecretion (the secretion of hormones by nerve cells). They suggested that specialized nerve cells might be able to manufacture and secrete true hormones, which would then be carried by the blood and would exert their effects in some target organ or tissue remote from their point of origin. The ability to manufacture hormones had traditionally been assigned to the endocrine glands: the thyroid, the gonads, the adrenals and so on. The suggestion that nerve cells could secrete hormones would endow them with a capacity far beyond their ability to liberate neurotransmitters such as epinephrine and acetylcholine at the submicroscopic regions (synapses) where they make contact with other nerve cells.

Even as these studies were in progress and these new concepts were being formulated other laboratories were reporting evidence that functions of the anterior lobe of the pituitary were somehow dependent on the structural integrity of the hypothalamic area and on a normal relation between the hypothalamus and the pituitary gland. For example, minute lesions of the hypothalamus, such as can be created by introducing small electrodes into the base of the brain in an experimental animal and producing localized electrocoagulation, were found to abolish the secretion of anterior pituitary hormones. On the other hand, the electrical stimulation of nerve cells in the same regions dramatically increased the secretion of the hormones [*see illustration below*].

Thus the question was presented: Precisely how does the hypothalamus regulate the secretory activity of the anterior pituitary? The results produced by electrocoagulation and electrical stimulation of the hypothalamus suggested some kind of neural mechanism. One objection to this theory was rather hard to overcome. Careful anatomical studies over many years had clearly established that there were no nerve fibers extending from the hypothalamus to the anterior pituitary. The only nerve fibers found in the pituitary stalk were those that terminate in the posterior lobe.

A way out of the dilemma was provided by an entirely different working hypothesis, suggested by the discovery in 1936 of blood vessels of a peculiar type that were shown to extend from the floor of the hypothalamus through the pituitary stalk to the anterior pituitary [*see bottom illustration on opposite page*]. If these tiny blood vessels were cut, the secretions of the anterior pituitary would instantly decrease. If the capillary vessels regenerated across the surgical cut, the secretions resumed.

Accordingly a new hypothesis was put forward about 1945 with which the name of the late G. W. Harris of the University of Oxford will remain associated. The hypothesis proposed that hypothalamic control of the secretory activity of the anterior pituitary could be neurochemical: some substance manufactured by nerve cells in the hypothalamus could be released into the capillary vessels that run from the hypothalamus to the anterior pituitary, where it could be delivered to the endocrine cells of the gland. On reaching these endocrine cells the substance of hypothalamic origin would somehow stimulate the secretion of the various anterior pituitary hormones.

The hypothesis that pituitary function is controlled by neurohormones originating in the hypothalamus was soon well established on the basis of intensive physiological studies in several laboratories. The next problem was therefore to isolate and characterize the postulated hypothalamic hormones. It was logical to guess that the hormones might be polypeptides of small molecular weight, since it had been well established that the two known neurosecretory products of hypothalamic origin, oxytocin and vasopressin, are each composed of eight amino acids. Indeed, in 1955 it was reported that crude aqueous hypothalamic extracts designed to contain polypeptides were able specifically to stimulate the secretion of ACTH, the pituitary hormone that controls the secretion of the steroid hormones of the adrenal cortex.

It was quickly demonstrated that none of the substances known to originate in the central nervous system (such as epinephrine, acetylcholine, vasopressin and oxytocin) could account for the ACTH-releasing activity observed in the extract of hypothalamic tissue. It therefore seemed reasonable to postulate the existence and involvement in this phenomenon of a new substance designated (adreno)corticotropin-releasing factor, or CRF. Several laboratories then undertook the apparently simple task of purifying CRF from hypothalamic extracts, with the final goal of isolating it and establishing its chemical structure. Seventeen years later the task still remains to be accomplished. Technical difficulties involving the methods

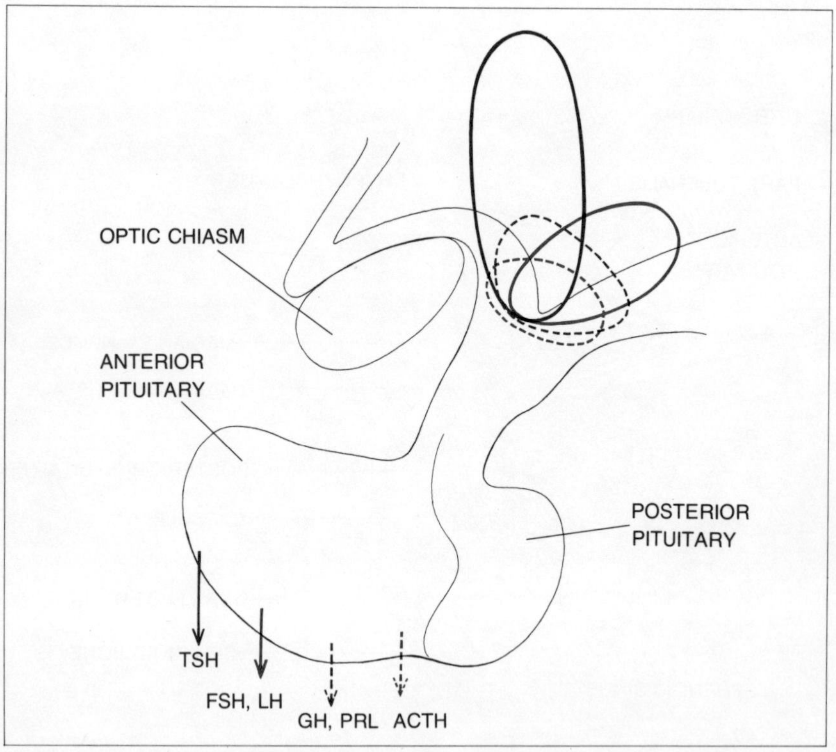

RELATION between the hypothalamus and anterior pituitary was established experimentally. Lesions in specific regions of the hypothalamus interfere with secretion by the anterior lobe of specific hormones; electrical stimulation of those regions stimulates secretion of the hormones. The regions associated with each hormone are mapped schematically.

of assaying for CRF, together with certain peculiar characteristics of the molecule, have defied the enthusiasm, ingenuity and hard work of several groups of investigators.

More rewarding results were obtained in a closely related effort. About 1960 it was clearly established that the same crude extracts of hypothalamic tissue were able to stimulate the secretion of not only ACTH but also the three other pituitary hormones mentioned above: thyrotropin (TSH) and the two gonadotropins (LH and FSH). TSH is the pituitary hormone that controls the function of the thyroid gland, which in turn secretes the two hormones thyroxine and triiodothyronine. LH controls the secretion of the steroid hormones responsible for the male or female sexual characteristics; it also triggers ovulation. FSH controls the development and maturation of the germ cells: the spermatozoa and the ova. In reality the way in which LH and FSH work together is considerably more complicated than this somewhat simplistic description suggests.

Results obtained between 1960 and 1962 were best explained by proposing the existence of three separate hypothalamic releasing factors: TRF (the TSH-releasing factor), LRF (the LH-releasing factor) and FRF (the FSH-releasing factor). The effort began at once to isolate and characterize TRF, LRF and FRF. Whereas it was difficult to find a good assay for CRF, a simple and highly reliable biological assay was devised for TRF. At first, however, the assays for LRF and FRF still left much to be desired.

With a good method available for assaying TRF, progress was initially rapid. Within a few months after its discovery TRF had been prepared in a form many thousands of times purer. Preparations of TRF obtained from the brains of sheep showed biological activity in doses as small as one microgram. A great deal of physiological information was obtained with those early preparations. For example, the thyroid hormones somehow inhibit their own secretion when they reach a certain level in the blood. This fact had been known for 40 years and was the first evidence of a negative feedback in endocrine regulation. Studies with TRF showed that the feedback control takes place at the level of the pituitary gland as the result of some kind of competition between the number of available molecules of thyroid hormones and of TRF. Other significant observations were made on the gonadotropin-releasing factors when

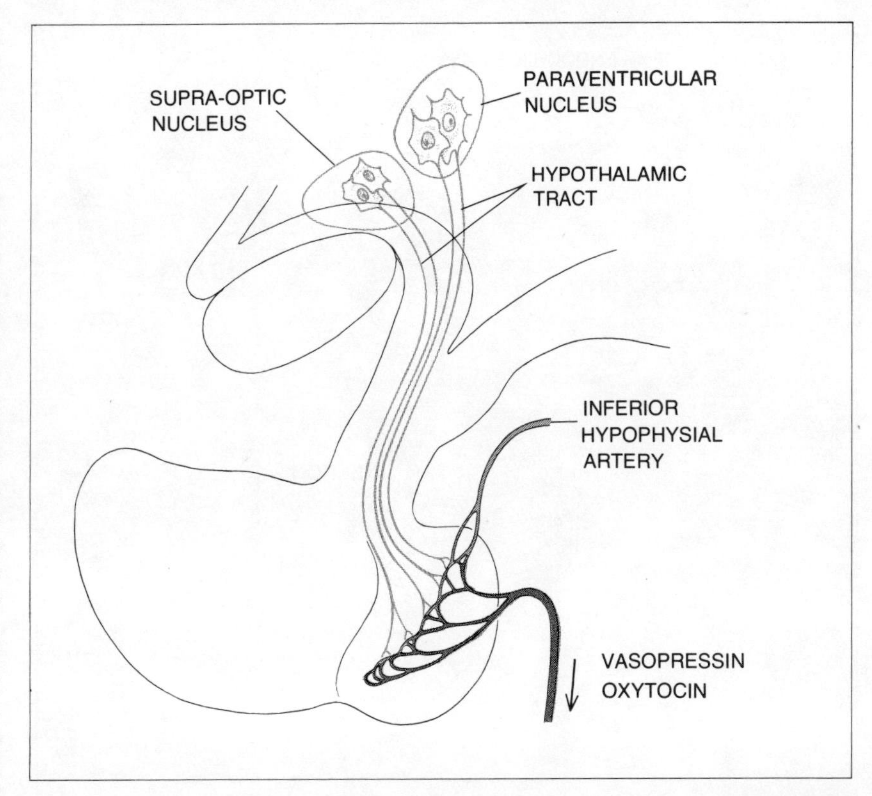

NEURAL CONNECTIONS could not explain the relation of the hypothalamus and the anterior lobe. The only significant nerve fibers connecting hypothalamus and pituitary run from two hypothalamic centers to the posterior lobe. They transmit oxytocin and vasopressin, two hormones manufactured in the hypothalamus and stored in the posterior lobe.

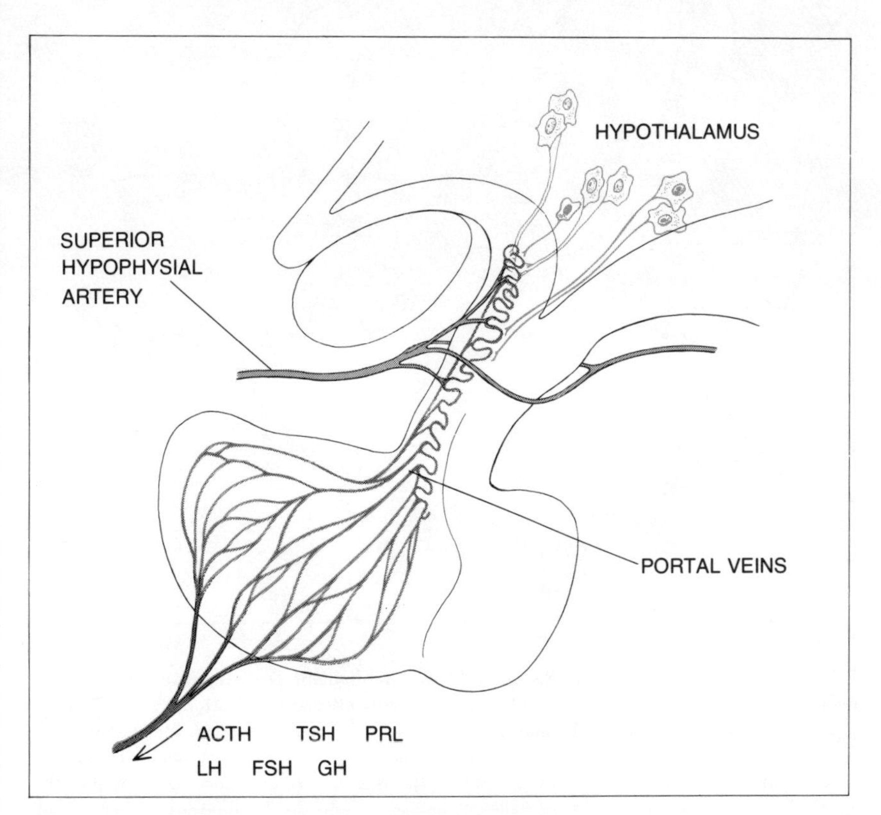

VASCULAR CONNECTIONS between hypothalamus and anterior lobe were eventually discovered: a network of capillaries reaching the base of the hypothalamus supplies portal veins that enter the anterior pituitary. Small hypothalamic nerve fibers apparently deliver to the capillaries releasing factors that stimulate secretion of the anterior-lobe hormones.

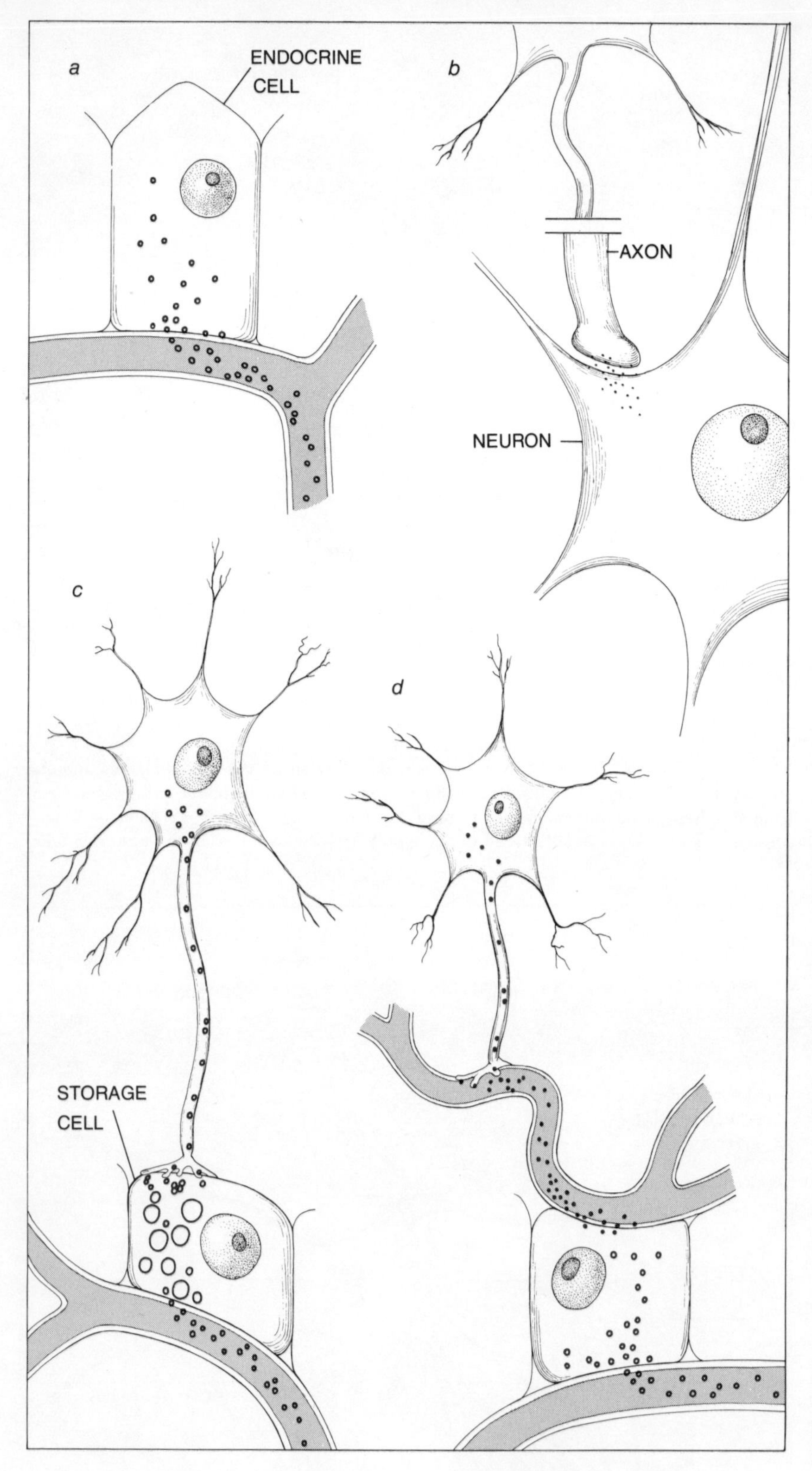

NEUROHUMORAL SECRETIONS involved in hypothalamic-pituitary interactions differ from classical hormone secretion and classical nerve-cell communication. A classical endocrine cell (such as those in the anterior pituitary or the adrenal cortex, for example) secretes its hormonal product directly into the bloodstream (*a*). At a classical synapse, the axon, or fiber, from one nerve cell releases locally a transmitter substance that activates the next cell (*b*). In neurosecretion of oxytocin or vasopressin the hormones are secreted by nerve cells and pass through their axons to storage cells in the posterior pituitary, eventually to be secreted into the bloodstream (*c*). Hypothalamic (releasing factor) hormones go from the neurons that secrete them into local capillaries, which carry them through portal veins to endocrine cells in the anterior lobe, whose secretions they in turn stimulate (*d*).

purified preparations, also active at microgram levels, were injected in experimental animals, for instance to produce ovulation.

It soon became apparent, however, that the isolation and chemical characterization of TRF, LRF and FRF would not be simple. The preparations active in microgram doses were chemically heterogeneous; they showed no clear-cut indication of a major component. It was also realized that each fragment of hypothalamus obtained from the brain of a sheep or another animal contained nearly infinitesimal quantities of the releasing factors. The isolation of enough of each factor to make its chemical characterization possible would therefore require the processing of an enormous number of hypothalamic fragments. Two groups of workers in the U.S. undertook this challenge: a group headed by A. V. Schally at the Tulane University School of Medicine and our own group, first at the Baylor University College of Medicine in Houston and then at the Salk Institute in La Jolla, Calif.

Over a period of four years the Tulane group worked with extracts from perhaps two million pig brains. Our laboratory collected, dissected and processed close to five million hypothalamic fragments from the brains of sheep. Since one sheep brain has a wet weight of about 100 grams, this meant handling 500 tons of brain tissue. From this amount we removed seven tons of hypothalamic tissue (about 1.5 grams per brain). Semi-industrial methods had to be developed in order to handle, extract and purify such large quantities of material. Finally in 1968 one milligram of a preparation of TRF was obtained that appeared to be homogeneous by all available criteria.

On careful measurement the entire milligram could be accounted for by the sole presence of three amino acids: histidine, glutamic acid and proline. Moreover, the three amino acids were present in equal amounts, which suggested that we were dealing with a relatively simple polypeptide perhaps as small as a tripeptide. In the determination of peptide sequences it is customary to subject the sample to attack by proteolytic enzymes, which cleave the peptide bonds holding the polypeptide chain together in well-established ways. Pure TRF, however, was shown to be resistant to all the proteolytic enzymes used. Since we could spare only a tiny amount of our precious one-milligram sample for studies of molecular weight, we could not obtain a

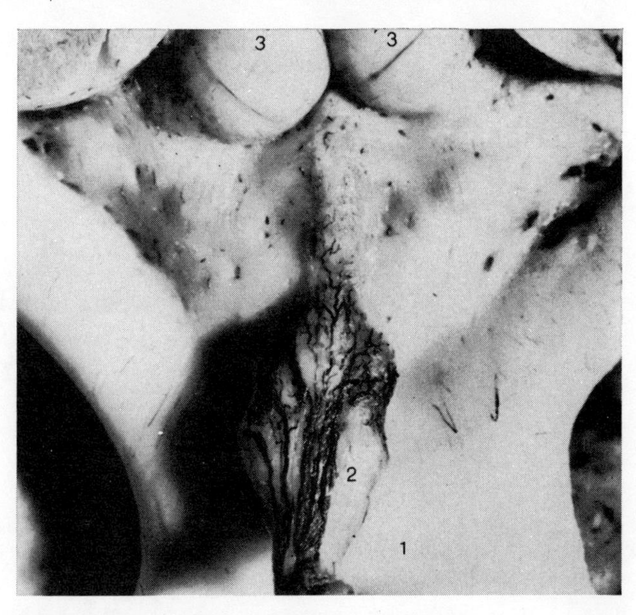

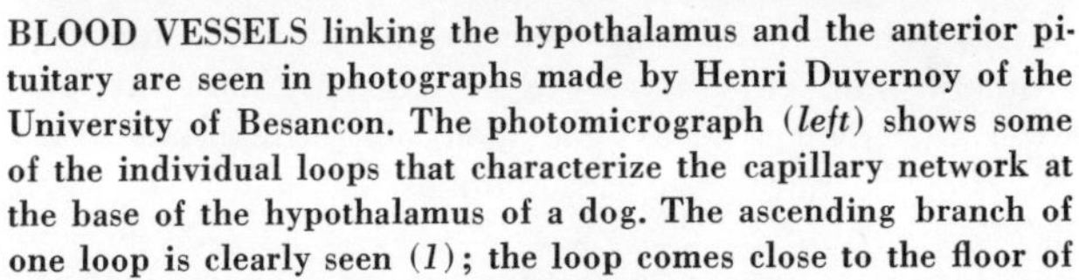

BLOOD VESSELS linking the hypothalamus and the anterior pituitary are seen in photographs made by Henri Duvernoy of the University of Besancon. The photomicrograph (*left*) shows some of the individual loops that characterize the capillary network at the base of the hypothalamus of a dog. The ascending branch of one loop is clearly seen (*1*); the loop comes close to the floor of the third ventricle (*2*) and then descends (*3*), carrying with it the releasing factors that are secreted by this region of the hypothalamus and entering the pars tuberalis of the anterior lobe (*4*). The photograph of the floor of the human hypothalamus (*right*) shows the optic chiasm (*1*), the posterior side of the pituitary stalk with its portal veins (*2*) and the mammillary bodies of the brain (*3*).

precise value for that important measurement. On the basis of inferential evidence, however, it seemed to be reasonable to assume that the molecular weight of TRF could not be more than 1,500.

With small molecules it is often possible to use methods based on the technique of mass spectrometry to obtain in a matter of hours the complete molecular structure of the compound under investigation. Because of the minute quantities of TRF available such efforts on our part were frustrated; the mass-spectrometric methods available to us in 1969 were not sensitive enough to indicate the structure of our unknown substance. Other approaches involve the use of infrared or nuclear magnetic-resonance spectrometry, which can provide direct insight into molecular structure. Here too the techniques then available were inadequate for providing clear-cut information about polypeptide samples that weighed only a few micrograms.

Confronted with nothing but dead ends, we decided on an entirely different approach to finding the structure of TRF. That approach was first to synthesize each of six possible tripeptides composed of the three amino acids known to be present in TRF: histidine (abbreviated His), glutamic acid (Glu) and proline (Pro). The six tripeptides were then assayed for their biological activity. None showed any activity when they were injected at doses of up to a million times the level of the active natural TRF.

Was this another dead end? Not quite. Our synthetic polypeptides all had a

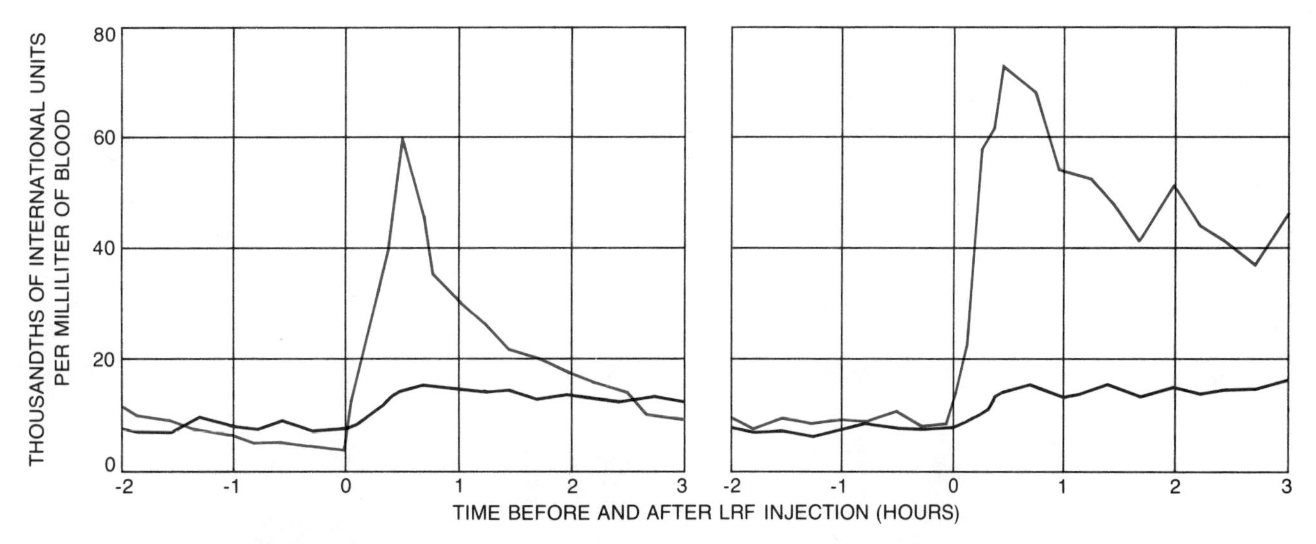

HYPOTHALAMIC HORMONES have clinical implications and applications. For example, women with no pituitary or ovarian defect respond to the administration of synthetic LRF by secreting normal amounts of the hormones LH and FSH. Curves show effect of LRF on secretion of LH (*color*) and FSH (*black*) in a normal woman on the third (*left*) and 11th (*right*) day of menstrual cycle.

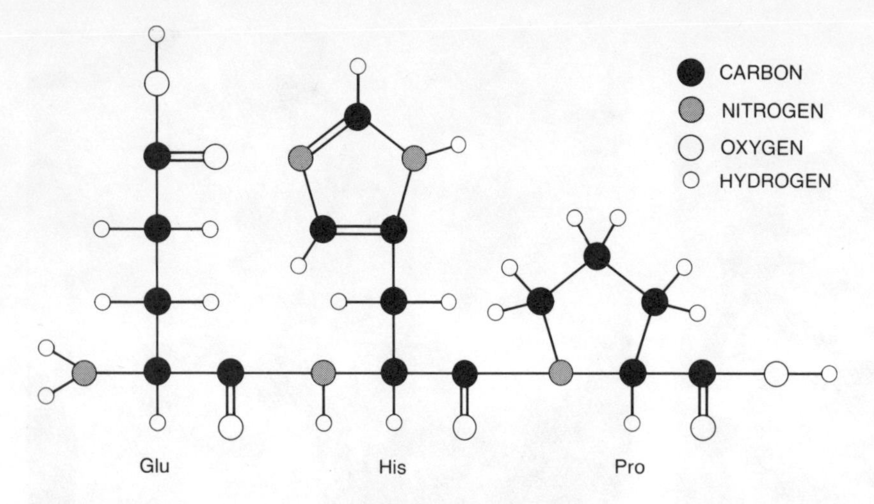

AMINO ACID CONTENT of TRF, the releasing factor for thyroid-stimulating hormone (TSH), was established: glutamic acid, histidine and proline in equal proportions. Each of six possible tripeptides was synthesized; one is diagrammed. None was active biologically.

free amino group (NH_2) at the end of the molecule designated the *N* terminus. We knew that in several well-characterized hormones the *N*-terminus end was not free; it was blocked by a small substitute group of some kind. Indeed, we had evidence from the small quantity of natural TRF that its *N* terminus was also blocked. To block the *N* terminus of our six candidate polypeptides was not difficult: we heated them in the presence of acetic anhydride, which typically couples an acetyl group (CH_3CO) to the *N* terminus. When these "protected" tripeptides were tested, the results were unequivocal. The biological activity of the sequence Glu-His-Pro, and that sequence alone, was qualitatively indistinguishable from the activity of natural TRF. Quantitatively, however, there was still a considerable difference between the synthetic product and natural TRF. Next it was shown that the protective effect of heating Glu-His-Pro with the acetic anhydride had been to convert the glutamic acid at the *N* terminus into a ring-shaped form known as pyroglutamic acid (pGlu).

We now had available gram quantities of the synthetic tripeptide pGlu-His-Pro-OH. (The OH is a hydroxyl group at the end of the molecule opposite the *N* terminus.) Accordingly we could bring into play all the methods that had yielded no information with the microgram quantities of natural TRF. Several of the techniques were modified, particularly with the aim of obtaining mass spectra of the synthetic peptide at levels of only a few micrograms.

Meanwhile, armed with knowledge about the structure of other hormones, we modified the synthetic pGlu-His-Pro-OH to pGlu-His-Pro-NH_2 by replacing the hydroxyl group with an amino group (NH_2) to produce the primary amide [*see top illustration on opposite page*]. This substance proved to have the same biological activity as the natural TRF. At length the complete structure of the natural TRF was obtained by high-resolution mass spectrometry. It turned out to be the structure pGlu-His-Pro-NH_2. The time was late 1969. Thus TRF not only was the first of the hypothalamic hormones to be fully characterized but also was immediately available by synthesis in amounts many millions of times greater than the hormone present in one sheep hypothalamus. TRF from pig brains was subsequently shown to have the same molecular structure as TRF from sheep brains.

Characterization of the hypothalamic releasing factor LRF, which controls the secretion of the gonadotropin LH, followed rapidly. Isolated from the side fractions of the programs for the isolation of TRF, LRF was shown in 1971 to be a polypeptide composed of 10 amino acids. Six of the amino acids are not found in TRF: tryptophan (Trp), serine (Ser), tyrosine (Tyr), glycine (Gly), leucine (Leu) and arginine (Arg). The full sequence of LRF is pGlu-His-Trp-Ser-Tyr-Gly-Leu-Arg-Pro-Gly-NH_2 [*see bottom illustration on these two pages*]. Although this structure is more compli-

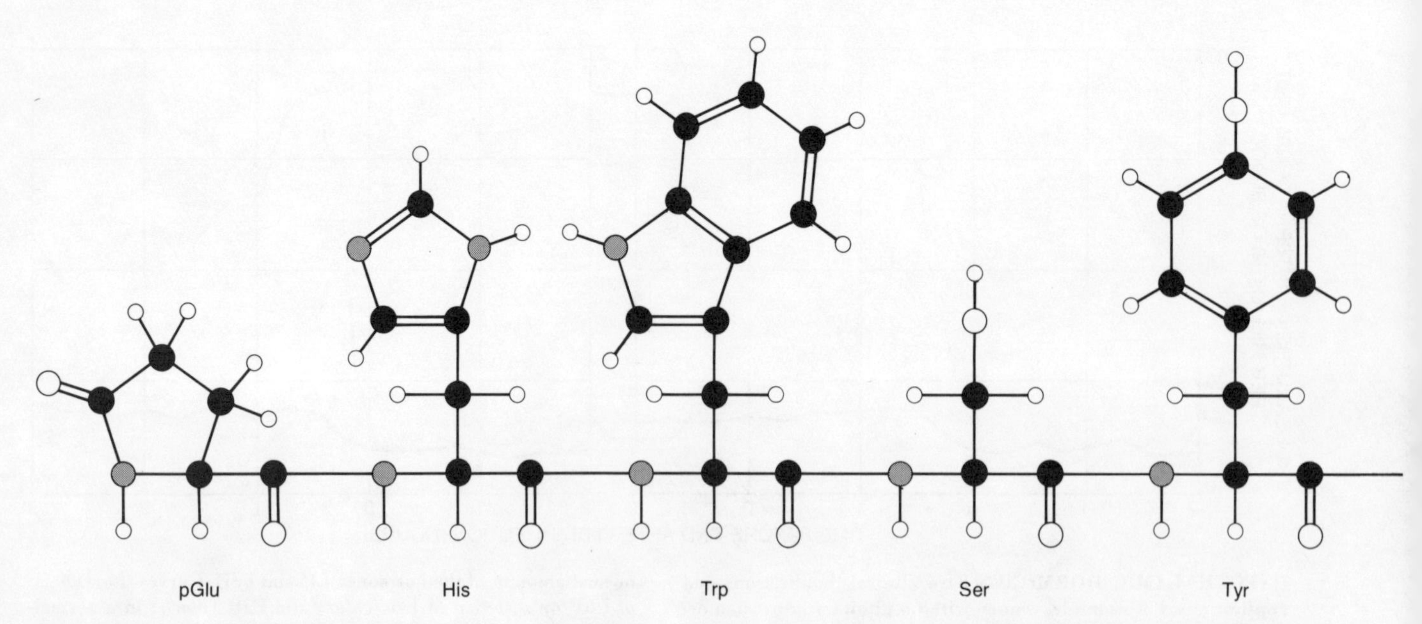

LRF, the releasing factor for the luteinizing hormone (LH), which affects the activity of the gonads, was characterized and synthesized soon after. First the hormone was isolated and its amino acid content was determined. Then their intramolecular sequence was es-

cated than the structure of TRF, it begins with the same two amino acids (pGlu-His) and has the same group at the other terminus (NH_2).

It turns out that LRF also stimulates the secretion of the other gonadotropin, FSH, although not as powerfully as it stimulates the secretion of LH. It has been proposed that LRF may be the sole hypothalamic controller of the secretion of the two gonadotropins: LH and FSH.

There is good physiological evidence that the hypothalamus is also involved in the control of the secretion of the other two important pituitary hormones: prolactin and growth hormone. Curiously, prolactin is as plentiful in males as in females, but its role in male physiology is still a mystery. The hypothalamic mechanism involved in the control of the secretion of prolactin or growth hormone is not fully understood. It is quite possible that the secretion of these two pituitary hormones is controlled not by releasing factors alone but perhaps jointly by releasing factors and specific hypothalamic hormones that somehow act as inhibitors of the secretion of prolactin or growth hormone. If it should turn out that inhibitory hormones rather than stimulative ones are involved in the regulation of prolactin and growth hormone, one should not be too surprised. The brain provides many examples of inhibitory and stimulative systems working in parallel.

The hypothalamic hormones TRF and LRF are both now available by synthesis in unlimited quantities. Both are highly active in stimulating pituitary functions in humans. TRF is already a powerful tool for exploring pituitary functions in several diseases characterized by the abnormality of one or several of the pituitary secretions. There is increasing evidence that most patients with such abnormalities (primarily children) actually have normally functioning glands, since they respond promptly to the administration of synthetic hypothalamic hormones. Evidently their abnormalities are due to hypothalamic rather than pituitary deficiencies. These deficiencies can now be successfully treated by the administration of the hypothalamic polypeptide TRF.

Similarly, an increasing number of women who have no ovulatory menstrual cycle and who show no pituitary or ovarian defect begin to secrete normal amounts of the gonadotropins LH and FSH after the administration of LRF. The administration of synthetic LRF should therefore be the method of choice for the treatment of those cases of infertility where the functional defect resides in the hypothalamus-pituitary system. Indeed, ovulation can be induced in women by the administration of synthetic LRF. On the other hand, knowledge of the structure of the LRF molecule may open up an entirely novel approach to fertility control. Synthetic compounds closely related to LRF in structure may act as inhibitors of the native LRF. Two such analogues of LRF, made by modifying the histidine in the hormone, have been reported as antagonists of LRF. It is therefore possible that LRF antagonists will be used as contraceptives.

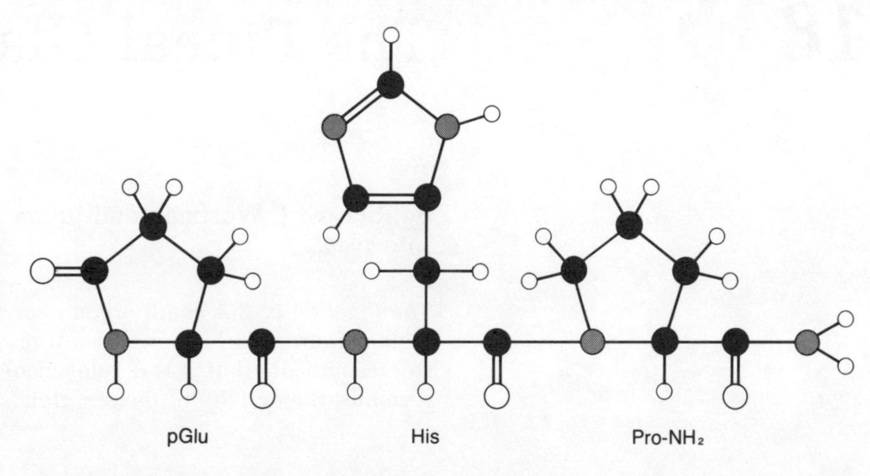

TRIPEPTIDES were modified in an effort to characterize the releasing factor. When the sequence glutamic acid–histidine-proline was modified by forming the glutamic acid into a ring and converting the proline end (*right*) to an amide, it was found to be TRF.

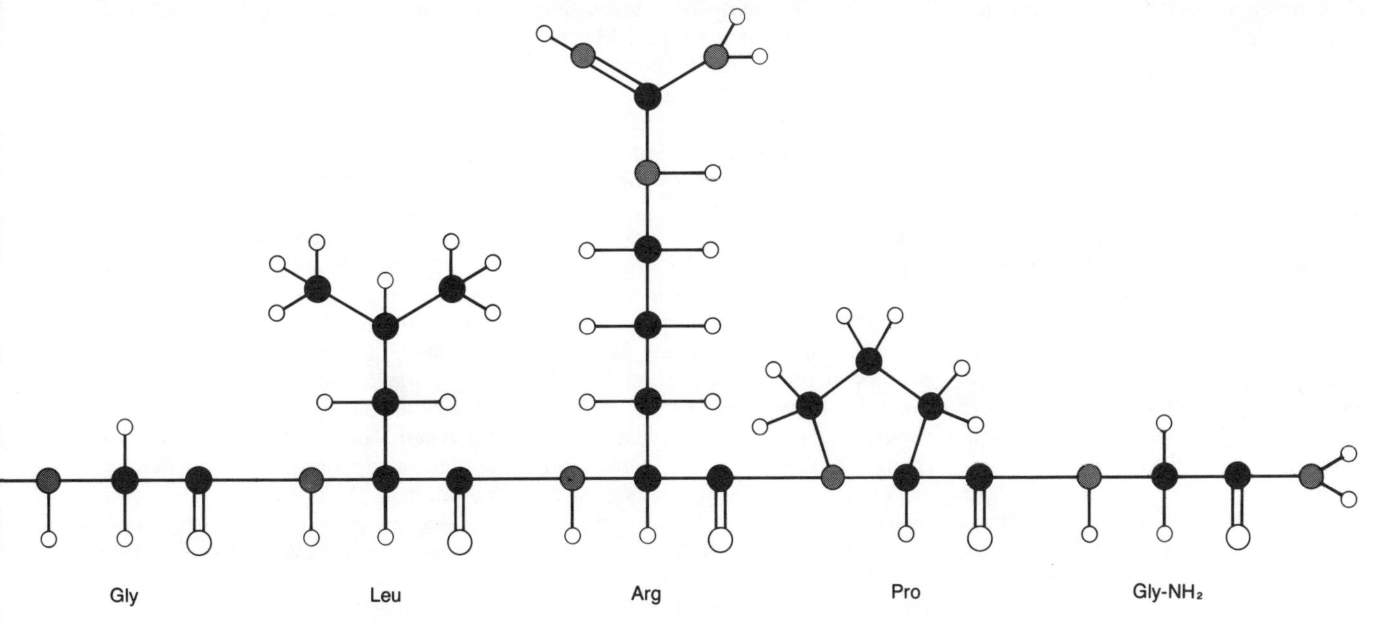

tablished and reproduced by synthesis; the synthetic replicate shown here was found to have full biological activity. In addition to stimulating LH activity, LRF also stimulates the secretion of another gonadotropic hormone, FSH, although not so powerfully.

18 The Pineal Gland

by Richard J. Wurtman and Julius Axelrod
July 1965

The function of this small organ near the center of the mammalian brain has long been a mystery. Recent studies indicate that it is a "biological clock" that regulates the activity of the sex glands

Buried nearly in the center of the brain in any mammal is a small white structure, shaped somewhat like a pinecone, called the pineal body. In man this organ is roughly a quarter of an inch long and weighs about a tenth of a gram. The function of the pineal body has never been clearly understood. Now that the role of the thymus gland in establishing the body's immunological defenses has been demonstrated, the pineal has become perhaps the last great mystery in the physiology of mammalian organs. This mystery may be nearing a solution: studies conducted within the past few years indicate that the pineal is an intricate and sensitive "biological clock," converting cyclic nervous activity generated by light in the environment into endocrine—that is, hormonal—information. It is not yet certain what physiological processes depend on the pineal clock for cues, but the evidence at hand suggests that the pineal participates in some way in the regulation of the gonads, or sex glands.

A Fourth Neuroendocrine Transducer

Until quite recently most investigators thought that the mammalian pineal was simply a vestige of a primitive light-sensing organ: the "third eye" found in certain cold-blooded vertebrates such as the frog. Other workers, noting the precocious sexual development of some young boys with pineal tumors, had proposed that in mammals the pineal was a gland. When the standard endocrine tests were applied to determine the possible glandular function of the pineal, however, the results varied so much from experiment to experiment that few positive conclusions seemed justified. Removal of the pineal in young female rats was frequently followed by an enlargement of the ovaries, but the microscopic appearance of the ovaries did not change consistently, and replacement of the extirpated pineal by transplantation seemed to have little or no physiological effect. Most experimental animals could survive the loss of the pineal body with no major change in appearance or function.

In retrospect much of the difficulty early workers had in exploring and defining the glandular function of the pineal arose from limitations in the traditional concept of an endocrine organ. Glands were once thought to be entirely dependent on substances in the bloodstream both for their own control and for their effects on the rest of the body: glands secreted hormones into the blood and were themselves regulated by other hormones, which were delivered to them by the circulation. The secretory activity of a gland was thought to be maintained at a fairly constant level by homeostatic mechanisms: as the level of a particular hormone in the bloodstream rose, the gland invariably responded by decreasing its secretion of that hormone; when the level of the hormone fell, the gland increased its secretion.

In the past two decades this concept of how the endocrine system works has proved inadequate to explain several kinds of glandular response, including changes in hormone secretion brought about by changes in the external environment and also regular cyclic changes in the secretion of certain hormones (for example, the hormones responsible for the menstrual cycle and the steroid hormones that are produced on a daily cycle by the adrenal gland). Out of the realization that these and other endocrine responses must depend in some way on interactions between the glands and the nervous system the new discipline of neuroendocrinology has developed.

In recent years much attention has centered on the problem of locating the nervous structures that participate in the control of glandular function. It has been known for some time that special types of organs would be needed to "transduce" neural information into endocrine information. Nervous tissue is specialized to receive and transmit information directly from cell to cell; according to the traditional view, glands are controlled by substances in the bloodstream and dispatch their messages to target organs by the secretion of hormones into the bloodstream. In order to transmit information from the nervous system to an endocrine organ a hypothetical "neuroendocrine transducer" would require some of the special characteristics of both neural and endocrine tissue. It should respond to substances (called neurohumors) released locally from nerve endings, and it should contain the biochemical machinery necessary for synthesizing a hormone and releasing it into the bloodstream. Three such neurosecretory systems have so far been identified. They are (1) the hypothalamus–posterior-pituitary system, which secretes the antidiuretic hormone and oxytocin, a hormone that causes the uterus to contract during labor; (2) the pituitary-releasing-factor system, also located in the hypothalamus, which secretes polypeptides that control the function of the pituitary gland, and (3) the adrenal medulla, whose cells respond to a nervous input by releasing adrenaline into the bloodstream.

The advent of neuroendocrinology has provided a conceptual framework

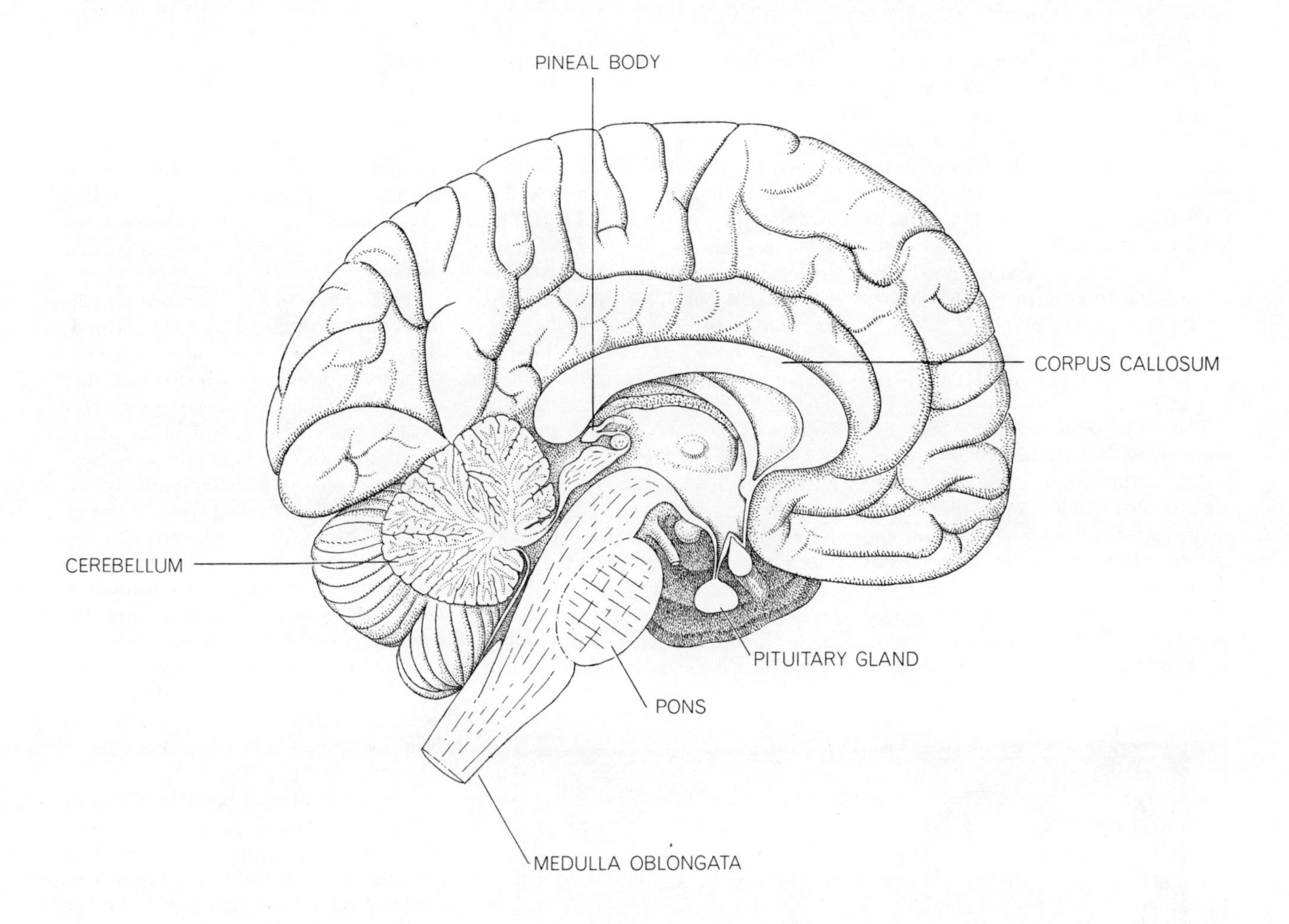

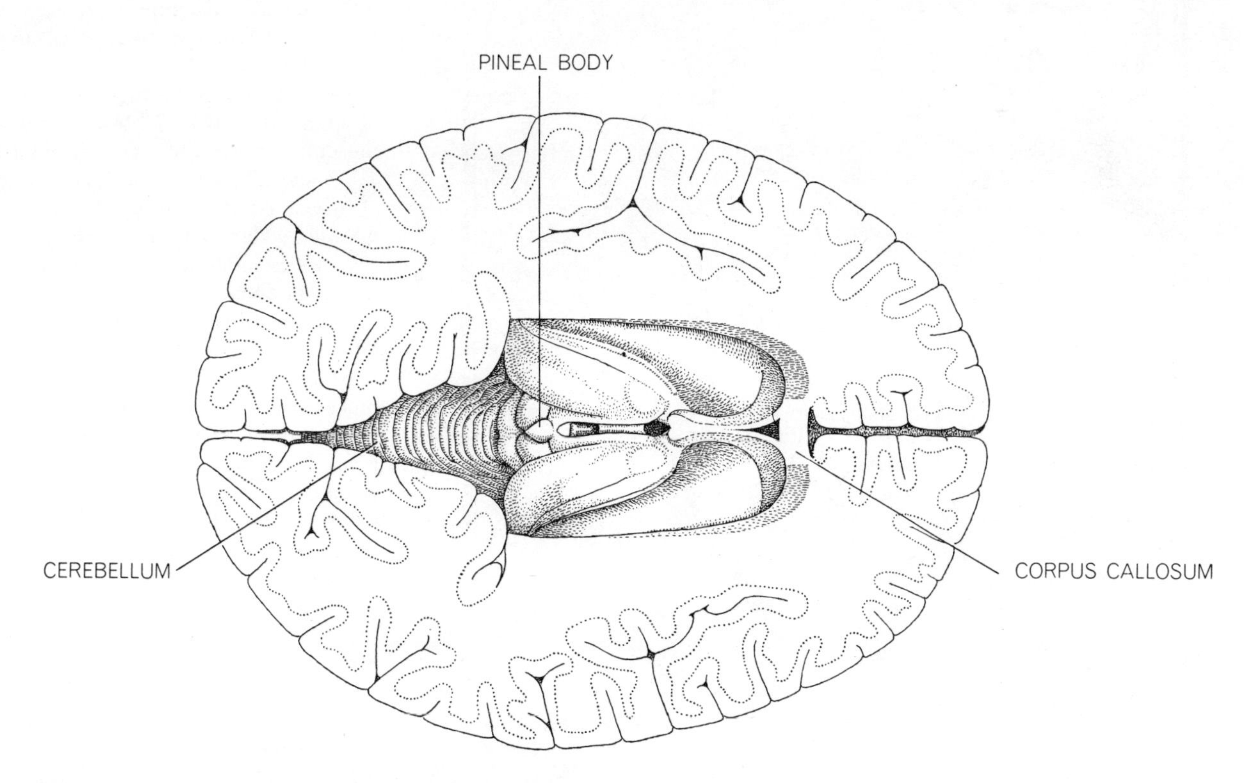

TWO VIEWS of the human brain reveal the central position of the pineal body. Section at top is cut in the median sagittal plane and is viewed from the side. Section at bottom is cut in a horizontal plane and is viewed from above; an additional excision has been made in this view to reveal the region immediately surrounding the pineal. In mammals the pineal is the only unpaired midline organ in the brain. The name "pineal" comes from the organ's resemblance to a pinecone, the Latin equivalent of which is *pinea*.

that has been most helpful in characterizing the role of the pineal gland. On the basis of recent studies conducted by the authors and their colleagues at the National Institute of Mental Health, as well as by investigators at other institutions, it now appears that the pineal is not a gland in the traditional sense but is a fourth neuroendocrine transducer; it is a gland that converts a nervous input into a hormonal output.

A Prophetic Formulation

The existence of the pineal body has been known for at least 2,000 years. Galen, writing in the second century A.D., quoted studies of earlier Greek anatomists who were impressed with the fact that the pineal was perched atop the aqueduct of the cerebrum and was a single structure rather than a paired one; he concluded that it served as a valve to regulate the flow of thought out of its "storage bin" in the lateral ventricles of the brain. In the 17th century René Descartes embellished this notion; he believed that the pineal housed the seat of the rational soul. In his formulation the eyes perceived the events of the real world and transmitted what they saw to the pineal by way of "strings" in the brain [*see illustration below*]. The pineal responded by allowing humors to pass down hollow tubes to the muscles, where they produced the appropriate responses. With the hindsight of 300 years of scientific development, we can admire this prophetic formulation of the pineal as a neuroendocrine transducer!

In the late 19th and early 20th centuries the pineal fell from its exalted metaphysical state. In 1898 Otto Heubner, a German physician, published a case report of a young boy who had shown precocious puberty and was also found to have a pineal tumor. In the course of the next 50 years many other children with pineal tumors and precocious sexual development were described, as well as a smaller number of patients whose pineal tumors were associated with delayed sexual development. Inexplicably almost all the cases of precocious puberty were observed in boys.

In a review of the literature on pineal tumors published in 1954 Julian I. Kitay, then a fellow in endocrinology at the Harvard Medical School, found that most of the tumors associated with precocious puberty were not really pineal in origin but either were tumors of supporting tissues or were teratomas (primitive tumors containing many types of cells). The tumors associated with delayed puberty, however, were in most cases true pineal tumors. He concluded that the cases of precocious puberty resulted from reduced pineal function due to disease of the surrounding tissue, whereas delayed sexual development in children with true pineal tumors was a consequence of increased pineal activity.

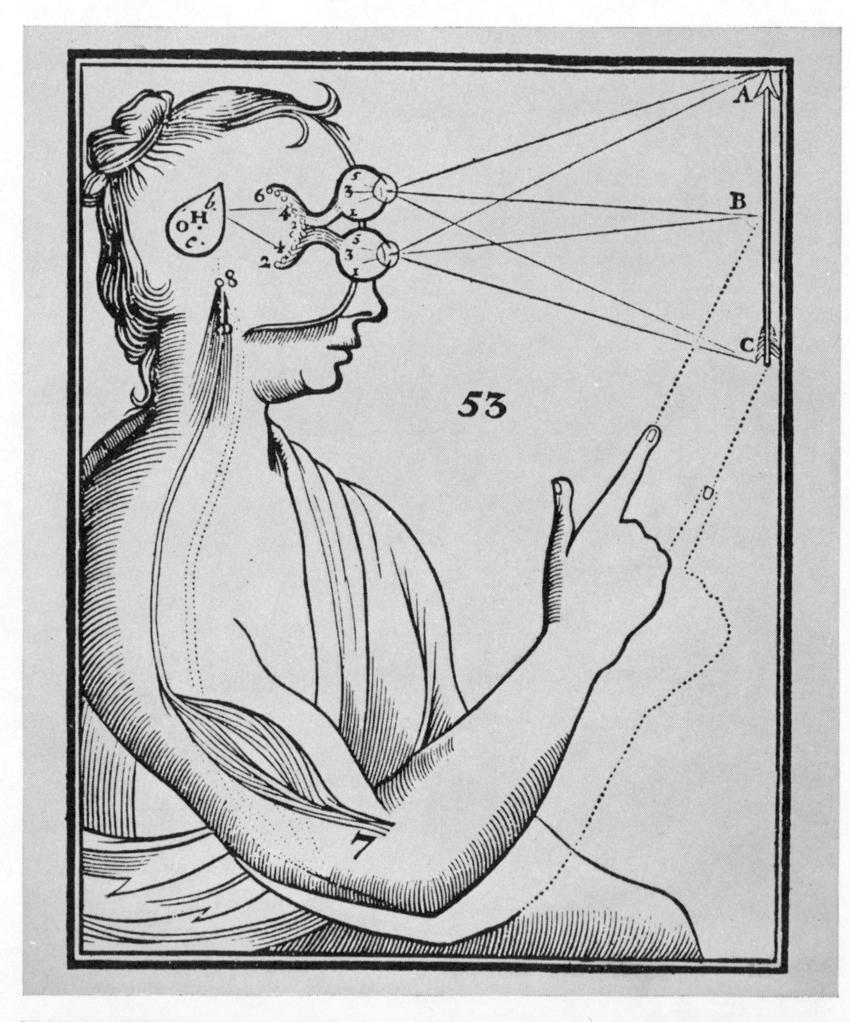

SEAT OF THE RATIONAL SOUL was the function assigned to the human pineal (*H*) by René Descartes in his mechanistic theory of perception. According to Descartes, the eyes perceived the events of the real world and transmitted what they saw to the pineal by way of "strings" in the brain. The pineal responded by allowing animal humors to pass down hollow tubes to the muscles, where they produced the appropriate responses. The size of the pineal has been exaggerated in this wood engraving, which first appeared in 1677.

The association of pineal tumors and sexual malfunction gave rise to hundreds of research projects designed to test the hypothesis that the pineal was a gland whose function was to inhibit the gonads. Little appears to have resulted from these early efforts. Later in 1954 Kitay and Mark D. Altschule, director of internal medicine at McLean Hospital in Waverly, Mass., reviewed the entire world literature on the pineal: some 1,800 references, about half of which dealt with the pineal-gonad question. They concluded that of all the studies published only two or three had used enough experimental animals and adequate controls for their data to be analyzed statistically. These few papers suggested a relation between the pineal and the gonads but did little to characterize it. After puberty the human pineal is hardened by calcification; this change in the appearance of the pineal led many investigators to assume that the organ was without function and further served to discourage research in the field. (Actually calcification appears to be unrelated to the pineal functions we have measured.)

As long ago as 1918 Nils Holmgren, a Swedish anatomist, had examined the pineal region of the frog and the dogfish with a light microscope. He was surprised to find that the pineal contained distinct sensory cells; they bore a marked resemblance to the cone cells of the retina and were in contact with nerve cells. On the basis of these observations he suggested that the pineal might function as a photoreceptor, or "third eye," in cold-blooded vertebrates. In the past five years this hypothesis has finally been confirmed by electrophysiological studies: Eberhardt Dodt and his colleagues in Germany have shown that the frog pineal is a wavelength discriminator: it converts light energy of certain wavelengths into nervous impulses. In 1927 Carey P. McCord and Floyd P. Allen, working at Johns Hopkins University, observed that if they made extracts of cattle pineals and added them to the media in which tadpoles were swimming, the tadpoles' skin blanched, that is, became lighter in color.

Such was the state of knowledge about the pineal as late as five or six years ago. It appeared to be a photoreceptor in the frog, had something to do with sexual function in rats and in humans (at least those with pineal tumors) and contained a factor (at least in cattle) that blanched pigment cells in tadpoles.

The Discovery of Melatonin

Then in 1958 Aaron B. Lerner and his co-workers at the Yale University School of Medicine identified a unique compound, melatonin, in the pineal gland of cattle [see "Hormones and Skin Color," by Aaron B. Lerner; Scientific American, July, 1961]. During the next four years at least half a dozen other major discoveries were made about the pineal by investigators representing many different disciplines and institutions. Lerner, a dermatologist and biochemist, was interested in identifying the substance in cattle pineal extracts that blanched frog skin. He and his colleagues prepared and purified extracts from more than 200,000 cattle pineals and tested the ability of the extracts to alter the reflectivity of light by pieces of excised frog skin. After four years of effort they succeeded in isolating and identifying the blanching agent and found that it was a new kind of biological compound: a methoxylated indole, whose biological activity requires a methyl group (CH_3) attached to an oxygen atom [*see illustration on next two pages*].

Methoxylation had been noted previously in mammalian tissue, but the products of this reaction had always appeared to lose their biological activity as a result. The new compound, named melatonin for its effect on cells containing the pigment melanin, appeared to lighten the amphibian skin by causing

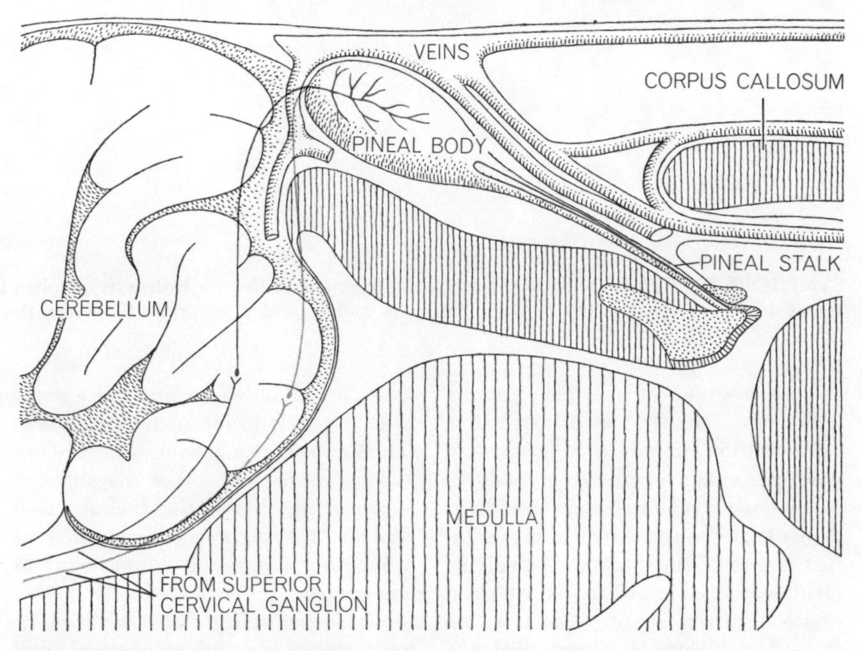

INNERVATION OF RAT PINEAL was the subject of a meticulous study by the Dutch neuroanatomist Johannes Ariëns Kappers in 1961. He demonstrated that the pineal of the adult rat is extensively innervated by nerves from the sympathetic nervous system. The sympathetic nerves to the pineal originate in the neck in the superior cervical ganglion, enter the skull along the blood vessels and eventually penetrate the pineal at its blunt end (*top*). Aberrant neurons from the central nervous system sometimes run up the pineal stalk from its base, but these generally turn and run back down the stalk again without synapsing. The pineal is surrounded by a network of great veins, into which its secretions probably pass. According to Ariëns Kappers, the innervation of the human pineal is quite similar.

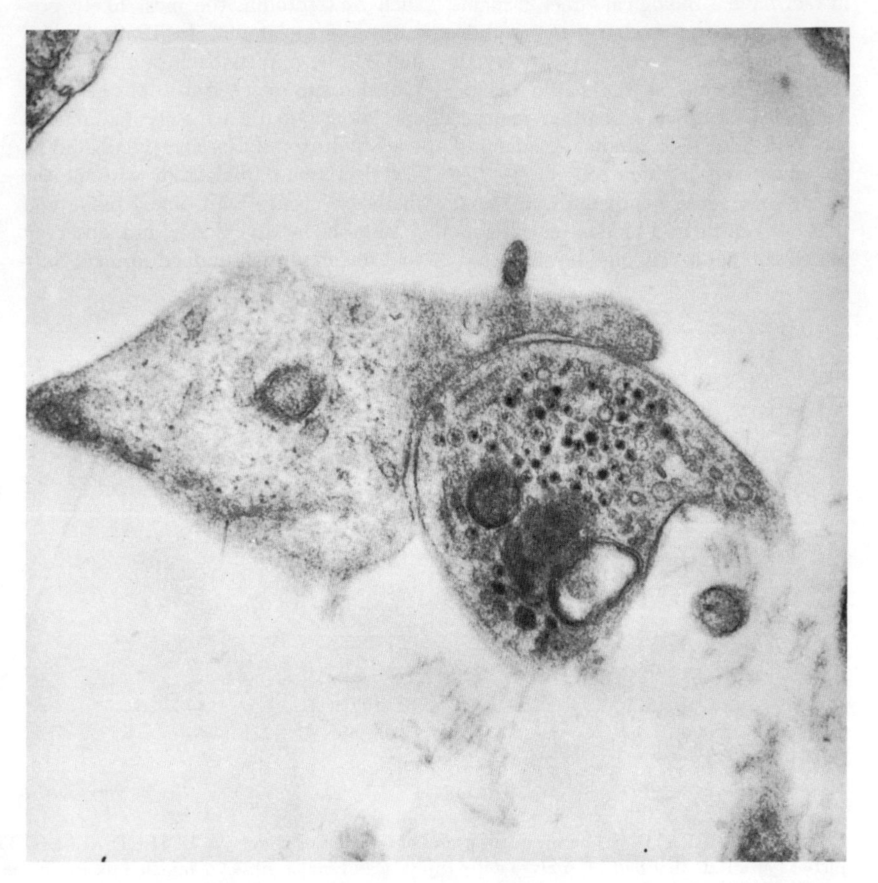

SYMPATHETIC NERVE terminates directly on a pineal cell, instead of on a blood vessel or smooth muscle cell, in this electron micrograph of a portion of a rat pineal made by David Wolfe of the Harvard Medical School. The nerve ending is characterized by dark vesicles, or sacs, that contain neurohumors. Magnification is about 12,500 diameters.

5-HYDROXYTRYPTOPHAN —DECARBOXYLATING ENZYME→ SEROTONIN —ACETYLATING ENZYME→

SYNTHESIS OF MELATONIN in the rat pineal begins with the removal of a carboxyl (COOH) group from the amino acid 5-hydroxytryptophan by the enzyme 5-hydroxytryptophan decarboxylase. Serotonin, the product of this reaction, is then enzymatically

the aggregation of melanin granules within the cells. It was effective in a concentration of only a trillionth of a gram per cubic centimeter of medium. No influence of melatonin could be demonstrated on mammalian pigmentation, nor could the substance actually be identified in amphibians, in which it exerted such a striking effect. It remained a biological enigma that the mammalian pineal should produce a substance that appeared to have no biological activity in mammals but was a potent skin-lightening agent in amphibians, which were unable to produce it!

Both aspects of the foregoing enigma have now been resolved. Subsequent research has shown that melatonin does in fact have a biological effect in mammals and can be produced by amphibians. Spurred by Lerner's discovery of this new indole in the cattle pineal, Nicholas J. Giarman, a pharmacologist at the Yale School of Medicine, analyzed pineal extracts for their content of other biologically active compounds. He found that both cattle and human pineals contained comparatively high levels of serotonin, an amine whose molecular structure is similar to melatonin and whose function in nervous tissue is largely unknown. Studies by other investigators subsequently showed that the rat pineal contains the highest concentration of serotonin yet recorded in any tissue of any species.

A year before the discovery of melatonin one of the authors (Axelrod) and his co-workers had identified a methoxylating enzyme (catechol-O-methyl transferase) in a number of tissues. This enzyme acted on a variety of catechols (compounds with two adjacent hydroxyl, or OH, groups on a benzene ring) but showed essentially no activity with respect to single-hydroxyl compounds such as serotonin, the most likely precursor of melatonin. In 1959 Axelrod and Herbert Weissbach studied cattle pineal tissue to see if it might have the special enzymatic capacity to methoxylate hydroxyindoles. They incubated N-acetylserotonin (melatonin without the methoxyl group) with pineal tissue and a suitable methyl donor and observed that melatonin was indeed formed. Subsequently they found that all mammalian pineals shared this biochemical property but that no tissue other than pineal could make melatonin. Extensive studies of a variety of mammalian species have confirmed this original observation that only the pineal appears to have the ability to synthesize melatonin. (In amphibians and some birds small amounts of melatonin are also manufactured by the brain and the eye.) Other investigators have found that the pineal contains all the biochemical machinery needed to make melatonin from an amino acid precursor, 5-hydroxytryptophan, which it obtains from the bloodstream. It was also found that circulating melatonin is rapidly metabolized in the liver to form 6-hydroxymelatonin.

Anatomy of the Pineal

While these investigations of the biochemical properties of the pineal were in progress, important advances were being made in the anatomy of the pineal by the Dutch neuroanatomist Johannes Ariëns Kappers and by several

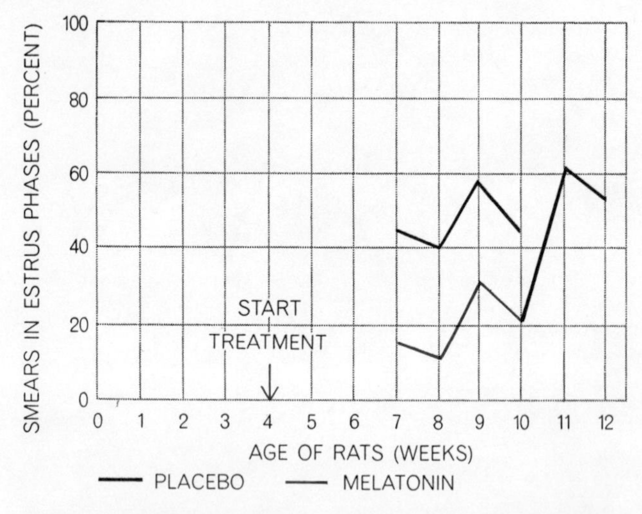

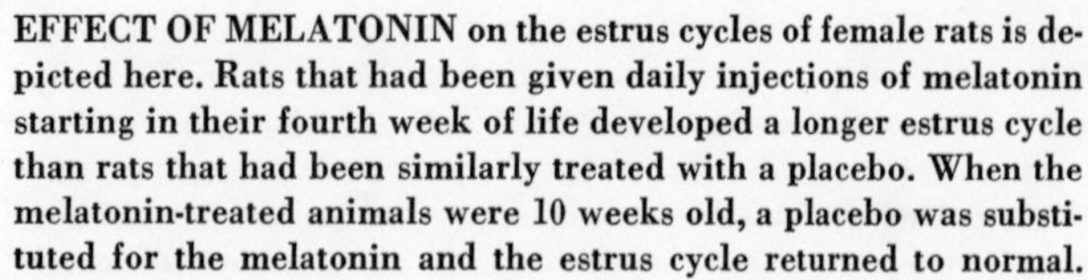

EFFECT OF MELATONIN on the estrus cycles of female rats is depicted here. Rats that had been given daily injections of melatonin starting in their fourth week of life developed a longer estrus cycle than rats that had been similarly treated with a placebo. When the melatonin-treated animals were 10 weeks old, a placebo was substituted for the melatonin and the estrus cycle returned to normal.

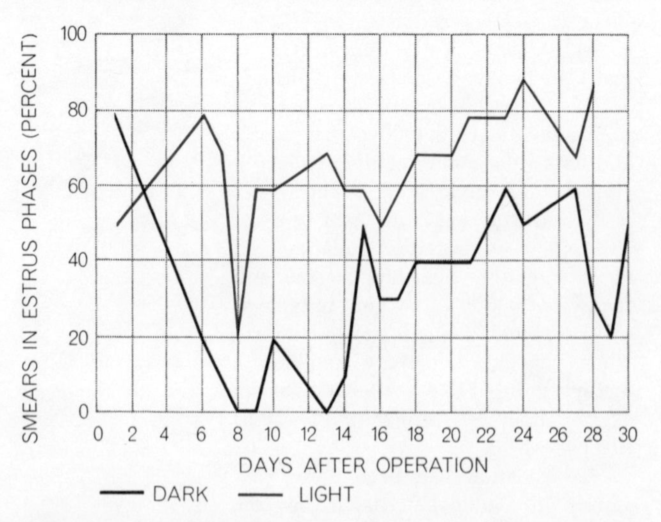

EFFECTS OF LIGHTING on the estrus cycles of three groups of female rats are shown in the graphs on these two pages. The groups, each consisting of about 20 rats, were subjected respectively to a sham operation (*left*), removal of their superior cervical ganglion (*middle*) and removal of their eyes (*right*). Each group was then further subdivided, with about half being placed in constant light

N-ACETYLSEROTONIN → METHOXYLATING ENZYME → MELATONIN

acetylated to form N-acetylserotonin. This compound in turn is methoxylated by the enzyme hydroxyindole-O-methyl transferase (HIOMT) to yield melatonin. In mammals HIOMT is found only in the pineal. Changes in basic molecule are indicated by color.

American electron microscopists, including Douglas E. Kelly of the University of Washington, Aaron Milofsky of the Yale School of Medicine and David Wolfe, then at the National Institute of Neurologic Diseases and Blindness. In 1961 Ariëns Kappers published a meticulous study of the nerve connections in the rat pineal. He demonstrated clearly that although this organ originates in the brain in the development of the embryo, it loses all nerve connections with the brain soon after birth. There is thus no anatomical basis for invoking "tracts from the brain" as the pathway by which neural information is delivered to the pineal.

Ariëns Kappers showed that instead the pineal of the adult rat is extensively penetrated by nerves from the sympathetic portion of the autonomic nervous system. The sympathetic nervous system is involuntary and is concerned with adapting to rapid changes in the internal and external environments; the sympathetic nerves to the pineal originate in the superior cervical ganglion in the neck, enter the skull along the blood vessels and eventually penetrate the pineal [*see top illustration on page 197*]. Electron microscope studies later showed that within the pineal many sympathetic nerve endings actually terminate directly on the pineal cells, instead of on blood vessels or smooth-muscle cells, as in most other organs [*see bottom illustration on page 197*]. Among endocrine structures the organization of nerves in the mammalian pineal appeared to be most analogous to that of the adrenal medulla, one of the three demonstrated neuroendocrine transducers.

Meanwhile electron microscope studies by other workers on the pineal regions of frogs had confirmed many of Holmgren's speculations. It was found that the pineal cells of amphibians contained light-sensitive elements that were practically indistinguishable from those found in the cone cells of the retina, but that the pineal cells of mammals did not contain such elements. By 1962 it could be stated with some assurance that the mammalian pineal was not simply a vestige of the frog "third eye," since the "vestige" had undergone profound anatomical changes with evolution.

The Melatonin Hypothesis

Even though the mammalian pineal no longer seemed to respond directly to light, there now appeared good evidence that its function continued to be related somehow to environmental light. In 1961 Virginia Fiske, working at Wellesley College, reported that the exposure of rats to continuous environmental illumination for several weeks brought about a decrease in the weight of their pineals. She had been interested in studying the mechanisms by which the exposure of rats to light for long periods induces changes in the function of their gonads. (For example, continous light increased the weight of the ovaries and accelerated the estrus cycle). At the same time one of the authors (Wurtman, then at the Harvard Medical School), in collaboration with Altschule and Willard Roth, was studying the conditions under which the administration

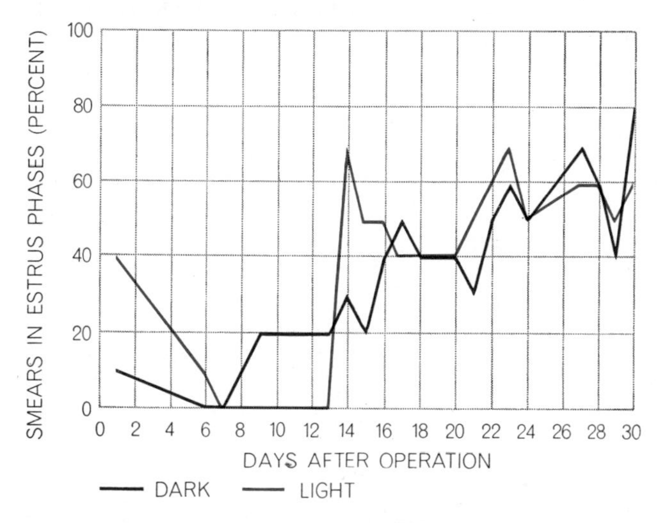

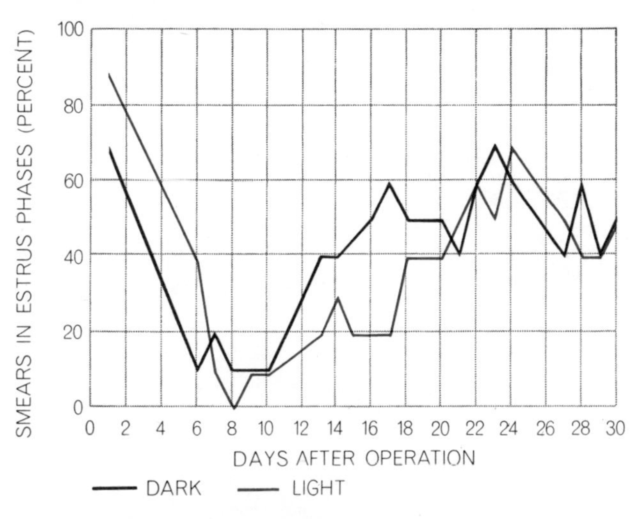

and the other half in constant darkness beginning one day after their respective operations. Daily vaginal smears were taken on the first day and on the sixth through the 30th days after the operations. Results were plotted as the percentage of all the smears in a treatment group showing estrus phases each day. In general it was found that interference with the transmission of light information to the pineal gland (either by blinding or by cutting the sympathetic nerves) also abolished most of the gonadal response to light. These findings supported the authors' melatonin hypothesis, which holds that one mechanism whereby light is able to accelerate the estrus cycle in normal animals is by inhibiting the synthesis in the pineal of melatonin, a compound that in turn inhibits estrus.

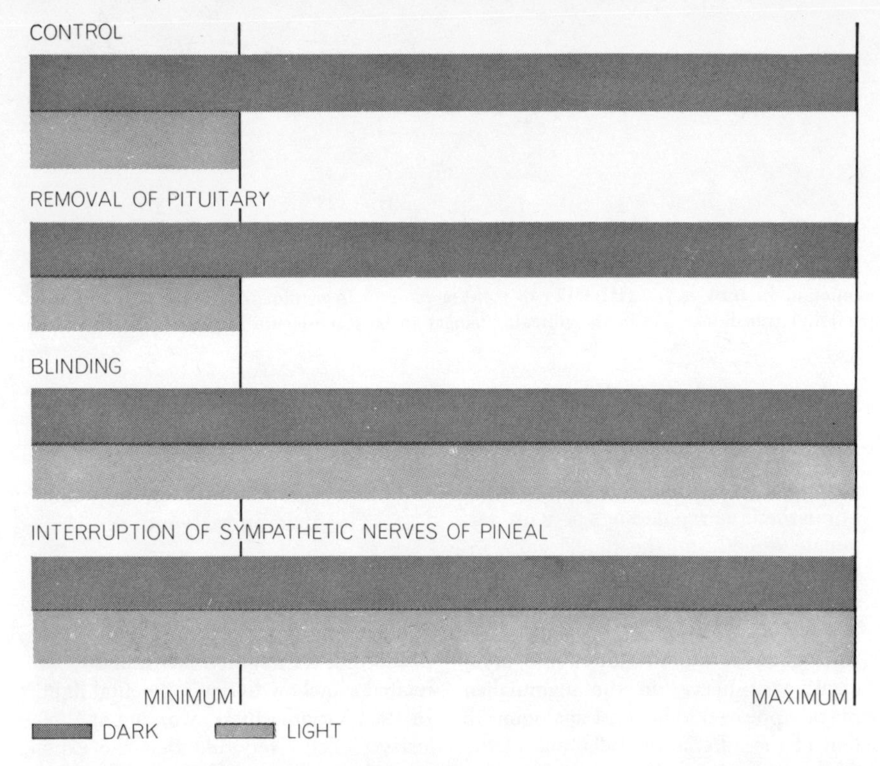

RESPONSE OF MELATONIN-FORMING ENZYME hydroxyindole-O-methyl transferase (HIOMT) to continuous light or darkness is shown under four different circumstances. In the control, or normal, animal continuous darkness induces an increase in HIOMT activity, whereas exposing the animal to continuous light has the opposite effect. The ability of the pineal gland to respond to environmental lighting is unaffected by the removal of the pituitary gland but is abolished following blinding or sympathetic denervation of the pineal.

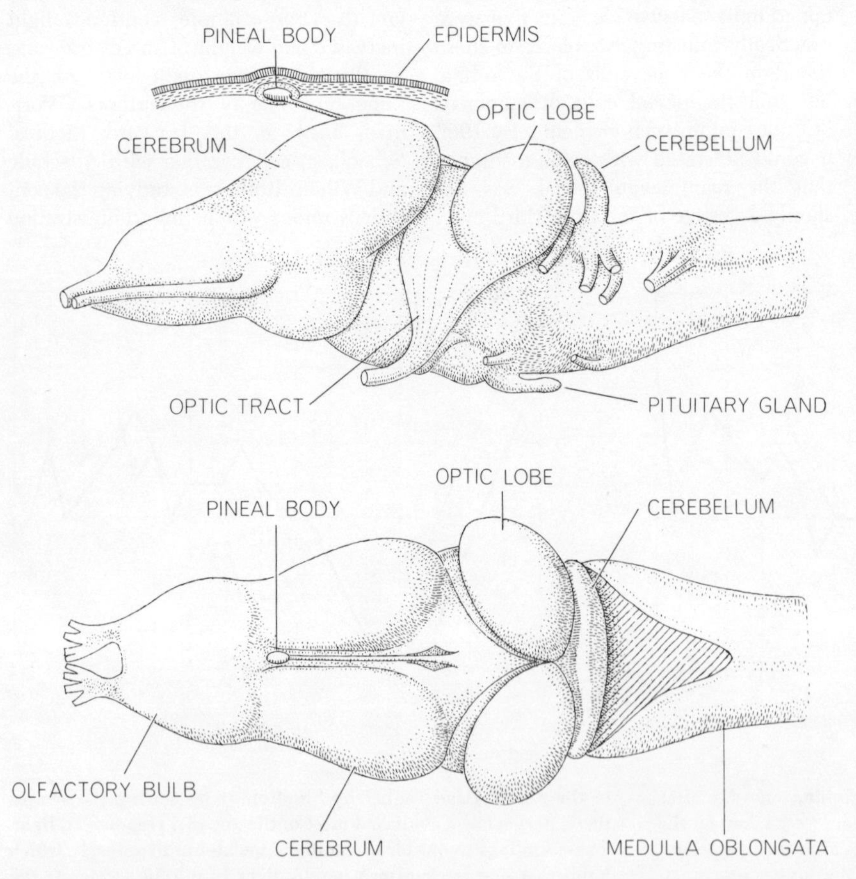

PINEAL EYE is a primitive photoreceptive organ found in certain cold-blooded vertebrates such as the frog. Frog's brain is shown from the side (*top*) and from above (*bottom*).

of cattle pineal extracts decreased ovary weight and slowed the estrus cycle.

We soon confirmed Mrs. Fiske's findings, and we were also able to show that the exposure of female rats to continuous light or the removal of their pineals had similar, but not additive, effects on the weight of their ovaries. These experiments suggested that perhaps one way in which light stimulates ovary function in rats is by inhibiting the action of an inhibitor found in pineal extracts. It now became crucial to identify the gonad-inhibiting substance in pineal extracts and to see if its synthesis or its actions were modified by environmental lighting.

In 1962 we began to work together on isolating the anti-gonadal substance present in pineal extracts. Our plan was to subject extracts of cattle pineal glands to successive purification steps and test the purified material for its ability to block the induction by light of an accelerated estrus cycle in the rat. Before undertaking the complicated and time-consuming procedure of isolating the active substance in the pineal glands of cattle, we first tested a mixture of all the constituents that had already been identified in this tissue. The mixture was found to block the effects of light on the estrus cycle.

Next we tested melatonin alone, since it was apparently the only compound produced uniquely by the pineal. To our good fortune we found that when rats were given tiny doses (one to 10 micrograms per day) of melatonin by injection, starting before puberty and continuing for a month thereafter, the estrus cycle was slowed and the ovaries lost weight—just as though the animals had been treated with pineal extracts. In later studies we found that this effect of melatonin was chemically specific: it was simulated by neither N-acetylserotonin, the immediate precursor of melatonin, nor 6-hydroxymelatonin, the major product of its metabolism. Moreover, it was possible to accelerate the estrus cycle by removal of the pineal and to block this response by the injection of melatonin.

On the basis of these studies, performed in collaboration with Elizabeth Chu of the National Cancer Institute, we postulated that melatonin was a mammalian hormone, since it is produced uniquely by a single gland (the pineal), is secreted into the bloodstream and has an effect on a distant target organ (the vagina and possibly also the ovaries). We were not able to identify the precise site of action of melatonin in affecting the gonads. The slowing of the estrus cycle could be produced by actions at any of several sites in the neuroendocrine apparatus, including the brain, the pituitary, the ovaries or the vagina itself. When melatonin was labeled with radioactive atoms and injected into cats, it was taken up by all these organs and was selectively concentrated by the ovaries.

William M. McIsaac and his colleagues at the Cleveland Clinic have confirmed the effects of melatonin on the estrus cycle and have identified another pineal methoxyindole—methoxytryptophol—that has similar effects. It appears likely that pineal extracts contain a family of hormones: the methoxyindoles, all of which have in common the fact that they can be synthesized by the methoxylating enzyme found only in the mammalian pineal.

We next set out to determine whether or not these effects of injected melatonin were physiological. Could the rat pineal synthesize melatonin and, if so, in what quantities? When rat pineal glands were examined for their ability to make melatonin, we were disappointed to find that the activity of the melatonin-forming enzyme (hydroxyindole-O-methyl transferase, or HIOMT) in the rat was much lower than in most other species; the maximum amount of melatonin that the rat could make was probably on the order of one microgram per day. Our disappointment was soon relieved, however, when we realized that the low activity of this enzyme made it likely that it was controlling the rate-limiting step in melatonin synthesis in the intact animal. Knowing that continuous exposure to light decreased pineal weight, as well as the amount of ribonucleic acid (RNA) and protein in the pineal, we next explored what effect illumination might have on HIOMT activity and thus on melatonin synthesis.

Since the rat pineal gland was so small (about a milligram in weight) and had so little enzymatic activity, it was necessary to devise extremely sensitive techniques to measure this activity. When rats were subjected to constant light for as short a period as a day or two, the rate of melatonin synthesis in their pineals fell to as little as a fifth that of animals kept in continuous darkness. Since this effect of illumination or its absence could be blocked by agents that interfered with protein synthesis, it appeared that light was actually influencing the rate of formation of the enzyme protein itself.

How was information about the state of lighting being transmitted to the rat pineal? Three possible routes suggested

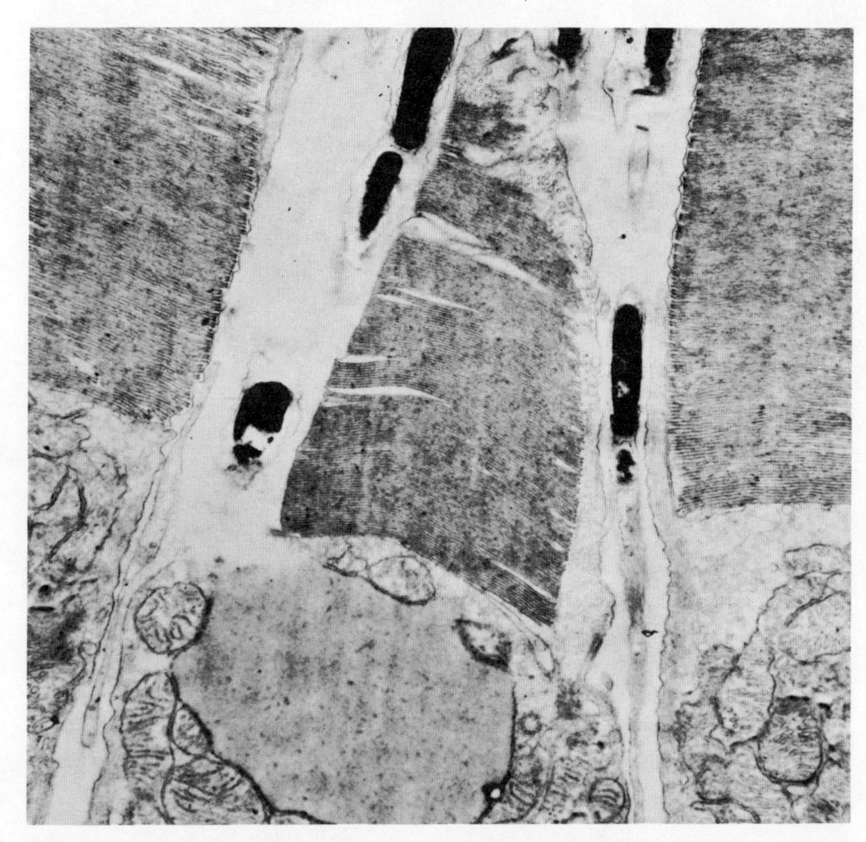

RETINAL CONE CELL from the eye of an adult frog is shown in this electron micrograph made by Douglas E. Kelly of the University of Washington. The photoreceptive outer segment of the cell (*top center*) consists of a densely lamellated membrane. Parts of two larger rod photoreceptors can be seen on each side of cone. Magnification is about 13,000 diameters.

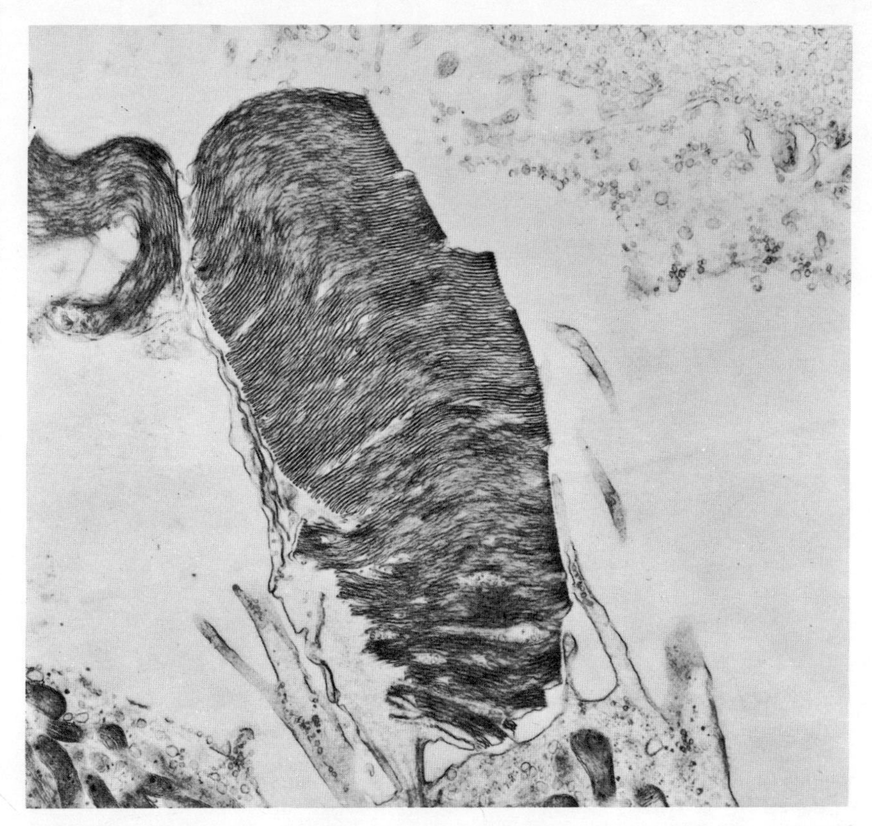

PINEAL CONE CELL from the pineal eye of an adult frog is shown in this electron micrograph made by Kelly at approximately the same magnification as the micrograph at top. The lamellated outer segment of the pineal cell is practically indistinguishable from that of the retinal photoreceptor. Part of the membrane has torn away from the cell (*top left*).

themselves. The first was that light penetrated the skull and acted directly on the pineal; W. F. Ganong and his colleagues at the University of California at Berkeley had already shown that significant quantities of light do penetrate the skulls of mammals. This hypothesis was ruled out, however, by demonstrating that blinded rats completely lost the capacity to respond to light with changed HIOMT activity; hence light had to be perceived first by the retina and was not acting directly on the pineal.

The second possibility was that light altered the level of a circulating hormone, perhaps by affecting the pituitary gland, and that this hormone secondarily influenced enzyme activity in the pineal gland. This hypothesis was also ruled out by demonstrating that the removal of various endocrine organs, including the pituitary and the ovaries, did not interfere with the response of pineal HIOMT to light.

The third possibility was that information about lighting was transmitted to the pineal by nerves. Fortunately Ariëns Kappers had just identified the nerve connections of the rat pineal as coming from the sympathetic nervous system. We found that if the sympathetic pathway to the pineal was interrupted by the removal of the superior cervical ganglion, the ability of melatonin-forming activity to be altered by light was completely lost. Thus it appeared that light was stimulating the retina and then information about this light was being transmitted to the pineal via sympathetic nerves. Within the pineal the sympathetic nerves probably released neurohumors (noradrenaline or serotonin), which acted on pineal cells to induce (or block the induction of) HIOMT; this enzyme in turn regulated the synthesis of melatonin.

Since one way light influences the gonads is by changing the amount of melatonin secreted from the pineal, we reasoned that the effects of light on the gonads might be blocked if the transmission of information about light to the pineal were interrupted. This could be accomplished by cutting the sympathetic nerves to the pineal—a procedure much less traumatic than the removal of the pineal itself. To test this hypothesis we placed groups of rats whose pineals had been denervated along with blinded and untreated animals in continuous light or darkness for a month. Vaginal smears were checked daily for evidence of changes in the estrus cycle, and pineals were tested for melatonin-synthesizing ability at the end of the experiment. It was found that interrupting the transmission of light information to the pineal (by cutting its sympathetic nerves—a procedure that does not interfere with the visual response to light) also abolished most of the gonadal response to light.

Incidentally, the observation that sympathetic nerves control enzyme synthesis in the pineal has provided, and should continue to provide, a useful tool for studies in a number of other biological disciplines. For example, studying the changes in brain enzymes produced by environmental factors offers a useful method for tracing the anatomy

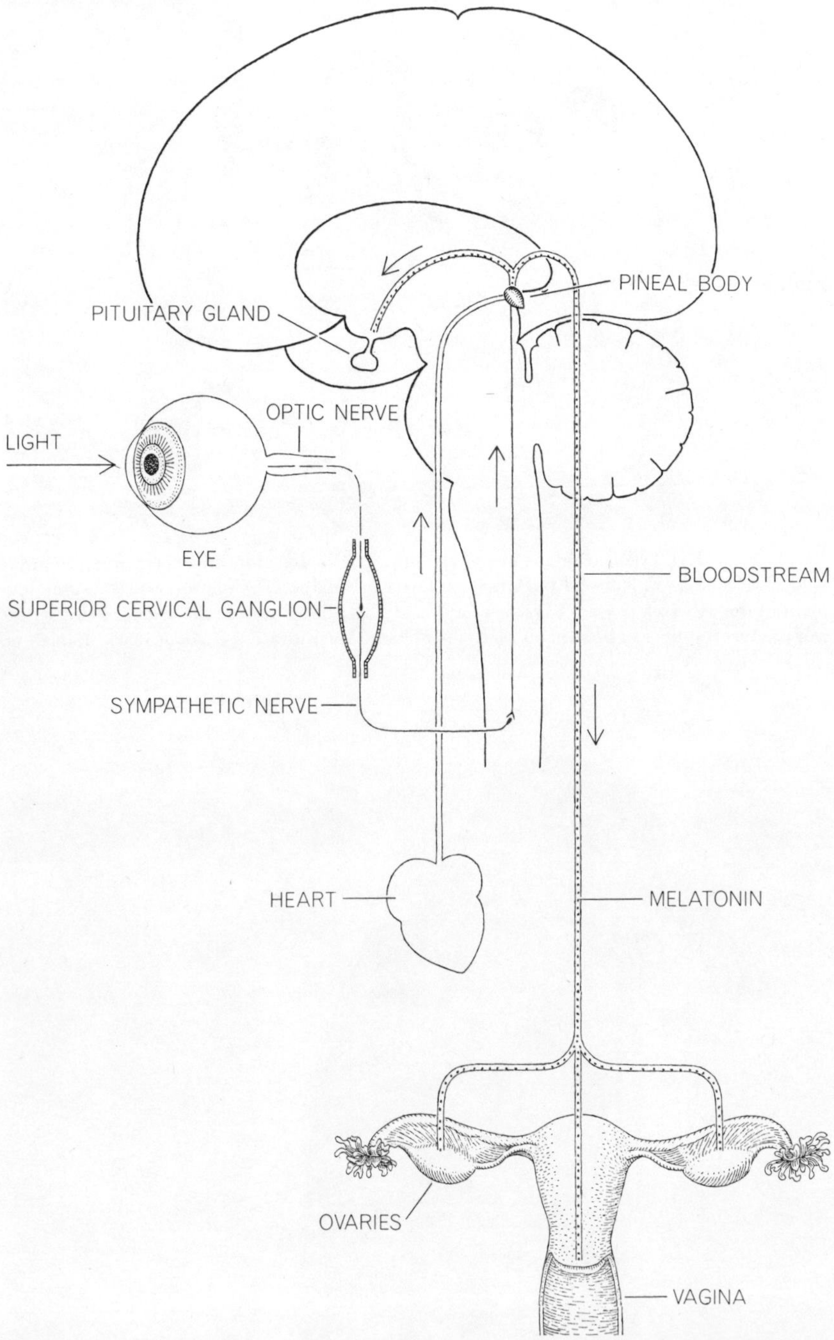

PINEAL BODY
PITUITARY GLAND
OPTIC NERVE
LIGHT
EYE
SUPERIOR CERVICAL GANGLION
SYMPATHETIC NERVE
BLOODSTREAM
HEART
MELATONIN
OVARIES
VAGINA

SUGGESTED PATHWAY by which light influences the estrus cycle in the rat is depicted in this schematic diagram. Light stimuli impinge on the retinas and cause a change in the neural output of the superior cervical ganglion by way of an unknown route. This information is then carried by sympathetic nerves to the pineal gland, where it causes a decrease in the activity of HIOMT and in the synthesis and release of melatonin. This decrease in turn lessens the inhibiting effect of the circulating melatonin on the rate of the estrus cycle. The precise site of action of melatonin in influencing the gonads is unknown; the slowing of the estrus cycle could be produced by actions at any one of several sites in the neuroendocrine apparatus, including the brain, the pituitary, the ovaries and the vagina.

of the nerve tracts involved. The observation that the activity of at least one part of the sympathetic nervous system (the superior cervical ganglion) is affected by environmental lighting raises the possibility that other regions of this neural apparatus are affected similarly. If so, physiological studies of the effects of light on other sympathetically innervated structures (for example the kidneys and fat tissue) may be profitable.

We have also found that light influences the serotonin-forming enzyme in the pineal gland but not in other organs. In contrast to HIOMT, the activity of this enzyme increases when rats are kept in constant light and decreases in darkness. When rats are blinded or when the sympathetic nerves to the pineal are cut, the effect of light and darkness on the serotonin-forming enzyme is also extinguished. Furthermore, certain drugs that block the transmission of sympathetic nervous impulses also abolish the effect of illumination on this enzyme. The fact that lighting influences pineal weight and at least two enzyme systems in this organ suggests that it may regulate many additional, undiscovered biochemical events in the pineal, via the sympathetic nervous system.

Diurnal and Circadian Rhythms

The pineal had been shown to respond and function under quite unusual conditions; for example, when an experimental animal was exposed to continuous light or darkness for several days. In nature, of course, animals that live in the temperate and tropical zones are rarely subjected to such conditions. It became important to determine if the pineal could also respond to naturally occurring changes in the environment.

In nature the level of light exposure changes with both diurnal and annual cycles. Except in polar regions every 24-hour day includes a period of sunlight and a period of darkness; the ratio of day to night varies with an annual rhythm that reaches its nadir at the winter solstice and its zenith on the first day of summer. Lighting cycles have been shown to be important in regulating several types of endocrine function: the increase in sunlight during the winter and spring triggers the annual gonadal growth and breeding cycles in many birds and some mammals that breed yearly, and the daily rhythm of day and night synchronizes a variety of roughly daily rhythms in mammals, such as the cycle of adrenal-steroid secretion. Such rhythms are called circadian, from the Latin phrase meaning "about one day." Could the pineal respond to natural diurnal lighting shifts? If so, it might function to synchronize the endocrine apparatus with these shifts.

In order to determine if normal lighting rhythms influenced the pineal, we kept a large population of rats under controlled lighting conditions (lights on from 7:00 A.M. to 7:00 P.M.) for several weeks and then tested their pineals for melatonin-forming ability at 6:00 A.M., noon, 6:00 P.M. and midnight. In the five hours after the onset of darkness (that is, by midnight) this enzymatic capacity increased between two and three times. Moreover, pineal weight also changed significantly during this period, again indicating that light was affecting many more compounds in the pineal than the single enzyme we were measuring.

All circadian rhythms studied up to this stage had in common the ability to persist for some weeks after animals were deprived of environmental lighting cues (by blinding or being placed in darkness). These rhythms no longer showed a period of precisely 24 hours, but they did fall in a range between 22 and 26 hours and hence were thought to be regulated by some internal mechanism not dependent on, but usually synchronized with, environmental lighting. Such endogenous, or internally regulated, circadian rhythms in rodents include motor activity and rectal temperature, as well as the rhythm in adrenal-steroid secretion. When we blinded rats or placed them in continuous light or darkness, the pineal rhythm in melatonin-forming activity was rapidly extinguished. If instead of turning off the lights at 7:00 P.M. illumination was

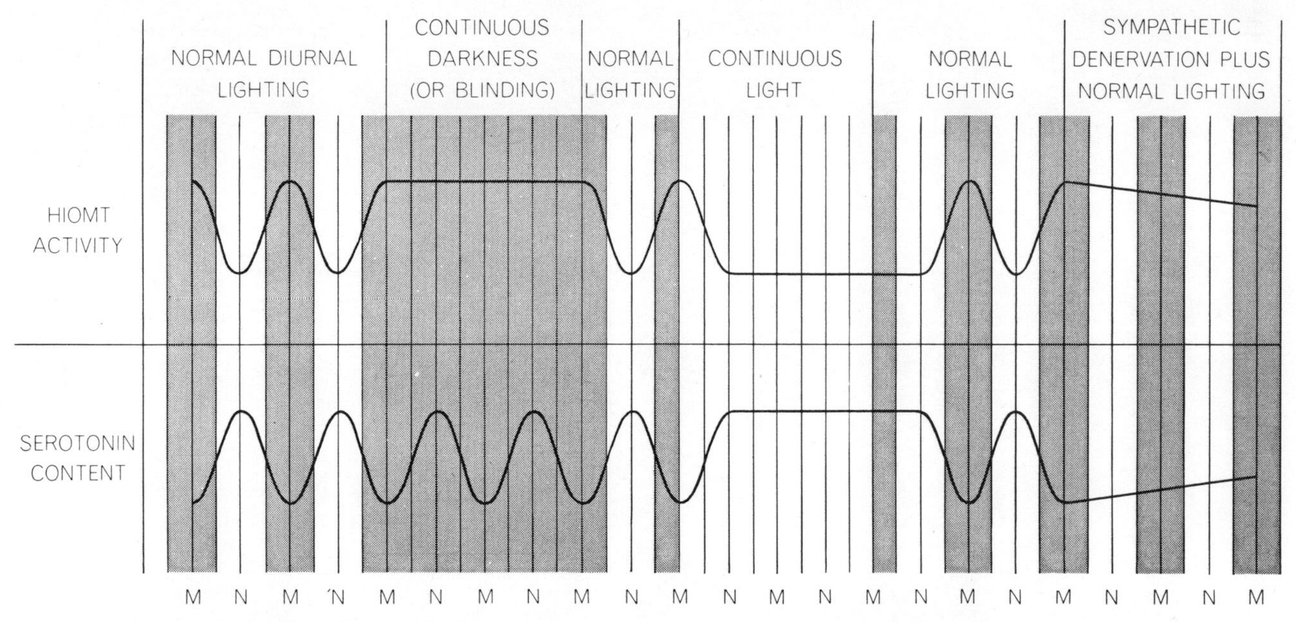

BIOCHEMICAL RHYTHMS in the pineal gland of the rat were recorded under various lighting and other conditions. Normally both the content of serotonin and the activity of the melatonin-forming enzyme (HIOMT) vary with a 24-hour cycle. The serotonin content is greatest at noon (*N*), whereas the HIOMT activity is greatest at midnight (*M*). The HIOMT cycle is completely dependent on environmental lighting conditions: it disappears when animals are kept in continuous light or darkness, or when they are blinded. The serotonin cycle persists in continuous darkness or after blinding but can be abolished by keeping the rats in continuous light. Both cycles are depressed when the sympathetic nerves to the pineal gland are cut (*extreme right*). Gray areas signify darkness.

continued for an additional five hours and pineals were examined as usual at midnight, the expected rise in melatonin-forming activity was completely blocked. This pineal rhythm in HIOMT activity thus appears to be truly exogenous, or externally regulated, and is entirely dependent on shifts in environmental lighting. Hence this enzyme rhythm may be more important in carrying information about light to the glands than other circadian rhythms that do not depend on light for their existence.

Recently Wilbur Quay of the University of California at Berkeley has found that the content of serotonin in the rat pineal also undergoes marked circadian rhythms. The highest levels of this amine are found in pineals at noon and the lowest levels at midnight. Serotonin content falls rapidly just at the time that melatonin-forming activity is rising. In collaboration with Solomon Snyder we studied the mechanism of the serotonin cycle. When rats are kept in continuous light, the serotonin cycle is extinguished. To our surprise, however, when rats are kept continuously in darkness or blinded, this rhythm persists, unlike the rhythm in the melatonin-forming ezyme. When the sympathetic nerves to the pineal are cut, the serotonin and HIOMT cycles are both suppressed. When the nerves from the central nervous system to the superior cervical ganglion are interrupted, the serotonin rhythm is also abolished [*see illustration on previous page*]. Hence the serotonin rhythm in the pineal gland is similar to most other circadian rhythms (and differs from the HIOMT cycle) in that it is endogenous and depends on environmental light only as an external synchronizer. The mechanism that controls the serotonin rhythm appears to reside within the central nervous system. The pineal gland thus contains at least two distinct biological clocks, one totally dependent on environmental lighting and the other originating within the brain but cued by changes in lighting.

At present little is known about what organs are dependent on the pineal clock for cues. The ability of melatonin to modify gonadal function suggests, but does not prove, that its secretion may have something to do with the timing of the estrus and menstrual cycles—two phenomena about whose mechanisms of control very little is known. One is tempted to argue teleologically that any control mechanism as complicated and sensitive as that found in the mammalian pineal gland must have some place in the economy of the body.

Calcitonin 19

by Howard Rasmussen and Maurice M. Pechet
October 1970

This recently discovered thyroid hormone plays an important role in metabolism: it inhibits the breakdown of bone and thus keeps the calcium in the blood from reaching an excessively high level

The endocrinology of mammals has been investigated so intensively and with such a wealth of discoveries over three-quarters of a century that one might suppose the subject by this time would have lost some of its novelty and excitement. In actuality, however, the story of the endocrine glands and their hormones retains an unending fascination. New mammalian hormones are currently being discovered at a high rate, as in the early rush of exploration during the first quarter of this century. The new hormone we shall discuss in this article turns out to be of great importance in the regulation of a crucial phase of human metabolism. Ironically, this hormone escaped attention for nearly 80 years although it was more or less constantly on stage throughout that time. It is a product of the thyroid gland, which over the years has been studied more intensively than any other endocrine organ. Discovered less than a decade ago, the long overlooked hormone, calcitonin, has evoked such interest among physiologists, biochemists and physicians that it has already been isolated in pure form, synthesized completely in the laboratory, investigated in detail as to its functions and used therapeutically in human disease.

The discovery of calcitonin grew out of a puzzling question regarding the regulation of the calcium level in the blood. A constant supply of calcium (and phosphate as well) is required in the circulating blood for the building of bone and the control of certain functions of cells. If the concentration of calcium ions in the blood plasma falls below normal, the nerve and muscle cells, for instance, begin to discharge spontaneously and the voluntary muscles go into continuous contraction—the condition known as tetany. It was known that two agents, vitamin D and a hormone of the parathyroid gland, were active in maintaining the plasma's calcium supply: vitamin D by assisting the uptake of calcium from food by the intestinal cells, the parathyroid hormone by causing the release of calcium to the blood from bone and by inducing the kidney tubules to capture, for return to the circulating blood, calcium that would otherwise be lost in the urine.

Experiments had established that the activity of the parathyroid was governed by a feedback system. When the calcium in the plasma dropped below the normal level, the gland increased its secretion of the hormone; when the calcium rose above normal, secretion of the hormone stopped. The question arose: Was this strictly negative control—the shutoff of the parathyroid hormone—sufficient to protect the animal against a dangerous rise in the blood's calcium concentration? Peter Sanderson and his associates at the Peter Bent Brigham Hospital in Boston and D. Harold Copp at the University of British Columbia began to look into the possibility that there was also some mechanism that exerted a positive control on the accumulation of calcium in the blood. They soon found evidence that such a mechanism was indeed at work.

Sanderson's group performed an elegant series of experiments involving both the thyroid and the parathyroid glands. They first tested the reactions of dogs, with the glands intact, to abrupt experimental changes in the plasma calcium level; they raised the level by injecting calcium salts or lowered it by injecting a chelating agent that bound and inactivated the calcium ions in the plasma. They found that in either case the animals' plasma calcium quickly returned to the normal level after the infusions were stopped. They then removed the thyroid and parathyroid glands surgically from the same animals and retested them with the injection treatments. This time there was no rapid return to normal; the calcium content of the plasma remained elevated or subpar for as long as 36 hours after the injection.

The failure of calcium to rise rapidly to normal in the cases where the level had been depressed could be explained by the absence of the stimulating parathyroid hormone (the source of the hormone having been removed). Obviously, however, lack of the hormone could not account for the fact that the calcium level remained high in the cases where the level had been elevated. It was clear that some other corrective agent must be missing in the animals with the thyroid and parathyroid glands removed—some positive agent that could cut back the delivery of calcium to the blood or speed up its removal. Copp, on the basis of experiments similar to Sanderson's, concluded that the agent must be a hormone that he named calcitonin, signifying that it participated in regulating the tone, or concentration, of calcium in the blood. Two groups of investigators, Philip F. Hirsch and Paul L. Munson at the Harvard Dental School and Iain MacIntyre and his colleagues at Hammersmith Hospital in London, soon located the source of the hormone. They found it not in the parathyroid, as Copp had suggested, but in the thyroid gland.

A. G. E. Pearse of the Postgraduate Medical School of London went on to discover the specific site where calcitonin is synthesized. It had been known for many years that the thyroid gland contains two types of cell: the type that produces the gland's classic hormone, thyroxine, and another type called *C* cells, which stain differently. Although

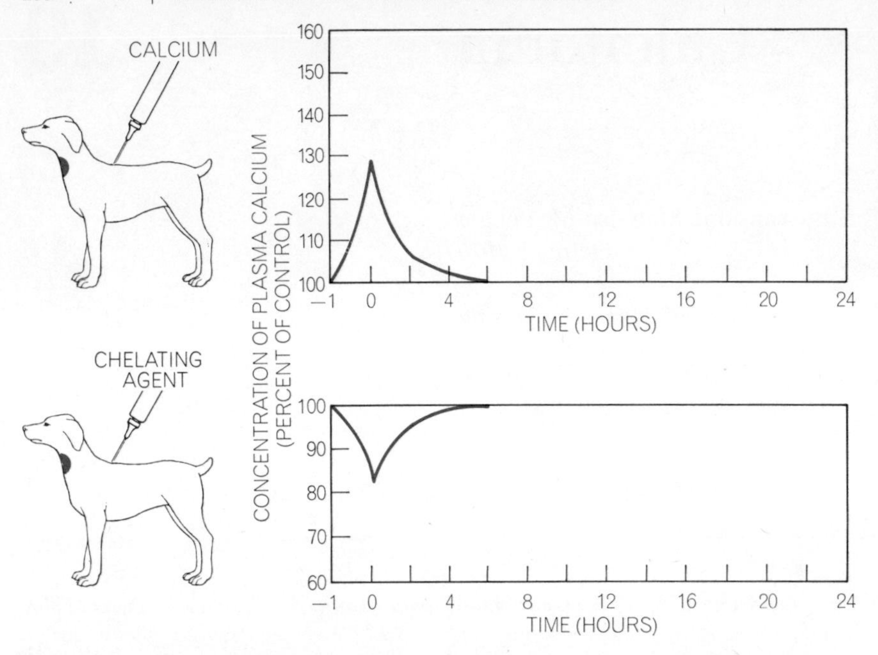

FIRST PHASE OF DOG EXPERIMENT conducted by Peter Sanderson and his colleagues at the Peter Bent Brigham Hospital indicates that there is a positive mechanism preventing excessively high levels of calcium in the blood. At top dog is injected with calcium. Graph at right shows that one hour before injection calcium concentration in the blood is normal. At time of injection concentration rises. About six hours later, however, concentration is normal. Therefore some agent is probably reducing the flow of calcium from the bone into the blood. At bottom same dog is injected with a chelating agent that captures calcium in blood and renders it inactive. The graph shows that one hour before the injection the calcium concentration is normal. After the injection it is depressed. About six hours after the injection the concentration is normal again, as is expected because a hormone secreted by the parathyroid gland is known to elevate the calcium concentration in the blood.

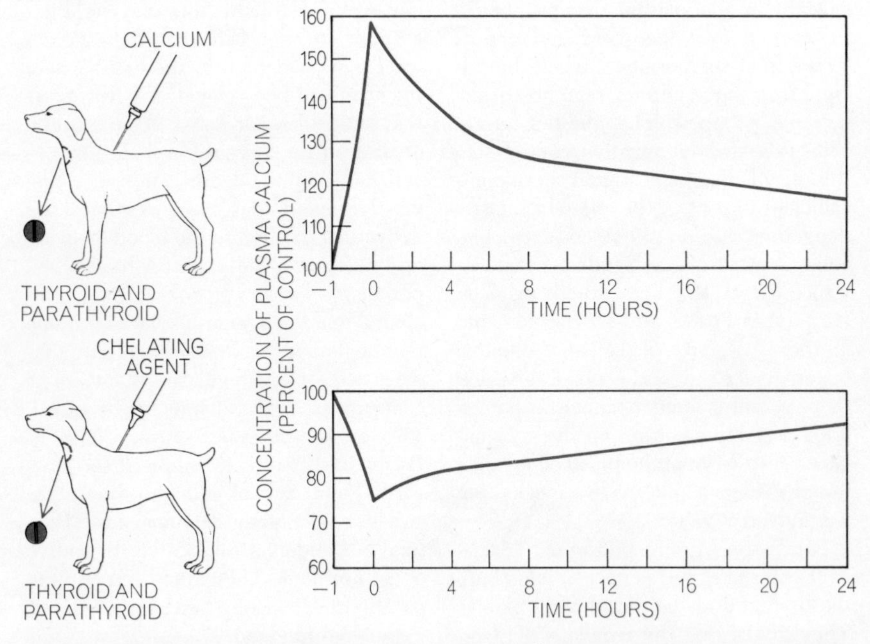

SECOND PHASE OF DOG EXPERIMENT demonstrates that there definitely is a controlling agent. At top a dog with thyroid and parathyroid glands removed is injected with calcium. The graph indicates that the calcium concentration, which was normal one hour before the injection, quickly rises. More than 24 hours later the concentration is still high, indicating that some agent that normally reduces the concentration is no longer present because the glands are missing. At bottom a chelating agent is administered. The graph at right indicates that the calcium concentration, normal one hour earlier, is depressed. More than 24 hours later the concentration is still low, because the parathyroid hormone, which keeps the concentration high, has been eliminated by removal of the gland. D. Harold Copp of the University of British Columbia demonstrated the significance of these experiments.

the *C* cells were described by Jose F. Nonidez of Cornell University in 1931, their function was unknown until Pearse showed that they produced calcitonin. These cells have had an interesting evolutionary history. In mammals the *C* cells arise in ultimobranchial glands that lose their distinct identity during development and merge with the thyroid. In fishes, amphibians, reptiles and birds, however, the ultimobranchial glands remain separate and persist as distinct organs in the adult. Stuart Tauber of the University of Texas Southwestern Medical School and Copp established that in these animals the ultimobranchial gland contains a high concentration of calcitonin but that the hormone does not show up at all in the thyroid gland.

Thus by 1967 the source of calcitonin in mammals was well established, and investigations of various aspects of its physiological activity were under way. One of the first aspects to be studied was the system that calls forth the secretion of calcitonin and its counterpart, the parathyroid hormone. Sanderson's and Copp's earlier studies had indicated, as we have seen, that the basic controlling factor was the calcium concentration in the blood. Direct measurements of the levels of calcium, calcitonin and the parathyroid hormone in the circulating blood of experimental animals now gave a quantitative picture of the control system. These measurements were made possible by newly developed assay techniques derived from the fundamental work of Solomon A. Berson and Rosalyn S. Yalow of the Veterans Administration Hospital in the Bronx. These investigators, working with John T. Potts, Jr., and Gerald D. Aurbach of the National Institutes of Health, had devised a radioimmunoassay for measuring the concentration of the parathyroid hormone; later Potts, and independently Claude Arnaud of the Mayo Clinic, had worked out a similar assay for calcitonin.

Using these assays, Potts and Arnaud found that in pigs infused with measured amounts of calcium the secretion of calcitonin increased in direct proportion to the rise in the blood's calcium content, and the secretion of the parathyroid hormone decreased in proportion to the calcium rise. Thus the results showed that the physiological control system responsible for keeping the blood's calcium supply at a stable level consists of two feedback loops: the parathyroid hormone operating to sustain the supply, calcitonin operating to prevent calcium from rising above the required level.

After the discoveries of calcitonin's

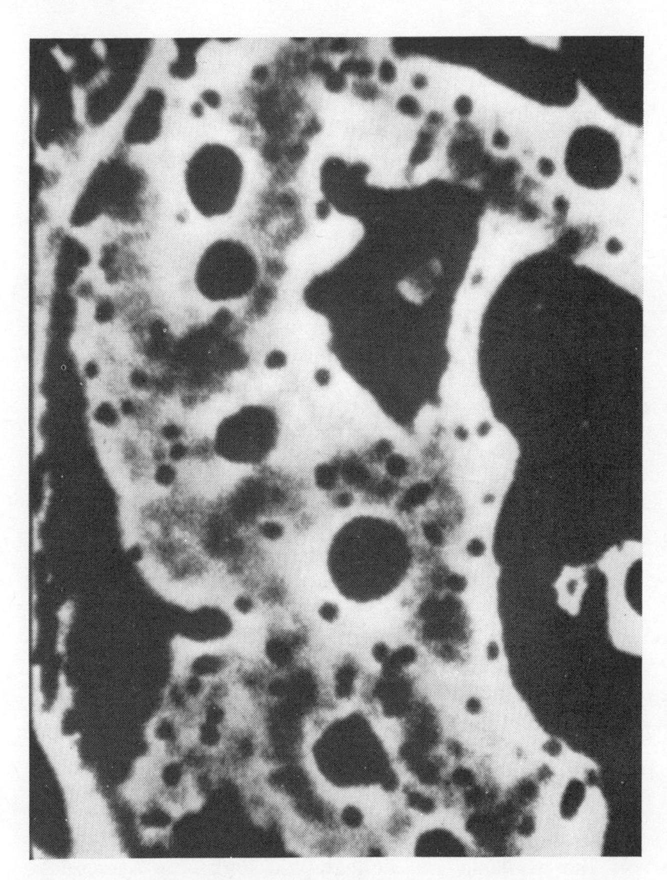

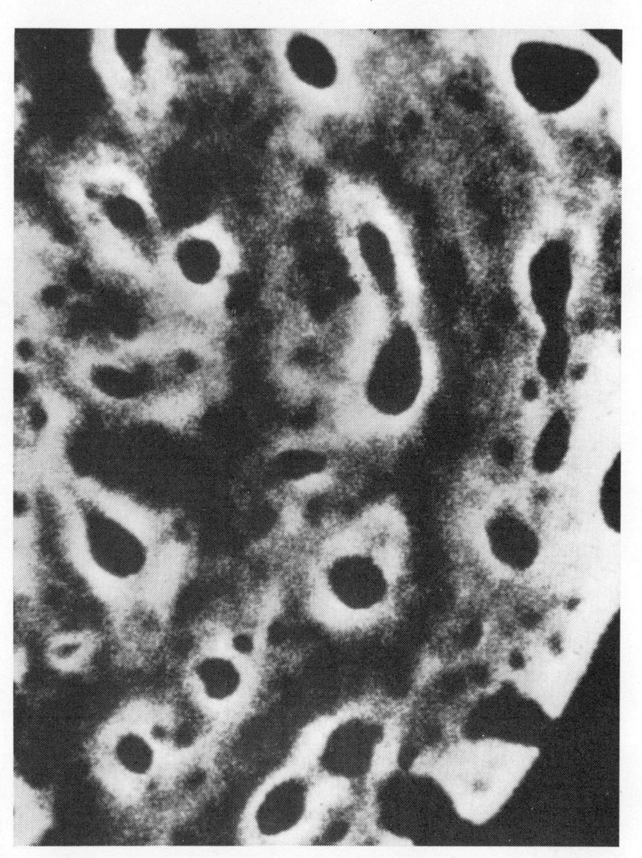

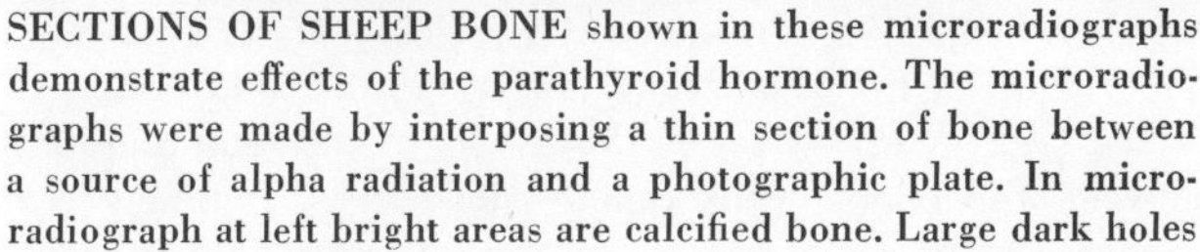

SECTIONS OF SHEEP BONE shown in these microradiographs demonstrate effects of the parathyroid hormone. The microradiographs were made by interposing a thin section of bone between a source of alpha radiation and a photographic plate. In microradiograph at left bright areas are calcified bone. Large dark holes are channels for blood vessels; small dark holes are occupied by osteocytes. The microradiograph at right was made after parathyroid hormone had caused osteocytes to resorb bone tissue. Resorption of bone tissue produces large blurred areas. The osteocyte spaces are enlarged and some of them have merged with one another.

role and the source that produced the hormone, the next challenge was to decipher its chemical nature, as a preliminary to learning how the hormone brings about its striking effect. A number of investigators set out to isolate the hormone, using various assay methods based on measurements of the potency of their preparations. Six laboratories reported late in 1967 and early in 1968 that they had extracted the hormone in pure form from pig thyroid glands. It turned out to be a polypeptide consisting of a single chain of 32 amino acids. Three laboratories soon determined the sequence of the amino acids in the molecule's structure, and shortly thereafter the Lederle Laboratories in Pearl River, N.Y., and the Ciba and Sandoz laboratories in Switzerland announced the synthesis of the hormone from its amino acids. The Ciba group, headed by R. Neher, also succeeded in isolating the human version of calcitonin (from tissue obtained from a patient with a carcinoma of the thyroid gland) and analyzed its amino acid sequence. This proved to be somewhat different from the structure of the pig hormone, indicating that the activity of calcitonin in curtailing the calcium content of blood depends on certain critical features of the amino acid sequence (common to both the pig and the human forms of the molecule) rather than on the structure as a whole.

The way was now open to investigate calcitonin's mode of action. How does the hormone function to lower the concentration of calcium in the blood plasma? Experiments with preparations of purified calcitonin in many laboratories in the U.S. and Europe soon showed that the hormone produces its effect by decreasing the release of calcium to the blood from bone. Calcitonin was found to act on the metabolism of bone in a way that inhibits bone resorption.

Bone is by far the most refractory and durable of all biological tissues, as is evident in the survival of fossil bones that have been buried for hundreds of thousands of years. Because of the apparent obdurability of bone it was supposed until very recently that once the skeleton of a vertebrate has been formed it ceases to partake of metabolism or to have any appreciable breakdown. Definitive evidence that this is by no means the case did not come until the availability of radioactive isotopes made it possible to examine the events actually occurring in the bone of living animals.

Tracer experiments with labeled isotopes of the main elements that go into the makeup of this tissue (calcium, phosphorus and the carbon of amino acids) showed that even in fully mature bone there is a constant turnover of these materials. It then became evident that metabolic activity in the skeleton is a vital necessity for at least two reasons. For one thing, such activity allows remodeling of the skeleton to enable it to deal with the developing mechanical stresses on the body. It is as if each bone were an elaborate Gothic structure in which a resident engineer, in response to changes in stresses, continually directs the replacement of supporting arches with new ones providing a slightly different center of thrust. The other vital function of metabolic activity in bone is that, by allowing an exchange of minerals (calcium and phosphate ions) with the blood, it provides a storehouse of these materials to help meet the body's general requirements. And in fact normally there is an overall balance of supply and de-

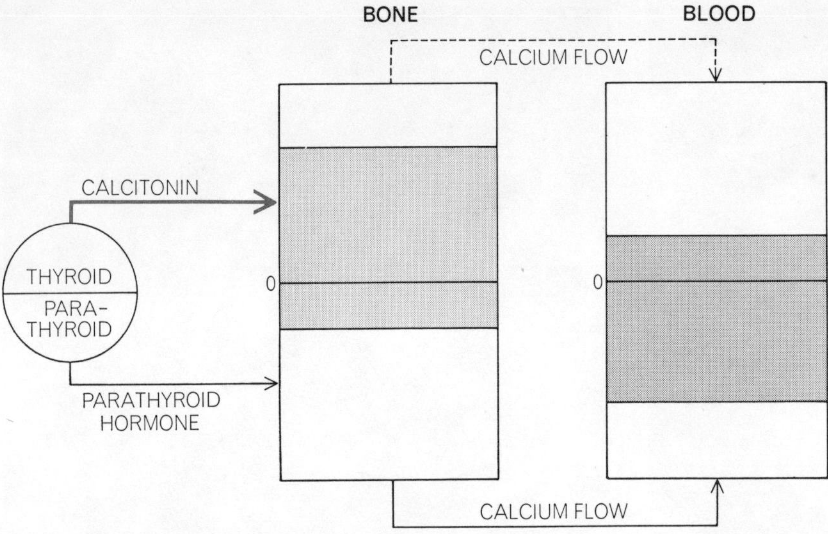

CONTROL LOOP regulating balance of calcium in bone and blood is established by calcitonin and the parathyroid hormone. Thyroid (*left*) secretes calcitonin (*colored arrow*) when blood calcium is high. Calcitonin prevents resorption, so that bone calcium level (*color in upper part of box in middle*) remains high and flow of calcium into blood is reduced (*broken arrow at top*). Therefore calcium level in the blood (*shading in upper part of box at right*) is low. When calcium level gets too low, parathyroid gland (*left*) increases it by secreting hormone (*black arrow*) so that more calcium enters blood (*solid arrow at bottom*).

mand for the bone minerals, new bone being laid down in some parts of the skeleton and old bone being resorbed elsewhere at the same rate.

To understand the operations of calcitonin and the parathyroid hormone in this system we need to look into the events in the formation and the resorption, or destruction, of bone. We are concerned here with what goes on at the surface of existing bone as new tissue is formed or old bone is dissolved away. The building on of new bone is initiated by a surface layer of osteoblasts, the cells that synthesize and extrude molecules of the fibrous protein collagen. The collagen molecules form thin, insoluble fibrils that pack themselves closely in a definite geometric array constituting a matrix, firmly anchored in the extracellular spaces of the bone. The matrix, which takes several days to form, then acts as a template for the deposit of crystals of mineral in an ordered arrangement. Studies by Melvin J. Glimcher and Stephen M. Krane of the Massachusetts General Hospital indicate that the buildup of the minerals begins with the attachment of phosphate groups on specific sites in the matrix. Once the first mineral crystals form, the depositing of phosphate and calcium proceeds rapidly, and within a few days the new bone contains 70 percent or more of all the mineral it will ever possess. As mineral is deposited, it displaces water and eventually occupies most of the available space in the matrix. It is then locked in so that (in an adult animal) most of the mineral is not free to take part in exchanges of mineral between the bone and the blood. This, of course, is an essential condition, because unrestricted removal of mineral from the bone could deprive it of its rigidity and mechanical strength.

Exactly where in the matrix is the mineral deposited? Is it encrusted around the collagen fibrils? If this were so, one would expect the mineral to play a large role in maintaining the structural form (shape and size) of the bone. Actually experiments have shown that when the mineral is extracted (by artificial means), the bone is not changed in shape or size but merely softens. A more plausible picture of the location of the mineral crystals, suggested by John A. Petruska and Alan J. Hodge of the California Institute of Technology and fitting the known facts, is that they are lodged in spaces between the ends of the fibrils, which are supposedly arranged in an overlapping array. This model implies that the mechanical properties of a bone are determined by the location of mineral in the matrix, and the bone's shape and size are determined by the three-dimensional configuration of the matrix, which is fixed by strong chemical bonds tying the fibrils together.

We come now to the opposite of the bone-building process: the resorption of bone that makes possible the continuous exchange of phosphate and calcium ions between bone and the blood. It turns out that the bone-forming cells play a role in resorption as well. As the osteoblasts complete their function of producing collagen, they become entrapped in these fibrils and are continually overlain by new osteoblasts growing on the surface. The buried osteoblasts are now converted into osteocytes and put out thin extensions, forming an extensive cellular network throughout the bone. When the bone has been fully formed, it is covered by a film of fluid topped by a thin layer of "resting" osteoblasts, which are no longer making collagen. These cells, together with the osteocytes, perform the important function of acting as sentinels that patrol the flow of mineral ions between the blood plasma and the special fluid bathing the bone. Were it not for this guarding network, a small change in the bone's chemistry might lead to a flood of calcium into or out of the blood, either of which could be lethal.

The involvement of osteocytes in bone resorption was clearly disclosed recently in radiographic studies by Leonard F. Bélanger of the University of Ottawa. It had long been supposed that bone was broken down only by the giant digesting cells called osteoclasts. Examining thin sections of bone by means of radiography (using alpha radiation and X rays), Bélanger produced pictures showing the matrix and indicating where it was calci-

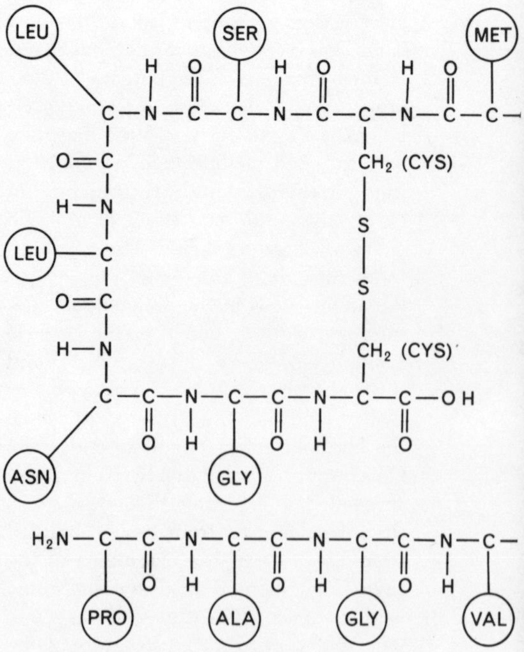

CALCITONIN MOLECULE in man consists of a specific sequence of 32 amino acid

fied and where it was not. He found that in bone that had been treated with the parathyroid hormone both the matrix and the calcium were dissolved away around each osteocyte [*see illustration on page 207*]. Apparently calcium is so tightly trapped in the matrix that it is not readily released unless the matrix structure is broken down.

The experiment thus demonstrated that the parathyroid hormone promotes bone resorption by acting on the osteocytes; in some way it incites these cells to engulf and destroy the bone tissue around them, with a consequent release of dissolved calcium into the bloodstream. We undertook a different set of experiments to probe the counteracting behavior of calcitonin. Conceivably this hormone might lower the blood's calcium level by speeding up the laying down of calcium in new bone, but it seemed much more likely that it acted by inhibiting bone resorption.

In looking into the effects of calcitonin we had the benefit not only of the radiographic techniques for examining what went on in bone tissue but also of a newly developed chemical measure of the rate of bone resorption in living animals. The collagen of bone is a protein rich in a peculiar amino acid: hydroxyproline. Darwin J. Prockop of the University of Pennsylvania School of Medicine had learned that hydroxyproline is not incorporated in a protein in the usual way—with the help of transfer RNA. There is no specific transfer RNA for this amino acid. A precursor molecule, protocollagen, is first synthesized and then specific proline units within it are acted on by an enzyme that adds the hydroxyl group to some of the prolines. In this way hydroxyproline is produced at the necessary positions on the collagen molecule. When the collagen molecule is broken down, hydroxyproline shows up in urine. It serves as a distinctive signal; the amount of hydroxyproline in the urine has in fact been found to reflect the rate of bone resorption in an animal. We found it useful in determining whether or not the resorption rate was influenced by calcitonin.

We removed the thyroid and parathyroid glands from rats and then injected the parathyroid hormone, so that the animals had an ample supply of this hormone but no calcitonin. The rats excreted large amounts of calcium and hydroxyproline in their urine, showing a high rate of bone resorption, and radiographic examination of sections from their bones confirmed that the osteocytes had destroyed bone tissue around them and had greatly swelled in size. When we carried out a parallel experiment with another set of rats, this time supplying them with injections of calcitonin as well as the parathyroid hormone after their thyroid and parathyroid glands were removed, the results were strikingly different. The animals' excretion of calcium and hydroxyproline in the urine *decreased*, and radiographic examination of bone sections showed no sign of bone breakdown around the osteocytes. It is also noteworthy that the osteoclasts, the large cells on the surface of bone that are involved in its remodeling, also increase their resorptive activity in response to parathyroid hormone and decrease it in response to calcitonin.

Here, then, was clear evidence that calcitonin performs its role of controlling the calcium content of the blood by inhibiting bone resorption. A simple further experiment demonstrated how vital this control is for the animal. In animals that had both glands removed and were then injected with a continuous infusion of calcium and the parathyroid hormone, the calcium concentration in the blood rose rapidly to a high level, urine excretion began to fail after 16 hours, and shortly afterward the animals died. The cause of death, as postmortem examination showed, was a massive accumulation of insoluble calcium phosphate in the kidneys, destroying their function. When, on the other hand, we removed only the parathyroid

ALA ALANINE
ASN ASPARAGINE
ASP ASPARTIC ACID
CYS CYSTINE
GLN GLUTAMINE
GLY GLYCINE
HIS HISTIDINE
ILE ISOLEUCINE
LEU LEUCINE
LYS LYSINE
MET METHIONINE
PHE PHENYLALANINE
PRO PROLINE
SER SERINE
THR THREONINE
TYR TYROSINE
VAL VALINE

LEU GLY THR TYR THR GLN ASP PHE ASN

LYS

PHE

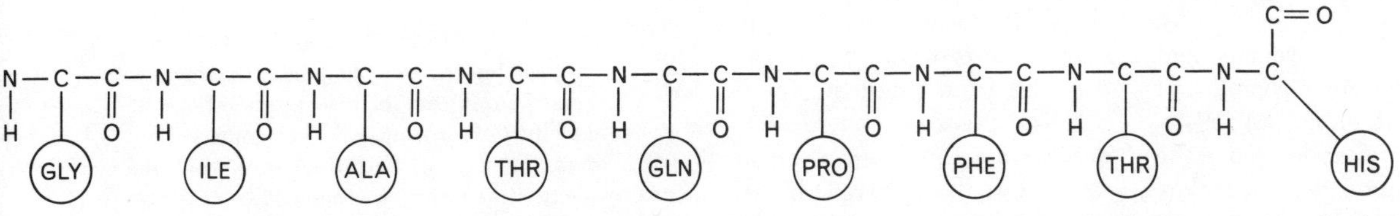

subunits. The identity of the individual subunits is indicated by the abbreviations that mark the amino acid side chains. The key at top right gives the full name of each of 17 different amino acid constituents of hormone. The sequence differs in other species.

gland, leaving the thyroid intact, the animals responded to the same infusion of calcium and parathyroid hormone without apparent ill effects. Clearly in this case the thyroid gland saved the day with an output of calcitonin that prevented a marked rise in the blood's calcium level, and we found no damaging deposits of calcium in the kidneys.

In view of calcitonin's now clearly established function as an inhibitor of bone resorption, it is curious to find that some primitive cartilaginous animals, such as the dogfish shark, produce calcitonin (in an ultimobranchial gland) although they have no bony skeleton. We can only speculate about what function the hormone may serve for them. Apparently calcitonin, like some other hormones, is an ancient biological agent that evolution converted from its original function to a new use in the vertebrates.

From the physiological study of calcitonin's action we have now gone on to try to learn how the hormone produces its effect in biochemical terms. This is difficult to get at, because bone tissue is structurally so complex and composed of such a diversity of cells that direct exploration by specific chemical experiments is almost out of the question. On the basis, however, of various items of evidence, mostly indirect, we have arrived at a tentative hypothesis about the biochemistry of calcitonin's control of metabolism in bone cells.

A starting point was provided by recent discoveries concerning the biochemical action of the parathyroid hormone. These discoveries have been made in the years since one of us (Rasmussen) wrote an article in *Scientific American* on that hormone ["The Parathyroid Hormone," by Howard Rasmussen; SCIENTIFIC AMERICAN Offprint 86]. In the urine of rats given parathyroid hormone L. R. Chase and Aurbach of the National Institutes of Health noted an increased content of the remarkable compound known as cyclic 3′5′ adenosine monophosphate (cyclic AMP). This substance, first discovered by Earl W. Sutherland, Jr., then at Washington University, had been found to serve as an intermediate in the action of many hormones—the hormone being called the "first messenger" and cyclic AMP the "second messenger."

Sutherland and his co-workers had established that cyclic AMP is produced from adenosine triphosphate (ATP) by an enzyme, adenyl cyclase, located at the surface of cells. From experiments they developed the hypothesis that many hormones interact with specific adenyl cyclases on the surface of specific cells and bring about an increase in the production of cyclic AMP within the cell. For instance, the hormone epinephrine triggers the synthesis of cyclic AMP in liver cells, the parathyroid hormone does so in kidney cells but not in liver cells, and so on. The effect of the second messenger varies according to the special character of the target cell; for example, in the liver it causes an increase in the synthesis and release of glucose; in the adrenal cortex (where cyclic AMP acts as the second messenger for ACTH) it stimulates the synthesis and release of steroid hormones. It is as if cyclic AMP conveyed the messages of the hormones in a generalized form, simply saying to each cell: "Do your thing."

The discovery by Chase and Aurbach that the amount of cyclic AMP in the urine of rats increases following the administration of parathyroid hormone offered an opening for analysis of calcitonin's action. Assume that the parathyroid hormone fosters production of cyclic AMP in bone cells and that the cyclic AMP brings about the resorption of bone. It could be supposed, then, that calcitonin might block the effect of the parathyroid hormone either by preventing the synthesis of cyclic AMP or by inactivating that compound in the cell. (It was known that cyclic AMP could be inactivated by phosphodiesterase, an enzyme in cells that converts it by hydrolysis to 5′ AMP.) Looking into the question experimentally, we first infused rats that had been deprived of their thyroid and parathyroid glands with cyclic AMP, using a dibutyryl form of the compound that had been synthesized by Theo Posternak of Case Western Reserve University and had been found to be invulnerable to hydrolysis by phosphodiesterase.

The urine of these animals showed that the injection of the cyclic AMP increased bone resorption just as injections of the parathyroid hormone did. We then found that injections of calcitonin blocked the bone-resorption effect of cyclic AMP, just as it blocked bone resorption induced by parathyroid hormone. The results of these experiments clearly indicated that neither of the hypotheses about calcitonin's action was correct: obviously it did not block bone resorption simply by preventing the parathyroid hormone from producing cyclic AMP, because it was effective against the action of that compound itself in the absence of the parathyroid hormone, and our use of the unhydrolyzable form of the cyclic AMP showed on the other hand that calcitonin did not inactivate cyclic AMP by causing its hydrolysis. Chase and Aurbach produced supporting evidence for these conclusions: they demonstrated that the parathyroid hormone brings about an increase in the concentration of cyclic AMP in bone cells and that this increase is not affected by a simultaneous presence of calcitonin.

Evidently the answer to the problem of how bone resorption was inhibited had to be sought not in some action on the cyclic AMP system itself but in metabolic events in the cell that presumably were influenced by cyclic AMP. Andre B. Borle of the University of Pitts-

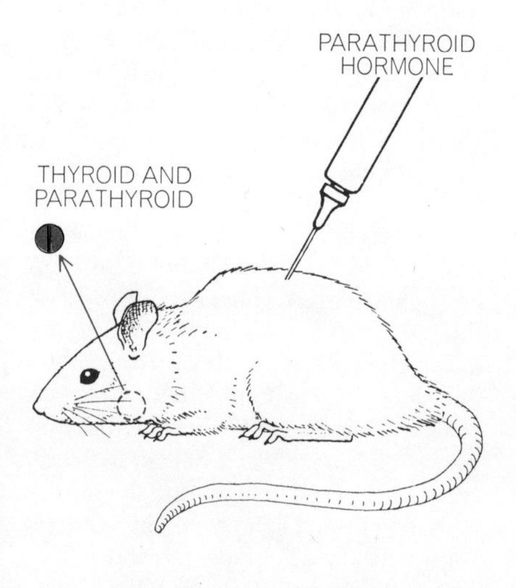

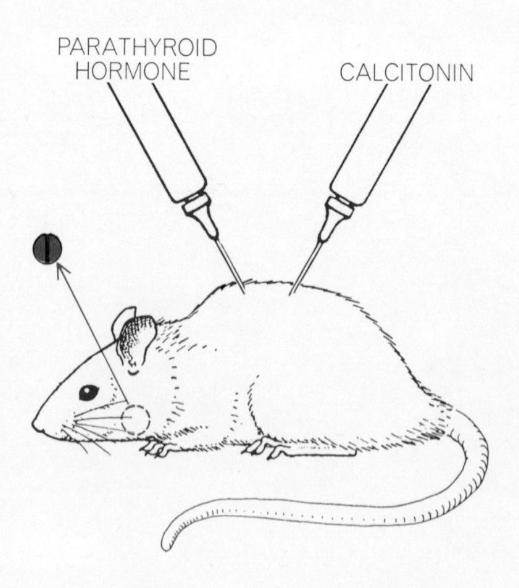

RAT EXPERIMENTS demonstrate clearly that calcitonin controls calcium concentration by inhibiting the resorption of bone induced by the parathyroid hormone. In first experiment (*top*) both the parathyroid and the thyroid glands are removed from the rat (*left*). First graph shows the level of hy-

burgh School of Medicine produced a clue that carried the search forward. Growing cells in a tissue culture, he found that when he added the parathyroid hormone to the culture medium, the cells increased their uptake of calcium. Did this effect also take place in cells growing normally in a living organism? Following up Borle's clue, one of us (Rasmussen) and a co-worker, Naokazu Nagata, established that the hormone did indeed affect calcium uptake in natural conditions, and we learned some details of the associated events.

Our first experiments were on whole animals. After removing the thyroid and parathyroid glands from rats, we injected some of the animals with the parathyroid hormone and others with calcium chloride. We then rapidly removed the kidneys, froze them to stop the many metabolic reactions going on in the cells, inactivated all the enzymes with cold perchloric acid and extracted the collection of metabolites, and then by a complex of techniques determined the concentrations of the various important metabolic intermediates.

We found that the infusion of calcium salt produced much the same metabolic picture as that resulting from the parathyroid hormone infusion. In order to obtain a more detailed picture under conditions enabling us to control some of the variables, we followed up with test-tube studies of kidney tissue, broken down with two enzymes, collagenase and hyaluronidase, that split apart the many thousands of small kidney tubules. The experiments on small segments of isolated tubules confirmed that the pattern of metabolism brought about by adding to the cells' calcium supply was almost identical with that produced by supplying the parathyroid hormone. This indicated that both treatments somehow operated on the same enzymes in the cell. Both, for example, speeded up the rate at which the cells convert lactic or malic acid to glucose. The rate depends on the cells' uptake of calcium, acting as a signal, and it appears that the amount of calcium entering the cells can be enhanced either by increasing

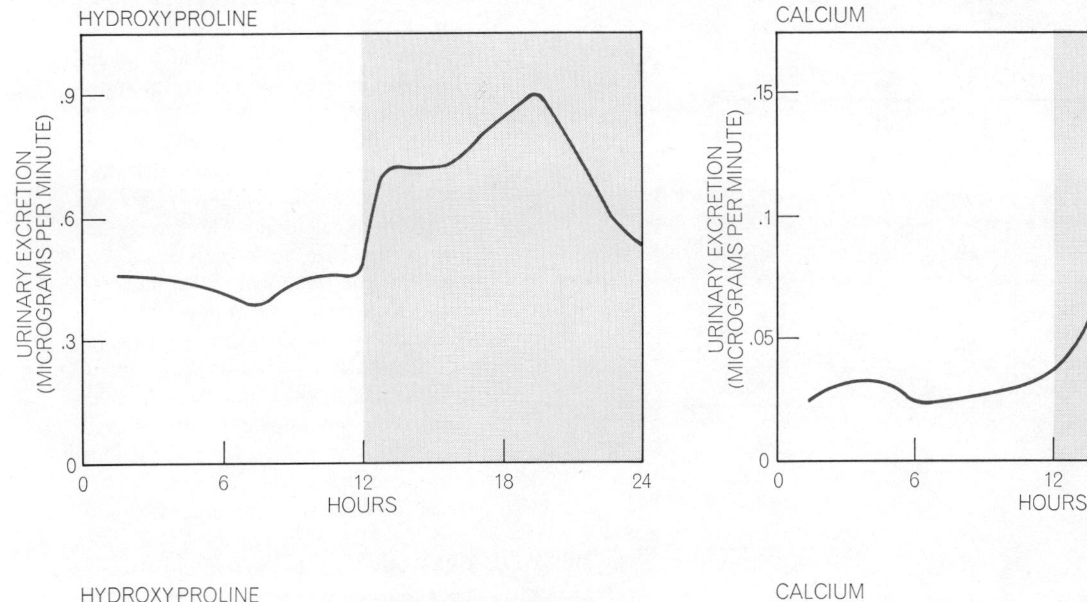

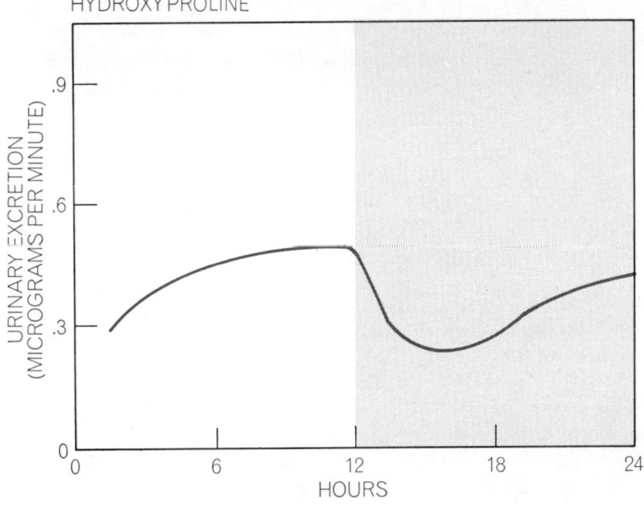

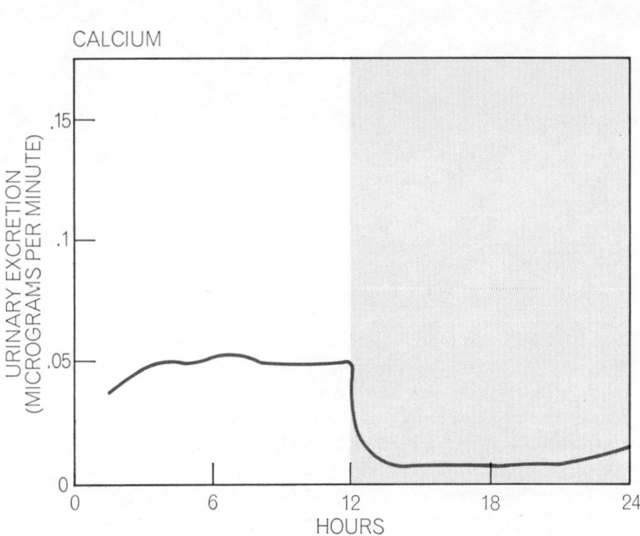

droxyproline, an amino acid produced by the breakdown of bone collagen, in the rat's urine. The second graph shows the calcium levels in the urine. The white area in each graph represents control period. The colored area represents the period during which the rat is constantly infused with the parathyroid hormone. During the control period in the first experiment, hydroxyproline levels and calcium levels are low. During the infusion period, however, both calcium and hydroxyproline levels are high, indicating considerable release of bone material. In the second experiment (*bottom*) glandless rat receives both parathyroid hormone and calcitonin. The first graph shows that the hydroxyproline level falls during infusions, indicating that calcitonin has blocked resorption of bone tissue. The second graph shows that calcium excretion also falls, confirming that the resorption of bone has been inhibited.

the supply of calcium outside the cell or by providing the parathyroid hormone. Thus in these cells calcium is an important second messenger relaying the hormonal, or first, message into changes in cellular activity.

How does the parathyroid hormone produce that effect on cell calcium? Here we may find an answer to the still mysterious question of cyclic AMP's function. Alan M. Tenenhouse of McGill University and one of us (Rasmussen) have suggested as a working hypothesis that the parathyroid hormone has two simultaneous effects on responsive cells: it increases the uptake of calcium and it stimulates the production of cyclic AMP. The increased intracellular concentration of cyclic AMP has two important effects related to calcium. It activates certain enzymes that now become calcium-sensitive, and it alters the intracellular distribution of calcium among various cell organelles.

In the light of this hypothesis and all the experimental findings, we can now see a possible explanation of the counteracting effect of calcitonin. That hormone may block the uptake of calcium by bone cells, a passive process and the one increased by parathyroid hormone, or alternatively it may somehow bring into play energy for active pumping of calcium out of the cells, thereby altering the changes induced by parathyroid hormone in calcium transport and bone resorption. The latter alternative seems more likely, and recently I. Radde of the University of Toronto reported early results in experiments on red blood cells that indicate calcitonin does stimulate the calcium pump in those cells.

The Ciba group of investigators announced several months ago that they had achieved total synthesis of the human form of calcitonin. This has opened the way for large-scale trials of the hormone for the treatment of various bone diseases. Some important uses in diseases characterized by derangement of calcium and bone metabolism had already been established. Unfortunately calcitonin does not appear promising as a cure for osteoporosis, by far the most common of the metabolic bone diseases. (It is particularly common in women over the age of 55 and in severe cases frequently leads to the fracture of bones because they cannot withstand ordinary mechanical stresses.) The discovery of calcitonin has, however, given a great impetus to the study of the fundamental processes of bone metabolism and turnover of the bone substances, and the rapid growth of knowledge in this field may soon bring forth a rational therapy for the disease. If so, the bringing to light of this remarkable new hormone will have served a most important catalytic function in medicine.

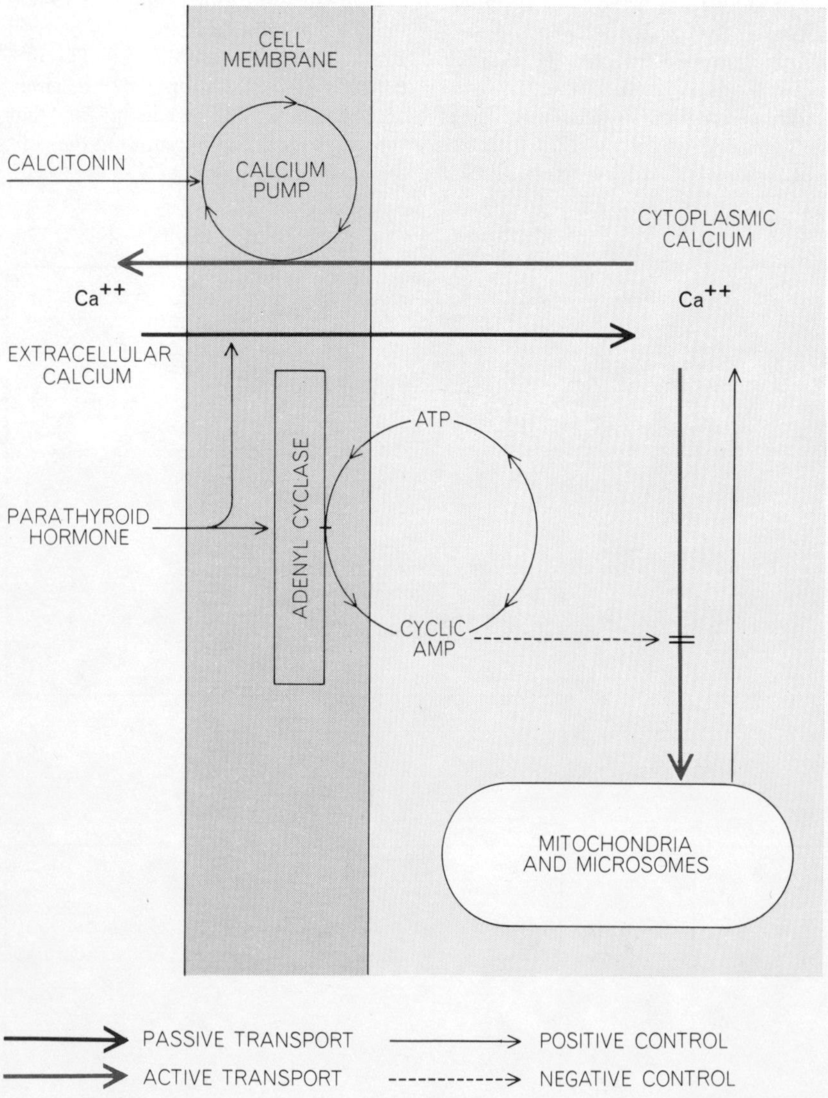

CELL MEMBRANE
CALCITONIN
CALCIUM PUMP
CYTOPLASMIC CALCIUM
Ca^{++}
Ca^{++}
EXTRACELLULAR CALCIUM
ATP
ADENYL CYCLASE
PARATHYROID HORMONE
CYCLIC AMP
MITOCHONDRIA AND MICROSOMES
PASSIVE TRANSPORT
POSITIVE CONTROL
ACTIVE TRANSPORT
NEGATIVE CONTROL

ACTION OF HORMONES ON CELLS controls body's calcium levels. When parathyroid hormone (*left*) reaches a cell wall (*dark colored area at left*), it stimulates the uptake of calcium by the cell from the extracellular space. At the same time the hormone activates adenyl cyclase, an enzyme in the membrane. Adenyl cyclase acts on ATP (adenosine triphosphate) to produce cyclic AMP (3′5′ adenosine monophosphate). As calcium enters the cytoplasm (*light colored area at right*), cyclic AMP (*center*) interferes with active movement of calcium out of the cytoplasm and into the mitochondria and other organelles (*right*). Since this active transport is interrupted, the calcium remains in the cytoplasm. Meanwhile calcium continues to flow passively from the organelles back into the cytoplasm. As a result calcium accumulates there. This accumulation is the key signal to the cell to begin resorbing the surrounding matrix of bone tissue. Calcitonin (*left*), however, counteracts this chain of events by activating a "calcium pump" represented by circular arrows at top left. This process pumps calcium out of the cell's cytoplasm, across the membrane and into the extracellular space, in effect "canceling" the message. Key at bottom shows that colored arrows indicate active transport of calcium and heavy black arrows passive transport. Small black arrows represent positive control; broken arrow shows negative control.

"*C*" CELLS that produce calcitonin are visible as luminous bodies in the micrograph on the opposite page. The tissue is from the thyroid gland of a dog. The *C* cells were made luminous by the injection of a fluorescent antibody that combines with calcitonin. The large dark areas are follicles that contain the principal thyroid hormone, thyroxin. Thyroxin is produced by cells forming grayish perimeter of each follicle. Small dark areas in some of the *C* cells are nuclei. Micrograph was made by A. G. E. Pearse of the Postgraduate Medical School of London.

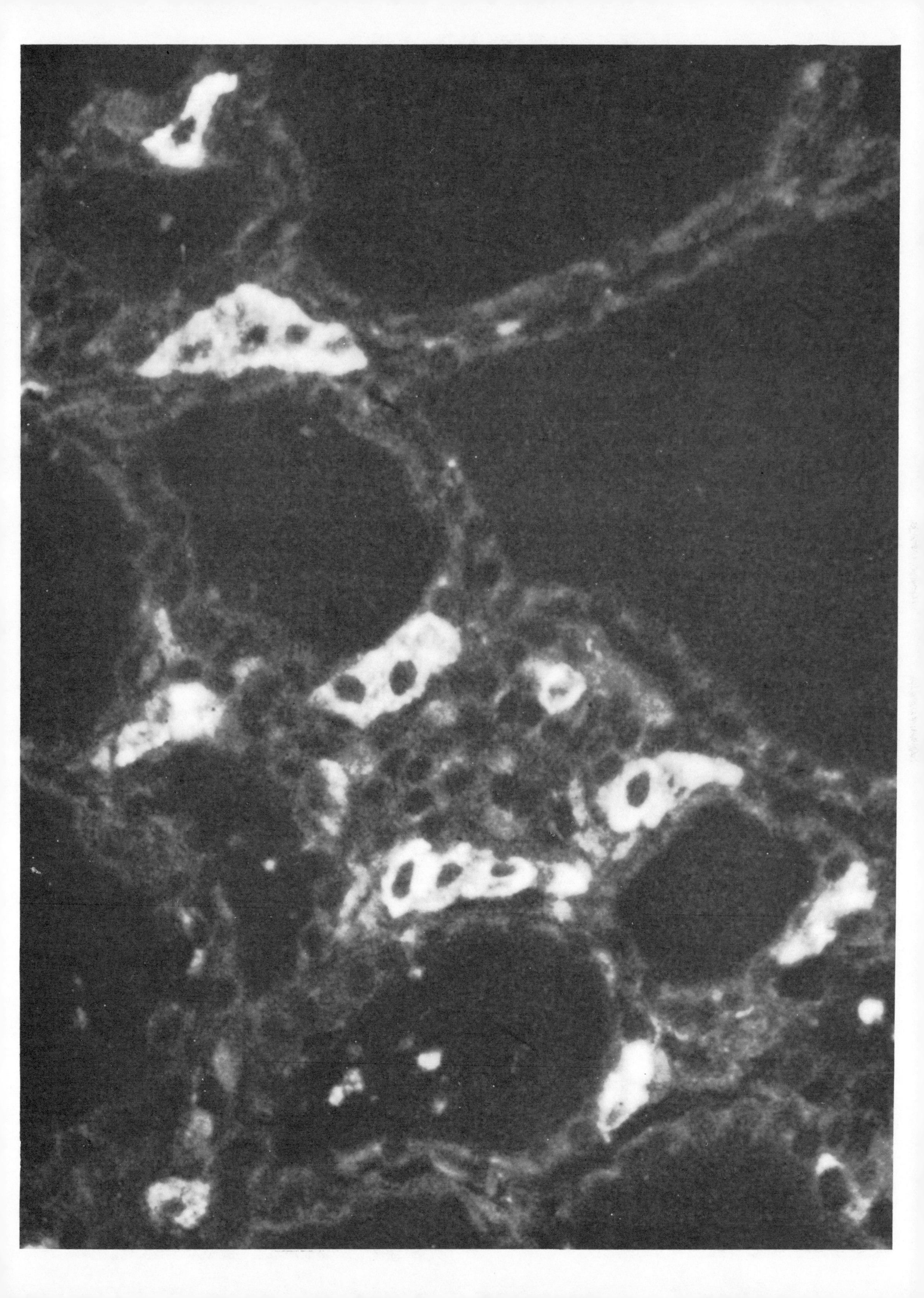

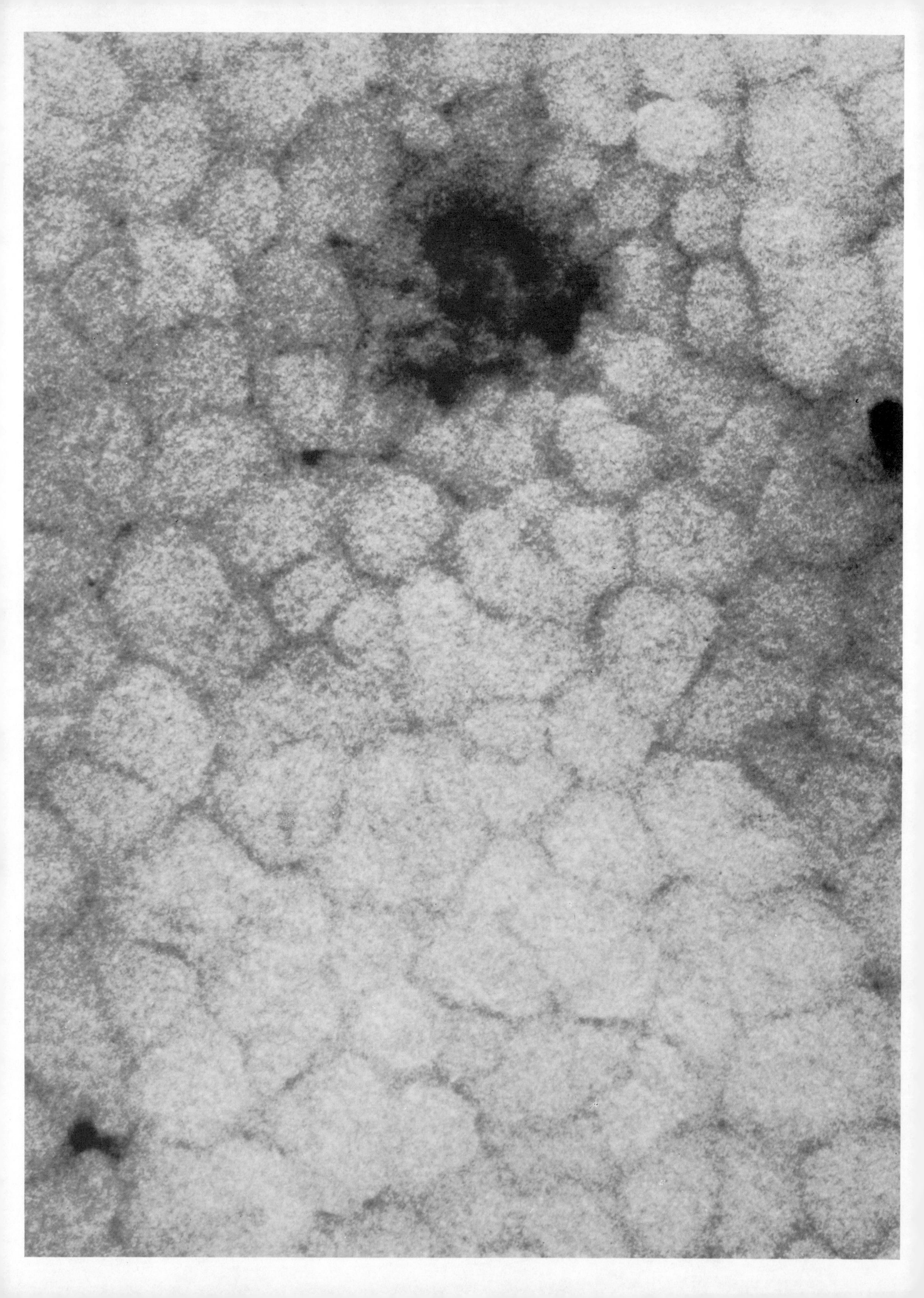

How An Eggshell Is Made

T. G. Taylor
March 1970

Eggshell is largely crystalline calcium carbonate. The calcium comes partly from the hen's bones, and when necessary the hen can mobilize 10 percent of her bone for this purpose in a day

To a housewife an egg is an article of food, and its shell serves to protect it from physical damage and to prevent the entry of dirt and microorganisms. To the hen an egg is a potential chick, and the shell serves not only as a protective covering but also as a source of calcium for the embryo and as a membrane through which the embryo respires. The eggshell performs its various functions with high efficiency, which is remarkable considering the number of eggs (five to seven a week) that the hen turns out. What is even more remarkable is the process whereby the hen obtains the substantial supply of calcium needed for the formation of the eggshells. The element comes in large part from her bones. Indeed, in extreme cases the hen can mobilize for this purpose as much as 10 percent of her total bone substance in less than a day! The physiology of this unusual process rewards close examination.

In its immature state the egg is one of many oöcytes, or unripened ova, in the ovary of the hen. Each oöcyte is encased in a membrane one cell thick; the entire structure is termed a follicle. At any one time follicles of various sizes, containing yolks at different stages of development, can be found in the ovary. Normally follicles ripen singly at a rate of one a day in hens that are laying regularly. There are occasional pauses. On the other hand, two follicles sometimes ovulate at the same time, giving rise to a double-yolk egg.

CHICKEN'S EGGSHELL consists mainly of columns of calcite, a crystalline form of calcium carbonate. They appear on the opposite page in an X-ray micrograph made through the thickness of a shell by A. R. Terepka of the University of Rochester; enlargement is 370 diameters. Large dark spot at top center is a "glassy" region of less opaque mineral; to its right is a pore.

Ovulation takes place within six or eight hours after the release of a high level of a hormone produced by the pituitary gland. The release of the hormone is related to the time of onset of darkness, and it normally occurs between midnight and about 8:00 A.M. It follows that the hen always ovulates in daylight. Moreover, since it takes about 24 hours after ovulation to complete the formation of the egg, the egg is also laid during the daylight hours.

Once the yolk is released from the ovary all the remaining stages of egg formation take place in the oviduct, which consists of several distinct regions: the infundibulum, the magnum, the isthmus, the shell gland (uterus) and the vagina [*see illustration on page 217*]. The oviduct, like the ovary, is on the left side of the hen's body; a vestigial ovary and a vestigial oviduct are sometimes found on the right side in a mature bird, but they normally degenerate completely during the development of the embryo. One can only speculate on the evolutionary reason for the disappearance of the right ovary and oviduct. A reasonable guess is that two ovaries were disadvantageous because of the problem of providing enough calcium for the shells of two eggs at once. Birds have enough of a job supplying calcium for one egg a day. Certain species of wild birds have retained two functional ovaries and oviducts. It is not known how ovulation is controlled in these species, but apparently wild birds do not lay two eggs in one day.

After the ovum is released from the follicle it is engulfed by the funnel-like infundibulum of the oviduct. It is here that the egg is fertilized in hens that have been mated. As the yolk passes along the oviduct, layers of albumen are laid down in the magnum. The proteins of the albumen, which constitute the egg white, are synthesized in the magnum from amino acids removed from the blood. The synthesis is continuous, and in the periods between the passage of yolks down the oviduct albumen is stored in the tissue of the magnum. The addition of the layers of albumen to the yolk takes about four hours.

The next stage in the formation of the egg is the laying down of two shell membranes, an inner one and an outer one, around the albumen. The membranes are formed in the thin, tubular isthmus. When the membranes are first laid down, they cover the albumen tightly, but they soon stretch. By the time the egg enters the shell gland they fit quite loosely.

The egg passes the next five hours in the process known as "plumping." This entails the entry of water and salts through the membranes until the egg is swollen. The plumping period appears to be an essential preliminary to the main process of shell calcification, which occupies the next 15 to 16 hours.

The shell is composed of calcite, which is one of the crystalline forms of calcium carbonate. A sparse matrix of protein runs through the crystals of the shell. The final stage in the formation of the egg is the deposition of a cuticle on the fully calcified shell; this is accomplished just before the egg is laid.

Let us now look at the structure of the eggshell in rather more detail. From the accompanying illustration [*bottom of page 219*] it will be seen that the shell is attached to the outer membrane by hemispherical structures known as mammillary knobs. Histochemical studies have shown that the cores of the knobs consist of a protein-mucopolysaccharide complex rich in acid groups, and that anchoring fibers run from the outer membrane into the knobs.

The cores of the mammillary knobs are laid down as the membrane-covered egg passes through the part of the oviduct called the isthmo-uterine junction; it is between the isthmus and the shell gland. It seems probable that the knobs are calcified soon after they are formed, before the egg enters the shell gland, and that they subsequently act as nuclei for the growth of the calcite crystals comprising the shell. Modern ideas on the mechanism of biological calcification—whether in bones, teeth, eggshells or any of the other places where calcium is deposited in animal bodies—emphasize the importance of crystal growth. Earlier theories seeking to explain the mechanism laid much stress on the role of precipitation of calcium salts from supersaturated solutions, but in the light of more recent evidence this concept no longer seems valid.

The mechanism whereby the mammillary knobs are calcified is not well understood. It is thought to involve the binding of calcium ions to the organic cores of the knobs by means of the sulfonic acid groups on the acid-mucopolysaccharide-protein material of which the cores are composed. It is suggested that the spatial arrangement of the bound calcium ions is the same as it is in the lattice of the calcite crystal, so that these oriented calcium ions act as seeds or nuclei for the growth of calcite crystals forming the shell. Some years ago my colleagues and I found that the isthmus contains extremely high concentrations of both calcium and citric acid, the former reaching a maximum of about 90 milligrams per 100 grams of fresh tissue and the latter about 360 milligrams. We concluded that the high level of calcium in this region may be of significance in the calcification of the mammillary knobs.

The main part of the shell was once known as the spongy layer but has more recently come to be called the palisade layer. It is composed of columns of tightly packed calcite crystals; the columns extend from the mammillary knobs to the cuticle. Occasional pores run up between the crystals from spaces formed where groups of knobs come together. The pores reach the surface in small depressions that are just visible to the unaided eye on the outside of the shell. It is through these pores that the embryo takes in oxygen and gives out carbon dioxide during the incubation of the egg.

The raw materials for the formation of the calcite crystals, namely the ions of calcium and carbonate, come from the blood plasma. The shell gland is provided with an extremely rich supply of blood. Careful measurements have shown that the level of plasma calcium falls as the blood passes through the gland when the calcification of a shell is in progress but does not fall when there is no egg in the gland.

Changes in the level of calcium in the blood of female birds during the breeding season have engaged the attention of many workers since 1926, when Oscar Riddle and Warren H. Reinhart of the Carnegie Institution of Washington discovered that breeding hen doves and pigeons had blood calcium levels more than twice as high as those found in cocks or nonbreeding hens. Adult males, nonbreeding females and immature birds of both sexes have plasma calcium levels of about 10 milligrams per 100 milliliters, whereas the level in females during the reproductive period is usually between 20 and 30 milligrams per 100 milliliters. For many years it was assumed that the high level of plasma calcium found in laying females was related to the trait of producing eggs with calcified shells, but it is generally recognized now that it is related to the production of large, yolky eggs. The extra calcium in the blood of laying birds (as compared with nonlaying ones) is almost entirely bound to protein. In contrast, the level of ionic calcium, which is the form of calcium mainly used in the formation of the eggshell, is about the same in laying and nonlaying hens.

The particular protein concerned in the binding of the increased plasma calcium is the phosphorus-containing protein phosvitin. It is the characteristic protein of the egg yolk. Phosvitin has a great affinity for calcium: the greater the amount of this phosphoprotein in the blood, the higher the level of plasma calcium. Phosvitin is synthesized in the liver under the influence of estrogen and is carried in the blood (in combination with lipid material) to the follicles developing in the ovary. Similar proteins are found in the blood of all animals that lay yolky eggs, including fishes, amphibians and reptiles, and yet neither fishes nor amphibians lay eggs with calcified shells, and among the reptiles only the Chelonia (turtles and tortoises) and the Crocodilia do so.

In the passage of blood through the shell gland there is a fall in both the protein-bound calcium (also termed nondiffusible calcium because the molecules of the protein to which it is bound are too large to diffuse through a semipermeable membrane) and in the diffusible calcium, the latter being mainly in the form of calcium ions. The two forms of calcium appear to be in equilibrium with each other. It seems likely that calcium

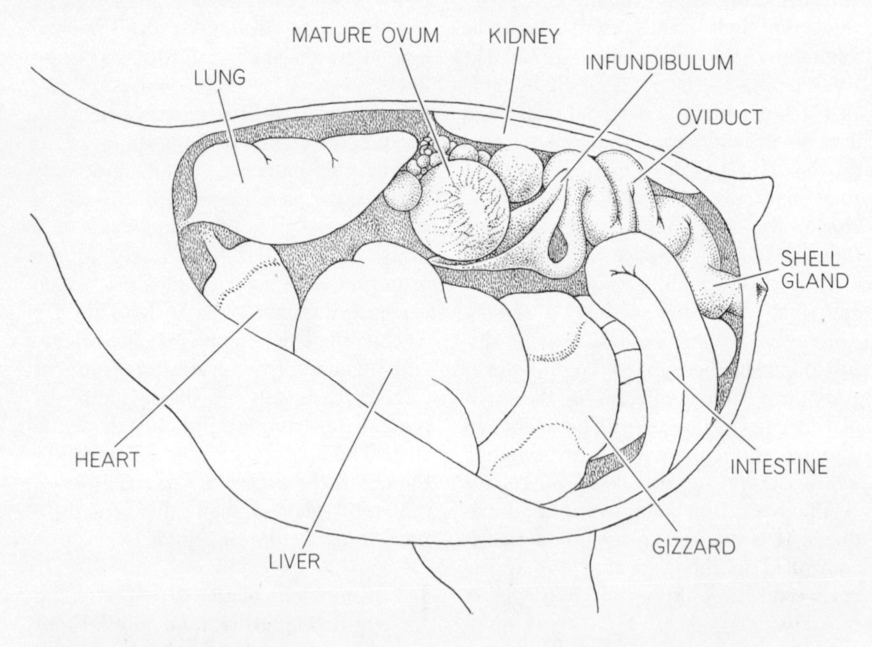

REPRODUCTIVE TRACT of the chicken is indicated in relation to the other organs in the body cavity. The single ovary and oviduct are on the hen's left side; an undeveloped ovary and an oviduct are sometimes found on the right side, having degenerated in the embryo.

in the form of ions is taken up from the plasma by the shell gland and that the level of ionic calcium is partly restored by the dissociation of a portion of the protein-bound calcium.

So much for the calcium ions. The origin of the carbonate ions is much harder to explain. At the slightly alkaline level of normal blood (*p*H 7.4) their concentration is extremely low, and it is the bicarbonate ion that predominates.

Theories to explain the formation of carbonate ions center on the enzyme carbonic anhydrase, which is present in high concentration in the cells lining the shell gland. One theory assumes that two bicarbonate ions are in equilibrium with a molecule of carbonic acid and a carbonate ion, with the equilibrium strongly in favor of the bicarbonate ions. The hypothesis is that the carbonic acid is continuously being dehydrated to carbon dioxide gas under the influence of the carbonic anhydrase, and that carbonate ions continuously diffuse or are pumped across the cell membranes into the shell gland, where they join calcium ions to form the calcite lattice of the growing crystals in the eggshell. An alternative theory, proposed by Kenneth Simkiss of Queen Mary College in London, is that the carbonate arises directly in the shell gland by the hydration of metabolic carbon dioxide under the influence of carbonic anhydrase.

The main evidence in support of the intimate involvement of carbonic anhydrase in eggshell formation is that certain sulfonamide drugs, which are powerful inhibitors of the enzyme, inhibit the calcification of shells. By feeding laying hens graded amounts of sulfanilamide, for example, it is possible to bring about a progressive thinning of the shells. Eventually, at the highest levels of treatment, completely shell-less eggs are laid.

On the average the shell of a chicken's egg weighs about five grams. Some 40 percent of the weight, or two grams, is calcium. Most of the calcium is laid down in the final 16 hours of the calcification process, which means that it is deposited at a mean rate of 125 milligrams per hour.

The total amount of calcium circulating in the blood of an average hen at any one time is about 25 milligrams. Hence an amount of calcium equal to the weight of calcium present in the circulation is removed from the blood every 12 minutes during the main period of shell calcification. Where does this calcium come from? The immediate source is the

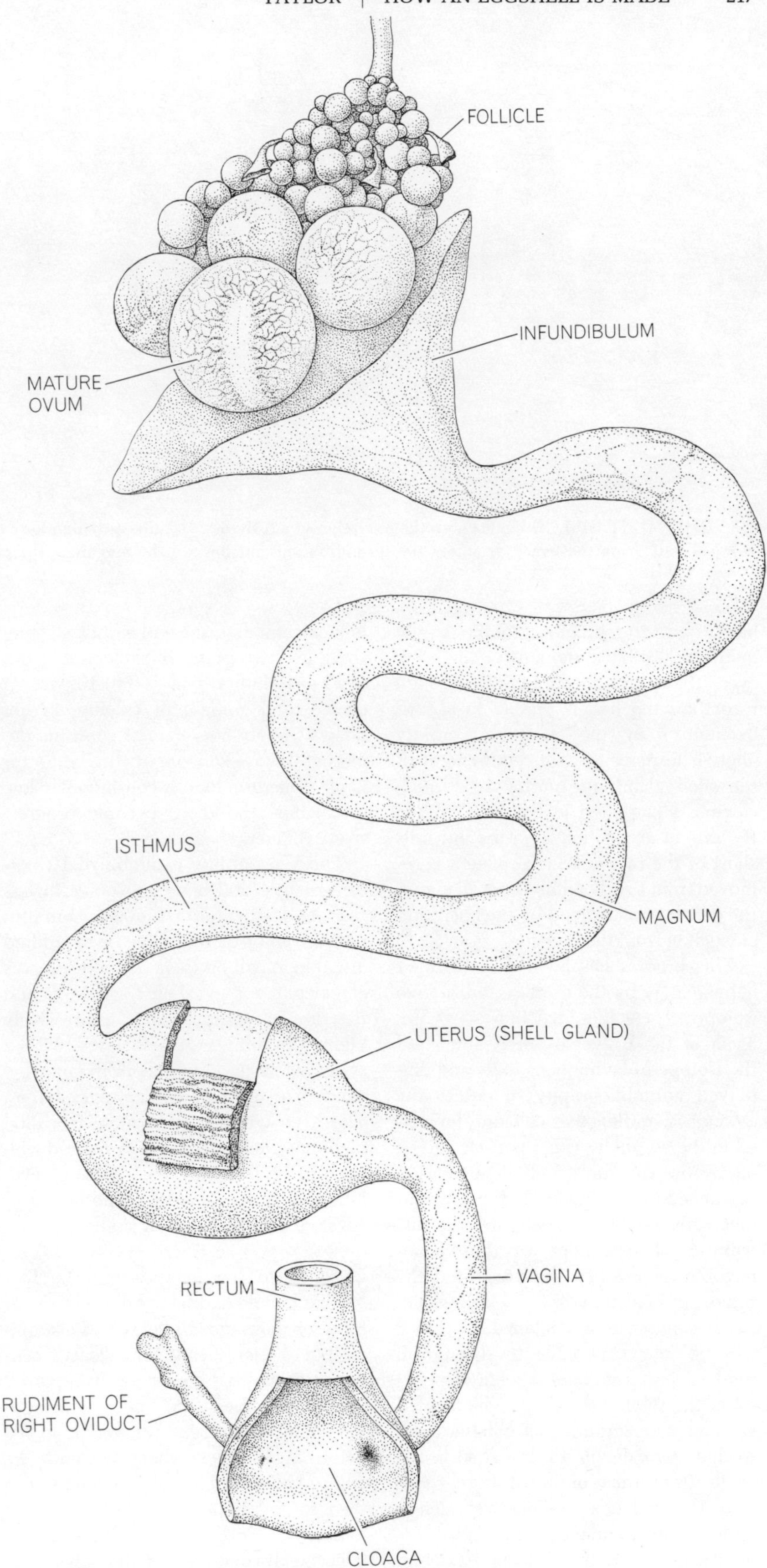

OVARY AND OVIDUCT of the chicken are involved in the formation of the egg. The shell is formed in the uterus, which is also called the shell gland. The principal steps in the formation of a chicken's egg are shown in the illustration at the top of the next two pages.

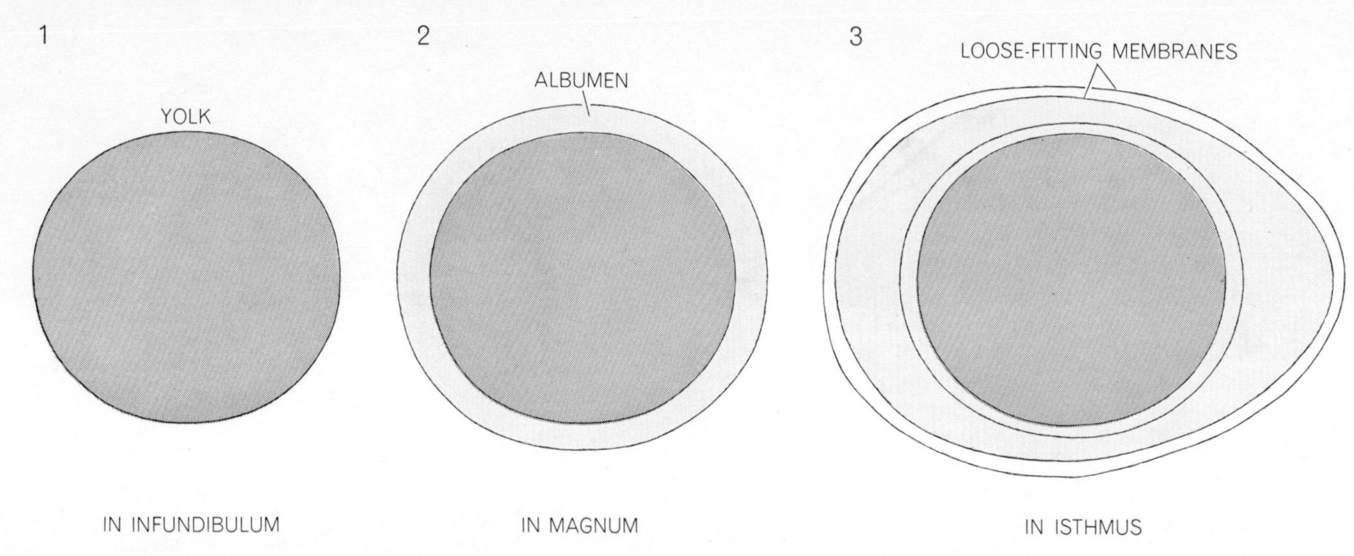

DEVELOPMENT OF EGG begins with the ovulation of a fully developed yolk from the ovary. It enters the infundibulum and begins moving along the oviduct. Layers of albumen are laid down in the magnum; the process takes about four hours. Two membranes

blood, but the ultimate source is the food. It has been demonstrated, however, that during the period of shell formation the hen is unable to absorb calcium from the intestines rapidly enough to meet the full requirement of the shell gland, no matter how much calcium is supplied in the food. When the rate of absorption from the gut falls short of the rate at which calcium is removed from the blood by the shell gland, the deficit is made good by the liberation of calcium from the skeleton.

This process has been demonstrated convincingly by the use of a radioactive isotope of calcium, calcium 45. Cyril Tyler of the University of Reading fed the isotope to laying hens daily and employed autoradiography to detect the amount of radioactive calcium deposited in the eggshells. (Beta particles given off by the calcium 45 of dietary origin blackened the X-ray film that was in contact with sections of shell, and the distribution of the isotope was thus visualized.) After the hens had been fed the radioactive calcium for a week the skeleton became intensely labeled, so that it was no longer possible to distinguish food calcium from bone calcium deposited in the shell. Accordingly the labeled calcium was withdrawn from the food, so that any calcium 45 deposited in the shells from then on must have come from the skeleton. Radioactive calcium appeared in abundance in the shells.

The mobilization of skeletal calcium for the formation of eggshell increases as the dietary supply of calcium decreases. When food completely devoid of calcium is fed, all the shell calcium comes from the bones. If a hen is fed a low-calcium diet, she will mobilize something like two grams of skeletal calcium in 15 to 16 hours. That is 8 to 10 percent of the total amount of calcium in her bones. Clearly hens cannot continue depleting their skeleton at this rate for long. When the food is continuously low in calcium, the shells become progressively thinner.

The hen's ability to mobilize 10 percent of her total bone substance in less than a day is quite fantastic but not unique: all birds that have been studied are able to call on their skeletal reserves of calcium for eggshell formation, and the rate of withdrawal is impressively high. This ability is associated with a system of secondary bone in the marrow cavities of most of the animal's bones. The secondary bone, which is called medullary bone, appears to have developed in birds during the course of evolution in direct relation to the laying of eggs with thick, calcified shells.

Strange to say, considering the fact that people had been killing birds for food for thousands of years and examining bones scientifically for at least a century, this unusual bone was not reported until 1916, when J. S. Foote of Creighton Medical College observed it in leg bones of the yellowhammer and the white pelican. The phenomenon was then forgotten until Preston Kyes and Truman S. Potter of the University of Chicago discovered it in the pigeon in 1934.

Medullary bone is quite similar in structure to the cancellous, or spongy, bone commonly found in the epiphyses (the growing ends) of bones. It occurs in the form of trabeculae, or fine spicules, which grow out into the marrow cavity from the inner surface of the structural bone. In males and nonbreeding females the marrow cavities of most bones are filled with red marrow tissue, which is involved in the production of blood cells. The spicules of medullary bone ramify through the marrow without interfering with the blood supply.

Medullary bone is found only in female birds during the reproductive period, which in the domestic fowl lasts many months. (In wild birds it lasts only a few weeks.) Medullary bone is never found in male birds under normal conditions, but it can be induced in males by injections of female sex hormones (estrogens). In hen birds medullary bone is produced under the combined influence of both estrogens and male sex hormones (androgens). It is thought that the developing ovary produces both kinds of hormone.

The formation and breakdown of medullary bone have been studied more closely in the pigeon than in any other bird. Pigeons lay only two eggs in a clutch; the second egg is laid two days after the first one. A pigeon normally lays the first egg about seven days after mating. The medullary bone is formed during this prelaying period. By the time the first egg is due to be provided with its shell, the marrow cavities of many bones of the skeleton are almost filled with bone spicules, which have grown steadily since the follicles developing in the ovary first started to secrete sex hormones.

About four hours after the egg enters the shell gland marked changes begin in

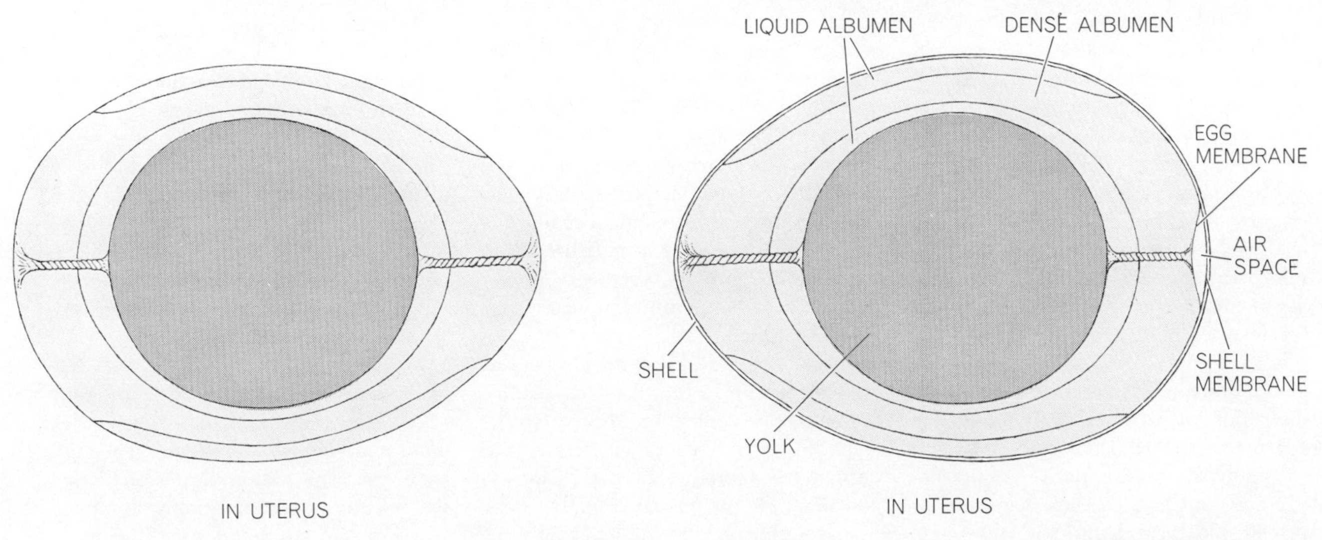

are added in the isthmus. At first they fit tightly, but by the time the egg enters the shell gland they have stretched so that the egg can undergo a five-hour process called plumping. The formation of the shell occupies the 15 to 16 hours needed to complete the egg.

the medullary bone. Within a few hours its cellular population has been transferred from one dominated by osteoblasts, or bone-forming cells, to one dominated by osteoclasts, or bone-destroying cells. The phase of bone destruction continues throughout the period of shell calcification. The calcium released from the bone mineral is deposited on the shell as calcium carbonate, and the phosphate liberated simultaneously is excreted in the urine.

The breakdown of the medullary bone persists for a few hours after the egg is laid. Then, quite suddenly, another phase of intense bone formation begins. This phase lasts until the calcification of the shell of the second egg starts; at that time another phase of bone destruction begins. No more bone is formed in this cycle. Resorption of the medullary bone continues after the second egg of the clutch is laid until, a week or so later, all traces of the special bone structure have disappeared and the marrow cavity regains its original appearance.

What mechanism might account for the rapid change from bone formation to bone destruction and vice versa? One suggestion is that variations in the level of estrogen control the cyclic changes in the medullary bone. There can be little doubt that the high level of estrogen plus androgen in the blood plasma is primarily responsible for the induction of medullary bone during the prelaying period; the drop in the level of estrogen or androgen or both after the second egg of the clutch is laid might well give rise to the bone destruction. It is difficult to see, however, how the fine degree of control necessary to induce bone destruction when calcification of the first eggshell is due to start, and to

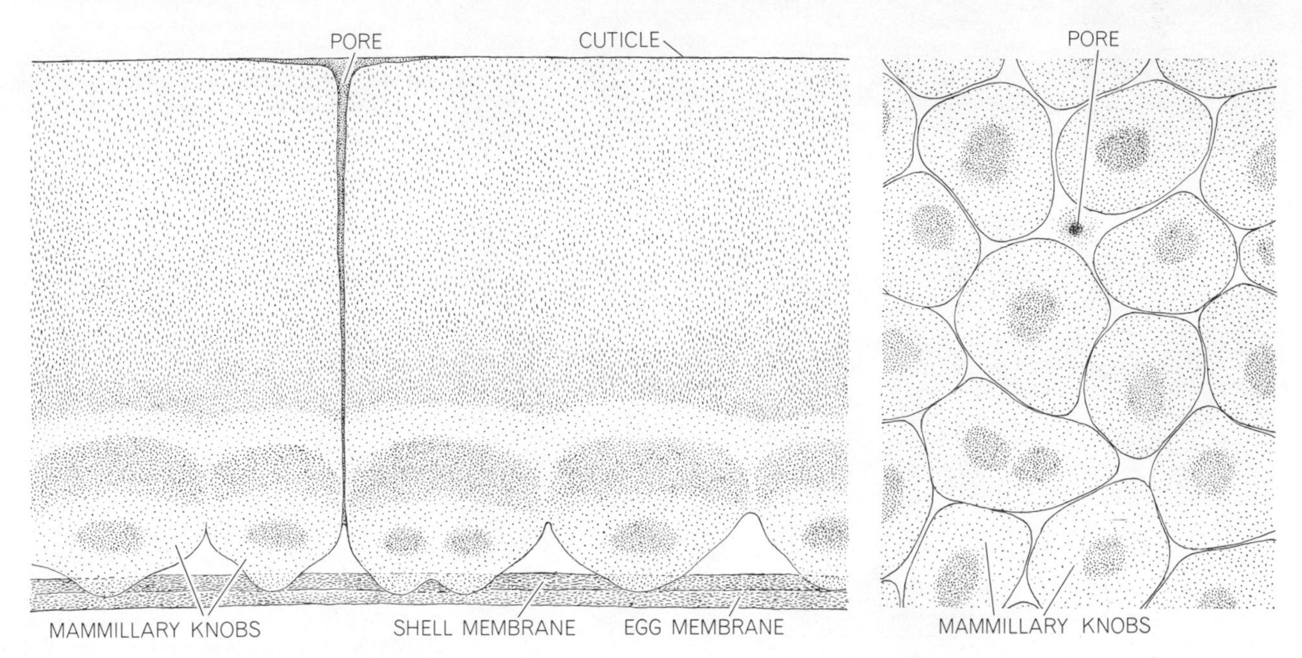

STRUCTURE OF EGGSHELL is portrayed in cross section (*left*) and in a tangential section (*right*) made through the layer containing the hemispherical mammillary knobs. The knobs, which attach the shell to the outer membrane of the egg by fibers, have an organic core; the rest of the structure is made of oriented ions of calcium that apparently act as seeds for the shell's calcite crystals.

reverse the process soon after it is completed, can be exercised by changes in the secretion of sex hormones, presumably from the single follicle present in the ovary and possibly from the recently ruptured follicle.

The control mechanism that my colleagues and I consider more likely is one mediated by the parathyroid gland. The role of this gland is to regulate the level of calcium ions in the blood. A drop in the level of plasma calcium causes the release of parathyroid hormone from the gland, and the hormone brings about a resorption of bone tissue through the agency of the bone cells (osteoclasts and enlarged osteocytes). Both organic matrix and bone mineral are removed together, and the calcium and phosphate are released into the blood. The level of plasma calcium is thus restored; the phosphate is excreted.

Bone resorption under the influence of parathyroid hormone is largely due to an increase in the number and activity of osteoclasts. The histological picture observed in the medullary bone of pigeons at the height of eggshell calcification bears a strong resemblance to the resorption of bone in rats and dogs following the administration of parathyroid hormone. Leonard F. Bélanger of the University of Ottawa and I have recently shown that the histological changes in the medullary bone of hens treated with parathyroid hormone were very similar to those occurring naturally during eggshell formation.

It has been shown that the level of diffusible calcium in the blood drops during eggshell calcification in the hen; the stimulus for the release of parathyroid hormone is therefore present. The hypothesis that the parathyroid hormone is responsible for the induction of bone resorption associated with shell formation is also consistent with the time lag between the end of the calcification of the eggshell of the first egg in the pigeon's clutch and the resumption of medullary bone formation.

When hens are fed a diet deficient in calcium, they normally stop laying in 10 to 14 days, having laid some six to eight eggs. During this period they may deplete their skeleton of calcium to the extent of almost 40 percent. It is interesting to inquire why they should stop laying instead of continuing to lay but producing eggs without shells. Failure to lay is a result of failure to ovulate; once ovulation takes place and the ovum enters the oviduct, an egg will be laid, with or without a hard shell.

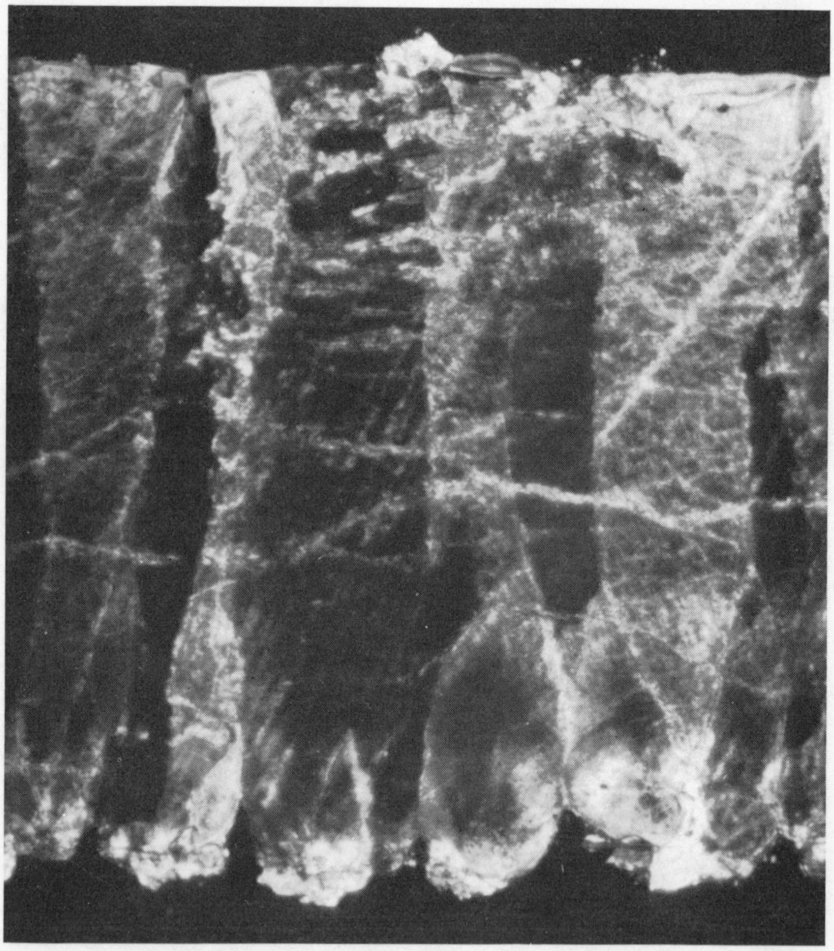

The question therefore becomes: Why do hens cease to ovulate when calcium is withheld from their diet? The most probable answer seemed to us to be that the release of gonadotrophic hormones from the anterior pituitary gland is reduced under these conditions. To test this hypothesis we placed six pullets, which had been laying for about a month, on a diet containing only .2 percent calcium—less than a tenth of the amount normally supplied in laying rations.

After five days on the deficient diet, when each hen had laid three or four eggs, we administered daily injections of an extract of avian pituitary glands to three of the experimental birds. During the next five days each of these hens laid an egg a day, whereas two of the untreated hens laid one egg each during the five days and the third untreated hen laid three eggs. We concluded that the failure to produce eggs on a diet deficient in calcium is indeed due to a reduction in the secretion of pituitary gonadotrophic hormones.

The mechanism of pituitary inhibition under these conditions has not been established. It is possible that the severe depression of the level of plasma calcium inhibits the part of the brain known as the hypothalamus, which is known to be sensitive to a number of chemical influences. The secretion of gonadotrophins in mammals is brought about by hormone-like factors released by the hypothalamus, but it is not known if the same mechanism operates in birds.

Plainly the laying of eggs with highly calcified shells has profound repercussions on the physiology of the bird. The success of birds in the struggle for existence indicates that they have been able to meet the challenge imposed on them by the evolution of shell making. Many facets of the intricate relations between eggshell formation, the skeletal mobilization of calcium, the ovary and the parathyroid and anterior pituitary glands await elucidation, but the general picture is now clear.

COLUMNAR STRUCTURE of an eggshell stands out in a photomicrograph also made by Terepka, using polarized light. The shell is seen in cross section at an enlargement of 325 diameters. The lumpy structures at bottom are the mammillary knobs of the eggshell.

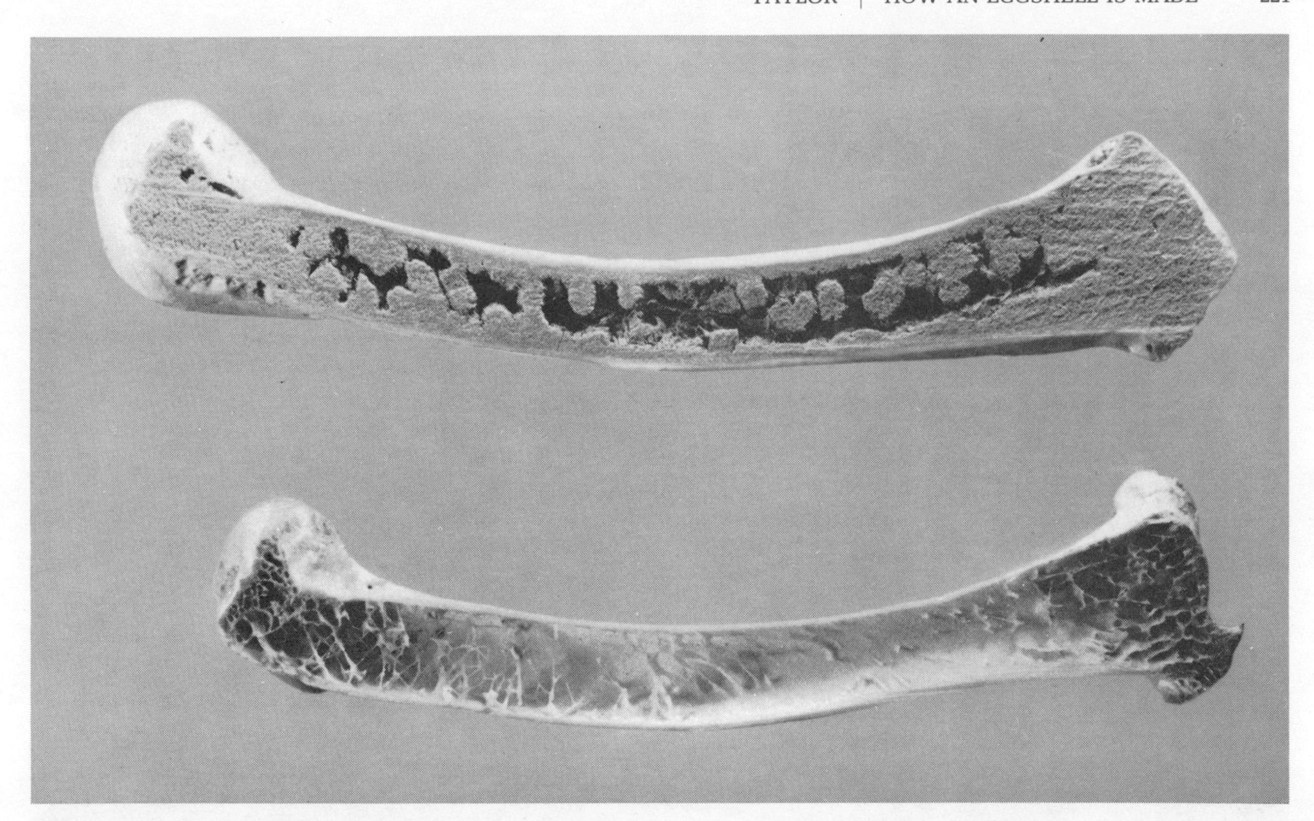

MEDULLARY BONE of a laying chicken contains the reserve of calcium that the hen draws on in forming the eggshell. Medullary bone in the femur of a laying chicken is shown at top; the structure consists of trabeculae, or fine spicules, of bone that grow into the marrow cavity from the inside of the structural bone. The femur of a nonlaying bird (*bottom*) shows no medullary bone.

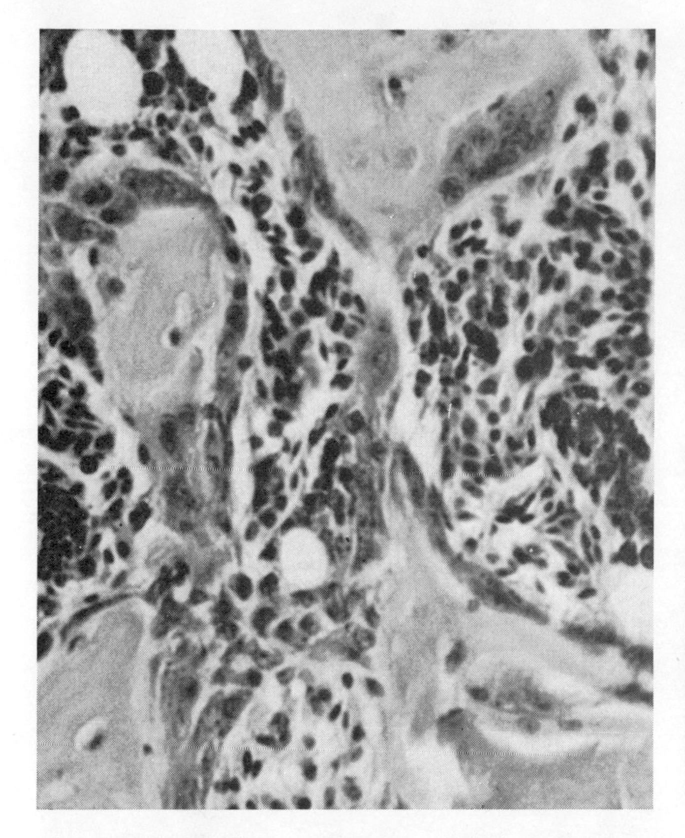

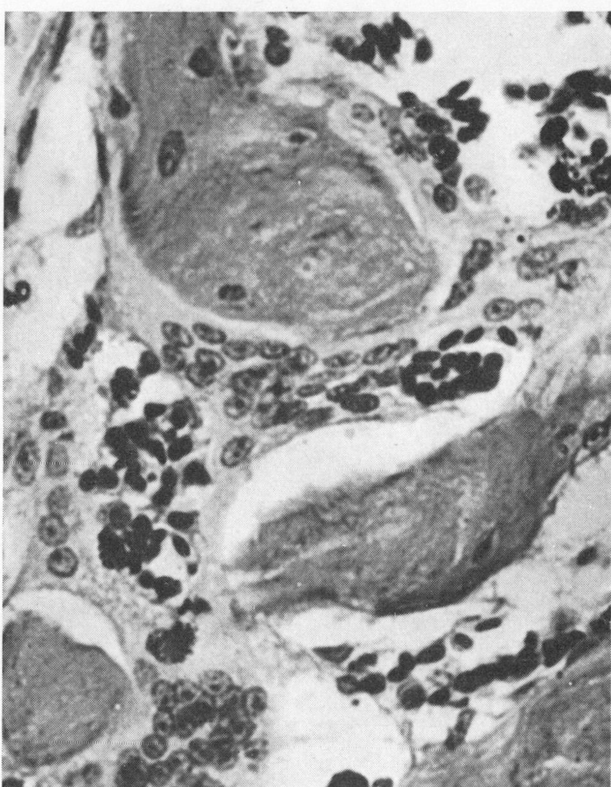

CELL POPULATION of medullary bone differs according to whether the bone is being built up or broken down. The bone is dominated by osteoblast cells (*left*) when the hen is accumulating a reserve of calcium and by osteoclast cells (*right*) when she is drawing on the reserve. The cells are the small, dark objects; the larger, gray objects are trabeculae. The bone is femur of a chicken; enlargement is 600 diameters. These micrographs were made by Werner J. Mueller and A. Zambonin of Pennsylvania State University.

DARK ALLEY in Glasgow, photographed in about 1870, is typical of the environment in which rickets was once endemic. In such a setting children received little ultraviolet radiation, which is necessary for the synthesis of the hormone that prevents rickets. Along the left side of the alley are two groups of children. Images are blurred because the children moved during the time exposure.

Rickets

21

by W. F. Loomis
December 1970

Although it is still widely regarded as a dietary-deficiency disease resulting from a lack of "vitamin D," it results in fact from a lack of sunlight. In smoky cities it was the first air-pollution disease

The discovery of the cause and cure of rickets is one of the great triumphs of biochemical medicine, and yet its history is little known. Indeed, it is so little known that even today most textbooks list rickets as a dietary-deficiency disease resulting from a lack of "vitamin D." In actual fact rickets was the earliest air-pollution disease. It was first described in England in about 1650, at the time of the introduction of soft coal, and it spread through Europe with the Industrial Revolution's pall of coal smoke and the increasing concentration of poor people in the narrow, sunless alleys of factory towns and big-city slums. This, we know now, was because rickets is caused not by a poor diet but by a deficiency of solar ultraviolet radiation, which is necessary for the synthesis of calciferol, the calcifying hormone released into the bloodstream by the skin. Without calciferol not enough calcium is laid down in growing bones, and the crippling deformities of rickets are the consequence. Either adequate sunlight or the ingestion of minute amounts of calciferol or one of its analogues therefore prevents and cures rickets, and so the disease has been eradicated.

That seems a clear enough story, and yet the textbooks speak of diet and vitamin D. How can that be? What happened is that the investigation of rickets proceeded along two quite independent lines. Intuitive folk medicine and then medical studies pointed in the direction of sunlight and calciferol. At the same time, however, common assumptions about poverty and poor nutrition and then studies by nutritionists pointed in the direction of diet and vitamin D. Now, with the inestimable advantage of hindsight, it is possible to trace these two chains of thought, to disentangle them and set the historical record straight.

Now that one knows what to look for, the evidence of a climatic influence on rickets can be discerned quite early. In the early 19th century G. Wendelstadt published *The Endemic Diseases of Wezlar*, a German town of 8,000 population with exceptionally narrow streets and dark alleys. The town was infamous for rickets, he wrote, with entire streets where in house after house individuals crippled by rickets could be found. "The children must sit indoors... which ends in death or if they continue to live, they develop thick joints, cease to be able to walk or have deformed legs. The head becomes large and even the vertebral column bends. It comes to pass that such children sit often for many years without being able to move; at times they cease to grow and are merely a burden to those about them." This terrible picture of an entire town afflicted with severe rickets leads one to guess that many of William Hogarth's sketches of frightfully deformed men and women may have depicted the crippling effects of rickets in London in the 18th century.

As early as 1888 the English physician Sir John Bland-Sutton found unmistakable evidence of rickets in animals in the London zoo—chimpanzees, lions, tigers, bears, deer, rabbits, lizards, ostriches, pigeons and many other species. He noted that "in spite of every care and keeping them in comfortable dens" lions in London developed rickets, whereas "in Dublin, Manchester, and some other British towns, lions can be reared successfully in captivity." It is clear in retrospect that the pall of coal smoke over London was the causative factor.

The geographical relation between rickets and cities was clearly noted in 1889 by the British Medical Association. After a survey of the incidence of the disease in the British Isles the association published maps [*see illustration on page 225*] that supported its major conclusion: There was widespread and severe rickets "in large towns and thickly peopled districts, especially where industrial pursuits are carried on," whereas rickets was almost totally absent in rural districts. Specifically, the report added, "almost the whole of London and the greater number of its outlying suburbs" reported severe rickets among rich and poor alike.

Solar ultraviolet may be blocked by many means, among them being the industrial smog in London and the sunless alleys of Wezlar, but beneath such specific industrial and urban conditions there is a major underlying factor: the far northern location of the entire European land mass. The area is made habitable by the benign influence of the Gulf Stream, yet its winter sun, hanging low in the sky, is almost without potency in effecting the crucial conversion of 7-dehydrocholesterol into calciferol. Elsewhere in the world, lands as northerly as Europe are largely uninhabited—the Aleutian Islands, for example, or Labrador or northern Siberia. The long, dark winters of Europe therefore powerfully predisposed European infants toward rickets during the winter months.

The seasonal variation was noted as early as 1884 by M. Kassowitz in Germany, who attributed it to the prolonged confinement of infants indoors during the winter. Then in 1906 D. Hansemann noted that nearly all German children who were born in the fall and died in the spring had rickets; those who were born in the spring and died in the fall were free of the disease. Noting the progressive rise of rickets during the winter months, he concluded that rickets was primarily a disease of "domestication,"

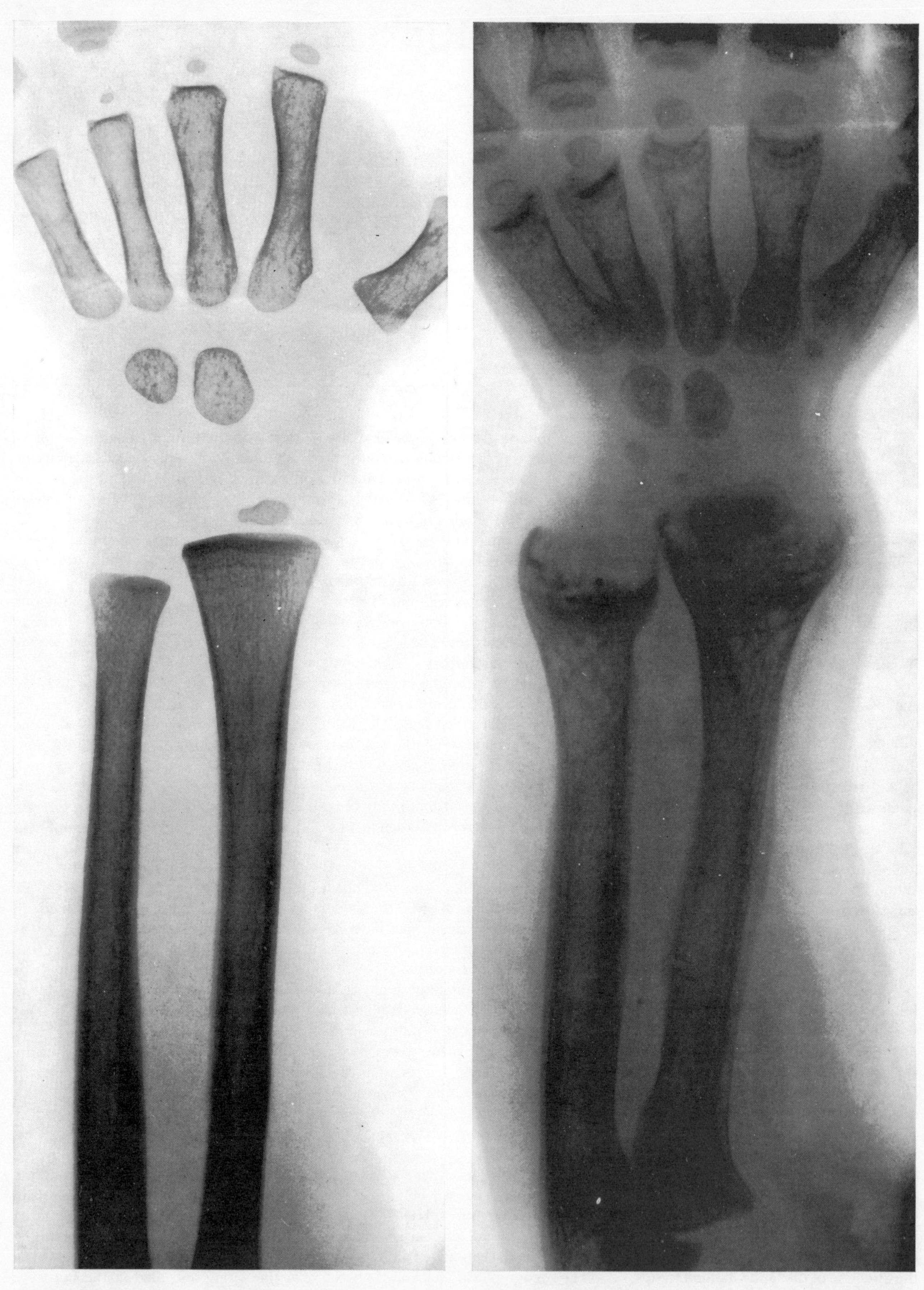

EFFECT OF RICKETS is deformation of bone for want of the calcifying hormone calciferol, which is synthesized on ultraviolet radiation. An X ray of normal arm and hand bones in an 18-month-old child (*left*) is compared with an X ray of the bones of a child of the same age with rickets (*right*). The disease can be prevented or cured by sunlight or the ingestion of small amounts of calciferol.

for "I have learned that rickets never exists in wild tribes or in animals [that] live in complete freedom. Once caught, however, most of these formerly wild animals—and especially monkeys—show great disposition towards rickets. Hardly one young captured animal can avoid this danger. By observing rickets in people, who do not get this disease to such a degree as monkeys, one can also see that it is a sickness of *domestication*. We can say that in living locked indoors, with thick, heavy walls and windows facing brick walls in other houses, the natural habitat of a child is being disturbed—namely the outdoors." In 1909 G. Schmorl strongly documented this marked seasonal variation in the frequency of rickets with a series of 386 postmortem examinations carried out on children under four years old.

Perhaps the most brilliant investigation into the nature of rickets was made in 1890 by Theobald Palm, an English medical missionary who went to Japan and "was struck with the absence of rickets among the Japanese as compared with its lamentable frequency among the poor children of the large centres of population in England and Scotland." He wrote to other medical missionaries around the world, collated the results and was amazed to find that rickets was essentially confined to northern Europe and was almost totally absent from the rest of the world.

Dugald Christie wrote him from Mukden, for example, as follows: "I have met with not a single case of rickets during a residency of six years in Manchuria," and this in spite of the fact that there were "no sanitary conditions whatever" and the only articles of diet were millet, rice, pork and vegetables. C. P. Smith reported from Mongolia that he had not seen any rickets. "We have 10 months in the year of almost constant sunshine. In summer the children go practically naked, and even in winter, with the rivers frozen into a solid mass of ice, I have seen children running about almost naked, that is during the day while the sun is shining." From Java a Dr. Waitz reported what was a known fact there: European children suffering from rickets recovered from the malady within a few months of moving to Java and without any medical treatment.

From data such as this Palm deduced that rickets was caused by the absence of sunlight. "It is in the narrow alleys, the haunts and playgrounds of the children of the poor, that this exclusion of sunlight is at its worst, and it is there that the victims of rickets are to be found in abundance." He proceeded to recommend "the systematic use of sun-baths as a preventive and therapeutic measure in rickets."

The first successful attempt to induce rickets experimentally in animals was made at the University of Glasgow in 1908 by Leonard Findlay. He published conclusive pictures of puppies that had been confined in cages and developed rickets; unconfined animals did not become rachitic. His results convinced him that the cause of rickets was not any

CORRELATION OF RICKETS with industrial areas and smoke from the burning of coal appeared in data assembled in 1889 by the British Medical Association. The map, which shows in gray the principal concentrations of rickets, is based on maps of England and Scotland prepared by the association. Since diets in these areas were in general better than those in poorer surrounding areas, the distribution of rickets is not what one would expect if the disease were of dietary origin. In actuality the cause was smoke that obscured sunlight.

defect in the diet, but he did not come to quite the right conclusion; he suggested that rickets was caused by "confinement, with consequent lack of exercise." More accurate was a brilliant experiment by Jan Raczynski of Paris in 1912. Raczynski pointed to lack of sunlight as the principal etiological factor in rickets. Two puppies, "newborn in the month of May from the same mother, were reared for six weeks, the first in sunlight from morning to evening, the second in deep shade in a large, well-ventilated cage. Both were fed in the same manner, that is exclusively on the milk of their mother." After six weeks the puppy kept out of the sunlight was markedly rachitic, a diagnosis confirmed by chemical analysis of its bones, which were found to contain 36 percent less calcium phosphate than the bones of the puppy that had been raised in the sun. Raczynski concluded that sunlight played a principal role in the etiology of the disease.

In 1918 Findlay returned to the problem with the assistance of Noel Paton. They did experiments with 17 collie puppies from two litters and reported that "all those kept in the laboratory showed signs of rickets to a greater or less degree. One which had been confined and had had butter was most markedly affected. It was unable to walk. Another of the confined animals which had had no butter was least affected." They concluded: "Pups kept in the country and freely exercised in the open air, although they had actually a smaller amount of milk fat than those kept in the laboratory, remained free of rickets, while the animals kept in the laboratory all became rickety." Findlay and Paton fed butter to some of their animals to check on the effect of diet, since the idea that rickets was a dietary-deficiency disease was already taking hold and milk fat was known as an important source of vitamin A. Their results argued against such a theory, of course. Moreover, it is known today that the adverse effect of butter they observed was due to the fact that "florid," or severe, rickets develops best in well-fed puppies; a poorly fed animal develops only mild rickets since the defect in calcification does not have as much effect in an animal whose bones are not growing. Rather than being a dietary-deficiency disease, therefore, florid rickets required a good diet, complete with vitamins A, B and C; only then could the puppy grow rapidly and hence develop incapacitating rickets.

Findlay's group had by now become known as the "Glasgow school," as opposed to the "London school" of nutritionists. Their competing theories led to two important studies of human rickets, one in Scotland and one in India.

Margaret Ferguson studied 200 families living in Glasgow among whom marked rickets existed and decided that inadequate air and exercise appeared to be the most potent factors. "Over 40 percent of the rachitic children had not been taken out, while only 4 percent of the nonrachitic children had been confined indoors." It is clear now that being out of doors was the chief variable, for both sets of children were free to exercise at will.

The most clear-cut investigation was conducted by Harry S. Hutchinson in Bombay. He found no rickets at all among poor Hindus who subsisted on a pitifully inadequate diet but who worked outdoors all day "and while at work left their young infants at some nearby point in the open air." In contrast, he found that rickets was exceedingly common among the well-fed Moslems and upper-caste Hindus, whose women usually married at the age of 12 and entered purdah, where the ensuing infants usually remained with their mother for the first six months of life in a semidark room in the interior of the house. Hutchinson found that infants of both sexes kept in purdah suffered severely from rickets; the girls, who entered purdah when they were married, recontracted the disease then. He concluded that "the most important etiological factor in the production of rickets is lack of fresh air, sunlight, and exercise." He then proceeded to cure 10 such cases of purdah-induced rickets by taking the patients out into the open air, "showing that removal of the cause removes the effect. All other factors remained constant and no medicine was given."

Although it was becoming increasingly clear by 1919 to many physicians that sunlight had the power both to prevent and to cure rickets, no method of providing summer sunlight during European winters was available. Not only were such winters generally cloudy, with an ineffective sun less than 30 degrees from the horizon, but also the cold usually required exposure of children to the sun in glassed-in solariums whose windowpanes, it is now known, effectively filtered out the required ultraviolet rays. Folk custom had taught northern European mothers to put their infants out of doors even during January for "some fresh air and sunshine." The trouble was that in large cities with narrow streets even this became ineffective because of the intervening buildings and the pall of smoke.

With natural sunlight ineffective, doctors such as E. Buchholz in Germany turned to artificial illumination such as that provided by the carbon-filament electric bulb. Since the ultraviolet component of such light is very small, the treatments did little good. Then in 1919 a Berlin pediatrician, Kurt Huldschinsky, tried the light from a mercury-vapor quartz lamp, which includes the ultraviolet wavelengths, on four cases of advanced rickets in children. He obtained complete cures within two months.

Huldschinsky's discovery of the subtle fact that it is the invisible portion of the sun's rays that prevents rickets solved the problem of this disease for all time. In addition to providing a truly effective method of curing the disease, he proceeded to show that an endocrine hormone must be involved. He irradiated one arm of a rachitic child with ultraviolet. Then he showed, with X-ray pictures, that calcium salts were deposited not only in the irradiated arm but in the other arm as well. This proved that on irradiation the skin released into the bloodstream a chemical that had the needed power to induce healing at a distance—in other words, a hormone.

After World War I, Huldschinsky's findings were extended by Alfred F. Hess in New York. He showed that sunlight alone had the power to cure rickets in children. He then showed that this was true also of rats that had been made artificially rachitic by means of a low-phosphate diet. In June, 1924, Hess found that ultraviolet irradiation rendered linseed or cottonseed capable of curing rickets. Similar results were obtained on whole rat rations later that year by Harry Steenbock. Hess proceeded to show that a crude cholesterol and plant sterols, as well as the skin, acquired the property of curing rickets when irradiated by ultraviolet light. In 1927 Otto Rosenheim and Thomas A. Webster showed that the plant sterol ergosterol (derived from ergot, a fungus) became enormously antirachitic when irradiated with ultraviolet light. This is the process that has now become routine: Some .01 milligram per quart of ergocalciferol—or what is called "vitamin D_2"—is added to almost all the milk sold in the U.S. and most European countries.

A description of the nature of the skin hormone naturally released by irradiated skin was finally provided in 1936 by Adolf Windaus of the University of Göt-

FOOD	WINTER	SUMMER
MILK	0	.025
BUTTER	0	4
CREAM	0	1
EGG YOKE	4	1.2
OLIVE OIL	0	0
CALF LIVER	0	0

CALCIFEROL IN FOOD is charted. The substance, which is often called vitamin D, is essentially absent from foodstuffs other than fish, particularly in winter. The numerals give the percentage of the minimum daily protective dose of calciferol in a gram of each food.

tingen. He demonstrated that 7-dehydrocholesterol is the natural prehormone that is found in the skin and showed how it becomes calciferol on ultraviolet irradiation.

The hormonal nature of calciferol had been recognized to some degree as early as 1923 by such an authority as the American pediatrician Edwards A. Park, who wrote a careful summary of the history of rickets in *Physiological Reviews.* He summarized his view of the complex situation by saying that rickets is best compared to the endocrine-deficiency disease diabetes rather than to the genuine vitamin-deficiency diseases such as scurvy, pellagra, xerophthalmia and beriberi. Hess shared this view. In the first sentence of a 1929 monograph he stated that rickets "must be regarded as essentially a climatic disorder."

How, then, has the London school's view that rickets is due to a deficiency of "vitamin D" prevailed even up to the present day? Why is the error almost universally found in modern textbooks of endocrinology, physiology, biochemistry and medicine and further propagated by the words printed on every carton of milk sold in the U.S. and many other countries: "400 U.S.P. units vitamin D added per qt."? The remainder of this article will attempt to explain briefly the origin of the mistake.

Modern studies such as those of G. A. Blondin of Clark University support the long-suspected fact that fish, unlike birds and mammals, are able to synthesize calciferol enzymatically without ultraviolet light. Shielded by water, fish receive essentially no ultraviolet (290-to-320-millimicron) radiation, and yet the bluefin tuna has up to a milligram of calciferol per gram of liver oil—enough to provide a daily protective dose of calciferol for 100 children. Cod-liver oil contains less than 1 percent as much, enough to protect against rickets if it is consumed in amounts equal to four grams per day. It is an effective antirachitic medicine because of calciferol's unusual stability: an oil or fat containing the hormone preserves its efficacy for a long time.

In the north of Europe fish has always been a staple of diet, and so the normal diet tended to protect children against rickets. Slowly, over the years, the people of Scandinavia and the Baltic regions became aware of the specific therapeutic value of cod-liver oil as a preventive and even as a cure for rickets. By the end of the 19th century this therapy had come to the attention of physicians, but it was not generally accepted because a number of variables made the evaluation difficult: the advent of spring, chance exposure to sunlight or some unrelated retardation of growth that reduced the severity of the rickets could mask the effect of the cod-liver oil. It remained for Hess to make the unequivocal demonstration. In 1917 he conducted a controlled test with Negro children in New York City, among whom rickets was severe and almost universal, and proved the prophylactic value of routine administration of cod-liver oil.

It was a significant finding but it helped to turn investigators away from sunlight and toward diet. In 1919, the very year in which Huldschinsky pointed directly to ultraviolet radiation as the crucial factor in preventing rickets, the British nutritionist Edward Mellanby re-

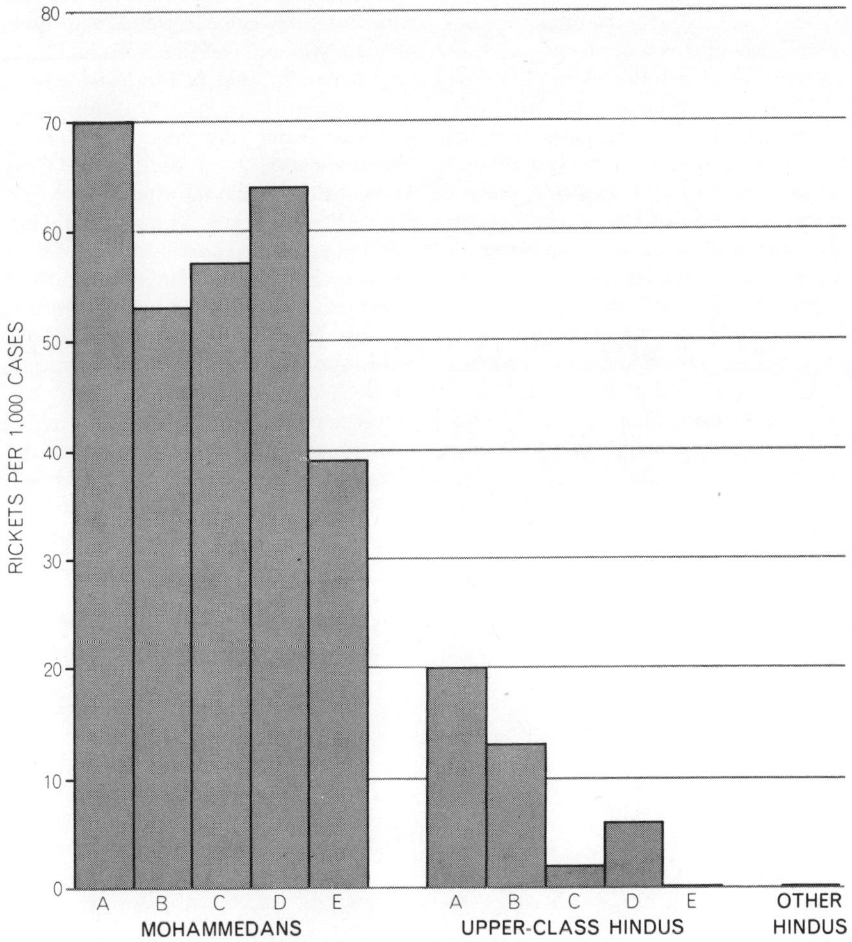

RICKETS PER 1,000 CASES
MOHAMMEDANS A B C D E
UPPER-CLASS HINDUS A B C D E
OTHER HINDUS

A UNDER 20
B 20 TO 24
C 25 TO 29
D 30 TO 34
E 35 AND OVER

INDOOR LIVING had a marked effect on rickets in India, according to a survey conducted by Harry S. Hutchinson in Bombay. He found a high incidence (*left*) among rich, well-fed Moslems, whose married women entered purdah and whose infants remained indoors; less among well-to-do Hindus (*middle*), whose children got outdoors more; none among poor Hindus (*right*), who had bad diets but who worked (and whose babies played) outdoors.

ported on what he called a "rachitogenic" diet. Working in London, under its pall of industrial smoke, Mellanby found that puppies would rapidly develop florid rickets on a diet reinforced with yeast, milk and orange juice—a natural finding, considering that Bland-Sutton had shown that almost all the animals in the zoo in industrial London suffered from rickets. Mellanby's announcement of the "production" of rickets in dogs by means of a particular diet was in line with the new "vitamine" theories of Frederick Gowland Hopkins and Casimir Funk; Mellanby suggested that the efficacy of cod-liver oil was "most probably" due to vitamin A, which was presumably missing in his "rachitogenic" diet. The diet finding was greeted with enthusiasm—even though Findlay had produced rickets in dogs 11 years earlier by simply keeping them indoors.

Park has pointed out that the report of Mellanby's "first experiments was meagre and would probably have awakened little interest, had not the British Medical Research Committee endorsed the work and publicly committed itself to the view that rickets was a deficiency disease due to a lack in the diet of 'antirachitic factor.'" It is clear enough today that Mellanby's idea that the cod-liver oil factor was probably vitamin A was wrong, but it is not generally recognized that *no* specific medicine such as cod-liver oil can be called a dietary vitamin unless it is present in normal foods in significant amounts. (Orally administered thyroid, for example, cannot be regarded as a "vitamin T" even though patients with insufficient thyroid of their own are cured by thyroid extract. Both endocrine secretions require external factors, incidentally: ultraviolet radiation in the case of calciferol and iodine in the case of thyroxine.)

In 1921 the American nutritionist Elmer V. McCollum, who had just accomplished the separation of fat-soluble vitamin A from water-soluble vitamin B, turned his attention to the rickets problem, putting his laboratory rats on Mellanby's "rachitogenic" diet. At first he could not produce rickets; being nocturnal animals, rats have become adapted to survival without direct sunlight and are resistant to rickets. Eventually Henry C. Sherman and Alwin M. Pappenheimer came on the trick of artificially giving rats a diet low in phosphate. Under this artificial stress the bones of young rats failed to calcify—unless they were placed in direct sunlight or were given cod-liver oil.

McCollum went on to establish the difference between the active factor in cod-liver oil and vitamin A in 1922 by showing that, after having been aerated and heated, cod-liver oil could still cure rickets but had lost its ability to cure xerophthalmia, which is due to lack of vitamin A. On this basis he called the cod-liver oil factor "vitamin D." Final recognition of the uniqueness of fish-liver oils came from the finding that animal fats such as butter and lard have essentially no calciferol, particularly in winter; the conclusion was clear that no nonfish diet of any kind could protect against rickets in a sunless environment. It was quite clear then that cod-liver oil was a medicine and not a food.

Nevertheless, McCollum had called it "vitamin D," and in the flush of enthusiasm for these new-found dietary factors the name acquired general acceptance. Semantic confusion now entered the picture in overwhelming force. Circular verbal proof made it evident that if "vitamin D" cured rickets, then rickets was a vitamin-deficiency disease! All the careful work demonstrating that rickets was primarily a climatic disorder was forgotten in the enthusiasm for the latest "vitamin." Chemists such as Windaus set about the task of deciphering its chemical structure. When Windaus received the Nobel prize in chemistry in 1928, it was "for his researches into the constitution of the sterols and their connection with the vitamins"—a curious citation in view of the fact that all biologically active sterols are manufactured by the body and are hormonal in character, whereas none of the known vitamins have a steroid structure. Even the discovery that calciferol was produced naturally in the skin in the presence of ultraviolet did not wipe out its classification as a vitamin or the definition of rickets as a dietary problem. Meanwhile the addition of ergocalciferol to milk had essentially eradicated the disease in Europe and America. Ironically, its effectiveness tended to buttress the dietary concept of the disease.

It took time for the correct view to emerge. In 1927 the chairman of the American Medical Association's section on the diseases of children remarked that "cod-liver oil is our civilization's excellent, economical and practical substitute—at least during the colder and darker half of the year—for exposure to sunlight. Is it not strange that the established vitamin deficiencies such as xerophthalmia, beriberi and scurvy are so rare in infants fed human milk from mothers and that rickets is so common? The great primary importance of the actinic [chemically active] rays to normal growth is evidenced by the fact that rickets occurs most severely and most frequently at the end of winter, and especially in those infants whose skins are pigmented. These observations strongly suggest that in human infants vitamins do not play a primary role in the development of rickets....

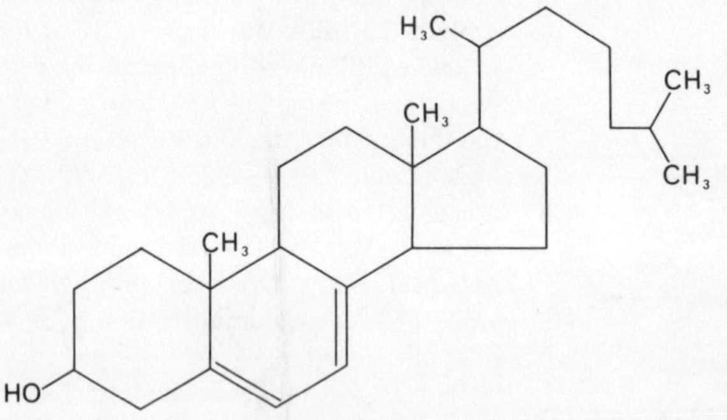

7-DEHYDROCHOLESTEROL

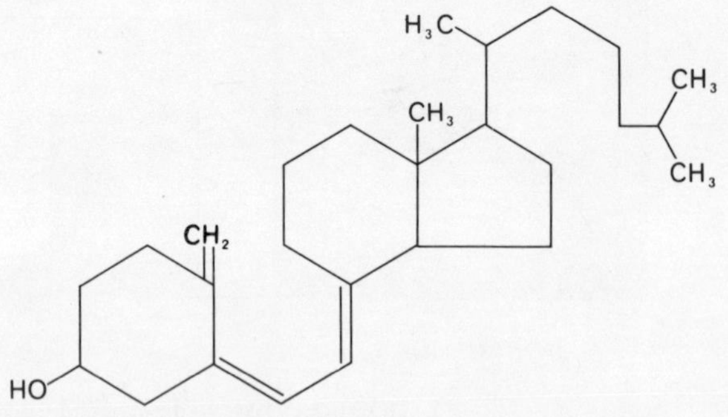

CALCIFEROL

CHEMICAL STRUCTURES of calciferol (*right*) and its precursor, 7-dehydrocholesterol (*left*), are closely related. The precursor is in the skin, and ultraviolet radiation from the sun is crucial in effecting its conversion to calciferol, which then enters the blood.

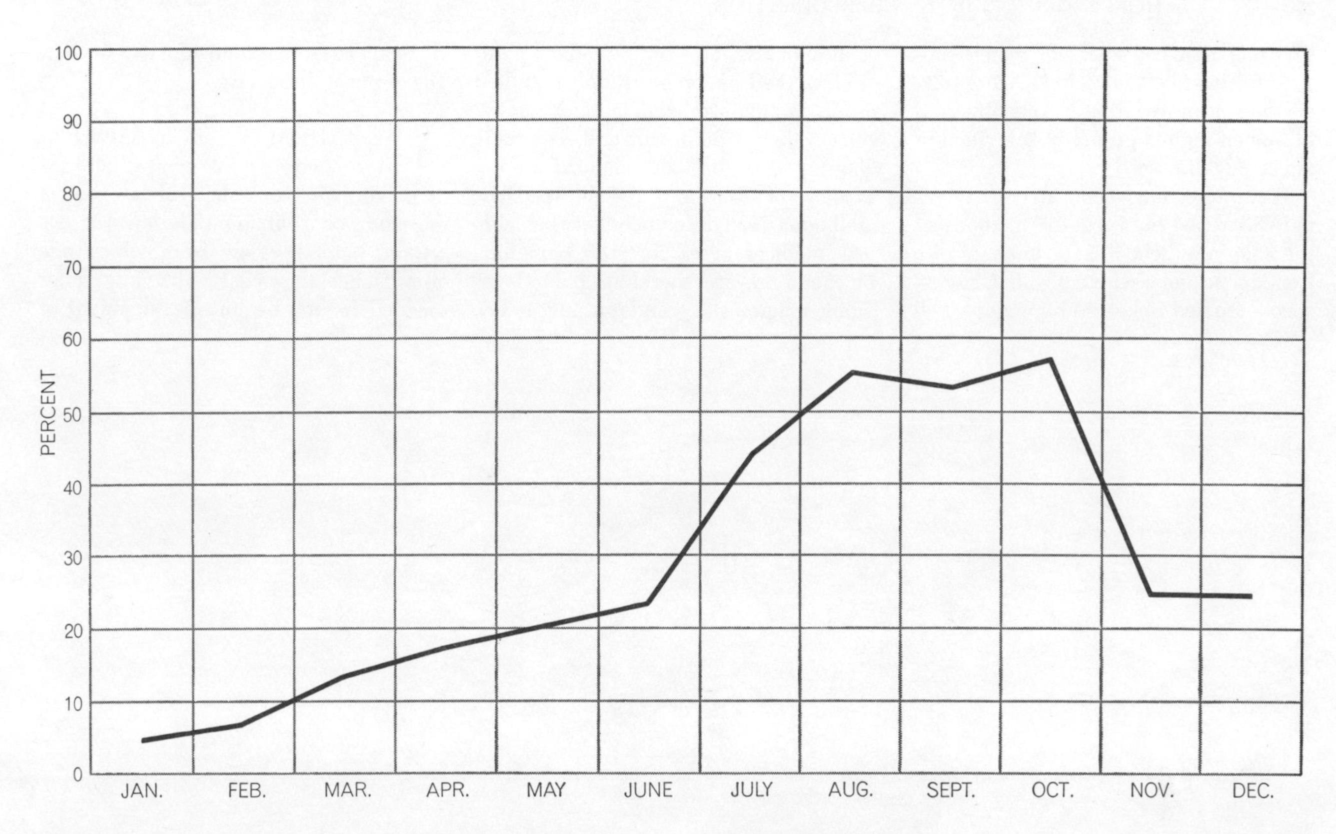

SEASONAL VARIATION in the frequency of rickets appeared in a series of 386 postmortem examinations of children with rickets conducted by G. Schmorl in 1909. Children were classified according to whether they had an active case of rickets at the time of death (*color*) or a "healing" case (*black*). It was clear that the severity of the disease increased in the fall and decreased with spring.

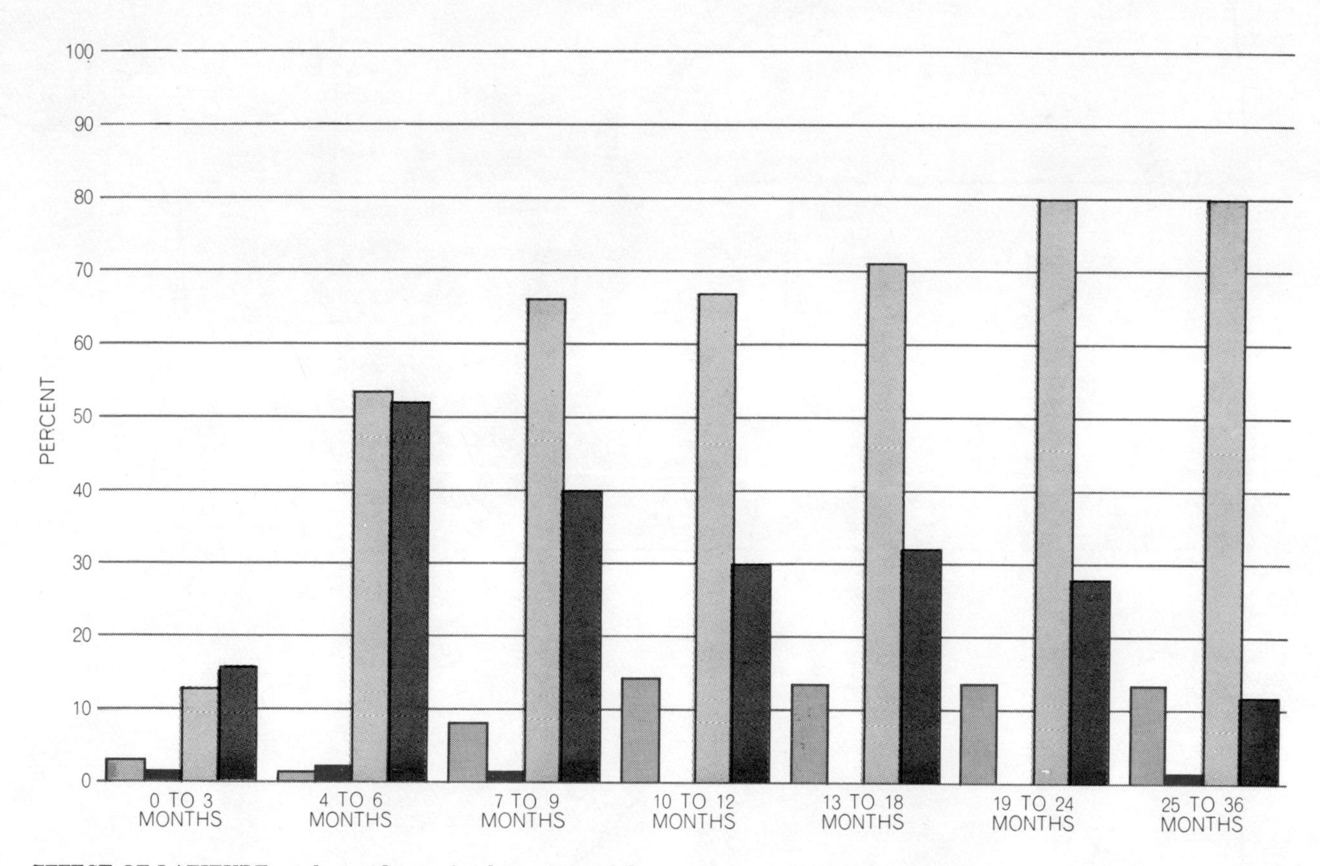

EFFECT OF LATITUDE on the incidence of rickets among children of various ages in Puerto Rico (*gray*) and New Haven (*color*) was demonstrated in a survey that was reported in 1933. The incidence is apparently related to the amount and strength of sunlight at 18 degrees of latitude and at 42 degrees. The light bars indicate clinical diagnosis of rickets, dark bars X-ray diagnosis.

The fact that cod-liver oil, which contains the so-called 'vitamin D,' cures rickets does *not* prove that rickets observed in human infants primarily is a vitamin-deficiency disease."

It is interesting to consider the essential difference between the methods of the Glasgow school and those of the London school. Whereas the Glasgow school studied rickets in humans as well as in animals, and from a medical point of view, the London nutritionists studied it only in animals, believing only the results of their experiments and essentially ignoring such brilliant medical studies as those of Palm and Hutchinson. The methodological differences between clinical medical research and beginning biochemistry are therefore behind the whole tangled story, and it is only today, 50 years later, that hindsight can explain the errors of those days.

A word should be said in answer to those who may ask what difference it makes whether calciferol is called a hormone or a vitamin. The answer lies in the point of view from which one approaches this vital calcifying factor needed for the healthy development of

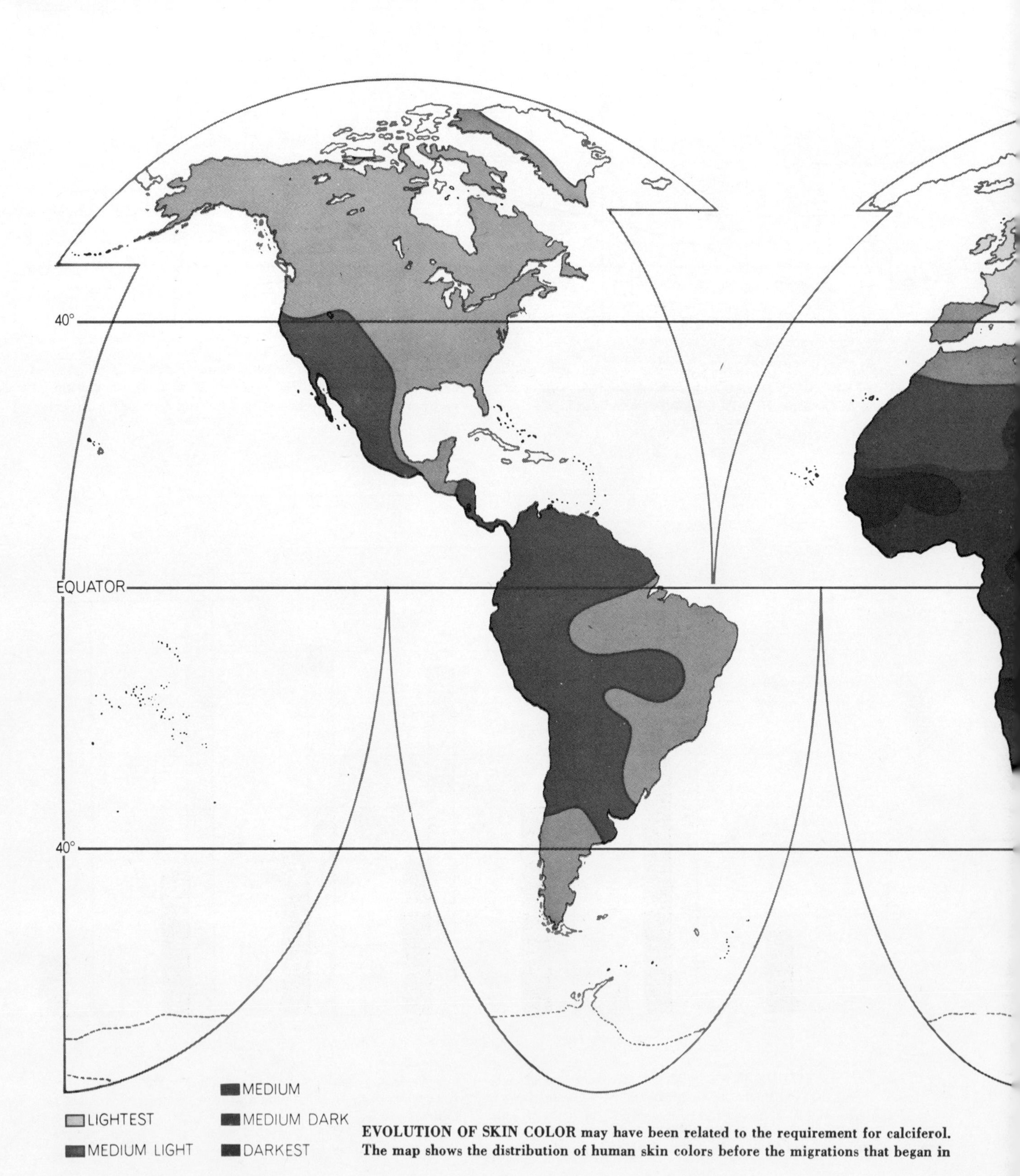

EVOLUTION OF SKIN COLOR may have been related to the requirement for calciferol. The map shows the distribution of human skin colors before the migrations that began in

the skeleton. Calling calciferol "vitamin D" at least suggests that it forms the nucleus of some cellular coenzyme, as is the case with many vitamins. Calling calciferol a hormone, on the other hand, explains why three hormones, calciferol, thyrocalcitonin and the parathyroid hormone are linked together in the delicate control of the level of calcium in the blood [see "Calcitonin," by Howard Rasmussen and Maurice M. Pechet; *beginning on page 205*]. Since no other cases of hormones and vitamins working together are known to medicine, it should not be surprising that calciferol turns out to be a steroid hormone whose production rate is under physiological control rather than being left to the vagaries of diet.

Other leads opened up by the hormonal view include the evolutionary development of the hormone. Fish synthesize it without ultraviolet light. Amphibians, reptiles, birds and mammals each apparently have some ultraviolet-receptive area of the body where the hormone is made, such as the ears of rabbits and the feet of birds. By and large, northern animals avoid rickets by living out of doors and by bearing their offspring in

the 16th century. Dark skin presumably protected against overproduction of calciferol. In Europe, much of which is at very high latitudes, man needed all the ultraviolet he could get, particularly in winter, and was presumably selected for unpigmented skin.

the spring, so that they are exposed to the summer sun during their growing period. Truly arctic animals, such as the polar bear and the seal, that live the year round in an area of deficient ultraviolet obtain their calciferol orally from their staple diet of fish. (The same was true of the Eskimos, who were entirely free of rickets until they were placed on a European diet by missionaries—when they too rapidly developed the disease.)

The recognition of calciferol as an ultraviolet-dependent hormone gives fresh meaning to a number of seemingly unrelated physiological and cultural adaptations. Tropical man probably avoids the dangers of too much calciferol production by virtue of his dark skin; the melanin granules in the outer layers protect the lower layers of the skin. European man, on the other hand, needed to use all the scanty ultraviolet light available, and consequently was gradually selected for an unpigmented skin such as is present in extreme degree in the blond-haired, blue-eyed, fair-skinned and rosy-cheeked infants of the English, north German and Scandinavian peoples. Indeed, the northern European idea of female beauty fits this picture: a girl with trim ankles, straight legs, fair skin and rosy cheeks must never have suffered from rickets and hence would probably bear strong sons and daughters if chosen as a mate. The very phrase "a fair young damsel" implies that a girl who is beautiful in the eyes of northern beholders is the possessor of an unpigmented skin! Delicate wrists are further proof of the absence of a history of rickets, as is a free-swinging walk, which is only possible in the absence of the pelvic deformities of rickets that would later endanger the process of childbirth.

June weddings tend to bring the first baby in the spring; an infant born in the fall was almost certain to have rickets by the time he was six months old. The fish-on-Friday tradition was as adaptive as the scurvy-preventing eating of an apple a day. Taking the baby out of doors even in the middle of winter for "some fresh air and sunshine" became a northern folk custom. Pink cheeks are visible evidence of the thinness of the unpigmented skin in the one area left uncovered in babies wrapped up warmly and placed out of doors in winter. The ability of outdoor-living Europeans to become deeply bronzed by the sun prevents the synthesis of too much calciferol in summer; it is significant that this seasonal pigmentation is induced by the identical 290-to-320-millimicron radiation that produces calciferol. Clearly northern European man, bronzed in summer but crocus white in winter, has an epidermis well adapted to the seasonal variation in ultraviolet radiation. Only with the advent of industrial smog did rickets appear in England and northern Europe.

Milk

22

by Stuart Patton
July 1969

The fluid made by the mammary gland is a remarkable blend of complex biological molecules. How the gland does its work is the subject of active investigation

According to the census of manufactures taken in 1967 the production, distribution and sale of dairy products constituted the seventh largest industry in the U.S., exceeded in value of shipments only by the motor vehicle, steel, aircraft, meat, petroleum and industrial-chemical industries. The basic product of the dairy industry is of course milk, which in the year of the census was produced in the amount of 118,769,000,000 pounds by 15,198,000 cows. As one might expect, the basic product has been studied intensively: nearly every state university has a program of dairy research, and the industry itself maintains a substantial research program. Notwithstanding these activities, much remains to be learned about milk. One reason is that milk is a remarkably complex substance; another is that the cellular processes whereby it is produced in the mammary tissue are highly intricate.

Milk's role, as a nearly complete food, in sustaining life processes is well known. Of equal importance is its role as a product of life processes. Milk is a record of the exquisite functioning of a cell, a fascinating cell that might be described as a factory, but a factory with the unusual property of becoming, to a certain extent, a product. Indeed, the lactating mammary cell ranks second in importance only to the photosynthesizing cell as a factor in sustaining life. For these reasons I shall focus here on the biology of milk, dealing to a lesser extent with its physical and chemical properties.

One's senses readily ascertain that cow's milk is a white, opaque liquid with characteristics of odor and flavor that are normally quite faint; a taste shows that it is slightly sweet and just perceptibly salty. One might go a step further and reason that rather large particles or molecules must be suspended in milk, because if all the constituents were fairly small dissolved molecules, milk would be as clear a solution as water is. Milk does have large components in suspension; they are mainly globules of fat and particles of protein.

Constituents of Milk

These observations indicate the gross composition of cow's milk: it is about 3.8 percent fat, 3.2 percent protein, 4.8 percent carbohydrate, .7 percent minerals and 87.5 percent water. Such an analysis, however, greatly oversimplifies cow's milk. For example, milk contains a large number of trace organic substances, some that pass through the mammary gland directly from the blood and others that result from the synthesis of milk in the mammary tissue. Moreover, the fat globules contain thousands of different molecules and are enclosed by a complex membrane acquired at the time of secretion. Milk protein was originally thought to have three components: casein, albumin and globulin. It is now known that there are four caseins, each with a number of genetic variants, and that albumin and globulin are actually a complex group of proteins known as the whey proteins. The number of proteins eventually discovered in this group probably will be limited only by the patience of the investigators and the sensitivity of the methods they apply. Only lactose, the sugar of milk, seems to be a pure and relatively simple compound.

The statements that can be made about cow's milk do not apply uniformly to the milk of other mammals, because there are large variations in the composition of milk. For example, the pinnipeds (the group of aquatic mammals including seals, sea lions and walruses) have milk that is often like heavy cream, containing 40 to 50 percent fat. In addition, depending on the species, pinniped milk contains little or no lactose. These variations can be explained in terms of their value in assisting the survival of the young of the species. A young pinniped is in special need of fat (in the form of blubber) as insulation against its cold environment, as an aid to buoyancy, as a source of energy and as a source of metabolic water in a salty environment.

A Closer Look

With the request that the reader keep in mind the important fact that milk differs substantially among mammalian species, I shall now be discussing milk in terms of cow's milk. Because of its commercial importance as a food and a raw material for foods, more is known about it than about the milk of other mammals. Moreover, the mechanisms of the synthesis of milk by the cow have been investigated closely because of the cow's importance as a unit of agriculture.

In addition to the major constituents already described, milk contains a large number of substances that occur in small amounts, ranging from .1 percent or so down to parts per billion. Among them are fatty acids, amino acids, sugars and sugar phosphates, proteoses, peptones, nitrogenous bases, gases and other volatiles. Many of these substances, such as the vitamins and minerals, play a key role in nutrition. Nonetheless, the most important components of milk are the lipids (fats), the proteins and the carbohydrate (lactose). A more precise description of them will lay a foundation for considering the remarkable processes of their synthesis and secretion by the cells of the mammary gland.

The term lipid specifies a broad group of fatty (greasy, waxy or oily) substances found in biological systems, including those used for food. The term fat is often used interchangeably with lipid, but in

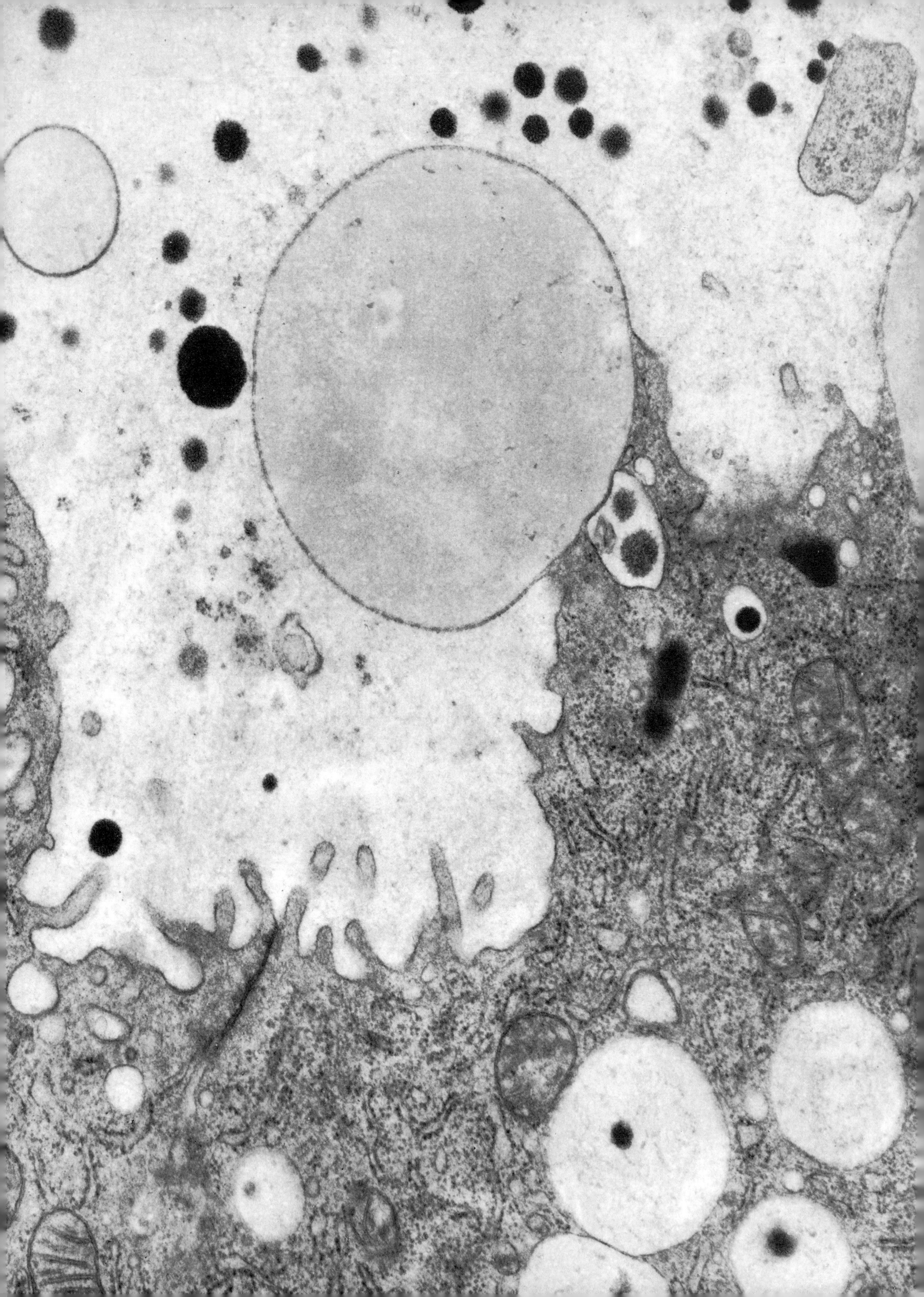

fact fat refers more narrowly to edible oils or the characteristically fatty tissue of the animal body. The lipids in milk are sometimes called its butterfat content; they exist as minute droplets or globules that under proper conditions will rise to form a layer of cream. The process known as homogenization reduces the globules in size and stabilizes their suspension, so that they no longer form a layer of cream. Agitation, in the form of churning, causes the globules to aggregate into granules that can be gathered and worked into butter. Butter is about 80 percent fat, and the part that is not fat is mainly water. If butter is melted, clarified and dried, it yields a product that is almost 100 percent fat and is known as butter oil; it is used commercially in the making of candy and baked goods.

The fat droplets in milk have an average diameter of three to four microns (about .00014 inch). A droplet consists of a membranous coat about .02 micron thick and a core that is virtually pure glyceride material. A glyceride is the ester, or product, resulting from the combination of glycerol with fatty acid. Because a molecule of glycerol has three reactive sites, it is possible to have monoglycerides, diglycerides or triglycerides, depending on how many molecules of fatty acid react with each molecule of glycerol. The lipids of milk are mostly triglycerides.

The fatty acids that are esterified with glycerol to form glycerides can vary in many ways. At least 150 different fatty acids can be found in the glycerides of milk, but only 10 of them occur consistently in amounts larger than 1 percent of the total. The principal ones are oleic acid, palmitic acid and stearic acid, which are also common in the glycerides of many other natural fats. The fat of cow's milk is unusual in that it contains the short-chain fatty acids, including butyric acid and caproic acid. Short-chain fatty acids are also found in the milk fat of other ruminants, such as the sheep and the goat. As I shall describe more fully below, the rumen, or first stomach, in these animals has a profound effect on their metabolism and on the composition of their milk fat. Another point to note in passing is that the short-chain fatty acids are highly odorous, and when they are released from the glycerides by the enzymes known as lipases, they contribute significantly to the flavor of many kinds of cheese.

The membrane that forms the surface of the milk-fat droplet is derived from the outer membrane of the lactating mammary cell at the time of secretion. It also appears to include materials that were at the surface of the droplet while it was still in the cell. The structure and composition of the membrane are the subject of intensive study. It is known that the portion of lipids not accounted for in the triglyceride fraction is involved in the membrane. The membrane lipids include part of the milk's cholesterol, phospholipids and glycolipids and most if not all of the vitamin A and carotene (a yellow pigment). The membrane also comprises unique proteins and enzymes, and its structure seems to be an aggregate of lipoprotein subunits. All in all, the milk-fat globule—a droplet of fat wrapped in a membrane—is a remarkable biological package.

Proteins of Milk

As with proteins in general, the proteins of milk are fundamentally chains of amino acid units. Since there are 18 common amino acids, the number of protein chains that could be formed is very large indeed. About 80 percent of the protein in milk, however, is casein. No other natural protein is quite like it. One aspect of its uniqueness is that it contains phosphorus; it is known as a phosphoprotein. In milk the molecules of casein are marshaled in aggregates called micelles, which are roughly spherical in shape and average about 100 millimicrons in diameter.

Casein occurs in four distinct types called alpha, beta, gamma and kappa, respectively representing about 50, 30, 5 and 15 percent of this protein. The four types differ in molecular weight and in a number of other characteristics. Kappa casein is unique in that it contains a carbohydrate, sialic acid. Little is known about the internal organization of the casein micelle and its subunits. It is assumed, however, that each subunit contains each of the four caseins.

The alpha, beta and gamma caseins can be made to aggregate by calcium ions, but kappa casein is highly resistant to such aggregation. Hence kappa casein serves as a protective colloid that keeps the casein micelles themselves from aggregating, which would give milk a curdlike consistency. In the making of cheese the enzyme rennin is added to milk to promote the formation of curds; the enzyme splits from kappa casein a peptide containing sialic acid, thereby destabilizing the casein micelles and giving rise to the formation of curds. After the resulting whey, or watery portion, has been drawn off, the curd can be used to make cheese.

Another protein unique to milk is beta-lactoglobulin, which accounts for about .4 percent of milk and is found in two common forms, *A* and *B*, and two uncommon ones, *C* and *D*. Beta-lactoglobulin contains a comparatively high proportion of the amino acid cysteine, which bears a reduced sulfur group (—SH). When milk is heated, these groups (starting at a temperature of about 70 degrees Celsius) are released from the protein as hydrogen sulfide; this is the source of the cooked flavor in heated milk.

Beta-lactoglobulin has much practical importance for the processing properties of milk proteins. If it is denatured beforehand, evaporated milk is stabilized against coagulating during sterilization by heat. On the other hand, if milk for cottage cheese is overheated so that the beta-lactoglobulin is denatured, an unsatisfactory soft curd is formed. Presumably the denatured protein is adsorbed on the surface of the casein micelles, thus hindering the action of rennin and the coalescence of the casein into a curd.

Enzymes are of course proteins too, and freshly secreted milk contains a great abundance of them. Robert D. McCarthy, working with radioactively labeled substances in our laboratory at Pennsylvania State University, has shown that enzymes in milk can incorporate fatty acids into glycerides and phospholipids and can convert stearic acid into oleic acid. It is also known that milk can synthesize lactose from added glucose. Such activities make it appropriate to describe milk as an unstructured tissue, in many ways resembling the enzymatically active solid tissues of the body.

The Carbohydrate of Milk

The substance responsible for the slightly sweet taste of milk is lactose, a

LACTATING CELL secretes a droplet of milk fat in the electron micrograph on the opposite page. The fat droplet is the large circular object at top center. The dark region from which it is emerging is the cell; the light region the droplet is entering is the lumen, or hollow portion, of an alveolus, one of the many pear-shaped structures that are basic units in lactation. The small dark circles visible in several places are granules of protein. The electron micrograph, which is of mammary tissue of a mouse, was made by S. R. Wellings of the University of Oregon Medical School; the enlargement is about 48,000 diameters.

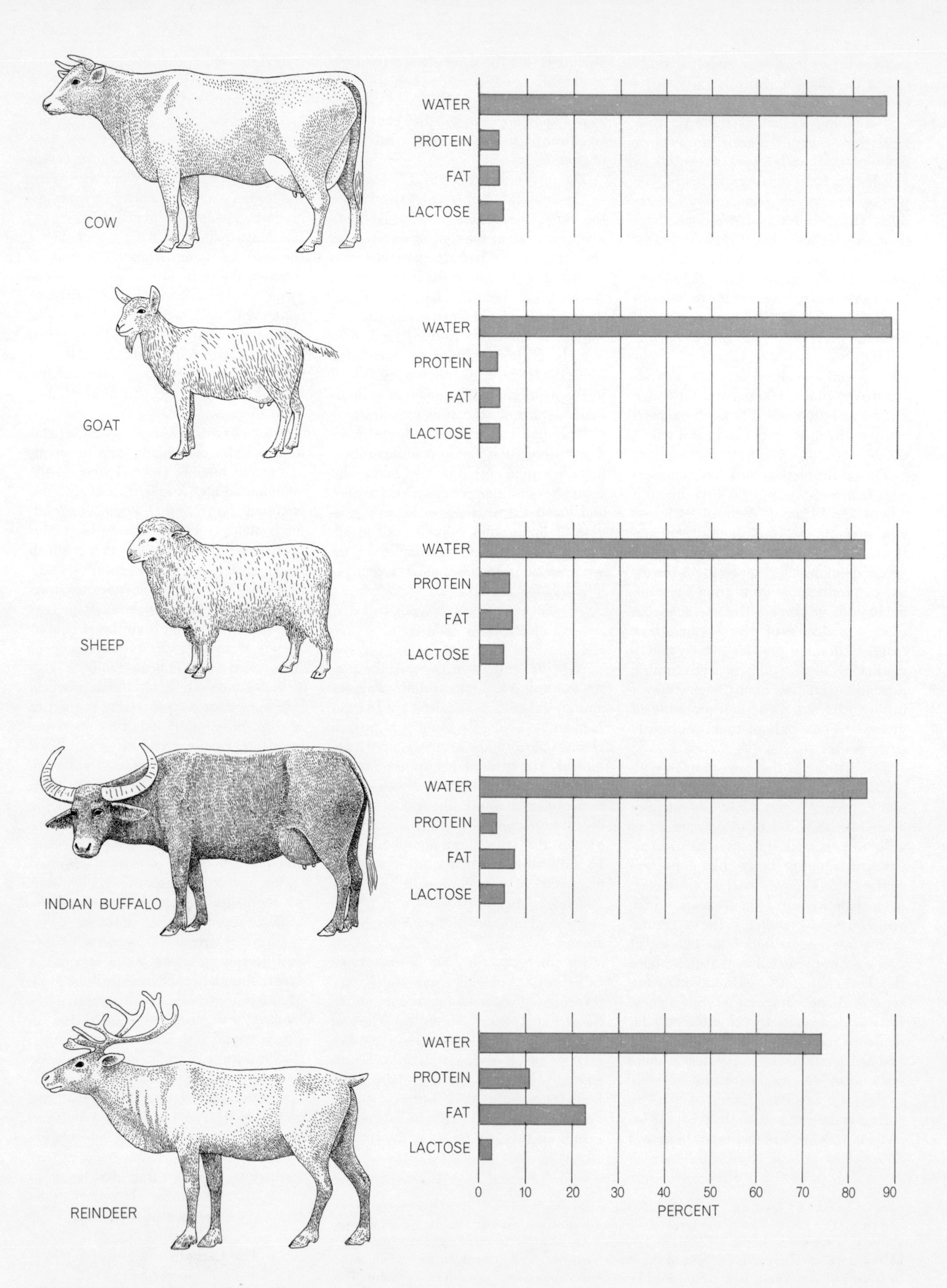

COMPOSITION OF MILK yielded by five kinds of animal is compared. In a few species the variation is even more marked; for example, the pinniped group, which includes seals and walruses, has milk that is about 50 percent fat and contains little or no lactose.

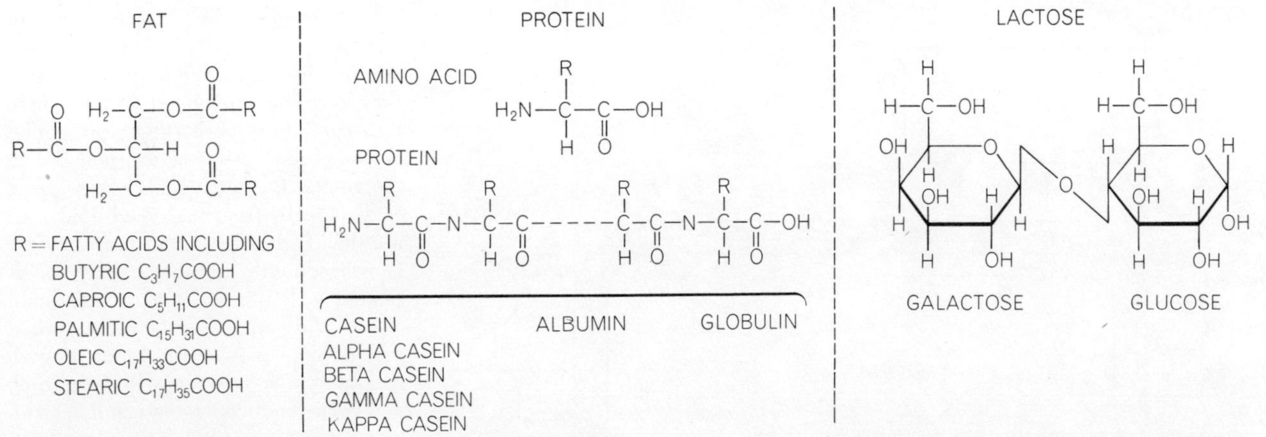

FAT

R = FATTY ACIDS INCLUDING
BUTYRIC C_3H_7COOH
CAPROIC $C_5H_{11}COOH$
PALMITIC $C_{15}H_{31}COOH$
OLEIC $C_{17}H_{33}COOH$
STEARIC $C_{17}H_{35}COOH$

PROTEIN

AMINO ACID

PROTEIN

CASEIN
ALPHA CASEIN
BETA CASEIN
GAMMA CASEIN
KAPPA CASEIN

ALBUMIN

GLOBULIN

LACTOSE

GALACTOSE

GLUCOSE

CONSTITUENTS OF COW'S MILK, exclusive of water and minerals, are portrayed with diagrams of basic chemical structure of its fat, protein and lactose. Casein is the major protein, and lactose the carbohydrate, of milk. Milk has many other fats and proteins.

carbohydrate with about a fifth the sweetness of ordinary sugar. Like casein, lactose is found only in milk. Lactose is composed of a molecule of galactose combined with a molecule of glucose, the simple sugar of the blood.

Since lactose is found only in milk, and only in the milk of certain species, one wonders why it is there. Indeed, it is reasonable to ask why milk contains carbohydrate of any kind, inasmuch as the most obvious role of lactose—providing a source of energy for the newborn—is filled by fat in milk from species such as the pinnipeds. The synthesis of lactose in the mammary gland does lock up molecules of glucose drawn from the blood, and since glucose is a highly active metabolite that might otherwise go elsewhere in the body or be metabolized in a different way, it may be that the synthesis of lactose provides a means of ensuring that glucose remains in the lactating cell and so becomes a part of the milk. Another possible role for lactose arises from its solubility; soluble molecules are important to the osmotic relations of cells, and lactose, which accounts for approximately 5 percent of milk, probably affects the osmotic relations of the lactating cell. Lactose may also be the carbohydrate of milk because it encourages certain desirable bacteria, which form lactic acid, to thrive in the intestine. Lactic acid is thought to promote the absorption of the calcium and phosphorus the young animal needs for the formation of bone. In any event, it would appear that the net effect of lactose from an evolutionary point of view was to promote the survival of the young, and so its synthesis was favored by natural selection.

A factor in the synthesis of lactose is the enzyme lactose synthetase, which is composed of two proteins. One of them, the *B* protein, was identified by Urs Brodbeck and Kurt E. Ebner of Oklahoma State University as alpha-lactalbumin. Thus for the first time a metabolic function for one of the principal milk proteins was identified. Then it was shown by a group at Duke University that the *A* protein is an enzyme that normally incorporates galactose into glycoproteins. In the presence of alpha-lactalbumin the enzyme has its specificity changed to the promotion of the reaction of galactose with glucose to form lactose. This seems to be the only case known where such a protein modifies the specificity of an enzyme. In subsequent work led by Roger W. Turkington the Duke group showed that organ cultures of mouse mammary gland, when pretreated with the hormones insulin and hydrocortisone, would produce both *A* and *B* proteins after treatment with the hormone prolactin. These three hormones are known to be necessary for the synthesis of milk in the mouse to begin.

In sum, the synthesis of lactose depends on enzymes, and the synthesis of the enzymes depends in turn on several hormones, which ultimately are also regulating the synthesis of the other components of milk. It is particularly interesting that lactose synthetase not only figures in the synthesis of lactose but also is present in the milk. This is evidence for my earlier observation that in milk production the factory becomes to a certain extent the product.

The Lactating State

In considering the synthesis of milk it is necessary to recognize that the process is related to all the other processes going on within the animal. It is an integral part of the animal's total metabolism. A case in point is the relation of milk fat to the other fats in the animal. Milk fat is immediately derived from two main sources: lipids circulating in the blood and synthesis from simple metabolites in the mammary gland. The origin of the simple metabolites traces back to the blood, to various sites in the body and ultimately to the food. The lipids circulating in the blood arise from all the many locations of lipid synthesis, transformation and storage in the body [*see top illustration on page 238*].

It is also necessary to consider metabolic activity at the cellular level. Clearly the lactating mammary cell, which is continuously turning out fat, protein, carbohydrate and many other substances, is not a resting cell—a cell that is simply maintaining itself. It is a busy place, with substances constantly moving in through the basal parts of the cell and out through the secreting parts. Some of the substances are used to maintain the cell and others are merely transported through the cell, but most of them are used by the lactating cell in the synthesis of the major constituents of milk.

Another consideration in the cow is the rumen, which is in effect a large fermentation tank ranging in capacity from 30 to 60 gallons depending on the size of the animal. Plant materials eaten by the cow are broken down in the rumen by a large and highly diverse population of bacteria and protozoa. The changes in the food are of major importance. Cellulose, which man cannot digest, is readily broken down in the rumen, and the products—acetate, propionate and butyrate—are prime metabolites in the bovine metabolic economy. Another sig-

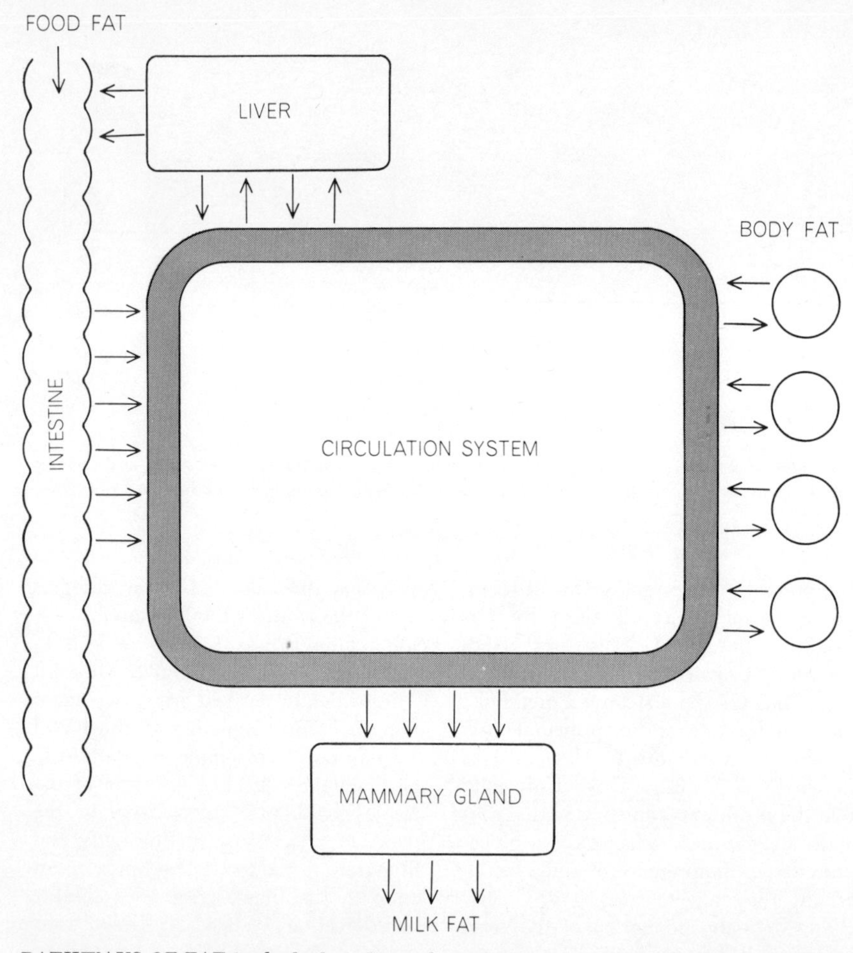

PATHWAYS OF FAT in the body indicate the sources of the fat in milk and thereby the mechanisms by which milk is synthesized. Fat from food enters the bloodstream through the intestines and also from the body's reserves of fat and from the liver. The mammary gland draws on these sources of raw material for the synthesis of the droplets of fat in milk.

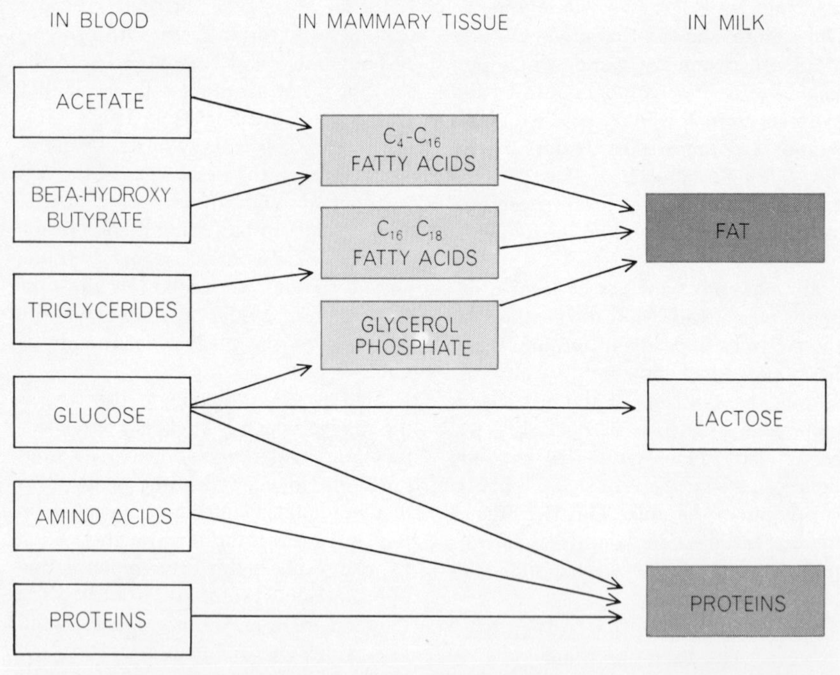

RAW MATERIALS of milk are depicted in a scheme showing the materials from the blood that are used by the mammary tissue in the synthesis of milk. Information on the materials going into the synthesis is obtained with radioactive tracers and by measuring changes in concentration of a substance between arterial and venous blood in mammary gland.

nificant change taking place in the rumen involves lipids. The lipids of plants are highly unsaturated, meaning that they have many free bonding sites where they can add more hydrogen molecules or form new chemical compounds. In the rumen the fatty acids are released enzymatically from such feed lipids and are then hydrogenated, so that they are converted into saturated fatty acids (mainly palmitic acid and stearic acid). These acids are subsequently absorbed and become part of the lipids of both meat and milk, which is why meat, milk and milk products of the ruminant animal contain saturated fats. In contrast, the milk fat of animals with a single stomach (such as human beings) will readily reflect dietary unsaturated fatty acids.

Another interesting fact about the rumen is that the microorganisms involved in fermentation become part of the milk as a result of subsequent digestion. Mark Keeney of the University of Maryland has estimated that at least 10 percent of the fatty acids of bovine milk are derived from the bacteria and protozoa in the rumen. Similarly, the amino acids used in the synthesis of milk proteins originate partly with microbes in the rumen.

Lactogenesis, the process that sets in motion the synthesis and secretion of milk, depends on the action of hormones. Hormonal changes in the female following conception lead to proliferation and differentiation of certain mammary cells. The organelles of the cell increase in size and quantity. Enzymes required to synthesize the various milk constituents appear in the cells, some gradually and some rather suddenly at about the time the animal gives birth. It is probably conservative to say that 100 enzymes are newly formed or greatly intensified in activity during the lactogenic transformation of tissue.

The mode of action of a hormone at the molecular level has not been established with certainty. As a result of work by Yale J. Topper and his colleagues at the National Institutes of Health, however, considerable progress has been made in determining what hormones are involved in lactogenesis and what effects they have at the cellular level. Working in vitro with mammary tissue from virgin mice, Topper's group has shown that the hormones insulin, hydrocortisone and prolactin, acting synergistically, are required to stimulate the synthesis of milk by the mammary tissue.

The hormone progesterone has an inhibitory effect on the differentiation of

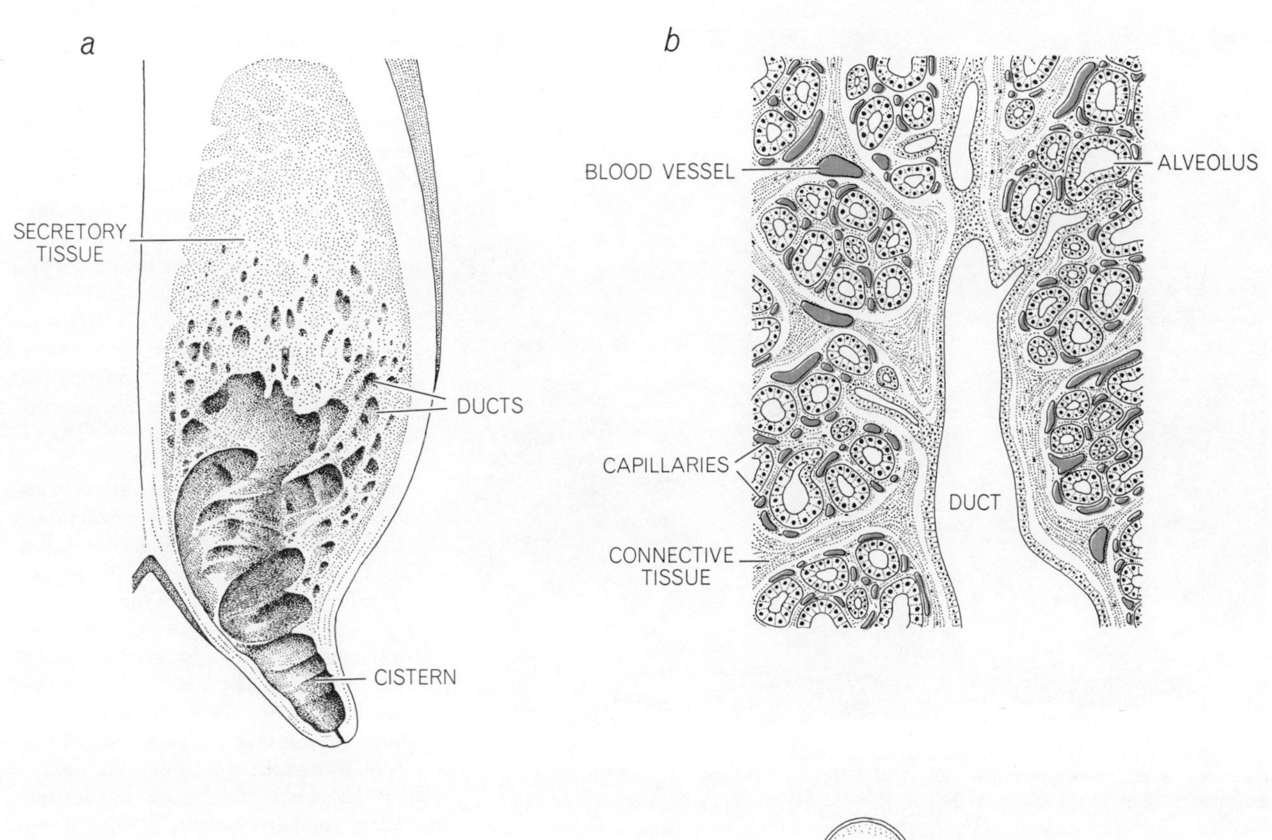

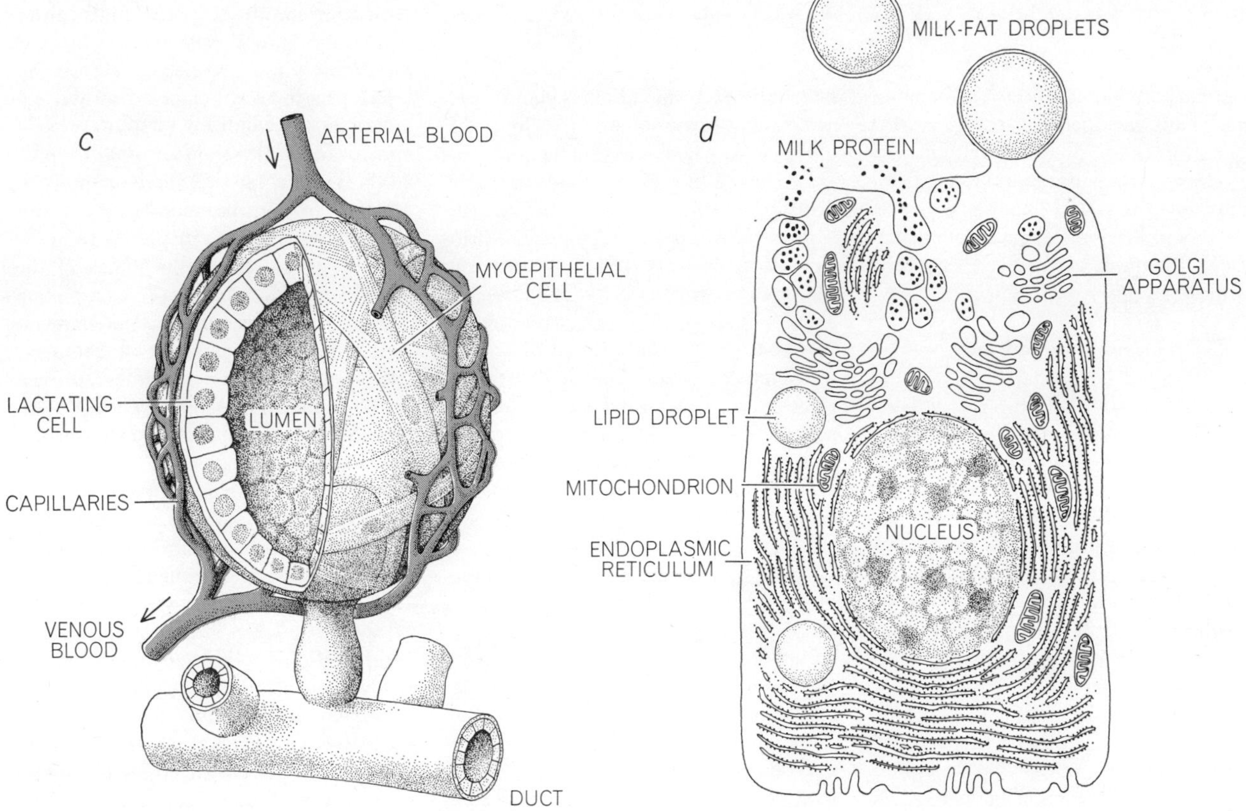

MILK-PRODUCING TISSUE of a cow is shown at progressively larger scale. At (*a*) is a longitudinal section of one of the four quarters of the mammary gland. The boxed area is reproduced at (*b*), where the arrangement of the alveoli and the duct system that drains them is apparent. A single alveolus (*c*) consists of an elliptical arrangement of lactating cells surrounding the lumen, which is linked to the duct system of the mammary gland. A lactating cell (*d*), similar to the one in the electron micrograph on page 114, is shown as it discharges a droplet of fat into the lumen. Part of the cell membrane apparently becomes the membranous covering of the fat droplet. Dark circular bodies in the vacuoles of Golgi apparatus are granules of protein, which are discharged into the lumen.

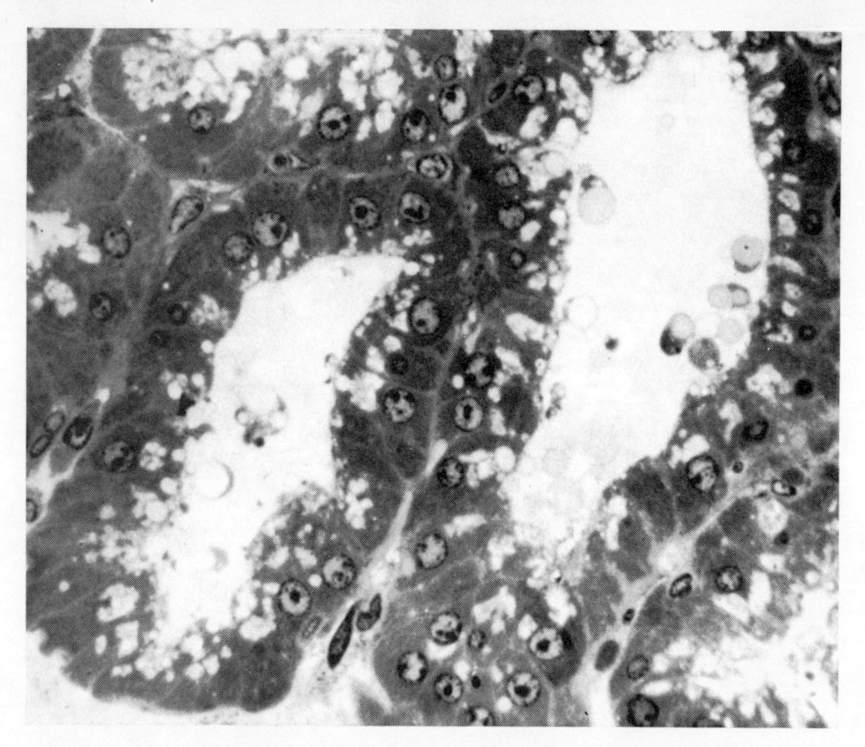

TWO ALVEOLI in the mammary gland of a lactating rat appear in a photomicrograph made by C. W. Heald in the laboratory of R. G. Saacke at the College of Agriculture of the Virginia Polytechnic Institute. The light areas are the lumens, which are surrounded by lactating cells. Between the two alveoli is a capillary. The enlargement is 2,500 diameters.

mammary tissue and the secretion of milk. Thus lactogenesis results in part from suppression of the activity of progesterone. Another important regulator of milk synthesis and secretion is the removal of milk from the mammary gland. Unless the milk is removed regularly, synthesis stops.

The Synthesis of Milk

Milk is produced by the vast number of cells that make up the mammary gland. The cells are formed into billions of pear-shaped, hollow structures called alveoli [*see illustration on previous page*]. Each cell in an alveolus discharges its milk into the lumen, which is the hollow part of the structure. When an alveolus is full, its outer cells contract under the influence of the hormone oxytocin, causing the alveolus to discharge its milk into a duct system that carries it to the cistern, or sac, that is the main collecting point.

In the present state of knowledge little more can be said about the precise mechanisms taking place within the lactating cell. It is possible, however, to describe to a certain extent the raw materials the cell draws from the blood and how they get from the blood into the cell. Much of the knowledge about the raw materials or metabolites comes from painstaking experiments by John M. Barry of the University of Oxford and J. L. Linzell and E. F. Annison of the British Agricultural Research Council's Institute of Animal Physiology. The principle of their work has been that any compounds being used by the mammary gland will show a drop in concentration between the entering arterial blood and the departing venous blood. The British workers have measured such arteriovenous differences and also have amplified their findings through use of radioactive tracers. The metabolites include acetate, triglycerides, glucose, amino acids and proteins.

In order to reach the cell the metabolites must move through the walls of the blood-carrying capillaries, across the endothelium lining the capillaries, through intercellular spaces and into the alveolar epithelium where the milk is synthesized. For the transport of small ions from the blood there is a selective system, as is shown by the fact that compared with the blood serum milk is much reduced in sodium and chloride ions and much elevated in potassium ions. Molecules such as the amino acids, glucose and acetate could in principle simply diffuse through the system. In the light of the rapid, selective and continuing uptake of these substances, however, it seems likely that here too there are special transport systems. The same is true for certain large molecules of protein that appear to pass unchanged from the blood into the milk. They include serum albumin and the immune globulins. Unless there are extremely large discontinuities in the cellular structure all the way from the blood to the milk, one is almost obliged to invoke special transport mechanisms for these large molecules. (Most of the milk proteins are made within the cell by the usual method of transcription by ribonucleic acid on the ribosomes of the cell.)

This question of how large particles can migrate through cell membranes without destroying the integrity of the cells arises for the exports of the lactating cells as well as for its imports. Indeed, the fat droplets steadily secreted by the cell are at times nearly as large as the cell itself. The first evidence of the remarkable biophysical mechanisms required for this task was supplied in 1959 by electron microscopy done in Germany by Wolfgang L. Bargmann and his coworkers at the University of Kiel. From their micrographs of mouse and hamster mammary tissue it was deduced that the fat droplets are secreted as a result of being progressively engaged in and enveloped by the limiting membrane at the apex of the cell [*see illustration on page 238*]. When a fat droplet is completely surrounded by the membrane in this way, it is effectively displaced from inside the cell out into the lumen of the alveolus. All that remains for completion of its secretion is for the slender membrane bridge to be pinched off. From our calculation of the sizable forces (as much as 100 atmospheres) involved in the attraction of the membrane to the surface of the droplet it is possible that once the droplet makes a close approach to the membrane it is quickly and forcefully snapped through.

Bargmann and his colleagues reported that their electron micrographs suggested the existence of milk-protein granules inside cell vacuoles that arise from the Golgi apparatus [*see illustration on previous page*]. According to the German workers the contents of the vacuoles were secreted by being emptied through the cell membrane. Our group believes the membrane processes in the two mechanisms of secretion (for fat droplets and protein granules) are related. The membrane around the Golgi vacuole carrying protein granules becomes continuous with the cell membrane at the time the vacuole empties its contents into the

STORAGE TANKS in a milk and ice cream plant of the Borden Company in West Allis, Wis., hold 48,000 quarts each. Raw milk delivered to the plant from dairy farms is stored in the tanks at a temperature of about 37 degrees Fahrenheit until it is processed. The five slender pipes at bottom are used in loading and unloading the tanks. An inspection port and a gauge are at top of each tank.

lumen. The cell membrane then is engaged by the milk-fat droplet and becomes the membrane around the secreted milk-fat globule. Evidence from both electron microscopy and biochemical studies is tending to substantiate these mechanisms of secretion for the lipids and proteins of milk, but many questions of both a gross and a refined nature remain to be answered.

Further Questions

The basal portion of the lactating cell, as distinguished from its apical, or secreting, end, contains extensive membranous processes known as the endoplasmic reticulum. The evidence is convincing that these membranes are the sites of synthesis for the major constituents of milk: triglycerides, proteins and lactose. The means whereby the components are gathered for secretion is not known. For example, it is not understood how all the triglyceride molecules are gathered into a droplet.

The precise operation of the Golgi apparatus is not established. The apparatus is defined as an organelle that accomplishes the differentiation of membranes and the "packaging" of materials for secretion, but just how it does these things in the lactating cell is unclear. It is now a reasonable assumption that alpha-lactalbumin, the *B* protein of lactose synthetase, joins the *A* protein in the Golgi apparatus, thus allowing the synthesis of lactose at that site. The lactose, milk protein and other constituents of milk serum are then packaged into Golgi vacuoles for secretion. One wonders, however, if the Golgi vacuoles are the vehicle of secretion for all the milk proteins. Electron micrographs show clearly that the granules the vacuoles contain have the appearance of casein micelles. Perhaps the other proteins are also present in the vacuoles but are not evident because they are transparent to electrons. We have suggested that these vacuoles, since they must carry some of the fluid of milk, may provide the vehicle for the secretion of milk-serum constituents such as lactose. Investigations of these questions and inferences are needed.

In sum, the lactating mammary cell, like all cells, is an almost incredible unit of organization and action. We are beginning to gain an understanding of this cell, which is so important to mammalian life, but the detailed revelation of its elegant mechanisms is still to come. The findings may lead to more and better food products. At the very least they will mean a deeper understanding of cellular processes and of life itself.

23

The Hormonal Control of Behavior in a Lizard

by David Crews
August 1979

The green anole, often mistakenly called a chameleon, is a good subject for examining how the sex glands and the brain interact in orchestrating the sexual behavior of both males and females

Generations of American children have become familiar with the green anole lizard (*Anolis carolinensis*) as the chameleon, since it is the animal usually sold by that name in pet shops. Recently biologists have become familiar with *A. carolinensis* as an excellent animal for laboratory studies of the interaction of behavior and hormones: the "chemical messengers" in the body that act at a distance from the site of their manufacture. It was a boyhood interest in the green anole (which I then thought was a chameleon) that led me as a behavioral biologist to work with the animal as a means of investigating the bases of reproductive behavior. The findings among other things provide a new outlook on the adaptive function of the relation between behavior and hormones that is seen in animals of quite different kinds.

The true chameleon and the green anole have much in common. They are both lizards. Most species live in trees or bushes, subsisting mainly on insects. Both can change color, although the anole's ability to do so is considerably more limited than the chameleon's. This is the trait that has made chameleons and anoles popular as pets, but the anole's color change is not (as many people think) related to the color of the background. Instead it is determined by such factors as light and temperature or by such emotions as fright, triumph and defeat. The chameleon is an animal of the Old World, whereas the anoles are found in the warmer regions of North and South America. The chameleon lays from two to 40 eggs at a time, the anole only a single egg.

The particular value of the green anole as an experimental animal is that it is abundant and that under the appropriate conditions it will establish in the laboratory the same social system and behavior it displays in its natural environment. A typical population might be found in southern Louisiana. There the anoles are reproductively inactive and essentially dormant from late September through late January. During that time they cluster in groups behind the loose bark of dead trees or under rocks and fallen logs. Beginning late in January or early in February the males emerge and establish breeding territories. The females begin to be active about a month later; by May each female is laying her single egg in the ground at intervals of from 10 to 14 days. The breeding season, which extends through August, is followed by about a month in which both males and females are refractory, that is, insensitive to the environmental and social stimuli that in the spring induced them to begin breeding.

Research on the reproductive behavior of a number of kinds of animal, most notably the work of the late Daniel S. Lehrman with the ring dove, has shown that the behavior is the consequence of two interlocking systems, one system represented by the animal and its internal state and the other by the animal and its environment. For example, among many birds and mammals the increasing amount of daylight in the spring acts on the brain of males to stimulate the pituitary gland to secrete increasing quantities of gonadotropic hormones. In lizards the primary stimulatory cue is the increase in temperature. Paul Licht of the University of California at Berkeley has shown that lizards also differ from birds and mammals in secreting one gonadotropic hormone rather than two of them.

The gonadotropin is transported by the bloodstream to the testes, where it stimulates the production of sperm and the secretion of androgens (the collective term for male hormones), specifically testosterone. The rising concentration of testosterone in the blood feeds back to specific areas of the brain to modulate further secretion of gonadotropin by the pituitary and to activate male sexual behavior. The male's behavior then becomes an important feature of the environment that stimulates the pituitary of the female to secrete gonadotropin, which rapidly induces ovarian development, steroid-hormone secretion and female sexual behavior. As the female responds to the male, her behavior influences his behavior and indirectly his physiology, and so the cycle begins anew.

The green anole has a rich repertory of behavior. A sexually active male will patrol his territory, stopping at prominent perches to execute an "assertion display." It is characterized by a bobbing movement that is typical of the species and is coordinated with the extension of a red dewlap at the throat. If a strange male green anole enters the territory, the resident male responds with a "challenge display." It is identified by an extreme lateral compression of the body and a highly stereotyped bobbing movement. If the intruding animal does not immediately respond with a "submission display" (a rapid nodding of the head) or responds aggressively to the challenge, the resident approaches the intruder and a fight ensues. During such a fight the dewlap is not extended but the throat region is engorged by a lowering of the tongue-supporting apparatus. As the fight progresses a crest is erected along the back and neck and a black spot forms directly behind the eye. It is not uncommon for males to lock jaws while they circle each other, each one trying to throw the other off the perch. After a fight the winning male (almost always the resident) is likely to climb to a prominent perch and execute a series of assertion displays.

In a "courtship display" a male advances toward a female, pausing for a series of bobbing movements and exhibitions of the dewlap. The number of bobbing movements varies greatly among individual males; the variation may help the female to distinguish one male from another. If the female is sexu-

ally receptive, she will allow the male to approach her and grip her neck. The male then mounts the female and curves his tail under hers. This action brings the cloacal regions into apposition so that the male can evert one of his two penises. (Lizards and snakes are unique among the vertebrates in having two of these organs, each of which is called a hemipenis.)

Much has been learned about the physiological control of the *A. carolinensis* male's reproductive behavior. As one might expect, male sexual behavior depends on the gonads. Castration leads to a rapid decline in courtship activity, but the administration of androgen reinstates it. Aggressive behavior appears to be less affected by castration. It is strongly affected, however, by environmental factors. For example, if a male

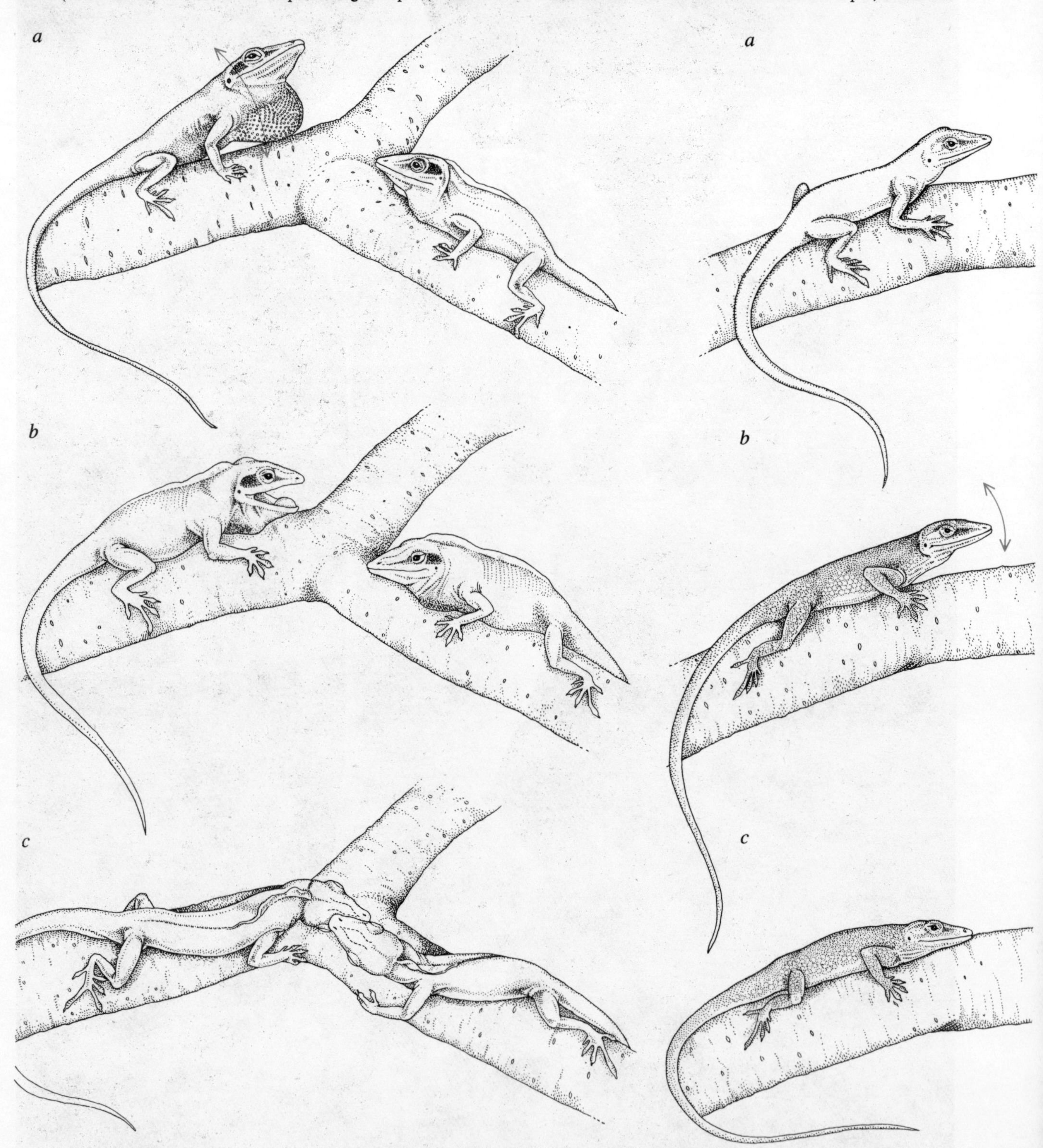

CHALLENGE DISPLAY given by a resident male (*a*) when a strange male intrudes on his territory involves an extension of the dewlap, a pronounced lateral compression of the body and a stereotyped bobbing of the body. If the intruder does not respond with a submission display, the resident moves toward the intruder (*b*), often with mouth agape, and they fight. During the fight a crest is erected along the back and neck and a black spot appears behind the eye. The males may lock jaws (*c*). The fight ends when one lizard throws the other off the perch.

CHALLENGE AND SUBMISSION are portrayed. A sequence begins (*a*) when a strange male, here the one on the left, intrudes on the territory of a resident male. The resident responds with a challenge display. This time, however, the intruder does not react aggres-

is returned to his home cage after castration, his aggressive behavior declines slowly or not at all, whereas if he is put in an unfamiliar cage, aggressiveness decreases quickly.

It has long been assumed that in vertebrate animals a male's reproductive behavior is triggered by the action of androgens on the brain. An alternative hypothesis was recently proposed by Frederick Naftolin of the Yale University School of Medicine. In essence it is that the "male" hormone testosterone is converted by enzymes in the brain into a "female" hormone, estradiol, which then activates the male's sexual behavior. The process that characterizes the structural change in the hormone is known as aromatization. In stimulating secondary sex structures, however, testosterone is believed to act directly or

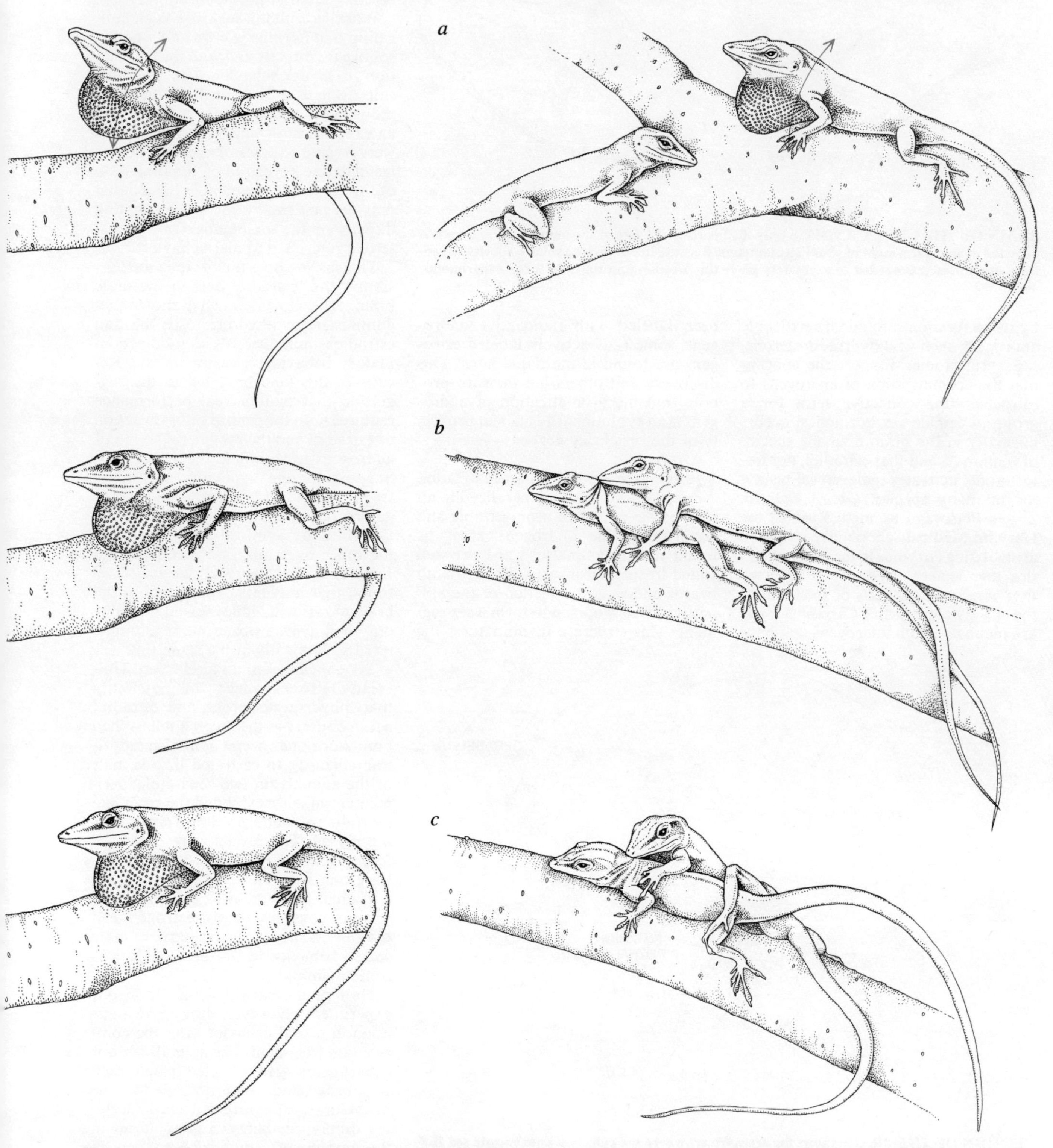

sively but instead gives a submission display (*b*), which entails a rapid nodding of the head, and turns from green (*light*) to brown (*dark*). With this response the resident male does not erect a crest and a black spot does not form behind his eye (*c*). In time the intruder leaves.

COURTSHIP AND MATING of the green anole begin (*a*) when a sexually active male sees a female and responds with a courtship display, which includes repeated extension of the dewlap (in the manner shown in the photograph on page 243) and the individually characteristic series of head-bobbing movements. If the female is sexually receptive, she arches her neck and allows the male to approach and grip it (*b*). In mounting (*c*) the male everts one of his two penises, each termed a hemipenis; if he is on the right side of the female, it is the left one, and vice versa.

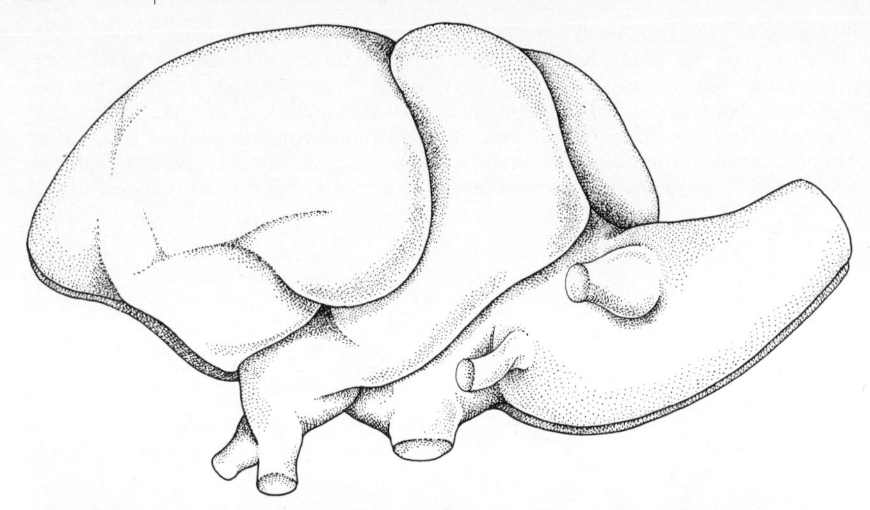

BRAIN OF THE GREEN ANOLE is only .003 percent the size of the human brain; here it is depicted at an enlargement of about 20 diameters. Because the brain is so small a special stereotaxic apparatus was needed to accurately place tiny needles and tubes in it for experiments.

by its conversion into nonaromatizable androgens such as dihydrotestosterone.

Several studies support the concept that the aromatization of androgens to estrogens (the collective term for a group of female sex hormones) occurs normally in the brain of many species of mammals and that estradiol, not testosterone, activates male sexual behavior in many species. Gloria Callard, Zoltan Petro and Kenneth Ryan of the Harvard Medical School have identified aromatizing enzymes in the brain of turtles and snakes. They have reported that when specific parts of the brain of the freshwater turtle *Chrysemys picta* are incubated with androgens that have been labeled with radioactive androgens, some radioactively labeled estrogens are found in the tissue later. This discovery and others led them to propose that the aromatization of androgens is an evolutionarily ancient property of the vertebrate nervous system.

Before the androgen-aromatization hypothesis can be generalized to all reptiles the effects of aromatizable and nonaromatizable androgens must be tested in living animals. Lizards are well suited to such an investigation. In both lizards and snakes a section of the kidney has been modified to form a sex segment. This structure manufactures the seminal fluid and is similar to the prostate and seminal vesicles of mammals. The secretory activity of the kidney sex segment, which is highly sensitive to stimulation by androgen, can be determined by measuring increases in the height of the epithelial cells of the kidney tubules after hormonal stimulation.

Working with this measure as an indication that hormones were affecting the sex organs directly and with the stimulation of sexual behavior such as courtship displays as an indication that hormones were acting on the brain, we did an experiment in which castrated males were treated with testosterone, dihydrotestosterone or estradiol. Examination of the kidney sex segment indicated that only the first two hormones were acting directly on the sex organs. Only testosterone reinstated sexual behavior.

This finding does not exclude the possibility that estradiol acts in the male brain. Several workers have shown that administering dihydrotestosterone and estradiol simultaneously stimulates copulatory behavior in castrated rats. Results of this kind have led to the suggestion that male sexual performance requires both the central or brain action of estradiol and the peripheral action of androgens to stimulate the sex organs. Support for the hypothesis has come from the work of two groups of investigators, one group headed by Julian M. Davidson of Stanford University and the other by Paula Davis and Ronald J. Barfield of Rutgers University. They found that implants of estradiol in the brain of rats will induce sexual behavior only if dihydrotestosterone is administered systemically at the same time.

My colleagues and I at Harvard University also examined the possibility that dihydrotestosterone and estradiol act in concert in the green anole. When both hormones were administered simultaneously to castrated lizards, half of the animals (in two separate experiments) suddenly exhibited the complete reproductive pattern within a single testing period. The results were quite different from the outcome when only testosterone is given; it typically induces a gradual restoration of behavior over several days. Evidently estrogen can play as vital a role in the control of male sexual behavior in lizards as it does in some mammals.

Half of the castrated lizards in our experiments, however, showed no alteration in sexual behavior after the combination treatment. The animals for our experiments were collected from a variety of sites, and so it is possible that the behavioral variability reflects underlying differences between populations in the amount of aromatizing enzyme in the reptilian brain. The behavioral variability may also reflect genetic differences in an individual's sensitivity to hormones.

The individual variability in response

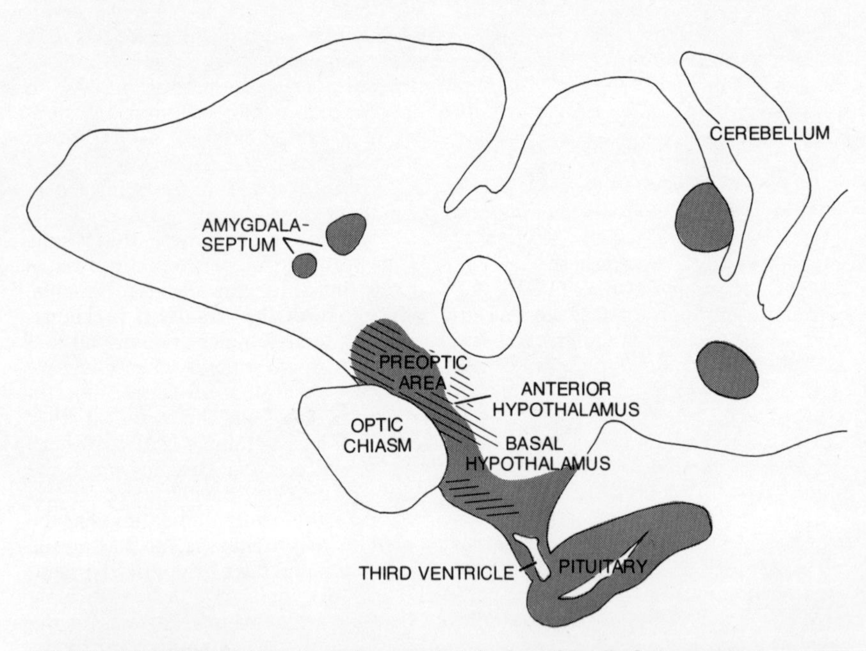

SECTION OF THE BRAIN shows the areas (*color*) where the cells that concentrate sex steroid hormones are situated, as revealed by autoradiography. In this technique male and female lizards from which the gonads have been removed are injected with radioactively labeled hormones; after the animal has been killed sections of its brain are put on slides coated with photographic emulsion. When the emulsion is developed, the areas exposed by radioactivity show where the hormones have been concentrated. Slanted lines indicate where brain lesions abolish male sexual and aggressive behavior, horizontal lines where lesions decrease pituitary function.

makes one wonder about the wisdom of relying on highly inbred stocks of animals for tests of drugs. Many of the inbred stocks of rodents result from repeated brother-sister matings, so that each strain is more like a single individual and does not reflect the normal spectrum of variability found in most natural populations. This important point is often overlooked in the interpretation of data on drug effects and in the determination of appropriate clinical treatments.

Hormones modulate both behavior and the function of the pituitary gland through their influence on the central nervous system. Our group has recently begun to study the interaction of hormones, the brain and behavior in the green anole.

Although much information is available about the areas of the mammalian brain that participate in the control of reproductive behavior and physiology, I was not sure that the information could be extrapolated to reptiles, and so I joined forces with Joan I. Morrell and Donald W. Pfaff of Rockefeller University. We first identified the sites in the brain of *A. carolinensis* that concentrate sex steroid hormones. Our technique was autoradiography. Male and female lizards from which the gonads had been removed were injected with a radioactively labeled hormone: estradiol, testosterone or dihydrotestosterone. After two hours the animals were killed; the brain and the pituitary gland were quickly removed and frozen. Then the brains were cut in thin sections in a darkroom, and the sections were placed on slides that had been coated with photographic emulsion. Half of these preparations were stored in lightproof boxes for six months, half for nine months. When the slides were removed and the emulsion was developed, the areas having a dense grouping of black grains over cell nuclei indicated where the radioactively labeled hormone had concentrated.

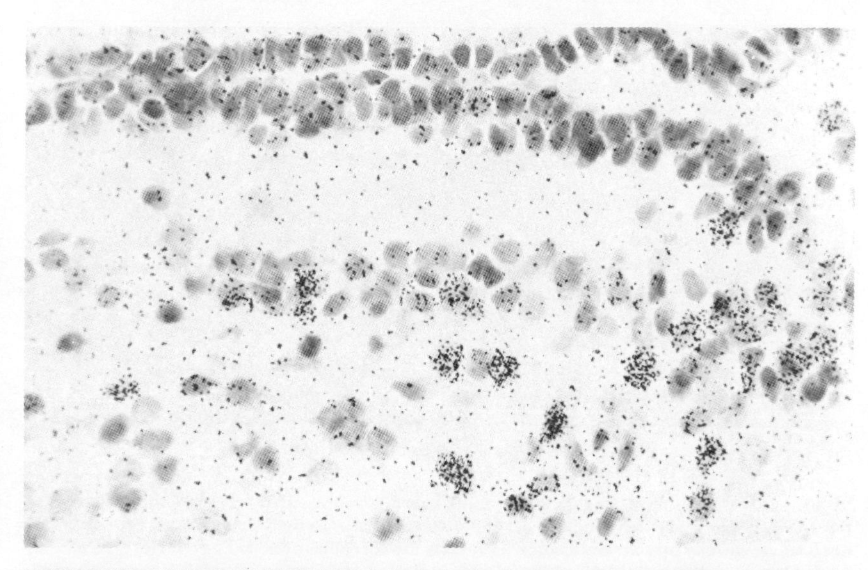

CONCENTRATION OF ESTRADIOL in the brain of a female *A. carolinensis* is indicated by the black grains in this autoradiograph of a section through the periventricular nucleus of the hypothalamus. The enlargement is 550 diameters. The injected estradiol was radioactively labeled; hence it exposed grains of silver in the emulsion on which the brain section was placed.

The results of this study show that in the lizard, as in all other vertebrate species that have been investigated, sex steroid hormones are concentrated in specific areas of the brain: the preoptic area, the anterior and basal hypothalamus and the limbic system. These structures have been found to play a crucial role in reproduction in mammals and birds. An examination of their function in reptiles, which have a more primitive level of organization, would shed light on the evolution of the mechanisms that control reproduction.

Since the brain of an adult male *A. carolinensis* is quite small, being only .003 percent the size of the human brain, we had to develop special techniques in order to investigate the function of the steroid-concentrating areas. Our efforts were greatly aided by the work of Neil B. Greenberg of the University of Tennessee. He devised a stereotaxic apparatus with which one can precisely direct the tip of an ultrafine needle or tube three-dimensionally to a particular place in a small brain, and he also prepared an atlas of the forebrain of *A. carolinensis*. Working with this atlas as a "road map" of the brain, we have begun to study the role the different hormone-sensitive areas play in the control of the animal's reproductive behavior and physiology.

A large body of research with mammals and birds indicates that the anterior hypothalamus-preoptic area is a major center for the integration of behavior. Since it is also a major site of steroid uptake in the lizard, we first examined its role in the regulation of male reproductive behavior. Destruction of the area in intact, sexually active males results in the immediate cessation of sexual behavior, whereas in sexually inactive castrated males the implantation of minute amounts of testosterone directly in the preoptic area reinstates sexual behavior.

When either the area immediately forward of the preoptic area or the anterior basal hypothalamus is destroyed, the release of gonadotropin from the pituitary is shut off and the testes soon collapse. In the first instance the effect is probably on the cells that produce the releasing hormones; cells of this kind have been found in the forebrain of a large variety of animals. The anterior basal hypothalamus is where these hormones are transferred to the pituitary circulation.

It should be kept in mind that an animal's behavior is not solely a consequence of the action of hormones on discrete areas of the brain. A number of workers have demonstrated that in mammals and birds the feedback of peripheral sensations from the secondary sex structures is important in the coordination and completion of behavior patterns. The relation is evident in the mounting behavior of male lizards.

It is possible to determine which hemipenis a male lizard or snake is using by noting the direction in which the tail is turned. If the male is on the left side of the female, he will intromit the right hemipenis by curving his tail to the right, and vice versa. Experiments have shown that sensations from the hemipenises play an important role not only in the control of the initial orientation of the male during copulation but also in the termination of copulation. The finding is in agreement with a large number of studies of mammals showing that sensory feedback from the penis is necessary for normal mating behavior.

The behavior of the male that results from hormone-brain interactions has a major influence on the reproductive physiology of the female. If reproductively inactive females are exposed to a springlike environment when they are housed either alone or with other females, the growth of the ovaries is stimulated. If such a female is also exposed to a sexually active, courting male, the rate of ovarian development is increased significantly, indicating that the behavior of the male is facilitating the stimulative effects of the environment. Indeed, one finds not only that male courtship behavior is necessary for the normal secretion of gonadotropin by the pituitary in the female but also that the amount of gonadotropin secreted is correlated with the amount or frequency of male courtship behavior to which the female is exposed.

On the other hand, aggressive behav-

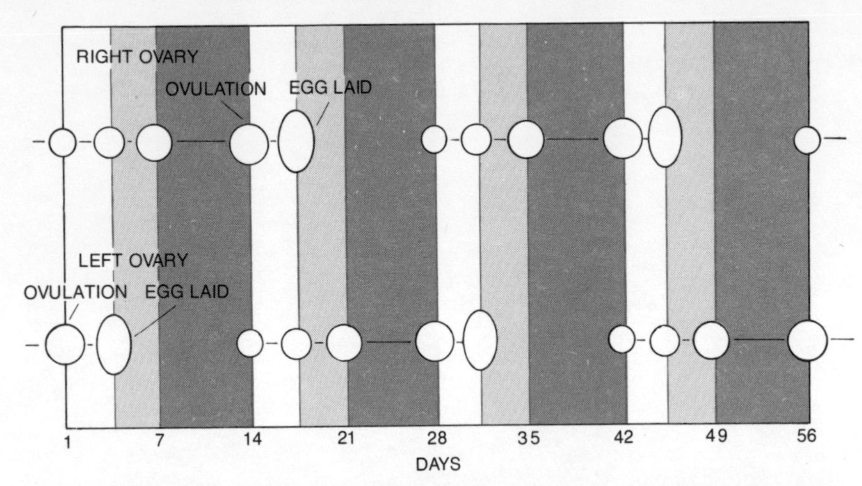

SEXUAL CYCLE of the female green anole lizard is represented schematically. The white regions indicate when the female is sexually unreceptive, the light color shows the period of onset of receptivity and the dark color represents the receptive period. The ovaries alternate in producing the follicle; its state of maturity is indicated by the size at which it is represented.

ior by males not only fails to facilitate the effects of the changing environment but also completely inhibits them. Females housed with males that are constantly fighting among themselves show little or no ovarian growth. To my knowledge this finding is the first demonstration that it is possible to turn female fertility off as well as on by varying the sociobehavioral environment.

In the female during the breeding season a single ovarian follicle develops and is ovulated about every two weeks. As in many human females, the ovaries alternate in the production of the follicle. Hence a sexually active female anole will typically have in one ovary a follicle that is in a more advanced state than the largest follicle in the other ovary. Moreover, like many mammals, a female *A. carolinensis* goes through cycles of sexual receptivity that are correlated with the pattern of follicle growth. The cycles presumably reflect corresponding fluctuations of sex steroids in the blood.

As one would expect, removal of the ovaries abolishes sexual receptivity in the female green anole. The administration of estrogen to such a female restores the behavior in a dose-related manner. The threshold dose appears to be .6 microgram of estradiol benzoate. A single injection of .8 microgram will induce receptivity in 85 percent of the females in 24 hours and in all of them in 48 hours.

In many mammals progesterone is important in the regulation of female receptivity. It does not function alone, but when it is administered at an appropriate time after the animal has been "primed" with estrogen, the two hormones act synergistically to facilitate receptivity. Studies of the pattern of hormone secretion by the ovaries in these animals reveal that this facilitation by progesterone coordinates the receptivity of the female with ovulation.

Progesterone appears to serve a similar function in lizards. For example, a female from which the ovaries have been removed will not be receptive to male courtship after a subthreshold dose of estradiol (.4 microgram) unless that dose is followed 24 hours later by 60 micrograms of progesterone.

As in mammals, progesterone can have a quite different effect if it is administered simultaneously with estrogen or sufficiently long afterward. Then it inhibits the stimulatory effect of estrogen. In the lizard a single injection of 160 micrograms of progesterone 48 hours after an above-threshold dose (.8 microgram) of estradiol completely inhibits the effect of the estrogen.

With the female as with the male both central and peripheral stimuli are involved in the regulation of receptivity. When a female that is preovulatory or has been primed with estrogen is put in a cage containing a male, the male will immediately begin to court her. Less than a minute after mating, however, the female stops being receptive to male courtship. Experiments have demonstrated that intromission of the hemipenis is the trigger of the transformation. A preovulatory female that is courted and mounted but not mated continues to be receptive. The duration of intromission also seems to be important, since females interrupted during mating continue to be receptive until they are allowed to mate to completion.

We have found that the presence of the ovaries is critical in the turning off of receptivity. If females lacking ovaries but primed with estrogen are mated, they will be receptive again within 24 hours. Indeed, preliminary experiments indicate that such females are unreceptive for only about six hours after mating. Normal females, on the other hand, continue to be unreceptive for several days, suggesting that some change in the production of ovarian hormones is the key to the long-term inhibition of receptivity.

Just as the behavior of the male influences the reproductive physiology of the female, so the presence of females has a strong effect on the male. Males housed with females have a rapider testicular growth and sperm development than males housed with other males.

In sum, the sequence of events in the

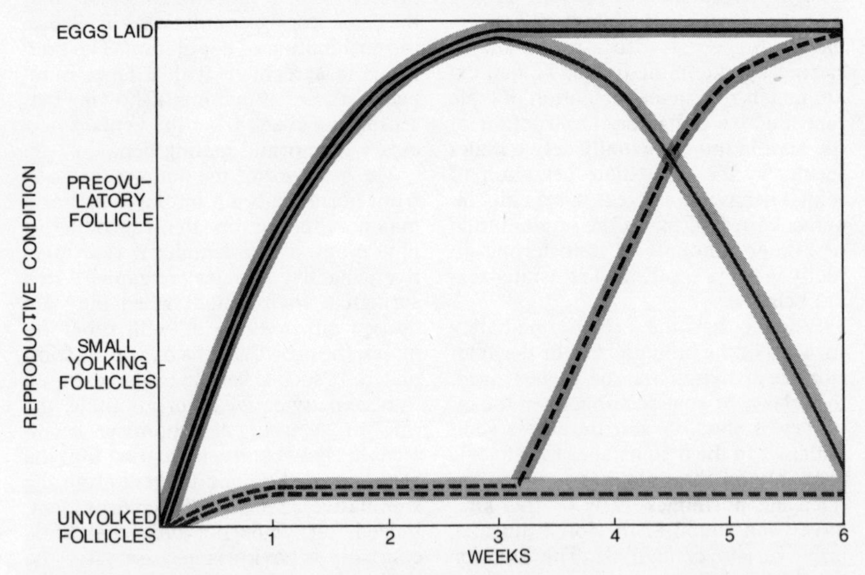

INTERACTION of male behavior and female reproductive status in *A. carolinensis* is charted. Four groups of females were involved. One of them (*bottom curve*) was exposed only to aggressive male behavior. Another (*top curve*) was exposed only to male courtship. The curve that ascends after three weeks represents a group that was exposed to three weeks of aggression and three weeks of courtship; curve that descends after three weeks represents a group that received the opposite treatment. Periods of courtship are in color, periods of aggression in gray. Follicle condition (and hence the reproductive status of the females) appears at the left.

reproductive cycle of the lizard begins with environmental stimulation of testicular activity. Hormones produced by the testes as a result of this stimulation act on discrete areas of the brain to influence the secretion of gonadotropin by the pituitary gland and also to activate sexual behavior in the male. Depending on the male's behavior, the secretion of gonadotropin in the female, and hence ovarian growth, is either stimulated or inhibited. The sexual behavior of the female, which is dependent in part on ovarian hormones and in part on external stimuli, then influences male behavior and testicular activity.

The experiments I have described show *A. carolinensis* to be a convenient animal for investigating biomedically important phenomena that in mammals are difficult to isolate and dissect into their component parts, particularly for studies of the neurological pathologies underlying sexual dysfunction. The work has also revealed a certain degree of conservatism in the neural and endocrine control of reproductive behavior in vertebrates. For example, we have found that the areas of the brain concentrating steroid hormones are similar in lizards and mammals and have found that the function of these areas in regulating reproductive behavior is similar too. We have also found that progesterone has a distinctive role in the hormonal control of female sexual receptivity (either stimulating it or inhibiting it) in lizards as in many mammals, suggesting that this biphasic effect of progesterone is ancient in evolutionary terms.

It is noteworthy also that the many unique characteristics of lizards and the integration of laboratory findings with observations in the field lead to a clearer understanding of how behavior interacts with physiology in the adaptation of an animal to its environment. For example, the work suggests that the typical time sequence observed in many species between aggression and courtship, and the physiological mechanisms underlying these actions, may reflect common evolutionary origins. The opposite effects on the female of male aggression and courtship may in turn account for the pattern often seen in vertebrates of the Temperate Zone in which males establish territories in the breeding grounds before the females arrive. Females arriving early would be prevented from reproducing by the aggressive behavior among the males, whereas females arriving later would be rapidly stimulated by the courtship behavior of the territorial males. Finally, the termination of sexual receptivity by mating has several advantages for the female. She locks in her choice of a mate and decreases her vulnerability to predators by limiting the time spent in mating, thereby increasing the chance that she will survive to leave a large number of offspring.

BIBLIOGRAPHIES

These bibliographies have been revised by Norman K. Wessels.

INTRODUCTORY ESSAYS

An Essay on Vertebrates

Alexander, R. M., THE CHORDATES. Cambridge University Press, Cambridge, 1975.

Barrington, E. J. W., THE BIOLOGY OF HEMICHORDATA AND PROTOCHORDATA. Oliver and Boyd, London, 1965.

Barrington, E. J. W., and Jefferies, L. P. S., PROTOCHORDATES. Academic Press, New York, 1975.

Beer, A. E., and Billingham, R. E., THE IMMUNOBIOLOGY OF MAMMALIAN REPRODUCTION. Prentice-Hall, Englewood Cliffs, N.J., 1976.

Diamond, I. T., and Hull, W. C., "Evolution of the neocortex." *Science* **164**, 1969, pp. 251–62.

Dobzhansky, T., MANKIND EVOLVING. Yale University Press, New Haven, 1962.

Gordon, M. S., ANIMAL PHYSIOLOGY: PRINCIPLES AND ADAPTATIONS. Macmillan, New York, 1972.

Halstead, L. B., THE PATTERN OF VERTEBRATE EVOLUTION. W. H. Freeman and Company, San Francisco, 1969.

Hardisty, M. W., and Potter, I. C., THE BIOLOGY OF LAMPREYS. Academic Press, New York, 1971.

Hoar, W. S., and Randall, D. J., FISH PHYSIOLOGY, Vols. I–VI. Academic Press, New York, 1969–71.

Hochachka, P. W., and Somero, G. N., STRATEGIES OF BIOCHEMICAL ADAPTATIONS. Saunders, Philadelphia, 1973.

Hogarth, P. J., BIOLOGY OF REPRODUCTION. Wiley, New York, 1978.

Jenkins, F. A., Jr., PRIMATE LOCOMOTION. Academic Press, New York, 1974.

LeGros Clark, W. E., THE ANTECEDENTS OF MAN. Quadrangle Books, Chicago, 1971.

Lewis, O. J., "Brachiation and early evolution of the Hominoidea." *Nature* **230**, 1971, pp. 577–78. But see also: *Nature* **237**, 1972, pp. 103–4.

McFarland, W. N., Pough, F. H., Cade, T. J., and Heiser, J. B., VERTEBRATE LIFE. MacMillan, New York, 1979.

Stahl, B. J., VERTEBRATE HISTORY: PROBLEMS IN EVOLUTION. McGraw Hill, New York, 1974.

Thomson, K. S., "Adaptation and evolution of early fishes." *Quart. Rev. Biol.* **46**, 1971, p. 139.

Tuttle, R. H., PRIMATE FUNCTIONAL MORPHOLOGY AND EVOLUTION. Mouton, Paris, 1975.

I Vascular System Biology

Baez, S., "Microcirculation." *Ann. Rev. Physiol.* **39**, 1977, p. 391.

Blaxter, J. H. S., and Tyler, P., "Physiology and function of the swimbladder." *Adv. Comp. Physiol. Biochem.* **7**, 1978, p. 311.

Nicoll, P. A., and Taylor, A. E., "Lymph formation and flow." *Ann. Rev. Physiol.* **39**, 1977, p. 73.

Oberg, B., "Overall cardiovascular regulation." *Ann. Rev. Physiol.* **38**, 1976, p. 537.

Perutz, M., "Hemoglobin structure and respiratory transport." *Sci. Am.* **239**(6), Dec. 1978, p. 92. Offprint 1413.

Riggs, A., "Functional properties of hemoglobins." *Physiol. Revs.* **45**, 1965, pp. 619–73.

Schmidt-Nielsen, K., and Taylor, C. R., "Red blood cells: why or why not?" *Science* **162**, 1968, pp. 274–75. See also *Science* **162**, 1968, pp. 275–77.

Seymour, R. S., "Dinosaurs, endothermy, and blood pressure." *Nature* **262**, 1976, p. 207.

II Gas Exchange and the Lungs

Bouverot, P., "Control of breathing in birds compared with mammals." *Physiol. Revs.* **58,** 1978, p. 604.

Hughes, G. M., RESPIRATION OF AMPHIBIOUS VERTEBRATES. Academic Press, New York, 1976.

Pattle, R. E., "Surface lining of lung alveoli." *Physiol. Revs.* **45,** 1965, pp. 48–79.

White, F. N., "Comparative aspects of vertebrate cardiorespiratory physiology." *Ann. Rev. Physiol.* **40,** 1978, p. 471.

III Water Balance and Its Control

Bonaventura, J., Bonaventura, C., and Sullivan, B., "Urea tolerance as a molecular adaptation of elasmobranch hemoglobin." *Science* **186,** 1974, p. 57.

Katz, A. I., and Lindeimer, M. D., "Action of hormones on the kidney." *Ann. Rev. Physiol.* **39,** 1977, p. 97.

Norman, A. W., "The mode of action of vitamin D." *Biol. Revs.* **43,** 1968, pp. 97–137.

Peaker, M., and Linzell, J. L., SALT GLANDS IN BIRDS AND REPTILES. Cambridge University Press, Cambridge, 1975.

Potts, W. T. W., "Osmotic and ionic regulation." *Ann. Rev. Physiol.* **30,** 1968, pp. 73–104.

Poulson, T. L., "Countercurrent multipliers in avian kidneys." *Science,* **148,** 1965, pp. 389–91.

Reid, I. A., Morris, B. J., and Ganong, W. F., "The renin-angiotensin system." *Ann. Rev. Physiol.* **40,** 1978, p. 377.

Salt, G. W., "Respiratory evaporation in birds." *Biol. Revs.* **39,** 1964, pp. 113–136.

Shoemaker, V. H., and Nagy, K. A., "Osmoregulation in amphibians and reptiles." *Ann. Rev. Physiol.* **39,** 1977, p. 449.

IV Temperature Adaptations

Augee, M. L., "Monotremes and the evolution of homeothermy." *Aust. Zool.* **20,** 1978, p. 111.

Bakker, R. T., "Anatomical and ecological evidence of endothermy in dinosaurs." *Nature* **238,** 1972, p. 81.

Bartholomew, G. A., "Body temperature and energy metabolism." *In* M. S. Gordon, Ed., ANIMAL FUNCTION: PRINCIPLES AND ADAPTATIONS. Macmillan, New York, 1968.

Graham, J. B., "Heat exchange in the black skipjack, and the blood-gas relationship of warm bodied fishes." *Proc. Natl. Acad. Sci.* **70,** 1973, p. 1964.

Griffiths, M., ECHIDNAS. Pergamon Press, New York, 1968.

Harrison, C. J. O., "Feathering and flight evolution in *Archaeopteryx.*" *Nature* **263,** 1976, p. 762.

Himms-Hagen, J., "Cellular thermogenesis." *Ann. Rev. Physiol.* **38,** 1976, p. 315.

Hochachka, P. W., "Organization of metabolism during temperature compensation." *In* C. L. Prosser, Ed., MOLECULAR MECHANISMS OF TEMPERATURE ADAPTATION. Washington, D.C., American Association for the Advancement of Science, 1967, pp. 177–204.

Johnston, P. A., "Growth rings in dinosaur teeth." *Nature* **278,** 1979, p. 635.

Licht, P., "Thermal adaptation in the enzymes of lizards in relation to preferred body temperatures." *In* C. L. Prosser, Ed., MOLECULAR MECHANISMS OF TEMPERATURE ADAPTATION. Washington, D.C., American Association for the Advancement of Science, 1967, pp. 131–46.

McClean, D. M., "A terminal Mesozoic 'greenhouse': lessons from the past." *Science* **201,** 1978, p. 401.

McNab, B. K. and Auffenberg, W., "The effect of large body size on the temperature regulation of the komodo dragon, *Varanus komodoensis.*" *Comp. Biochem. Physiol.* **55A,** 1976, p. 345.

Pough, F. H. *In* McFarland, W. N., Pough, F. H., Cade, T. J., and Heiser, J. B., VERTEBRATE LIFE. MacMillan, New York, 1979.

Smith, R. E., and Horwitz, B. A., "Brown fat and thermiogenesis." *Physiol. Rev.* **49,** 1969, pp. 330–425.

Stonehouse, B., and Gilmore, D., THE BIOLOGY OF MARSUPIALS. University Park Press, Baltimore, 1977.

V Hormones and Internal Regulation

Bergland, R. M., and Page, R. B., "Pituitary-brain vascular relations: a new paradigm." *Science* **204,** 1979, p. 18.

Gorbman, A., and Bern, H. A., A TEXTBOOK OF COMPARATIVE ENDOCRINOLOGY. Wiley, New York, 1962.

Jost, J. P., and Rickenberg, H. V., "Cyclic AMP." *Ann. Rev. Biochem.* **40,** 1971, pp. 741–74.

Packard, G. C., Tracy, C. R., and Roth, J. J., "The physiological ecology of reptilian eggs and embryos, and the evolution of viviparity within the class Reptilia." *Biol. Rev.* **52,** 1977, p. 71.

Seif, S. M., and Robinson, A. G., "Localization and release of neurophysins." *Ann. Rev. Physiol.* **40,** 1978, p. 345.

Labrie, F., et al., "Mechanism of action of hypothalamic hormones in the adenohypophysis." *Ann. Rev. Physiol.* **41,** 1979, p. 555.

Nottebohm, F., "Ontogeny of bird song." *Science* **167,** 1970, p. 950.

Sutherland, E. W., "Studies on the mechanism of hormone action." *Science* **177,** 1972, pp. 401–408.

Wallis, M., "The molecular evolution of pituitary hormones." *Biol. Rev.* **50,** 1975, p. 35.

ARTICLES

1. The Heart

PHYSIOLOGY IN HEALTH AND DISEASE. Carl J. Wiggers. Lea & Febiger, 1949.

THE MOTION OF THE HEART: THE STORY OF CARDIOVASCULAR RESEARCH. Blake Cabot. Harper & Brothers, 1954.

2. The Microcirculation of the Blood

THE ANATOMY AND PHYSIOLOGY OF CAPILLARIES. August Krogh. Yale University Press, 1929.

GENERAL PRINCIPLES GOVERNING THE BEHAVIOR OF THE MICROCIRCULATION. B. W. Zweifach in *The American Journal of Medicine,* Vol. 23, No. 5, pages 684–696; November, 1957.

3. The Lymphatic System

THE LYMPHATIC SYSTEM WITH PARTICULAR REFERENCE TO THE KIDNEY. H. S. Mayerson in *Surgery, Gynecology & Obstetrics,* Vol. 116, No. 3, pages 259–272; March, 1963.

LYMPHATICS AND LYMPH CIRCULATION. István Rusznyák, Mihály Földi and György Szabó. Pergamon Press Ltd., 1960.

OBSERVATIONS AND REFLECTIONS ON THE LYMPHATIC SYSTEM. H. S. Mayerson in *Transactions & Studies of the College of Physicians of Philadelphia,* Fourth Series, Vol. 28, No. 3, pages 109–127; January, 1961.

4. A Brain-Cooling System in Mammals

STUDIES OF THE CAROTID RETE AND ITS ASSOCIATED ARTERIES. P. M. Daniel, J. D. K. Dawes and Marjorie M. L. Prichard in *Philosophical Transactions of the Royal Society, Series B.* Vol. 237, pages 173–208; 1954.

INFLUENCE OF THE CAROTID RETE ON BRAIN TEMPERATURE IN CATS EXPOSED TO HOT ENVIRONMENTS. M. A. Baker in *Journal of Physiology,* Vol. 220, No. 3, pages 711–728; February, 1972.

RAPID BRAIN COOLING IN EXERCISING DOGS. M. A. Baker and L. W. Chapman in *Science,* Vol. 195, No. 4280, pages 781–783; February 25, 1977.

5. The Physiology of the Giraffe

CIRCULATION OF THE GIRAFFE. Robert H. Goetz, James V. Warren, Otto H. Gauer, John L. Patterson, Jr., Joseph T. Doyle, E. N. Keen and Maurice McGregor in *Circulation Research,* Vol. 8, No. 5, pages 1049–1058; September, 1960.

CARDIORESPIRATORY DYNAMICS IN THE OX AND GIRAFFE, WITH COMPARATIVE OBSERVATIONS ON MAN AND OTHER MAMMALS. John L. Patterson, Jr., et al. in *Annals of the New York Academy of Sciences,* Vol. 127, Article 1, pages 393–413; September 8, 1965.

BLOOD PRESSURE RESPONSES OF WILD GIRAFFES STUDIED BY RADIO TELEMETRY. Robert L. Van Citters, William S. Kemper and Dean L. Franklin in *Science,* Vol. 152, No. 3720, pages 384–386; April 15, 1966.

CEREBRAL HEMODYNAMICS IN THE GIRAFFE. Robert L. Van Citters, Dean L. Franklin, Stephen F. Vatner, Thomas Patrick and James V. Warren in *Transactions of the Association of American Physicians,* Vol. 82, pages 293–303; 1969.

6. Air-Breathing Fishes

AIR BREATHING IN THE TELEOST SYMBRANCHUS MARMORATUS. Kjell Johansen in *Comparative Biochemistry and Physiology,* Vol. 18, No. 2, pages 383–395; June, 1966.

CARDIOVASCULAR DYNAMICS IN THE LUNGFISHES. Kjell Johansen, Claude Lenfant and David Hanson in *Zeitschrift für vergleichende Physiologie,* Vol. 59, No. 2, pages 157–186; June 5, 1968.

GAS EXCHANGE AND CONTROL OF BREATHING IN THE ELECTRIC EEL, ELECTROPHORUS ELECTRICUS. Kjell Johansen, Claude Lenfant, K. Schmidt-Nielsen and J. A. Petersen in *Zeitschrift für vergleichende Physiologie* (in press).

OBSERVATIONS ON THE AFRICAN LUNGFISH PROTOPTERUS AETHIOPICUS, AND ON EVOLUTION FROM WATER TO LAND ENVIRONMENT. Homer W. Smith in *Ecology,* Vol. 12, No. 1, pages 164–181; January 1931.

RESPIRATORY PROPERTIES OF BLOOD AND PATTERN OF GAS EXCHANGE IN THE LUNGFISH. Claude Lenfant, Kjell Johansen and Gordon C. Grigg in *Respiration Physiology,* Vol. 2, No. 1, pages 1–22; December, 1966–1967.

7. The Lung

THE MECHANISM OF BREATHING. Wallace O. Fenn in *Scientific American*, Vol. 202, No. 1, pages 138–148; Januray, 1960.

8. How Birds Breathe

BIRD RESPIRATION: FLOW PATTERN IN THE DUCK LUNG. William L. Bretz and Knut Schmidt-Nielsen in *The Journal of Experimental Biology*, Vol. 54, No. 1, pages 103–118; February, 1971.

A PRELIMINARY ALLOMETRIC ANALYSIS OF RESPIRATORY VARIABLES IN RESTING BIRDS. Robert C. Lasiewski and William A. Calder, Jr., in *Respiration Physiology*, Vol. 11, No. 2, pages 152–166; January, 1971.

RESPIRATORY PHYSIOLOGY OF HOUSE SPARROWS IN RELATION TO HIGH-ALTITUDE FLIGHT. Vance A. Tucker in *The Journal of Experimental Biology*, Vol. 48, No. 1, pages 55–66; February, 1968.

STRUCTURAL AND FUNCTIONAL ASPECTS OF THE AVIAN LUNGS AND AIR SACS. A. S. King in *International Review of General and Experimental Zoology: Vol. II*, edited by William J. L. Felts and Richard J. Harrison. Academic Press, Inc., 1964.

9. Surface Tension in the Lungs

MECHANICAL PROPERTIES OF LUNGS. Jere Mead in *Physiological Reviews*, Vol. 41, No. 2, pages 281–330; April, 1961.

THE PHYSICS AND CHEMISTRY OF SURFACES. Neil Kensington Adam. Oxford University Press, 1941.

SOAP-BUBBLES: THEIR COLOURS AND THE FORCES WHICH MOLD THEM. C. V. Boys. Dover Publications, Inc., 1959.

10. The Diving Women of Korea and Japan

THE ISLAND OF THE FISHERWOMEN. Fosco Maraini. Harcourt, Brace & World, Inc., 1962.

KOREAN SEA WOMEN: A STUDY OF THEIR PHYSIOLOGY. The departments of physiology, Yonsei University College of Medicine, Seoul, and the State University of New York at Buffalo.

THE PHYSIOLOGICAL STRESSES OF THE AMA. Hermann Rahn in *Physiology of Breath-Hold Diving and the Ama of Japan*. Publication 1341, National Academy of Sciences—National Research Council, 1965.

11. The Eland and the Oryx

DESERT ANIMALS: PHYSIOLOGICAL PROBLEMS OF HEAT AND WATER. Knut Schmidt-Nielsen. Oxford University Press, 1964.

THE FIRE OF LIFE: AN INTRODUCTION TO ANIMAL ENERGETICS. Max Kleiber. John Wiley & Sons, Inc., 1961.

TERRESTRIAL ANIMALS IN DRY HEAT: UNGULATES. W. V. Macfarlane in *Handbook of Physiology, Section 4: Adaptation to the Environment*. The American Physiological Society, 1964.

12. Salt Glands

THE SALT GLANDS OF THE HERRING GULL. Ragnar Fänge, Knut Schmidt-Nielsen and Humio Osaki in *Biological Bulletin*, Vol. 115, pages 162–171; October, 1958.

SALT GLANDS IN MARINE REPTILES. Knut Schmidt-Nielsen and Ragnar Fänge in *Nature*, Vol. 182, No. 783–785; September 20, 1958.

13. Fishes with Warm Bodies

WARM-BODIED FISH. Francis G. Carey, John M. Teal, John W. Kanwisher, Kenneth D. Lawson and James S. Beckett in *American Zoologist*, Vol. 11, pages 137–145; 1971.

14. Adaptations to Cold

BODY INSULATION OF SOME ARCTIC AND TROPICAL MAMMALS AND BIRDS. P. F. Scholander, Vladimir Walters, Raymond Hock and Laurence Irving in *The Biological Bulletin*, Vol. 99, No. 2, pages 225–236; October, 1950.

BODY TEMPERATURES OF ARCTIC AND SUBARCTIC BIRDS AND MAMMALS. Laurence Irving and John Krog in *Journal of Applied Physiology*, Vol. 6, No. 11, pages 667–680; May, 1954.

EFFECT OF TEMPERATURE ON SENSITIVITY OF THE FINGER. Laurence Irving in *Journal of Applied Physiology*, Vol. 18, No. 6, pages 1201–1205; November, 1963.

TERRESTRIAL ANIMALS IN COLD: INTRODUCTION. Laurence Irving in *Handbook of Physiology, Section 4: Adaptation to the Environment*. American Physiological Society, 1964.

15. The Thermostat of Vertebrate Animals

TEMPERATURE REGULATION IN MAMMALS AND OTHER VERTEBRATES. J. Bligh. Elsevier North-Holland Publishing Company, 1973.

PROBING THE ROSTRAL BRAINSTEM OF ANESTHETIZED, UNANESTHETIZED, AND EXERCISING DOGS, AND OF HIBERNATING AND EUTHERMIC GROUND SQUIRRELS. Harold T. Hammel, H. Craig Heller and F. R. Sharp in *Federation Proceedings*, Vol. 32, pages 1588–1596; 1973.

THERMOREGULATION DURING SLEEP AND HIBERNATION. H. Craig Heller and S. F. Glotzbach in *International Review of Physiology*, Vol. 15, pages 147–187; 1977.

SLEEP AND HIBERNATION: ELECTROPHYSICAL AND THERMOREGULATORY HOMOLOGIES. H. Craig Heller, J. M. Walker, G. L. Florant, S. F. Glotzbach and R. J. Berger in *Strategies in Cold: Natural Torpidity and Thermogenesis*, edited by Lawrence Wang and Jack W. Hudson. Academic Press, Inc., 1978.

16. The Production of Heat by Fat

BROWN ADIPOSE TISSUE AND THE RESPONSE OF NEW-BORN RABBITS TO COLD. M. J. R. Dawkins and D. Hull in *The Journal of Physiology*, Vol. 172, No. 2, pages 216–238; August, 1964.

BROWN FAT: A REVIEW. Bengt Johansson in *Metabolism: Clinical and Experimental*, Vol. 8, No. 3, pages 221–240; May, 1959.

ON THE ACTION OF HORMONES WHICH ACCELERATE THE RATE OF OXYGEN CONSUMPTION AND FATTY ACID RELEASE IN RAT ADIPOSE TISSUE IN VITRO. Eric G. Ball and Robert L. Jungas in *Proceedings of the National Academy of Sciences*, Vol. 47, No. 7, pages 932–941; July, 1961.

THERMOGENESIS OF BROWN ADIPOSE TISSUE IN COLD-ACCLIMATED RATS. Robert E. Smith and Jane C. Roberts in *American Journal of Physiology*, Vol. 206, No. 1, pages 143–148; January, 1964.

17. The Hormones of the Hypothalamus

NEURAL CONTROL OF THE PITUITARY GLAND. G. W. Harris. The Williams & Wilkins Company, 1955.

THE HYPOTHALAMUS: PROCEEDINGS OF THE WORKSHOP CONFERENCE ON INTEGRATION OF ENDOCRINE AND NON ENDOCRINE MECHANISMS IN THE HYPOTHALAMUS. Edited by L. Martini, M. Motta and F. Fraschini. Academic Press, 1970.

CHARACTERIZATION OF OVINE HYPOTHALAMIC HYPOPHYSIOTROPIC TSH-RELEASING FACTOR. Roger Burgus, Thomas F. Dunn, Dominic Desiderio, Darrell N. Ward, Wylie Vale and Roger Guillemin in *Nature*, Vol. 226, No. 5243, pages 321–325; April 25, 1970.

STRUCTURE OF THE PORCINE LH- AND FSH-RELEASING HORMONE, I: THE PROPOSED AMINO ACID SEQUENCE. H. Matsuo, Y. Baba, R. M. G. Nair, A. Arimura and A. V. Schally in *Biochemical and Biophysical Research Communications*, Vol. 43, No. 6, pages 1334–1339; June 18, 1971.

SYNTHETIC POLYPEPTIDE ANTAGONISTS OF THE HYPOTHALAMIC LUTEINIZING HORMONE RELEASING FACTOR. Wylie Vale, Geoffrey Grant, Jean Rivier, Michael Monahan, Max Amoss, Richard Blackwell, Roger Burgus and Roger Guillemin in *Science*, Vol. 176, No. 4037, pages 933–934; May 26, 1972.

SYNTHETIC LUTEINIZING HORMONE-RELEASING FACTOR: A POTENT STIMULATOR OF GONADOTROPIN RELEASE IN MAN. S. S. C. Yen, R. Rebar, G. VandenBerg, F. Naftolin, Y. Ehara, S. Engblom, K. J. Ryan, K. Benirschke, J. Rivier, M. Amoss and R. Guillemin in *The Journal of Clinical Endocrinology and Metabolism*, Vol. 34, No. 6, pages 1108–1111; June 1972.

18. The Pineal Gland

MELATONIN SYNTHESIS IN THE PINEAL GLAND: EFFECT OF LIGHT MEDIATED BY THE SYMPATHETIC NERVOUS SYSTEM. Richard J. Wurtman, Julius Axelrod and Josef E. Fischer in *Science*, Vol. 143, No. 3612, pages 1328–1329; March 20, 1964.

STRUCTURE AND FUNCTION OF THE EPIPHYSIS CEREBRI. Edited by J. Ariëns Kappers and J. P. Schadé in *Progress in Brain Research*, Vol. X. Elsevier Publishing Company, 1965.

19. Calcitonin

THE AMINO ACID SEQUENCE OF PORCINE THYROCALCITONIN. J. T. Potts, Jr., H. D. Niall, H. T. Keutmann, H. B. Brewer, Jr., and L. J. Deftos in *Proceedings of the National Academy of Sciences*, Vol. 59, No. 4, pages 1321–1328; April 15, 1968.

CALCITONIN FROM ULTIMOBRANCHIAL GLANDS OF DOGFISH AND CHICKENS. D. H. Copp, D. W. Cockcroft and Yankoon Kueh in *Science*, Vol. 158, No. 3803, pages 924–925; November 17, 1967.

SYMPOSIUM ON THYROCALCITONIN. Edited by Maurice M. Pechet in *Thee American Journal of Medicine*, Vol. 43, No. 5, pages 645–726; November, 1967.

20. How an Eggshell is Made

CALCIFICATION AND OSSIFICATION. MEDULLARY BONE CHANGES IN THE REPRODUCTIVE CYCLE OF FEMALE PIGEONS. William Bloom, Margaret A. Bloom and Franklin C. McLean in *The Anatomical Record*, Vol. 81, No. 4, pages 443–475; December 26, 1941.

CALCIUM METABOLISM AND AVIAN REPRODUCTION. K. Simkiss in *Biological Reviews*, Vol. 36, No. 3, pages 321–367; August, 1961.

THE EFFECT OF PITUITARY HORMONES ON OVULATION IN CALCIUM-DEFICIENT PULLETS, T. G. Taylor, T. R. Morris and F. Hertelendy in *The Veterinary Record*, Vol. 74, No. 4, pages 123–125; January 27, 1962.

EGGSHELL FORMATION AND SKELETAL METABOLISM. T. G. Taylor and D. A. Stringer in *Avian Physiology,* edited by Paul D. Sturkie. Comstock Publishing Associates, 1965.

21. Rickets

THE ETIOLOGY OF RICKETS. Edwards A. Park in *Physiological Reviews,* Vol. 3, No. 1, pages 106–163; January, 1923.

RICKETS INCLUDING OSTEOMALACIA AND TETANY. Alfred F. Hess. Lea & Febiger, 1929.

INVESTIGATIONS ON THE ETIOLOGY OF RICKETS: VITAMIN D. Elmer Verner McCollum in *A History of Nutrition: The Sequence of Ideas in Nutritional Investigations.* Houghton Mifflin Company, 1957.

SKIN-PIGMENT REGULATION OF VITAMIN-D BIOSYNTHESIS IN MAN. W. Farnsworth Loomis in *Science,* Vol. 157, No. 3788, pages 501–506; August 4, 1967.

22. Milk

PRINCIPLES OF DAIRY CHEMISTRY. Robert Jenness and Stuart Patton. John Wiley & Sons, Inc., 1959.

MILK: THE MAMMARY GLAND AND ITS SECRETION. Edited by S. K. Kon and A. T. Cowie. Academic Press, 1961.

MILK PROTEINS. H. A. McKenzie in *Advances in Protein Chemistry,* edited by C. B. Anfinsen, Jr., M. L. Anson, John T. Edsall and Frederic M. Richards, Vol. 22, pages 55–234; 1967.

THE ROLE OF THE PLASMA MEMBRANE IN THE SECRETION OF MILK FAT. Stuart Patton and Frederick M. Fowkes in *Journal of Theoretical Biology,* Vol. 15, No. 3, pages 274–281; June, 1967.

23. The Hormonal Control of Behavior in a Lizard

PSYCHOBIOLOGY OF REPTILIAN REPRODUCTION. David Crews in *Science,* Vol. 189, No. 4208, pages 1059–1065; September 26, 1975.

BIOLOGICAL DETERMINANTS OF SEXUAL BEHAVIOR. Edited by John B. Hutchinson. John Wiley & Sons, Inc., 1977.

THE PARENTAL BEHAVIOR OF RING DOVES. Rae Silver in *American Scientist,* Vol. 66, No. 2, pages 209–215; March–April, 1978.

ENDOCRINE CONTROL OF REPTILIAN REPRODUCTIVE BEHAVIOR. D. Crews in *Endocrine Control of Sexual Behavior,* edited by Carlos Beyer. Raven Press, 1979.

INDEX